29th IUPAP Conference on Computational Physics (CCP 2017)

Journal of Physics: Conference Series
Volume 1136

Paris, France
9 – 13 July 2017

ISBN: 978-1-5108-8018-4
ISSN: 1742-6588

Printed from e-media with permission by:

Curran Associates, Inc.
57 Morehouse Lane
Red Hook, NY 12571

Some format issues inherent in the e-media version may also appear in this print version.

Printed by Curran Associates, Inc. (2018)

For permission requests, please contact the Institute of Physics at the address below.

Institute of Physics
Dirac House, Temple Back
Bristol BS1 6BE UK

Phone: 44 1 17 929 7481
Fax: 44 1 17 920 0979

techtracking@iop.org

Additional copies of this publication are available from:

Curran Associates, Inc.
57 Morehouse Lane
Red Hook, NY 12571 USA
Phone: 845-758-0400
Fax: 845-758-2633
Email: curran@proceedings.com
Web: www.proceedings.com

Table of contents

Volume 1136

XXIXth IUPAP Conference on Computational Physics CCP2017
9–13 July 2017, Paris, France

Accepted papers received: 13 November 2018
Published online: 24 December 2018

Preface

Papers

IOP Conf. Series: Journal of Physics: Conf. Series **1136** (2018) 011001 doi:10.1088/1742-6596/1136/1/011001

XXIX IUPAP Conference on Computational Physics (CCP2017)

Riccardo Spezia,[1] **A. Marco Saitta**[2] **and Rodolphe Vuilleumier**[3]

[1]Sorbonne Université, CNRS, Laboratoire de Chimie Théorique, LCT, 4, Place Jussieu, 75252 Paris Cedex 05 (France)
[2]Sorbonne Université, Muséum National d'Histoire Naturelle, UMR CNRS 7590, IRD, Institut de Minéralogie, de Physique des Matériaux et de Cosmochimie, IMPMC, F-75005 Paris, France
[3]PASTEUR, Département de Chimie, Ecole Normale Supérieure, PSL University, Sorbonne Université, CNRS 75005 Paris (France)

E-mail: [1]riccardo.spezia@sorbonne-universite.fr, [2]marco.saitta@sorbonne.universite.fr, [3]rodolphe.vuilleumier@ens.fr

Editorial

The XXIX IUPAP Conference in Computational Physics (CCP2017) was held for the first time in its history in Paris, at the Campus Jussieu of the former Université Pierre et Marie Curie (recently merged with other institutes in Sorbonne Université), ideally located in the center of Paris.

The conferences was on three and a half days of talks and poster sessions covering all fields of computational physics. These include nine plenary lectures in the main auditorium of the conference center. The Young Scientific Prize 2017 of the IUPAP C20 commission was awarded on this occasion to Glen Evenbly. Furthermore, the conference was followed by one half-day spin-off meeting on the history of simulation sponsored by CECAM.

The IUPAP Conference in Computational Physics (CCP) is a series of conferences held annually under the auspices of the International Union of Pure and Applied Physics (IUPAP) on the basis of endorsement by its Commission on Computational Physics (C20). The application of computation is ever increasing and far-reaching in all aspects of science and so in physics. It is a major building block of the modern day science. This XXIX edition presented the latest techniques and discoveries of the computational scientists working in various branches of physics and closely related areas. In particular the main topics of the conference were: Education; Astrophysics; Nuclear, Particle and Field Physics; Atomic, Molecular and Optical Physics; Quantum Many Body Physics; Classical Statistical Mechanics; Fluid Dynamics: from Macro- to Nano-fluidics; Chemical Physics; Soft Matter and Biophysics; Materials Science; Energy Storage and Production; Geosciences and Climate Modeling. The Proceedings of the conference are aimed to show most of these topics, giving an useful landmark on the status of computational physics.

Finally, we are grateful to different organizations and sponsors who made this event possible and a success: Université Pierre et Marie Curie (which since January 2018 is Sorbonne Université) for hosting and helping us in many technical issues, then IUPAP and American Physical Society for supporting us.

IOP Conf. Series: Journal of Physics: Conf. Series **1136** (2018) 011001 doi:10.1088/1742-6596/1136/1/011001

Conference Organization
Here we list the people involved in different aspects of the organization with the affiliation at the time of the event.

Chairs
A. Marco Saitta (Université Pierre et Marie Curie)
Riccardo Spezia (CNRS / Université d'Evry-Val-d'Essonne)
Rodolphe Vuilleumier (Université Pierre et Marie Curie/ENS Paris)

Local Organizing Committee
Théau Debroise (Université Pierre et Marie Curie)
Pauline Galy (ENS Paris)
Benoit Grosjean (ENS Paris)
Veronica Macaluso (Université d'Evry)
Mathieu Moog (Université Pierre et Marie Curie)
Félix Mouhat (Université Pierre et Marie Curie / ENS Paris)
Andrea Perez-Villa (Université Pierre et Marie Curie)
Danielle Raddas (CNRS / Université Pierre et Marie Curie)
Farshud Sorourifar (Université Pierre et Marie Curie)

Local Scientific Committee
Marc Baaden (CNRS / IPBC)
Silke Biermann (Ecole Polytechnique)
Lydéric Bocquet (CNRS / ENS Paris)
Marie-Laure Bocquet (CNRS / ENS Paris)
Matteo Cacciari (Université Paris Diderot)
Michele Casula (CNRS / Université Pierre et Marie Curie)
Philippe Depondt (Université Pierre et Marie Curie)
Frédéric Derue (CNRS / Université Pierre et Marie Curie)
Jean-Louis Dufresne (CNRS / Ecole Polytechnique)
Benjamin Fuks (Université Pierre et Marie Curie)
Marie-Pierre Gaigeot (Université d'Evry-Val-d'Essonne)
Paul Indelicato (CNRS / Université Pierre et Marie Curie)
Damien Laage (CNRS / ENS Paris)
Michele Lazzeri (CNRS / Université Pierre et Marie Curie)
Fabio Pietrucci (Université Pierre et Marie Curie)
Nicolas Rambaux (Université Pierre et Marie Curie/Observatoire de Paris)
Benjamin Rotenberg (CNRS / Université Pierre et Marie Curie)
Mathieu Salanne (Université Pierre et Marie Curie)
Emmanuel Trizac (Université Paris Sud)
Angela Vasanelli (Université Paris Diderot)
Pascal Viot (Université Pierre et Marie Curie)
Félix Werner (CNRS / ENS Paris)

International Scientific Committee
Hai-Qing Lin, Chair of C20 (Bejing Computational Science Research Center, China)
Joan Adler (Israel Institut of Technology, Israel)
Constantia Alexandrou (The Cyprus Institute, Cyprus)

Wanda Andreoni (Ecole Polytechnique Fédérale de Lausanne, Switzerland)
Norbert Attig (Julich Supercomputing Center, Germany)
Stefano Baroni (Scuola Internazionale di Studi Superiori Avanzati, Trieste, Italy)
Sitangshu Bikas Santra (Indian Institute of Technology Guwahati, India)
Marc Etienne Brachet (Ecole Normale Supérieure de Paris, France)
Nithaya Chetty (National Research Foundation, South Africa)
Giovanni Ciccotti (Università di Roma "La Sapienza", Italy)
Pere Colet (Consejo Superior de Investigaciones Científicas, Spain)
Ronald Dickman (Federal University of Minas Gerais, Brasil)
Hans Fangohr (University of Southampton, UK)
Daan Frenkel (University of Cambridge, UK)
James Gubernatis (Los Alamos National Laboratory, USA)
Alex Hansen (Norvegian University of Science and Technology, Norway)
Suklyun Hong (Sejong University, South Korea)
Masatoshi Imada (University of Tokyo, Japan)
Barry M. Klein (University of California Davis, USA)
Werner Krauth (Ecole Normale Supérieure de Paris, France)
Georg Kresse (University of Vienna, Austria)
David P. Landau (University of Georgia, USA)
Richard Liska (Czech Technical University in Prague, Czech Republic)
Priya Mahadevan (S.N. Bose Natinal Center of Basic Sciences, India)
Michele Parrinello (ETH Zurch, Switzerland)
Purusattam Ray (The Institute of Mathematical Sciences, India)
Mathieu Salanne (Université Pierre et Marie Curie, France)
Anders W. Sandvik (Boston University, USA)
Lev Shchur (Russian Academy of Science, Russia)
Jianwei Sun (Temple University, USA)
Hideake Takabe (Osaka University, Japan)
Maria-Roser Valent (Univerity of Frankfurt, Germany)
Wei-Min Wang (Chinese Academy of Sciences, China)

IUPAP Young Scientist Award

The 2017 IUPAP Young Scientist Prize for Computational Physics was awarded during CCP2017 by the C20 Commission to Professor Glen Brian Evenbly from the University of Sherbrooke (Canada) for his exceptional work in the development of simulation algorithms for quantum many-body systems using the tensor network formalism. More details can be found in the web-page of IUPAP C20 commission (http://iupap.org/commissions/computational-physics/c20-news/).

Plenary Sessions Keynote Speakers

- Rainer Blatt (Universtität Innsbruck, Austria).
- Manuela Campanelli (Rochester Institute of Technology, USA).
- Cecilia Clementi (Rice University, USA).
- Giulia Galli (University of Chicago, USA).
- Eberhard K. U. Gross (Max Planck Institute of Microstructure Physics, Halle, Germany).
- Ulf-G Meißner (Universität Bonn/Forschungszentrum Jülich, Germany).
- Sauro Succi (CNR, Rome, Italy).

IOP Conf. Series: Journal of Physics: Conf. Series **1136** (2018) 011001 doi:10.1088/1742-6596/1136/1/011001

Parallel and Poster Sessions

The keynote speakers of the different parallel sessions are listed below:

- Education: Richard M Martin (University of Illinois, USA).
- Astrophysics: Parameswaran Ajith (International Centre for Theoretical Sciences, India).
- Nuclear, Particle and Fields Physics: Juan Rojo (Vrije University, Amsterdam, The Netherlands).
- Atomic, Molecular and Optical Physics: Evgeni Kozik (King's College London, UK).
- Quantum Many Body Physics: Tao Xiang (Chinese Academy of Sciences, China) and Ryotaro Arita (RIKEN Center for Emergent Matter Science, Japan).
- Classical Statistical Mechanics: Ying Wai Li (Oak Ridge National Laboratory, USA).
- Fluid Dynamics: from Macro- to Nano- fluidics: Roland Netz (TU Berlin, Germany).
- Chemical Physics: Baron Peters (University of California Santa Barbara, USA) and Sundaram Balasubramanian (Jawaharlal Nehru Centre for Advanced Scientific Research, India).
- Soft Matter and Biophysics: Gerhard Hummer (Max Planck Institute of Biophysics, Frankfurt, Germany).
- Materials Science: Stefano Baroni (SISSA, Trieste, Italy) and Xavier Blase (CNRS & Grenoble-Alpes University, France).
- Energy Storage and Production: Eva Perlt (Bonn University, Germany).
- Geosciences and Climate Modeling: Christoph Schaer (ETH Zurich, Switzerland).

IOP Conf. Series: Journal of Physics: Conf. Series **1136** (2018) 011002 doi:10.1088/1742-6596/1136/1/011002

Peer review statement

All papers published in this volume of *Journal of Physics: Conference Series* have been peer reviewed through processes administered by the proceedings Editors. Reviews were conducted by expert referees to the professional and scientific standards expected of a proceedings journal published by IOP Publishing.

IOP Conf. Series: Journal of Physics: Conf. Series **1136** (2018) 012001 doi:10.1088/1742-6596/1136/1/012001

Nuclear Physics as Precision Science

Ulf-G. Meißner

Universität Bonn, HISKP and Bethe Center for Theoretical Physics, D-53115 Bonn, Germany
Forschungszentrum Jülich, IKP-3, IAS-4, JARA-HPC and JCHP, D-52425 Jülich, Germany

E-mail: meissner@hiskp.uni-bonn.de

Abstract. Theoretical Nuclear Physics has entered a new era. Using the powerful machinery of chiral effective Lagrangians, the forces between two, three and four nucleons can now be calculated with unprecedented precision and with reliable uncertainties. Furthermore, Monte Carlo methods can be adopted to serve as a new and powerful approach to exactly solve problems in nuclear structure and reactions. I discuss the foundations of these new methods and provide a variety of intriguing examples. Variations of the fundamental constants of Nature can also be investigated and the consequences for the element generation in the Big Bang and in stars are considered. This sheds new light on our anthropic view of the Universe.

1. Short introduction

Nuclear physics has entered a new era, triggered by the work of Weinberg [1, 2]. The powerful machinery of chiral effective Lagrangians has given a completely new and systematic approach to nuclear forces and external currents. When combined with standard many-body techniques, these forces can also be used to address systematically topics in nuclear structure and reaction physics, see e.g. [3, 4, 5, 6, 7, 8]. In this talk, I will discuss a complementary and novel approach to investigate nuclear physics based on combining effective field theory (EFT) methods with powerful Monte Carlo simulation methods, called nuclear lattice EFT (NLEFT) or nuclear lattice simulations. As I will show, systematic and precise calculations become possible, providing also the much needed theoretical uncertainty, which is very difficult to estimate based on meson-exchange models or alike. This is important because theory much like experiment must provide uncertainties.

Before discussing these new developments, let me summarize a few basic facts about nuclear physics that underlie these developments. Atomic nuclei are self-bound systems of fermions, namely protons and neutrons, collectively called nucleons. The nucleons are bound by the strong force, that is balanced by the repulsion between the positively charged protons. Nuclei constitute non-relativistic systems that can be described by a nuclear Hamiltonian $H_{\rm nucl} = T+V$, with T the kinetic energy operator and V the potential. The latter is given by a sum of terms, $V = V_{\rm NN} + V_{\rm NNN} + \ldots$. For small and medium mass nuclei, the two-nucleon potential $V_{\rm NN}$ provides the largest part of the binding, but a small three-nucleon force $V_{\rm NNN}$ is required, see e.g. the review [9]. Furthermore, it is important to realize that nuclear binding energies are much smaller than nuclear masses, at most $\sim 8.5\,$MeV per nucleon in the Fe mass region. Chiral EFT allows one to systematically analyze the nuclear Hamiltonian based on the symmetries of the underlying gauge theory of the strong interactions, QCD.

IOP Conf. Series: Journal of Physics: Conf. Series **1136** (2018) 012001 doi:10.1088/1742-6596/1136/1/012001

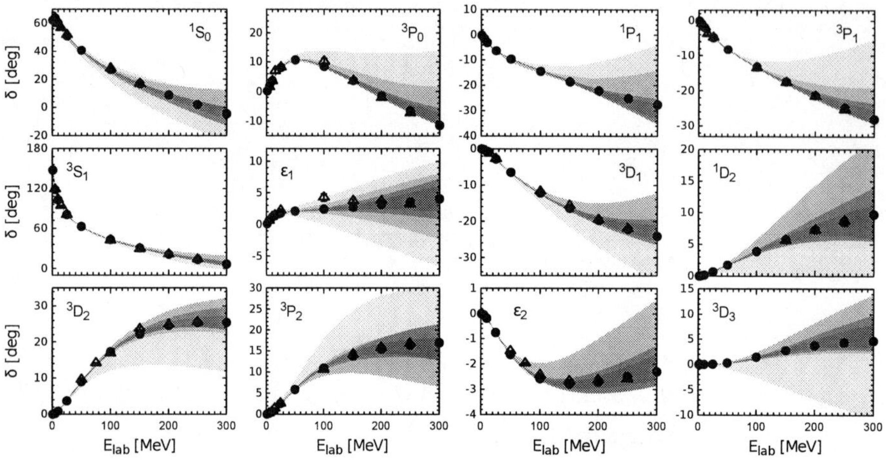

Figure 1. Phase shifts and mixing parameters for the neutron-proton interaction at NLO (yellow bands), N2LO (green bands), N3LO (blue bands) and N4LO (red bands).

2. Effective field theory for nuclear interactions

Let me first discuss a few new developments in the chiral EFT for nuclear forces and currents. For earlier work, I refer to the detailed review [10]. Following Weinberg, the nuclear potential is expanded in small momenta and/or pion masses making use of the power counting of chiral perturbation theory to leading order (LO), next-to-leading order (NLO) and so on. Three-nucleon forces appear at N2LO, which explains their smallness and four-nucleon forces start to contribute only at N3LO. There are two types of low-energy constants (LECs) that appear. The first set is related to pion-nucleon scattering. These LECs have recently been determined to high precision [11]. The other type of LECs that accompany the multi-nucleon contact interactions must be fixed by a fit to NN (and NNN) data. A refined analysis of the two-nucleon forces based on a combination of coordinate- and momentum space regulators at next-to-next-to-next-to-leading order (N3LO) in the expansion of the NN potential combined with a novel method to estimate the theoretical uncertainties was given in [12], for an application in few-nucleon systems, see [13]. The calculation of the forces was further refined to N4LO in [14]. At this order, no new LECs are involved and a very precise and systematically improved description of the NN phase shifts is obtained, c.f. Fig. 1. For related work on the peripheral phases, see [15]. It is obvious that a precise description with small uncertainties has been obtained. The more demanding task of working out the N3LO 3N forces based on [16, 17] and the N4LO forces from [18, 19] is presently under way by the LENPIC collaboration.

3. The framework of nuclear lattice simulations

NLEFT is a new method to tackle the nuclear many-body problem. Euclidean space-time is represented by a discrete hyper-cubic volume, $V = L\times L\times L\times L_t$, with spatial (temporal) length L (L_t) and corresponding lattice spacings a and a_t, respectively. The nucleons are placed on the lattice sites, see Fig. 2 (left panel). The interactions between the nucleons are given by the same chiral EFT potentials as in the continuum, simply adapted to the lattice formulation, see e.g. [20]. The Coulomb interaction between the protons can also straightforwardly be included [21]. A finite lattice spacing entails an UV cut-off, as the maximal momentum is the given by $p_{\max} = \pi/a$. For the most commonly used value of the lattice spacing, $a \simeq 2\,$fm, one has $p_{\max} = 314\,$MeV, which corresponds to a very soft interaction. However, it can be shown that the physics in the two-nucleon system does not depend on the lattice spacing if varied from 1 to 2 fm [22]. Monte Carlo methods can then be used to numerically exactly solve the A-body problem for a

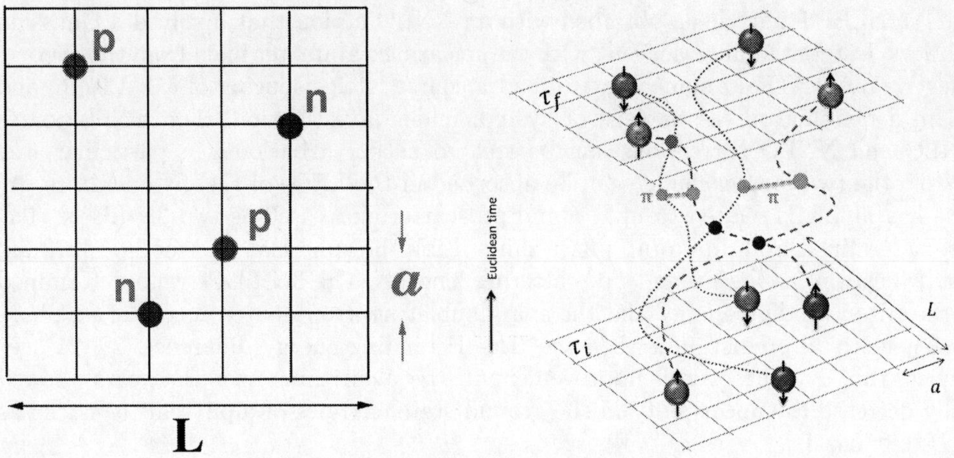

Figure 2. Left panel: Neutrons and protons on a space-time lattice with spatial length L and lattice spacing a. Right panel: Evolution of a ^{4}He nucleus in Euclidean time.

given set of NN and NNN interactions. A very important ingredient in these simulations is the approximate Wigner SU(4) symmetry of the nuclear interactions, that is crucial in suppressing the malicious sign oscillations that plague fermion MC studies at finite baryonic density [23, 24]. The remaining sign oscillations are caused by SU(4) non-symmetric contact terms as well as the one-pion-exchange. For more details, see the review [25].

The central object of NLEFT is the A-nucleon correlation function, $Z_A(t) = \langle\Psi_A|\exp(-tH)|\Psi_A\rangle$, with t the Euclidean time and Ψ_A a Slater determinant of A free nucleons or a more sophisticated correlated initial/final state. From the transient energy $E_A(t) = -d\ln Z_A(t)/dt$ one can infer the ground state energy of the A-nucleon system via $E_A^0 = \lim_{t\to\infty} E_A(t)$. Similarly, the expectation value of any normal-ordered operator follows from $Z_A^{\mathcal{O}} = \langle\Psi_A|\exp(-tH/2)\,\mathcal{O}\,\exp(-tH/2)\,|\Psi_A\rangle$ in the limit of infinite Euclidean time, $\lim_{t\to\infty}(Z_A^{\mathcal{O}}(t)/Z_A(t)) = \langle\Psi_A|\mathcal{O}\,|\Psi_A\rangle$. Excited state properties can also be extracted. In order to compute the low-lying excited states of a given nucleus, the Euclidean time projection method is generalized to a multi-channel calculation [26]. The Euclidean time evolution of a ^{4}He nucleus is depicted in the right panel of Fig. 2. Initial states are either properly antisymmetrized free standing waves of four particles or more complex correlated configurations. With the help of auxiliary fields, the multi-nucleon interactions and the pion exchanges can be mapped onto insertions on a single nucleon world-line, which makes such a computation most accessible for parallel computing. One major advantage of this approach is that all possible configurations are sampled, in particular also four nucleons on one lattice site. This already lets one suspect that clustering will emerge naturally in this approach.

4. Results from nuclear lattice simulations

4.1. General remarks

Before discussing results obtained using NLEFT, a few general remarks are in order. It is important to realize that nuclear structure and reactions dynamics should be treated on the same footing. This has important implications for the simulations. While originally all LECs have been determined in few-nucleon systems, which has led to a number of intriguing results, it was realized later that nucleus-nucleus collisions should also be used for determining some LECs as this appears to be advantageous in pinning down more precisely the three- and higher-body forces. Furthermore, the framework of nuclear lattice simulations could only be established as a novel quantum many-body method since one was able to solve problems that before could not be mastered in the well established schemes like e.g. the no-core-shell model.

IOP Conf. Series: Journal of Physics: Conf. Series **1136** (2018) 012001 doi:10.1088/1742-6596/1136/1/012001

Most results in NLEFT have been obtained with an NNLO action that involved a Gaussian smearing of the two LO contact interactions, with the smearing parameter fixed from the average S-wave np effective range. The canonical lattice had a coarse lattice spacing of $a = 1.97$ fm and $L \simeq 10 \ldots 16$ fm depending on the nucleus or system under investigation. For such a coarse lattice, the NLO and NNLO corrections can be treated as perturbations, in particular, the contribution from the two-pion exchange can be absorbed in the LECs of the 4N operators. At this order, one has 11 LECs related to np, nn and pp scattering as well as two 3N LECs. The 2N LECs were determined from fits to np phase shifts using the spherical wall method [27] and its refinement [28] as well as the nn and pp scattering lengths. The 3N LECs were determined from a fit to the triton binding energy and the spin-doublet neutron-deuteron scattering phase shift. The first non-trivial prediction is then the ^{3}He-^{3}H binding energy difference [29, 21], see the first entry in Tab. 1. The overall improvement in the algorithms over the years allowed to substantially decrease the uncertainty in the ground state energies of alpha-cluster nuclei as listed in Tab. 1, see [30, 31].

E [MeV]	^{3}He-^{3}H	^{4}He	^{8}Be	^{12}C	^{16}O	^{20}Ne	^{24}Mg	^{28}Si
NLEFT	0.78(5)	−28.3(6)	−55(2)	−92(3)	−131(1)	−166(1)	−198(2)	−234(3)
Exp.	0.76	−28.3	−56.5	−92.2	−127.6	−160.6	−198.3	−236.5

Table 1. Ground state energies at NNLO. The improvement in the simulations is shown in the much decreased uncertainty of the heavier nuclei based on the new algorithm compared to the older calculations for the lighter nuclei up to carbon [29, 21].

Similarly, excited states can be computed with similar accuracy. In Fig. 3 the LO calculation of the first two 0^+ states in ^{12}C is shown, starting from various initial states (plane waves and alpha cluster states) [32]. One set of these initial states directly gives the ground state (left panel), whereas the other set first traces out the first excitation with the same quantum numbers as shown by the intermediate plateau (right panel). This is the famous Hoyle state [33]. The thermalization of the various initial states with growing Euclidean time to almost the same energy gives a handle on the systematic uncertainties inherent to the simulations. For more details, see e.g. [31].

Using this framework, a number of interesting results has been obtained, such as the first *ab initio* calculation of the Hoyle state in ^{12}C [26, 32], the study of the triple-alpha process under variations of some fundamental constants [34, 35], the calculation of the ground state energies of the alpha-cluster nuclei up to ^{28}Si with an accuracy of about 1% [30], an *ab initio* calculation of the spectrum and structure of ^{16}O [36], and the first ever microscopic calculation of alpha-alpha scattering [37]. However, the employed NNLO action works well for alpha-type nuclei, but is less precise for other systems. Therefore, new forms of smearing including also the pion-exchange as well as a non-local distribution of lattice creation and annihilation operators have been employed to gain further insight. Based on these improved LO actions, it was found that nuclear physics is near a quantum phase transition from a Bose gas to the nuclear liquid, where first alpha-cluster nuclei are formed [38]. Furthermore, isotopic chains from H to O could be calculated and new insights into nuclear clustering was obtained recently, including also a new algorithm that for the first time allows to calculate density distributions in nuclei and the corresponding form factors [39]. Some selected topics from this rich spectrum of results will be discussed in what follows. Most of these results have been obtained on supercomputers like JUGENE and JUQUEEN at the Forschungszentrum Jülich. The CPU scaling is approximately quadratic in atomic number, so nuclei up to $A \simeq 30$ have been investigated. Going to larger nuclei requires more fine-tuned actions to suppress the remaining sign oscillations.

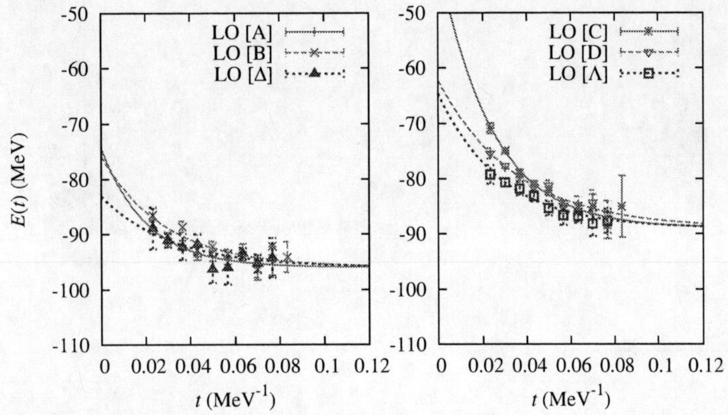

Figure 3. Results for the lowest 0^+ states in ^{12}C at LO. The left panel shows the results using various initial states, each of which approaches the ground state energy with increasing Euclidean time t. The right panel shows the results using other initial states. These trace out an intermediate plateau at an energy ~ 7 MeV above the ground state. Note that the respective energies are extracted from the large t behaviour, and not from any plateau as often done in lattice QCD.

4.2. Ab initio calculation of alpha-alpha scattering

Let us now consider α-α scattering as a prototypical nuclear reaction. This is related to the facts that processes involving α-type nuclei comprise a major part of stellar nucleosynthesis, and control the production of certain elements in stars. Also, ab initio calculations of scattering and reactions suffer from exponential or factorial scaling with the number of nucleons in the clusters, so therefore it was not possible so far to perform an *ab initio* calculation of α-α scattering. It is therefore a challenging task to use the lattice to tackle such type of processes. We note that on the lattice one only has discrete energy levels, and therefore a direct calculation of scattering processes appears impossible. This hurdle can be overcome by the so-called adiabatic projection method, that splits the problem of the calculation of scattering and inelastic reactions into two parts. First, using the Euclidean time projection method, one constructs an low-energy cluster Hamiltonian, called the adiabatic Hamiltonian. In the second step, one then computes scattering phase shifts or reaction amplitudes using this adiabatic Hamiltonian. The method was developed and refined in [40, 41, 42, 43, 44] and resembles in the methodology the Hamiltonian matrix approach combining the no-core-shell model with the resonating group method, see e.g. [45, 46, 47]. In more detail, the construction a low-energy effective theory for clusters proceeds as follows: One uses initial states as a direct product of two clusters located on the lattice, parameterized by the relative separation between the clusters, as shown in the left panel of Fig. 4

$$|\vec{R}\rangle = \sum_{\vec{r}} |\vec{r} + \vec{R}\rangle \otimes \vec{r}.$$

These are projected in Euclidean time with the chiral EFT Hamiltonian H, $|\vec{R}\rangle_\tau = \exp(-H\tau)|\vec{R}\rangle$. These so-called dressed cluster states include all possible interaction effects such a polarizations as well as deformations and, of course, the Pauli principle. The adiabatic Hamiltonian is then given by $[H_\tau]_{\vec{R}\vec{R}'} = {}_\tau\langle\vec{R}|H|\vec{R}'\rangle_\tau$. In general, this Hamiltonian needs to be normalized, which requires left and right multiplication with the corresponding norm matrices. What concerns the strong interactions, it can be shown that asymptotically, the adiabatic Hamiltonian is nothing but the free Hamiltonian for two clusters, eventually supplemented by infinite-range interactions as the Coulomb one. The underlying simulations can be simplified considerably by employing the so-called radial Hamiltonian based on the lattice version of

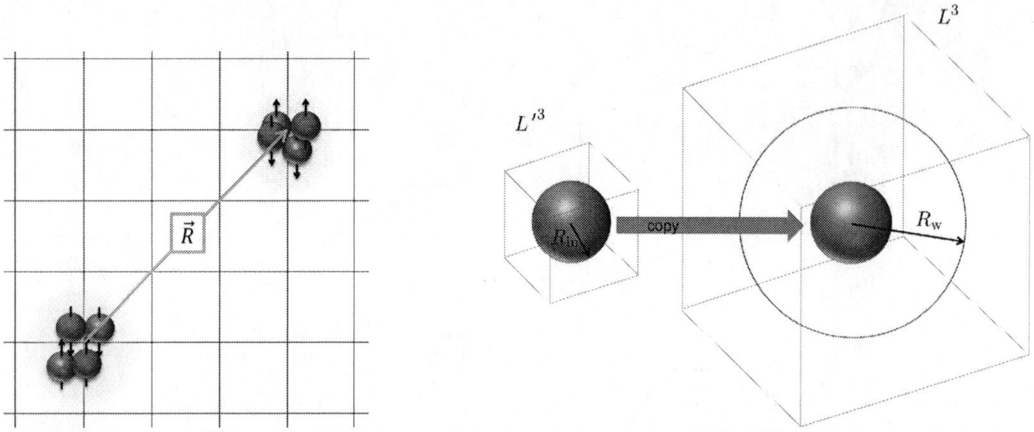

Figure 4. Left panel: A two-dimensional picture of the two-cluster initial state $|\vec{R}\rangle$ separated by the displacement vector $\vec{R}$. Right panel: A sketch of the lattices for the cluster-cluster calculations in the overlapping and in the noninteracting regions. $R_{\rm in}$ is the largest radial distance where the full adiabatic Hamiltonian is matched to the effective free cluster Hamiltonian without introducing any systematic errors. R_W indicates the radius of the spherical wall as discussed in the text.

angular momentum projection and binning the lattice points in rings of a given width. Further, the long-range Coulomb interaction can also be included exactly. For that, one performs first simulations in small box with a volume $L'^3 \sim (16\ {\rm fm})^3$, with all interactions switched on. This is the supplemented by a second set of simulations in a large box with a volume of about $L^3 \sim (120\ {\rm fm})^3$, where the strong interactions are turned off and the long-range Coulomb interaction is included by imposing Coulomb boundary conditions on a spherical wall with radius $R_W \simeq 40$ fm, see the right panel of Fig. 4. In that way, all effects of the strong and the electromagnetic interactions are included.

Using the same NNLO Hamiltonian as for the studies of the spectrum and structure of ^{12}C and ^{16}O, the S- and D-wave phase shifts have been computed in [37], as shown in Fig. 5. At LO in the employed counting, the Coulomb interaction is not included, so both the S- and D-wave phase shifts are off the data. This is visibly improved at NLO and further at NNLO for the D-wave. The small NNLO corrections in the S-wave are due to the coarse lattice spacing. Overall, one finds a good description of the scattering data. In the S-wave, we find a bound state corresponding to ^{8}Be that is bound by $-0.11(1)$ MeV, whereas in nature this nucleus is unbound by $+0.09$ MeV. This deviation of about 200 keV reflects the precision of the calculation. In the D-wave at NNLO, the resonance parameters are $E_R^{\rm NNLO} = 3.27(12)$ MeV and $\Gamma_R^{\rm NNLO} = 2.09(16)$ MeV, not far off the empirical data of 2.92(18) MeV and 1.35(50) MeV, respectively. May be the most significant result of this study is the fact that the computational time scales quadratically with the number of nucleons in the two clusters, $t_{\rm CPU} \sim (A_1 + A_2)^2$, with A_i the number of nucleons in cluster i $(i = 1, 2)$. This means that the computational time for the so-called holy grail of nuclear astrophysics, the radiative alpha capture on ^{12}C at stellar energies (given by the Gamov peak), $\alpha + ^{12}{\rm C} \to ^{16}{\rm O} + \gamma$, is in reach, requiring only 8 times as much CPU time as the computation of elastic α-α scattering (twice the number of nucleons and two channels). Before doing that, however, the chiral forces should be worked out to N3LO so as to reach the required accuracy.

4.3. New insights into nuclear clustering

Clustering in nuclei is an old but ever fascinating topic, introduced by Wheeler in 1937 in this seminal paper on "Molecular Viewpoints in Nuclear Structure" [49]. The most prominent type of

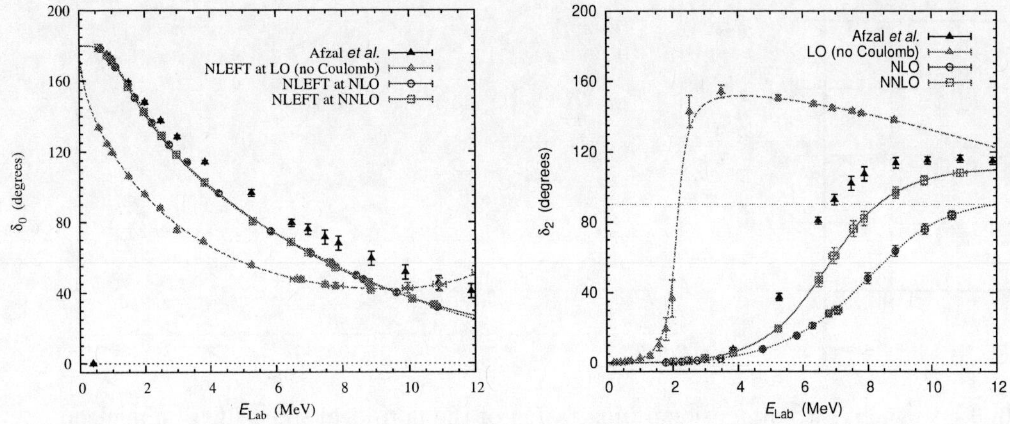

Figure 5. Left panel: S-wave α-α phase shift δ_0. Right panel: D-wave α-α phase shift δ_2. Shown are the NLEFT LO (green triangles), NLO (blue circles) and NNLO (red squares) results. The data (black triangles with error bars) are from [48].

clustering is the observation of α-particle substructures in light and medium heavy nuclei, and its eventual disappearance as the atomic number increases. There have been many works on alpha clustering, here I just mention recent work on alpha-clustering employing density functional methods by the Peking group [50] as well as work by the Paris-Zagreb group [51]. For a recent review, see e.g. Ref. [52].

As already mentioned above, alpha-clustering emerges naturally in NLEFT and a number of intriguing results on alpha-type nuclei and clustering have already been obtained, such as the first ab initio calculation of the Hoyle state or the observation that nuclear physics is close to a quantum phase transition from a Bose gas of α's to a nuclear liquid for α-type nuclei. However, when adding extra neutrons and/or protons, the precision of the calculations quickly deteriorates due to the remaining sign oscillations. To overcome this, a new LO action with smeared SU(4) local and non-local symmetric contact interactions as well as smeared one-pion exchange was constructed in [39]. The non-local smearing distributes any nucleon creation and/or annihilation operator over the six neighboring lattice sites as depicted in the left panel of Fig. 6, $a^{(\dagger)}_{\rm NL}(\mathbf{n}) = a^{(\dagger)}(\mathbf{n}) + s_{\rm NL}\sum_{\langle \mathbf{n}'\,\mathbf{n}\rangle} a^{(\dagger)}(\mathbf{n}')$, where $s_{\rm NL}$ is a real parameter, and the notation $\sum_{\langle \mathbf{n}'\,\mathbf{n}\rangle}$ represents the summation over nearest-neighbor lattice sites of the site $\mathbf{n}$. While this smearing was originally designed to just suppress the remaining sign oscillations when extra neutrons and/or protons are added to alpha-type nuclei, it turned out to work much better. For that, consider a LO action that is SU(4) symmetric with local and non-local smearing as well as smeared one-pion exchange. This action has three LECs, the strength of the SU(4)-symmetric contact term, the parameter related to the local smearing and the above-mentioned $s_{\rm NL}$. Fitting these to the average np S-wave scattering lengths and effective ranges and also the α-α S-wave scattering length, one can predict the isotopes chain from hydrogen to oxygen as shown in the right panel of Fig. 6. These have an accuracy of 0.7 MeV per nucleon or better. This is quite amazing given this highly simplified LO action. Clearly, NLO effects (and higher orders) need to be accounted for to achieve e.g. a better description of the $^1S_0$ np phase.

Using this action, one can also obtain deeper insight into nuclear clustering. For that, define as probes of alpha clusters the quantities $\rho_4 = \sum_{\mathbf{n}} : \rho^4(\mathbf{n})/4! :$ and $\rho_3 = \sum_{\mathbf{n}} : \rho^3(\mathbf{n})/3! :$. Here, ρ_4 couples to the center of the α-cluster while ρ_3 gets contributions from a wider portion of the alpha-particle wave function and thus these can be used for "measuring" cluster properties. Note that ρ_3 and ρ_4 depend on the regulator, the lattice spacing a, but not on the nucleus. However, the ratios $\rho_3/\rho_{3,\alpha}$ and $\rho_4/\rho_{4,\alpha}$ are free of short-distance ambiguities. If properly defined, the effective number of alpha clusters should be greater than or equal to N_α. A value equal to

IOP Conf. Series: Journal of Physics: Conf. Series **1136** (2018) 012001 doi:10.1088/1742-6596/1136/1/012001

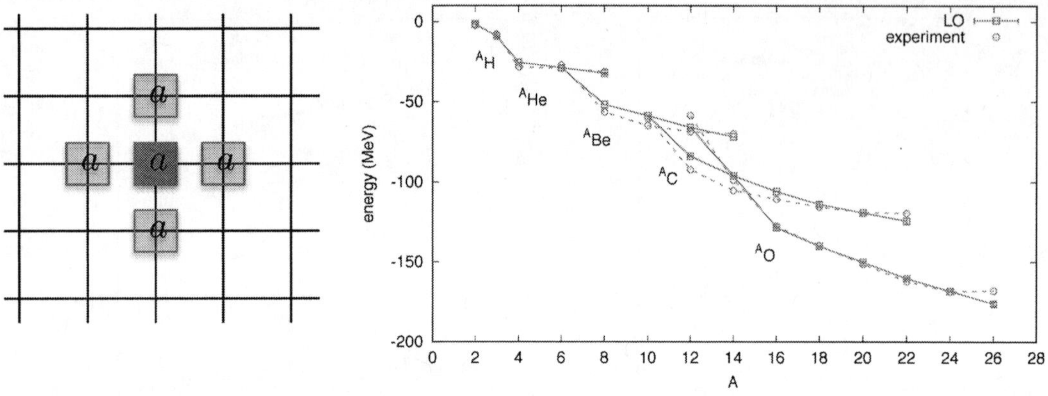

Figure 6. Left panel: Two-dimensional illustration of the non-local smearing of a nucleon creation/annihilation operator over the neighboring lattice sites. Right panel: Ground state energies versus number of nucleons A for the hydrogen, helium, beryllium, carbon, and oxygen isotopes (NLEFT: squares with error bars, experiment: circles). The errors are one-standard deviation error bars associated with the stochastic errors and the extrapolation to an infinite number of time steps.

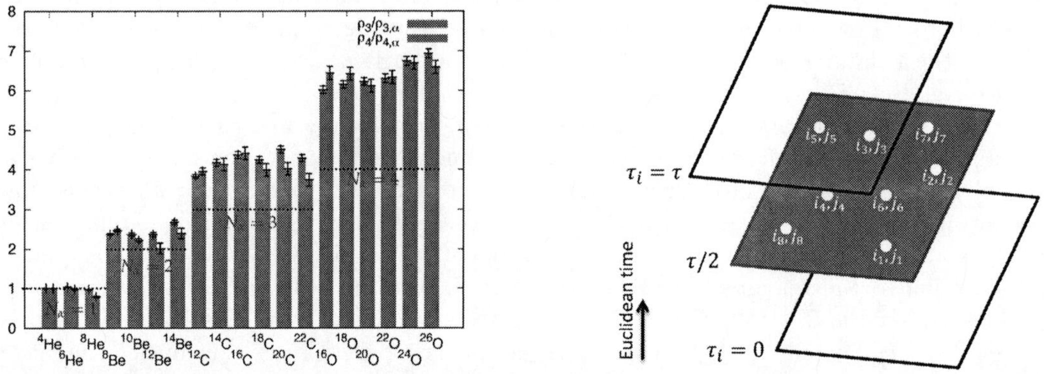

Figure 7. Left panel: The ratios $\rho_3/\rho_{3,\alpha}$ and $\rho_4/\rho_{4,\alpha}$ for the helium, beryllium, carbon and oxygen isotopes. For orientation, we also indicate by the dashed lines the expected number of α-clusters, N_α. Right panel: Sketch of the pinhole locations and spin-isospin indices at time $t = L_t a_t/2$.

N_α indicates that the alpha clusters are behaving as indivisible objects, and the nucleus can be regarded as a compound fluid of alpha particles and neutrons. If the effective number is significantly greater than N_α, then the description in terms of individual alpha clusters breaks down and the system behaves more as a nuclear liquid of protons and neutrons. The behavior is shown in the left panel of Fig. 7, where it is seen that for the oxygen isotope chain the entanglement between the clusters leads to expectation values of $\rho_3/\rho_{3,\alpha}$ and $\rho_4/\rho_{4,\alpha}$ much larger than 4. This shows that the transition from cluster-like states in light systems to nuclear liquid-like states in heavier systems should not be viewed as a simple suppression of multi-nucleon short-distance correlations, but rather as an increasing *entanglement* of the nucleons involved in the multi-nucleon correlations.

Another important development of Ref. [39] was the formulation of the so-called pinhole algorithm, see the right panel of Fig. 7. In general, auxiliary field quantum MC calculations involve states that are superpositions of many different center-of-mass positions, so a direct calculation of density distributions of nucleons in a nucleus is not possible. This can be overcome by inserting a screen with pinholes with spin and isospin labels that allows nucleons

IOP Conf. Series: Journal of Physics: Conf. Series **1136** (2018) 012001 doi:10.1088/1742-6596/1136/1/012001

with corresponding spin and isospin to pass. In that way, one measures the A-body density operator $\rho_{i_1,j_1,\cdots i_A,j_A}(\mathbf{n}_1,\cdots \mathbf{n}_A) = :\rho_{i_1,j_1}(\mathbf{n}_1)\cdots\rho_{i_A,j_A}(\mathbf{n}_A):$. MC sampling of the amplitude $A_{i_1,j_1,\cdots i_A,j_A}(\mathbf{n}_1,\ldots,\mathbf{n}_A,L_t) = \langle\psi(\tau/2)|\rho_{i_1,j_1,\cdots i_A,j_A}(\mathbf{n}_1,\ldots,\mathbf{n}_A)|\psi(\tau/2)\rangle$ then allows to measure the proton and neutron densities as well as more complicated two-, three- or higher-body correlations of nucleons within a given nucleus. This is because the pinhole sheet allows one to determine the center-of-mass (cm) of a given nucleus given simply by the minimal distance to all nucleons. Further, the resolution of this method is a/A because the cm position $\mathbf{r}_{\rm cm}$ is an integer $\mathbf{n}_{\rm cm}$ times a/A. Results for the proton and neutron distributions in the isotopes ^{12,14,16}C are shown in the left panel of Fig. 8. The proton size of $r_E^p = 0.84\,$fm is accounted for and asymptotic properties for the volume dependence of N-body bound states [54] have been used. Upon Fourier-transformation of these densities, one can obtain the corresponding elastic form factor. This is shown in the right panel of Fig. 8 for ^{12}C. Given the simplicity of the underlying Hamiltonian, the agreement is quite satisfactory. This paves the way for detailed nuclear structure studies.

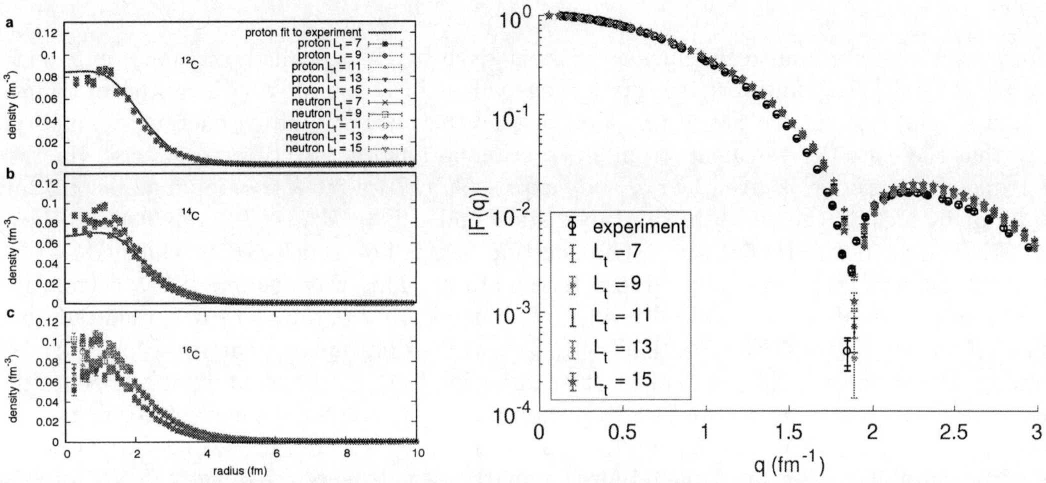

Figure 8. Left panel: Proton and neutron densities for the ground states of ^{12,14,16}C versus radial distance for MC data with $L_t = 9, 11, 13, 15$ time steps. The experimental data are from [53]. Right panel: Elastic form factor of ^{12}C. Data: black open symbols.

4.4. Fine-tunings and the multiverse

In nuclear physics, we observe a number of so-called fine-tunings, for some reviews and recent works, see e.g. [56, 57, 58, 59, 60]. A prominent example is the lightest nucleus, the deuteron. It is bound by a mere 2 MeV, just one percent of its total mass. Also, the aforementioned Hoyle state must be very closely placed to the triple-alpha threshold, in nature the energy difference is just 380 keV, much less than typical nuclear excitation energies of a few MeV. This close proximity is required so that in hot, old stars a sufficient amount of carbon and also oxygen is generated [33]. It is therefore natural to ask how much the fundamental parameters of the Standard Model can be detuned so that this resonance condition is no longer viable? First, however, we must find out what the relevant parameters are. Nuclear binding is a delicate balance between the attractive strong and the repulsive electromagnetic interactions. The latter are given in terms of Sommerfeld's fine-structure constant, $\alpha_{\rm EM} \simeq 1/137$. As concerns the strong interaction, the strong coupling constant α_S is intimately tied to the nucleon mass because of dimensional transmutation, and therefore the small light quark masses m_u, m_d are the relevant parameters that control nuclear binding. This appears at first counter-intuitive, as the major part of the nucleon mass is given by gluon field energy by means of the trace anomaly, while

IOP Conf. Series: Journal of Physics: Conf. Series **1136** (2018) 012001 doi:10.1088/1742-6596/1136/1/012001

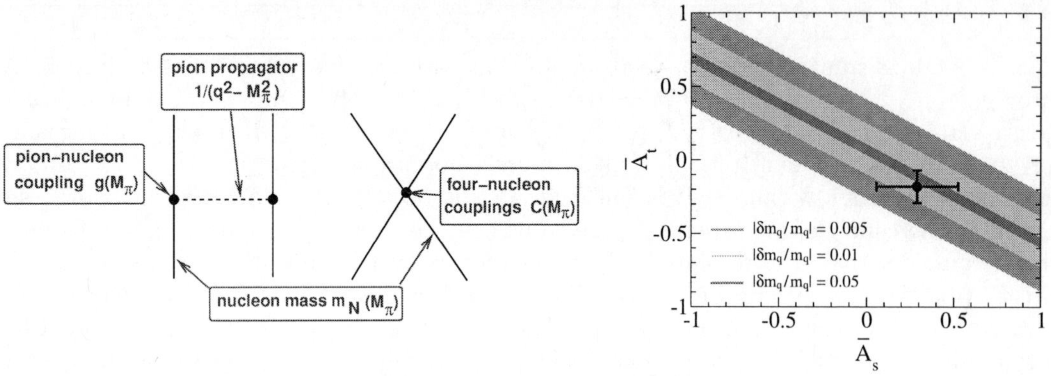

Figure 9. Left panel: Schematic drawing of the quark mass dependence in the two-nucleon force. Right panel: "Survivability bands" for carbon-oxygen based life due to 0.5% (broad outer band), 1% (medium band) and 5% (narrow inner band) changes in m_q in terms of the input parameters $\bar{A}_s$ and $\bar{A}_t$. The black cross denotes the results of the N2LO analysis from [66].

the light quark contribution to the nucleon mass is given by the so-called pion-nucleon σ-term, $\sigma_{\pi N} = 59(3)\,\text{MeV}$ [55]. However, the quark mass values of a few MeV (which are, of course, scale- and scheme-dependent) are of the same size as the nuclear binding energy per nucleon, E/A, so that these are the pertinent strong interaction parameters. To be more precise, the rate of the triple-alpha process is given by $r_{3\alpha} \sim \Gamma_\gamma \exp\left(-\Delta E/kT\right)$, with k the Boltzmann constant, T the temperature, Γ_γ the width of the Hoyle state and $\Delta E = E^\star_{12} - 3E_\alpha = 379.47(18)$ keV, with $E^\star$ the energy of the Hoyle state . The question now is how much ΔE be changed so that there is still enough ^{12}C and ^{16}O produced in the stars? This was answered in a calculation of the element generation in stars by varying ΔE but no other parameter. It turned out that the allowed variation is $\delta|\Delta E| \lesssim 100$ keV [61, 62], which does not appear to be any form of fine-tuning. However, one still has to make the connection to the fundamental parameters of the Standard Model. While is can be done for the electromagnetic interactions in cluster-type models as used e.g. in [61], the variation of the quark masses requires a more microscopic framework as provided by chiral EFT. This is depicted for the quark mass dependence of the LO NN force in the left panel of Fig. 9. Here, quark mass and pion mass dependence can be used synonymously, as the Gell-Mann–Oakes–Renner relation, $M_\pi^2 \sim m_u + m_d$, is fulfilled to better than 94% in QCD [63]. As can be seen from this figure, there are explicit (through the pion propagator) and implicit (through the pion-nucleon coupling, the nucleon mass and the four-nucleon couplings) pion mass dependences. All this can be accounted for systematically and precisely using chiral EFT. Coming back to the triple alpha-process, nuclear lattice simulations are the appropriate tool to study its dependence on the fundamental parameters, for details see [34, 35]. For that, one has to translate the condition $\delta|\Delta E| \lesssim 100$ keV into a constraint for the quark masses (and similarly for the fine-structure constant). For the quark masses, it reads (for fixed $\alpha_{\rm EM}$)

$$\left|\Big(0.571(14)\bar{A}_s + 0.934(11)\bar{A}_t - 0.069(6)\Big)\frac{\delta m_q}{m_q}\right| < 0.0015\ ,$$

with $m_q = (m_u + m_d)/2$ the average light quark mass (as strong isospin breaking plays no role here), $\bar{A}_{s,t} \equiv \partial a_{s,t}^{-1}/\partial M_\pi|_{M_\pi^{\rm phys}}$, where a_s, a_t denotes the singlet and the triplet NN scattering length, respectively. Independently of the precise values of these two quantities, it can be shown that the various fine-tunings in the triple-alpha process (the closeness of the ^{8}Be binding energy to the 2α threshold and the closeness of the Hoyle state to the 3α threshold) are indeed correlated. This had been speculated before [64] but could only be worked out precisely using NLEFT. Bounds on $\bar{A}_{s,t}$ had been obtained earlier based on resonance saturation of 4N operators [65] in [66] (see also [67]) as shown by the black cross in the right panel of Fig. 9. The fairly large uncertainty can eventually be overcome using lattice QCD to calculate these quantities. In the plane of $\bar{A}_s$-$\bar{A}_t$, varying the quark mass leads to diagonal bands whose widths depends on the assumed variations. This is shown for variations of $\delta m_q/m_q$ of 0.5, 1 and 5% by the three different bands. Clearly, a smaller variation leads to a broader band. If one focuses on the central value of $\bar{A}_{s,t}$, one finds that variations of $2-3\%$ of m_q are allowed so that the abovementioned condition is fulfilled. The large uncertainties in $\bar{A}_{s,t}$ do not allow for a more precise statement. For $\alpha_{\rm EM}$, no such uncertainties are present and it can be stated with certainty that it can be varied by at most 2.5%. Also, no other bounds are found if one varies both the quark masses and the fine-structure constant at the same time. This is clearly a stronger fine-tuning as for $|\Delta E|$ and its consequences for our anthropic view of the Universe are discussed in [68]. Lattice QCD can be used to tighten the bounds on $\bar{A}_{s,t}$, for the state-of-the-art see [69].

5. Summary and outlook

Let me briefly summarize the main messages of this talk:

- Chiral EFT for nuclear forces provides a precise framework for 2N and 3N forces with small uncertainties. It can also be formulated for varying strong and electromagnetic forces, which is a necessary requirement to study various fine-tunings in nuclear physics.
- Nuclear lattice simulations are a new quantum many-body approach that is based on the successful continuum nuclear chiral EFT. Already a number of intriguing results have been obtained based on NLEFT. In particular clustering emerges naturally and α-cluster nuclei are well described. Further, with an improved chiral action based on non-local smearing, neutron- and proton-rich nuclei can also be studied. With the invention of the pinhole algorithm, the calculation of charge densities and form factors has become possible. Furthermore, fine-tuning in nuclear reactions can be studied.
- Various bridges to lattice QCD studies need to be explored, in particular in pinning down some of the LECs related to multi-nucleon forces or the quark mass dependence of multi-nucleon operators.
- Finally, it must be said that many open issues in nuclear structure and reaction physics can now be addressed in a truly quantitative manner. For example, the "holy grail" of nuclear astrophysics [70], the *ab initio calculation* of the reaction ^{4}He+^{12}C $\to^{16}$O $+\,\gamma$, is in reach.

Acknowledgements

I thank the organizers for giving me the opportunity to present these thoughts at this wonderful conference. I am grateful to all my collaborators, who have contributed to my understanding of the issues discussed here. Partial financial support by the BMBF (05P15PCFN1), CAS (2017VMA0025), DFG (TRR 110), and VolkswagenStiftung is gratefully acknowledged.

References

[1] S. Weinberg, Phys. Lett. B **251** (1990) 288
[2] S. Weinberg, Nucl. Phys. B **363** (1991) 3
[3] P. Navratil, V. G. Gueorguiev, J. P. Vary, W. E. Ormand, A. Nogga, Phys. Rev. Lett. **99** (2007) 042501
[4] G. Hagen, T. Papenbrock, D. J. Dean and M. Hjorth-Jensen, Phys. Rev. Lett. **101** (2008) 092502
[5] T. Otsuka, T. Suzuki, J. D. Holt, A. Schwenk and Y. Akaishi, Phys. Rev. Lett. **105** (2010) 032501
[6] R. Roth, J. Langhammer, A. Calci, S. Binder, P. Navratil, Phys. Rev. Lett. **107** (2011) 072501
[7] A. Cipollone, C. Barbieri and P. Navrtil, Phys. Rev. Lett. **111** (2013) 062501
[8] J. E. Lynn, *et al.*, Phys. Rev. Lett. **113** (2014) 192501
[9] N. Kalantar-Nayestanaki, *et al.*, Rept. Prog. Phys. **75** (2012) 016301
[10] E. Epelbaum, H.-W. Hammer, U.-G. Meißner, Rev. Mod. Phys. **81** (2009) 1773
[11] M. Hoferichter, J. Ruiz de Elvira, B. Kubis and U.-G. Meißner, Phys. Rev. Lett. **115** (2015) 192301
[12] E. Epelbaum, H. Krebs and U.-G. Meißner, Eur. Phys. J. A **51** (2015) 53
[13] S. Binder *et al.* [LENPIC Collaboration], Phys. Rev. C **93** (2016) 044002 [arXiv:1505.07218 [nucl-th]].

[14] E. Epelbaum, H. Krebs and U.-G. Meißner, Phys. Rev. Lett. **115** (2015) 122301
[15] D. R. Entem, N. Kaiser, R. Machleidt and Y. Nosyk, Phys. Rev. C **91** (2015) 014002
[16] V. Bernard, E. Epelbaum, H. Krebs and U.-G. Meißner, Phys. Rev. C **77** (2008) 064004
[17] V. Bernard, E. Epelbaum, H. Krebs and U.-G. Meißner, Phys. Rev. C **84** (2011) 054001
[18] H. Krebs, A. Gasparyan and E. Epelbaum, Phys. Rev. C **85** (2012) 054006
[19] H. Krebs, A. Gasparyan and E. Epelbaum, Phys. Rev. C **87** (2013) 054007
[20] B. Borasoy, E. Epelbaum, H. Krebs, D. Lee, U.-G. Meißner, Eur. Phys. J. **A31** (2007) 105
[21] E. Epelbaum, H. Krebs, D. Lee and U.-G. Meißner, Eur. Phys. J. A **45** (2010) 335
[22] J. M. Alarcon *et al.*, Eur. Phys. J. A **53** (2017) 83
[23] E. Wigner, Phys. Rev. **51** (1937) 106
[24] J. W. Chen, D. Lee and T. Schäfer, Phys. Rev. Lett. **93** (2004) 242302
[25] D. Lee, Prog. Part. Nucl. Phys. **63** (2009) 117
[26] E. Epelbaum, H. Krebs, D. Lee, U.-G. Meißner, Phys. Rev. Lett. **106** (2011) 192501
[27] B. Borasoy, E. Epelbaum, H. Krebs, D. Lee, U.-G. Meißner, Eur. Phys. J. **A35** (2008) 343
[28] B. N. Lu, T. A. Lähde, D. Lee and U.-G. Meißner, Phys. Lett. B **760** (2016) 309
[29] E. Epelbaum, H. Krebs, D. Lee, U.-G. Meißner, Phys. Rev. Lett. **104** (2010) 142501
[30] T. A. Lähde, E. Epelbaum, H. Krebs, D. Lee, U.-G. Meißner and G. Rupak, Phys. Lett. B **732** (2014) 110
[31] T. A. Lähde, *et al.*, J. Phys. G **42** (2015) 034012
[32] E. Epelbaum, H. Krebs, T. A. Lähde, D. Lee and U.-G. Meißner, Phys. Rev. Lett. **109** (2012) 252501
[33] F. Hoyle, Astrophys. J. Suppl. **1** (1954) 121
[34] E. Epelbaum, H. Krebs, T. A. Lähde, D. Lee and U.-G. Meißner, Phys. Rev. Lett. **110** (2013) 112502
[35] E. Epelbaum, H. Krebs, T. A. Lähde, D. Lee and U.-G. Meißner, Eur. Phys. J. A **49** (2013) 82
[36] E. Epelbaum, *et al.*, Phys. Rev. Lett. **112** (2014) 102501
[37] S. Elhatisari, *et al.*, Nature **528** (2015) 111
[38] S. Elhatisari *et al.*, Phys. Rev. Lett. **117** (2016) 132501
[39] S. Elhatisari *et al.*, Phys. Rev. Lett. **119** (2017) 222505
[40] G. Rupak and D. Lee, Phys. Rev. Lett. **111** (2013) 032502
[41] M. Pine, D. Lee and G. Rupak, Eur. Phys. J. A **49** (2013) 151
[42] S. Elhatisari and D. Lee, Phys. Rev. C **90** (2014) 064001
[43] A. Rokash, M. Pine, S. Elhatisari, D. Lee, E. Epelbaum and H. Krebs, Phys. Rev. C **92** (2015) 054612
[44] S. Elhatisari, D. Lee, U. G. Meißner and G. Rupak, Eur. Phys. J. A **52** (2016) 174
[45] P. Navratil and S. Quaglioni, Phys. Rev. C **83** (2011) 044609
[46] P. Navratil, R. Roth and S. Quaglioni, Phys. Lett. B **704** (2011) 379
[47] P. Navratil and S. Quaglioni, Phys. Rev. Lett. **108** (2012) 042503
[48] S. A. Afzal, A. A. Z. Ahmad and S. Ali, Rev. Mod. Phys. **41** (1969) 247
[49] J. A. Wheeler, Phys. Rev. **52** (1937) 1083
[50] P. W. Zhao, N. Itagaki and J. Meng, Phys. Rev. Lett. **115** (2015) 022501
[51] J.-P. Ebran, E. Khan, T. Niksic and D. Vretenar, Nature **487** (2012) 341
[52] M. Freer, H. Horiuchi, Y. Kanada-En'yo, D. Lee and U.-G. Meißner, arXiv:1705.06192 [nucl-th], to appear in Rev. Mod. Phys.
[53] F. J. Kline, H. Crannell, J. T. O' Brien, J. Mccarthy and R. R. Whitney, Nucl. Phys. A **209** (1973) 381.
[54] S. König and D. Lee, Phys. Lett. B **779** (2018) 9
[55] M. Hoferichter, J. Ruiz de Elvira, B. Kubis and U.-G. Meißner, Phys. Rev. Lett. **115** (2015) 092301
[56] C. J. Hogan, Rev. Mod. Phys. **72** (2000) 1149
[57] R. L. Jaffe, A. Jenkins and I. Kimchi, Phys. Rev. D **79** (2009) 065014
[58] L. A. Barnes, Publ. Astron. Soc. Austral. **29** (2012) 529
[59] A. N. Schellekens, Rev. Mod. Phys. **85** (2013) 1491
[60] J. F. Donoghue, Ann. Rev. Nucl. Part. Sci. **66** (2016) 1
[61] H. Oberhummer, A. Csoto and H. Schlattl, Science **289** (2000) 88
[62] H. Schlattl, A. Heger, H. Oberhummer, T. Rauscher and A. Csoto, Astrophys. Space Sci. **291** (2004) 27
[63] G. Colangelo, J. Gasser and H. Leutwyler, Phys. Rev. Lett. **86** (2001) 5008
[64] S. Weinberg, "Facing Up", Harvard University Press, Cambridge, Massachusetts (2001).
[65] E. Epelbaum, U.-G. Meißner, W. Gloeckle and C. Elster, Phys. Rev. C **65** (2002) 044001
[66] J. C. Berengut, *et al.*, Phys. Rev. D **87** (2013) 085018
[67] P. F. Bedaque, T. Luu, L. Platter, Phys. Rev. **C83** (2011) 045803
[68] U.-G. Meißner, Sci. Bull. **60** (2015) 43
[69] V. Baru, E. Epelbaum and A. A. Filin, Phys. Rev. C **94** (2016) 014001
[70] W. A. Fowler, Rev. Mod. Phys. **56** (1984) 149

Transformed Lattice Rules for Feynman Loop Integrals on GPUs

E de Doncker[1], **A Almulihi**[1] **and F Yuasa**[2]

[1] Department of Computer Science, Western Michigan University, Kalamazoo MI 49008, U. S. A.

[2] High Energy Accelerator Research Organization (KEK), Oho 1-1, Tsukuba, Ibaraki, 305-0801, Japan

E-mail:
{elise.dedoncker, ahmedhassan.almulihi}@wmich.edu, fukuko.yuasa@kek.jp

Abstract. We focus on the application of rank-1 lattice rules with tanh and $\sin^m$ type transformations for the integration of functions which may have singularities on the boundaries of the integration domain. Using this technique we compute Feynman loop integrals for classes of 2-loop box and 3-loop self-energy Feynman diagrams that have up to eight internal lines with masses. The d-dimensional rule approximation with n points relies on a generator vector of length d with integer components, which is computed once and for all using the Component-by-Component method by Nuyens and Cools (2006). The generator vector is incorporated in the (CUDA) integration program that implements the lattice integration for fast execution on GPUs. Results are obtained efficiently and without special attention to specific problem characteristics.

1. Introduction

It is our goal to evaluate Feynman loop integrals accurately and efficiently. Thousands of Feynman diagrams may be associated with a particle interaction, each representing a possible configuration of virtual processes for the interaction and contributing a corresponding part to the total amplitude and thus to the probability or cross section of the interaction. The amplitude is expanded as a perturbation series, where the leading (lowest) order of approximation corresponds to tree level Feynman diagrams, and loop diagrams are needed for the higher orders. The increasing precision of parameter measurements in collider experiments necessitates the computation of loop integrals, underlying the theoretical predictions for present and future high-energy experiments. After the discovery of the Higgs boson at the Large Hadron Collider in 2012 [1, 2], new physics beyond the Standard model (SM) is expected but has not yet emerged. The motivation for the discovery of derivations from SM predictions increases, and the computation of higher order corrections including loop integrals in perturbative quantum field theory is of great importance. Detailed discussions on the precision levels are presented in [3, 4].

Analytic methods are generally not feasible for multi-loop integrals with a variety of masses, and the integral dimension increases with the number of loops and internal lines. Numerical computations need to deal with the dimension as well as integrand singularities and the large scale of the computations. We have targeted singularities with reduction, transformation and extrapolation methods [5, 6, 7, 8, 9, 10, 11, 12, 13, 14, 15, 16, 17, 18], and with adaptive and parallel strategies as further reported in [19, 20, 21, 22, 23, 24, 25, 26]. Accelerator boards such as GPUs can be utilized to parallelize approaches that involve a uniform computational structure. Product type approximations such as a product trapezoidal rule (typically applied with a transformation to relieve singularities at the boundaries of the domain) fall in this category but suffer from the dimensional effect. In this paper we explore GPU power for lattice rule integration combined with transformations for dealing with boundary singularities. In this context we investigate the accuracy and efficiency of $\tanh$ [27] and Sidi's $\sin^m$-transformations [28, 29].

Lattice integration was used with Sidi's transformations in programs for 2 and 3 dimensions in (*r2d2lri* [30], *elrint3d* [31]), with a product-wise implementation of the latter for 5-dimensional integration in *plrint5d* [32]. Lattice rules on GPU were recently applied to integrals resulting from sector decomposition in high energy physics [33]. In [34] we compared lattice rules using Sidi's $\sin^m$-transformations with $m = 2, 4, 6$ (and without transformation, $m = 0$) for Feynman integrals. We reported comparisons of lattice rules to Monte Carlo integration on GPUs in [35] for several classes of test functions in higher dimensions (up to dimension 25).

Importance sampling for variance reduction has been used extensively in Monte Carlo integration. The adaptive Monte Carlo integration and event generation package BASES/SPRING [36] uses a strategy similar to that of Vegas [37], and allows the user to select a set of important variables (coordinate directions) where the integrand function varies most. The implementation of Vegas in the Cuba package [38] incorporates sampling by Mersenne Twister and RANLUX pseudo-random numbers and by Sobol quasi-random (low discrepancy) sequences.

In subsequent sections we give some background for rank-1 lattice rules in Section 2, and outline transformations to treat singularities on the boundaries of the unit simplex domain in Section 3. Section 4 applies lattice rules with transformations to Feynman loop integrals for classes of 2-loop box and 3-loop self-energy diagrams, and compares GPU results for various numbers of points with respect to accuracy and time. Lattice generator vectors used for the application, as well as CUDA code for the integration kernel and for sample integrand device functions are given in appendices.

2. Lattice rules

A *Monte Carlo* approximation for the integral over the unit cube $\mathcal{C}_d = [0,1]^d$ in $\mathbb{R}^d$ takes the form

$$If = \int_{\mathcal{C}_d} f(\mathbf{x})\, d\mathbf{x} \approx \frac{1}{n}\sum_{j=0}^{n-1} f(\mathbf{x}_j), \tag{1}$$

where the $\mathbf{x}_j$ are uniform random. We consider *Quasi-Monte Carlo (QMC)* approximations of the same form as in Eq (1), but where the function evaluation points lie on a lattice within the half-open cube $[0,1)^d$. One direction to derive good rules is based on the theory of *optimal coefficients* or *good lattice points* introduced by Korobov [39, 40, 41] and Hlawka [42]. For overviews see, e.g., [43, 44, 45, 46].

A *rank-1 lattice rule* is represented as

$$Q(\mathbf{z}, n)f = \frac{1}{n}\sum_{j=0}^{n-1} f(\{\frac{j}{n}\mathbf{z}\}), \tag{2}$$

where $\mathbf{z}$ is an integer *generator* vector with components $z \in \mathcal{Z}_n = \{1 \le z < n,\ \gcd(z,n) = 1\}$, and $\{\mathbf{x}\}$ denotes the vector in $[0,1)^d$ obtained by taking the fractional part of each component of $\mathbf{x}$.

Assuming that $f(\mathbf{x})$ is one-periodic and its Fourier expansion

$$f(\mathbf{x}) = \textstyle\sum_{\mathbf{m}\in\mathbb{Z}^d} c_{\mathbf{m}} \exp(2\pi i\, \mathbf{m}\cdot\mathbf{x}) \quad \text{with} \quad c_{\mathbf{m}} = \int_{\mathcal{C}_d} \exp(-2\pi i\, \mathbf{m}\cdot\mathbf{x})\, f(\mathbf{x})\, d\mathbf{x}, \quad \mathbf{m}\in\mathbb{Z}^d$$

is absolutely convergent, and applying the rule of Eq (2) to the Fourier expansion leads to [44, 46]

$$Q(\mathbf{z}, n)f - If = \textstyle\sum_{\mathbf{m}\in\mathbb{Z}^d-\{0\}} c_{\mathbf{m}}\, Q(\mathbf{z}, n)\, \exp(2\pi i\, \mathbf{m}\cdot\mathbf{x}) = \sum'_{\mathbf{m}\cdot\mathbf{z}=0 (\mathrm{mod}\ n)} c_{\mathbf{m}}.$$

Here the prime on the sum indicates that the term for $\mathbf{m} = \mathbf{0}$ is omitted.
The class $\mathcal{E}^\alpha(\kappa)$, for $\alpha > 1$ and some $\kappa > 0$ is defined as the class of functions whose Fourier coefficients are bounded as

$|c_{\mathbf{m}}| \leq \kappa\, r_{\mathbf{m}}^{-\alpha}$, for $\mathbf{m} \neq \mathbf{0}$ and $r_{\mathbf{m}} = \prod_{j=1}^{d} \max\{1, |m_j|\}$.

Membership in the class $\mathcal{E}^{\alpha}(\kappa)$ characterizes the smoothness of the function f and allows for a bound on its integration error as

$$|Q(\mathbf{z}, n)f - If| \leq \kappa\, P_{\alpha}(\mathbf{z}, n), \tag{3}$$

where $P_{\alpha}(\mathbf{z}, n) = \sum'_{\mathbf{m}\cdot\mathbf{z}=0 (\mathrm{mod}\ n)} r_{\mathbf{m}}^{-\alpha}$. A sufficient condition for membership $f \in \mathcal{E}^{\alpha}(\kappa)$ is given by conditions of bounded variation on its mixed partal derivatives in the sense of Hardy and Krause [47, 46]. It is further noted that the bound can be reached (with an equality) in Eq (3).

With Korobov's definition [39], the components of the generator $\mathbf{z}$ are called *optimal components* (or $\mathbf{z}$ is called a *good lattice point* by Hlawka [42]) if, for $f \in \mathcal{E}^{\alpha}(\kappa)$,

$$|Q(\mathbf{z}, n)f - If| \leq \kappa\, \gamma(d, \alpha) \frac{(\log n)^{\beta(d,\alpha)}}{n^{\alpha}}, \tag{4}$$

with $\gamma(d, \alpha)$ and $\beta(d, \alpha)$ independent of n and $\alpha > 1$. Thus if $P_{\alpha}(\mathbf{z}, n) \leq \gamma(d, \alpha)\frac{(\log n)^{\beta(d,\alpha)}}{n^{\alpha}}$ in Eq (3), then $\mathbf{z}$ is a good lattice point according to Eq (4) (see [46]). Existence of good rank-1 rules was first proved by Korobov for n prime. Results by Niederreiter [43, 46] are used to show the existence of $\mathbf{z} \in \mathbb{Z}^d,\ d \geq 2,\ n \geq 2$ such that

$$P_{\alpha}(\mathbf{z}, n) \leq \frac{(2 \log n)^{\alpha d}}{n^{\alpha}} + \mathcal{O}(\frac{(\log n)^{\alpha d - 1}}{n^{\alpha}}). \tag{5}$$

Disney and Sloan [48] improved particularly on the factor $(2^{\alpha d})$ in the dominant term of Eq (5), by proving the existence of $\mathbf{z}$ such that $P_{\alpha}(\mathbf{z}, n)$ is asymptotically bounded by $\left(\frac{2e}{d}\right)^{\alpha d} \frac{(\log n)^{\alpha d}}{n^{\alpha}}$ for $d \geq 3$. For the sake of comparison consider the error bound of Monte Carlo integration with convergence rate of $\mathcal{O}(1/\sqrt{n})$ as $n \to \infty$.

The component-by-component (CBC) algorithm by Sloan and Reztsov [49] for the computation of the generator vector $\mathbf{z}$ runs in $\mathcal{O}(dn^2)$ time. For our implementation of the integration we pre-compute the generator vector with the fast CBC algorithm by Nuyens and Cools [50, 51], which runs in $\mathcal{O}(d n \log(n))$ time and $\mathcal{O}(n)$ space. Table A1 gives a list of the generator vectors we obtained for various numbers n of points and dimensions d. In the construction, weights can be associated with the coordinate directions (variables) according to their importance. However, the rules generated for the computations in this paper assumed equal weights. No a-priori information is provided for the integration problems.

The summation of Eq (2) is implemented in the CUDA kernel of Appendix B, with a structure similar to that of the dot product in [52]. In the `while` loop we compute the $\mathbf{x}$-components of the lattice evaluation points (in array `x`) from the generator given in array `z` on input, and perform the integrand evaluations. The array `cache` is shared by the threads over a block, and the reduction after the `while` loop sums the contributions into the `cache[0]` element for the block. The partial results over the blocks are returned in the array `q` and will be summed in the main program over the `blockdim` blocks. As the number of blocks is relatively small (e.g., `blockdim` = 64), this summation is carried out sequentially.

3. Transformations

Convergence results of lattice rules for functions in $\mathcal{E}^{\alpha}(\kappa)$ as those listed above motivate the use of periodizing transformations, such that the resulting function can be extended periodically. For the one-dimensional integral $I[0, 1]f = \int_0^1 f(x)\, dx$, the transformation $x = \varphi(u)$ yields

$I[0, 1]f = \int_0^1 f(\varphi(t))\, \varphi'(t)\, dt$.

Periodizing transformations include [44]

$\varphi(t) = 1 - 2|t - \frac{1}{2}|$, and

$\varphi(t) = \varphi_m(t) = (2m + 1)\binom{2m}{m} \int_0^t u^m (1 - u)^m\, du$ (Korobov's transformation [41]).

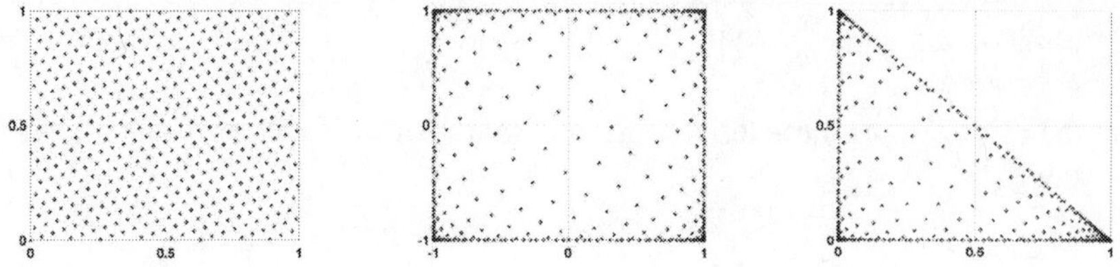

(a) (b) (c)

Figure 1. (a) Lattice points for $d = 2,\ n = 503,\ \mathbf{z} = (1, 192)$; (b) Transformed lattice points with tanh; (c) Transformed lattice points with tanh in unit triangle

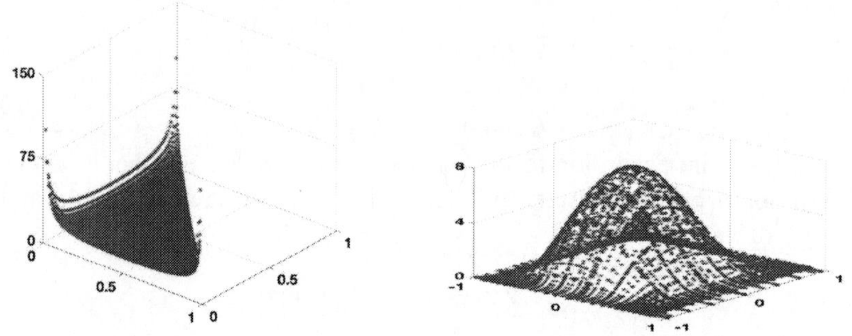

(a) (b)

Figure 2. (a) Function $1/\sqrt{x_1 x_2(1 - x_1 - x_2)}$ over unit triangle; (b) Transformed integrand $1/\sqrt{x_1 x_2(1 - x_1 - x_2)}$ × Jacobian with *tanh* transformation

We further consider the class of $\sin^m$-transformations by Sidi [28, 29],
$\varphi(t) = \Psi_m(t) = \frac{\theta_m(t)}{\theta_m(1)}$ where $\theta_m(t) = \int_0^t \sin^m(\pi u)\, du,\ m = 1, 2, \ldots,$
in particular,

$$\Psi_6(t) = t - (45\sin(2\pi t) - 9\sin(4\pi t) + \sin(6\pi t))/(60\pi) \text{ with } \Psi_6'(t) = \frac{16}{5}\sin^6(\pi t), \quad (6)$$

and the tanh transformation by Sag and Szekeres [27], given for $I[0,1]f = \int_{-1}^{1} f(\frac{t+1}{2})\, dt$ by

$$t = \varphi(u) = \tanh(\frac{cu}{1-u^2}), \text{ so that } I[0,1]f = \frac{c}{2}\int_{-1}^{1} f(\frac{\varphi(u)+1}{2})\frac{1+u^2}{(1-u^2)^2}\,\text{sech}^2(\frac{cu}{1-u^2})\, du \quad (7)$$

(where $c > 0$ is a constant). The Jacobian of the tanh transformation in Eq (7) and all its derivatives vanish at the end-points of the transformed integral. The latter property is also satisfied by the IMT [53] and DE (double exponential) [54, 55] type transformations. Korobov's and Sidi's transformations $\varphi(t) = \varphi_m(t)$ and $\varphi(t) = \Psi_m(t)$ satisfy that $\varphi'(t) = \mathcal{O}(t^m)$ as $t \to 0+$, and $\varphi'(t) = \mathcal{O}((1-t)^m)$ as $t \to 1-$ (see [28]). Whereas a low order Ψ_m such as $\Psi_1(t) = (1 - \cos(\pi t))/2$ may be used to transform a smooth nonperiodic function $f(x)$ into a smooth periodized function $f(\Psi_m(t))\Psi_m'(t)$, higher orders can help alleviate or smoothen boundary singularities more effectively through the occurrence of the Jacobian proportional with $\sin^m(\pi t)$.

Fig 1(a) illustrates the $n = 503$ lattice points of the 2D rule with $\mathbf{z} = (1, 192)$. Fig 1(b) and (c) show the lattice points transformed with the tanh-transformation (the latter to the unit triangle). Note that the transformed points spread out from the middle of the region, moving toward the edges. Especially for

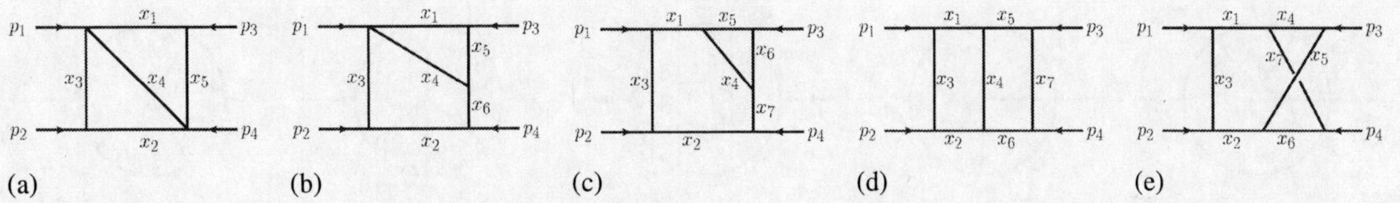

Figure 3. [2LB] Sample 2-loop box diagrams [56, 23]: (a) $N = 5$ (double-triangle), (b) $N = 6$ (tetragon-triangle), (c) $N = 7$ (pentagon-triangle), (d) $N = 7$ ladder, (e) $N = 7$ crossed box

rules with a large number of points this may lead to computational problems with over- or underflows when the integrand is evaluated at points very close to the boundaries of the domain.

With respect to Fig 2 consider the integral of $f(x_1, x_2) = \frac{1}{\sqrt{x_1 x_2 (1-x_1-x_2)}}$ over the unit triangle $\mathcal{S}_2$,

$$I(\mathcal{S}_2)f = \int_0^1 dx_1 \int_0^{1-x_1} dx_2\, f(x_1, x_2) \tag{8}$$

$$= \int_0^1 dx_1 (1-x_1) \int_0^1 dx_2\, f(x_1, (1-x_1)x_2) = \frac{1}{4}\int_{-1}^1 dt_1 \frac{1-t_1}{2} \int_{-1}^1 dt_2\, f(\frac{1+t_1}{2}, \frac{1-t_1}{2}\frac{1+t_2}{2})$$

$$= \frac{1}{8}\int_{-1}^1 du_1\, (1-\varphi(u_1)) \int_{-1}^1 du_2\, f(\alpha_1, \alpha_2)\, \varphi'(u_1)\, \varphi'(u_2), \tag{9}$$

with $\varphi(u) = \tanh(\frac{c\,u}{1-u^2})$, $c = 3$, and $\alpha_1 = \frac{1}{2}(1+\varphi(u_1))$, $\alpha_2 = \frac{1}{4}(1-\varphi(u_1))\,(1+\varphi(u_2))$. Then Fig 2(a) depicts the integrand $f(x_1, x_2)$ of Eq (8) as a function of x_1 and x_2 over $\mathcal{S}_2$, and Fig 2(b) plots the integrand of Eq (9) as a function of u_1 and u_2 over $[-1, 1]^2$. Even though the original integrand has singularities at the edges of the triangle, the transformed integrand is zero at the boundaries of the transformed domain. A similar behavior is depicted for Sidi's Ψ_4 transformation in [34]. The interest in the triangle region $\mathcal{S}_2$ is motivated by the computation of Feynman integrals over simplex domains.

4. Application to Feynman Diagrams

4.1. Feynman loop integrals

We represent a Feynman integral with L loops and N internal lines by $\mathcal{F} = (4\pi)^{-\nu L/2}\mathcal{F}_{L_N}$ with

$$\mathcal{F}_{N,L} = \Gamma(N - \frac{\nu L}{2})(-1)^N \int_{\mathcal{C}_N} \prod_{r=1}^{N} dx_r\, \delta(1 - \sum x_r) \frac{C^{N-\nu(L+1)/2}}{(D - i\varrho C)^{N-\nu L/2}}, \tag{10}$$

where $\nu = 4$ is the space-time dimension, and C and D are polynomials arising from determinants and depending on the topology of the Feynman diagram and physical parameters. C and D are homogeneous of degree L and $L+1$, respectively, in the integration variables x_r. The δ-function in Eq (10) can be eliminated by expressing one of the integration variables in terms of the other, leading to

$$\mathcal{F}_{L,N} = \Gamma(N - \frac{\nu L}{2})(-1)^N \int_{\mathcal{S}_{N-1}} \frac{C^{N-\nu(L+1)/2}}{(D - i\varrho C)^{N-\nu L/2}}\, d\mathbf{x}. \tag{11}$$

Here the domain is the $d = (N-1)$-dimensional unit simplex, $\mathcal{S}_d = \{\mathbf{x} \in \mathcal{C}_d \mid 0 \le \sum_{k=1}^{d} x_k \le 1\}$. The integral over $\mathcal{S}_{N-1}$ can then further be transformed to $\mathcal{C}_{N-1}$.

Subsequently we give numerical results for the 2-loop box integrals of Fig 3 and the 3-loop self-energy diagrams of Fig 4. We compute the C and D functions as outlined in [12]. Sample CUDA code for the integrand of Fig 4(e) (with $N = 8$) transformed from $\mathcal{S}_7$ with a tanh type transformation to the cube $\mathcal{C}_7$ is given in Appendix C.1. The integral approximation will be multiplied in the main program with the factor `pow(c*half,d)` $= (\frac{c}{2})^7$ (from the Jacobian of the tanh transformation in Eq (7)).

Similarly, Appendix C.2 lists the integrand for Fig 3(b) (with $N = 6$) as a CUDA device function using Ψ_6 after a transformation from the simplex $\mathcal{S}_5$ to the unit cube $\mathcal{C}_5$. Here the constant `con` $= (\frac{16}{5})^d = (3.2)^5$ arises from the Jacobian of the Ψ_6 transformation in Eq (6).

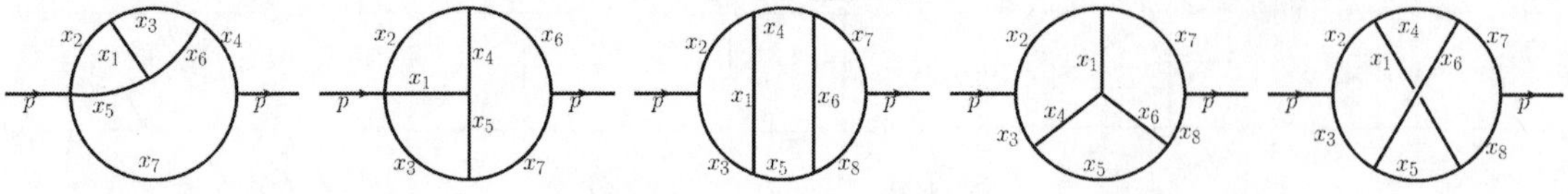

(a) (b) (c) (d) (e)

Figure 4. [3LS] Sample 3-loop self-energy diagrams with massive internal lines, cf., Laporta [56]: (a) $N = 7$, (b) $N = 7$, (c) $N = 8$, (d) $N = 8$, (e) $N = 8$.

Table 1. Results for 2-loop box diagrams, LR with tanh transformation

DIAG:	[2LB](a)			[2LB](b)			[2LB](c)			[2LB](d)			[2LB](e)		
EXACT:	0.9509235623171			0.276209225359			0.1723367907503			0.1036407209893			0.0853513981538		
	E_a		$T(s)$	E_a		$T(s)$	E_a		$T(s)$	E_a		$T(s)$	E_a		$T(s)$
n / d_R:	4	6		5	10		6	10		6	10		6	12	
10M	1.9e-09	4.0e-12	0.02	6.7e-08	1.5e-09$^{(7)}$	0.03	2.6e-06	2.4e-06	0.03	8.7e-08	3.3e-07	0.03	5.4e-09	8.5e-08$^{(10)}$	0.03
100M	4.3e-13	3.1e-13	0.20	9.3e-10	1.8e-11	0.26	6.8e-09	6.7e-09$^{(12)}$	0.31	1.0e-09	5.6e-11	0.30	4.0e-12	3.6e-10	0.32
200M	3.9e-13	1.9e-11	0.40	3.7e-12	1.7e-12	0.51	2.3e-09	2.5e-09	0.61	3.2e-08	1.4e-10	0.61	3.5e-10	4.2e-12	0.64
350M	2.6e-12	3.1e-12	0.71	1.4e-11	7.4e-12	0.90	7.1e-10	7.5e-09$^{(12)}$	1.09	1.0e-09	2.3e-11	1.07	4.3e-14	1.6e-14	1.12
400M	1.0e-13	4.1e-13	0.80	6.8e-13	2.1e-12	1.02	3.2e-10	5.9e-08	1.26	4.7e-12	1.0e-11	1.21	2.1e-11	1.5e-12	1.27

Table 2. Results for 2-loop box diagrams, LR with Ψ_6 transformation

DIAG:	[2LB](a)			[2LB](b)			[2LB](c)			[2LB](d)			[2LB](e)		
EXACT:	0.9509235623171			0.276209225359			0.1723367907503			0.1036407209893			0.0853513981538		
	E_a		$T(s)$	E_a		$T(s)$	E_a		$T(s)$	E_a		$T(s)$	E_a		$T(s)$
n / d_R:	4	6		5	10		6	10		6	10		6	12	
10M	4.0e-10	1.3e-11	0.02	1.4e-08	2.2e-09$^{(7)}$	0.03	1.4e-06	1.5e-06	0.03	1.5e-08	2.7e-07	0.03	1.3e-08	4.4e-08$^{(10)}$	0.03
100M	2.9e-14	1.7e-13	0.21	6.2e-11	1.4e-11	0.25	6.8e-09	1.7e-09$^{(12)}$	0.30	1.1e-10	8.1e-13	0.30	2.6e-13	3.7e-12	0.31
200M	2.7e-13	2.7e-12	0.41	3.1e-12	3.9e-12	0.51	3.0e-10	2.8e-10	0.60	3.3e-09	1.7e-11	0.60	2.6e-11	1.0e-12	0.62
350M	2.5e-12	2.5e-12	0.71	7.0e-12	3.8e-12	0.89	1.5e-10	1.9e-09$^{(12)}$	1.05	6.1e-13	8.2e-13	1.06	2.2e-12	5.2e-13	1.09
400M	1.3e-13	5.4e-13	0.82	1.8e-12	4.5e-12	1.01	1.2e-11	1.2e-08	1.20	7.6e-15	3.3e-12	1.20	5.2e-13	3.5e-12	1.24

4.2. GPU results

The numerical calculations are performed on a cluster node of the *thor* cluster at WMU. The node has dual 8-core, 2.6 GHz Intel Xeon E5-2670 processors, and 128 GB of memory, and furthermore an NVIDIA Tesla K20m GPU, which has 2496 CUDA cores and 4800 MB global memory.

Tables 1 and 2 report results for the 2-loop box diagrams of Fig 3 using lattice rules with $n = 10, 100, 200, 350$ and 400M (million) points, and with the tanh and Sidi's Ψ_6 transformation, respectively. We list the *exact* values from [56] – except for the crossed box box diagram of Fig 3(e), where the integral value resulted from an extensive computation with large numbers of integrand evaluations using the adaptive ParInt code in long double precision [16]. The integral dimensions for the diagrams [2lb](a-e) are $d = 4, 5, 6, 6, 6$. The lattice rule sums are generated from a **z**-vector for dimension $d_R = d$ and a **z**-vector for a dimension $d_R > d$; the rule for the latter is applied by using the first d components of the generator.

For each result the absolute error and the time in seconds are reported. Note that, where the accuracy is listed as $E_a{}^{(d'_R)}$ in a column labeled d_R, the rule for dimension d'_R was used. In a number of cases the accuracy of the higher (d_R)-dimensional rule is comparable to (and occasionally better than) that of the d-dimensional rule. For each integral, the times for the same number of points n are very close for different d_R; in any case the average time is given. The timings for corresponding cases in Table 1 (with tanh) and Table 2 (with Ψ_6) are generally very similar. Overall the accuracies are somewhat better with

Table 3. Results for 3-loop self-energy diagrams, LR with tanh transformation

DIAG:	[3LS](a)			[3LS](b)			[3LS](c)			[3LS](d)			[3LS](e)		
EXACT:	1.326448208272			1.341399241447			0.2796089232826			0.1826272375392			0.1480133039584		
	E_a		$T(s)$	E_a		$T(s)$	E_a		$T(s)$	E_a		$T(s)$	E_a		$T(s)$
n / d_R:	6	10		6	10		7	10		7	10		7	10	
10M	5.0e-04	5.3e-05$^{(12)}$	0.03	3.2e-04	2.1e-04$^{(7)}$	0.03	2.6e-04	9.8e-06$^{(12)}$	0.04	8.9e-05	7.5e-06	0.04	5.5e-05	4.6e-06	0.04
100M	9.8e-06	1.1e-05$^{(8)}$	0.31	8.9e-06	7.9e-06$^{(8)}$	0.31	1.7e-05	2.1e-06$^{(8)}$	0.37	4.0e-06	4.5e-06$^{(12)}$	0.37	5.3e-07	7.7e-07	0.37
200M	1.5e-05*	2.5e-06	0.63	3.1e-04	2.0e-07$^{(12)}$	0.62	8.9e-07	5.2e-07	0.73	1.4e-07	1.8e-06	0.75	7.5e-07	3.1e-07	0.75
350M	1.1e-05	5.7e-07	1.11	1.8e-05*	3.0e-07	1.09	5.5e-07	2.7e-07$^{(8)}$	1.28	4.6e-07	5.5e-07	1.31	2.3e-07	1.8e-07$^{(12)}$	1.31
400M	3.2e-06	5.7e-06$^{(8)}$	1.26	3.5e-06	3.8e-06$^{(8)}$	1.24	3.0e-07	2.4e-08	1.46	2.4e-08	5.1e-08	1.49	1.9e-10	9.1e-08$^{(12)}$	1.49

Table 4. Results for 3-loop self-energy diagrams, LR with Ψ_6 transformation

DIAG:	[3LS](a)			[3LS](b)			[3LS](c)			[3LS](d)			[3LS](e)		
EXACT:	1.326448208272			1.341399241447			0.2796089232826			0.1826272375392			0.1480133039584		
	E_a		$T(s)$	E_a		$T(s)$	E_a		$T(s)$	E_a		$T(s)$	E_a		$T(s)$
n / d_R:	6	10		6	10		7	10		7	10		7	10	
10M	3.2e-06	9.9e-05$^{(7)}$	0.03	1.0e-05	5.9e-05$^{(12)}$	0.03	8.7e-05	4.0e-06$^{(10)}$	0.04	3.6e-06	2.2e-05	0.04	2.1e-05	1.8e-06	0.04
100M	2.5e-07	6.4e-07$^{(7)}$	0.31	2.7e-06	1.4e-06$^{(7)}$	0.31	5.3e-07	1.6e-08$^{(12)}$	0.37	2.3e-06	7.5e-07$^{(12)}$	0.37	6.9e-08	8.9e-08$^{(8)}$	0.36
200M	3.6e-07	6.6e-07$^{(8)}$	0.63	2.3e-05	6.1e-07$^{(7)}$	0.63	7.5e-08	5.0e-09	0.73	1.3e-07	2.3e-08$^{(8)}$	0.73	2.3e-07	2.7e-08$^{(8)}$	0.72
350M	1.0e-06	4.4e-07$^{(12)}$	1.10	2.7e-07	1.9e-07	1.10	1.4e-07	2.8e-08	1.31	4.3e-08	4.6e-09	1.28	4.8e-09	3.0e-08	1.26
400M	1.1e-07	6.9e-07$^{(8)}$	1.25	9.3e-06	3.5e-08$^{(8)}$	1.25	5.9e-08	7.8e-09	1.49	3.1e-08	5.6e-07$^{(12)}$	1.46	4.2e-08	2.2e-09	1.44

the Ψ_6 transformation. We also obtained results with Ψ_4 (not shown), which were however less accurate than with Ψ_6. Preliminary tests with Ψ_8 often yielded less accuracy, and some runs resulted in overflows; therefore Ψ_8 was not further considered.

Tables 3 and 4 give results for the 3-loop self-energy diagrams of Fig 4, for which the integrals are of dimensions $d = 6, 6, 7, 7, 7$. In the columns labeled d_R of these tables, somewhat better results obtained with generators for dimensions d'_R are listed (as $E_a{}^{(d'_R)}$), which are more accurate than the corresponding d-dimensional approximations in some cases. For the tanh results in Table 3, two results appear with a superscript *. In these cases the tanh transformation with constant $c = 3$ resulted in an overflow, and $c = 2.5$ was used instead. Otherwise, overflows were not problematic using the type of coding shown in Appendix C.1. For most of the integrals the accuracy improved from the 10M to the 350 or 400M rule.

Parallel performance results are given in Table 5 for the [2lb(e)] (crossed box) diagram, and lattice rules for dimension 6. In each case the sequential time T_{seq} and the parallel (GPU) time T are reported in milliseconds (ms), as well as the parallel speedup $S = T_{seq}/T$, for rules with $n =$ 10M, 100M, 200M and 400M points. The speedup measures the improvement in time when the program is executed in parallel, versus the sequential running time.

For the different numbers of points it emerges that speedups are achieved of around 500 for the lattice rules applied with the Ψ_6 transformation, and around 400 with the tanh transformation. The reason for the difference in speedup lies in the sequential times, which are somewhat higher for the Ψ_6 transformation than for tanh, whereas the parallel times are similar. As a consequence the speedup ratio of the sequential to the parallel time is higher for the Ψ_6 transformation. Excellent speedups are also obtained for other diagrams of the problem sets considered.

The approach combining lattice rules and transformations for these loop integrals performs well compared to other numerical methods. For the 2-loop crossed box problem using parallel adaptive integration in [23] with 600M to 300B function evaluations, the absolute accuracy estimate ranged from 2.9e-08 to 1.1e-11, obtained in execution times of about 1 second to 587 seconds, with 64 MPI (Message Passing Interface) processes running on four cluster nodes (16 per node) – compare to the GPU times up to the order of a second. Iterated adaptive integration, also explored in [23] and implemented as a multi-threaded program on one node, took considerably longer than the adaptive method. Even though

IOP Conf. Series: Journal of Physics: Conf. Series **1136** (2018) 012002 doi:10.1088/1742-6596/1136/1/012002

Table 5. Speedup results of LR with tanh or Ψ_6 transformation

n	tanh			Ψ_6		
	$T_{seq}(ms)$	$T(ms)$	S	$T_{seq}(ms)$	$T(ms)$	S
10M	12652.1	32.39	391	15840.4	31.55	502
100M	125939	317.3	397	156570	311.7	502
200M	251200	636.1	395	313278	623.6	502
400M	503561	1271	396	630059	1247	505

the adaptive strategies are more versatile and may further improve the accuracy (up to a certain point) by increasing the number of region subdivisions, the lattice rule program for this application is far simpler and currently runs on one node plus GPU.

5. Conclusions

We presented an approach for the evaluation of loop integrals on GPUs for classes of 2-loop box and 3-loop self-energy Feynman diagrams. Lattice rules are particularly suited for GPU computations, and can be applied with transformations to deal with integrand singularities at the boundaries of the integration region. For the target application in this paper, Sidi's $\sin^m$-transformation (with $m = 6$) overall attained better accuracy than the tanh-transformation, and the combined lattice rule and transformation techniques achieved excellent accuracy and parallel efficiency.

The goal of this work is to establish high speed numerical methods to derive precise results for important classes of Feynman loop integrals. As such it contributes to the development of computer simulations in particle physics, and has potential for extensions in this and other physics areas, such as in condensed matter physics where Feynman integrals play an important role.

Acknowledgments

We acknowledge the support from the National Science Foundation under Award Number 1126438, which funded the equipment used for the computations in this paper, as well as the funding by Grant-in-Aid for Scientific Research (17K05428) of JSPS, and by the Large Scale Simulation Program No.16/17-21 of KEK. We further express our appreciation for the interesting comments and suggestions by the reviewers of this paper.

Appendix A. Lattice generator vectors

Table A1 lists the generator vectors **z** used for the computations in this paper.

Appendix B. CUDA kernel for rule sum

```
__global__ void lattice_kernel( int *z, double *q ) {
    // Input: generator z
    // Output: array q contains partial lattice rule sums over the blocks

    __shared__ double cache[threadsPerBlock]; // cache array is shared over block

    int tid = threadIdx.x + blockIdx.x * blockDim.x; // thread ID
    int cacheIndex = threadIdx.x;
    double temp = 0;
    double x[dim], temp1, temp2, intpart;
    int i;

    while (tid < n) { // add function evaluations for thread tid into temp

        for( i = 0; i < dim; i++) { // compute evaluation point x
          temp1 = (double) tid * z[i];
```

IOP Conf. Series: Journal of Physics: Conf. Series **1136** (2018) 012002 doi:10.1088/1742-6596/1136/1/012002

Table A1. Lattice rule generators

d	n	z components
4	10M	(1, 2928962, 1859617, 3250721)
4	100M	(1, 38278307, 43388112, 5988368)
4	200M	(1, 76384079, 90104983, 20251157)
4	350M	(1, 146940205, 127904721, 59228284)
4	400M	(1, 154642716, 171950363, 128667067)
5	10M	(1, 3675449, 4456704, 3864450, 4934306)
5	100M	(1, 38278307, 43388112, 48206750, 45950883)
5	200M	(1, 76384079, 93229505, 12830863, 56635299)
5	350M	(1, 133694234, 154078417, 9896546, 125929561)
5	400M	(1, 154642716, 171950363, 128667067, 118799181)
6	10M	(1, 2928962, 1859617, 4304839, 2689131, 2592686)
6	100M	(1, 38278307, 34519177, 26711202, 33749025, 1281880)
6	200M	(1, 53716462, 63701826, 51044020, 15378683, 37953591)
6	350M	(1, 146940205, 127904721, 27682752, 91275310, 62729638)
6	400M	(1, 155987582, 55105452, 187223455, 170818423, 163800605)
7	10M	(1, 2928962, 1859617, 4304839, 4474033, 273363, 1426318)
7	100M	(1, 41883906, 22682973, 44229424, 29466837, 15518263, 42112409)
7	200M	(1, 76384079, 93229505, 12830863, 56635299, 54395013, 90891159)
7	350M	(1, 146940205, 127904721, 27682752, 91275310, 62729638, 70372598)
7	400M	(1, 155987582, 55105452, 187223455, 109010593, 87245157, 73239020)
10	10M	(1, 3675449, 2181023, 834814, 4235339, 3173487, 171907, 735463, 2446815, 1818771)
10	100M	(1, 41883906, 22682973, 44229424, 29466837, 8176047, 49462874, 1162485, 46871525, 36107330)
10	200M	(1, 53716462, 63701826, 51044020, 67272988, 11162620, 9925795, 63388095, 54986324, 19264518)
10	350M	(1, 146940205, 127904721, 108579162, 30890237, 166719091, 28591254, 114681466, 54322685, 136274288)
10	400M	(1, 155987582, 105723490, 76332612, 3135442, 168381052, 61518902, 153655949, 14227594, 127700487)
12	10M	(1, 3675449, 2181023, 2356124, 3034497, 3444697, 2127900, 1305196, 1184663, 525861, 4284855, 708123)
12	100M	(1, 41883906, 22682973, 44229424, 2504352, 2579113, 5421496, 10190424, 23617111, 19238982, 34419287, 47805957)
12	200M	(1, 76384079, 93229505, 19725574, 77435876, 14972381, 48799902, 61957549, 61297601, 4060651, 6876378, 79938247)
12	350M	(1, 146940205, 64635382, 75928369, 106876745, 170290273, 20400538, 155050145, 29237893, 70053721, 114745098, 75194860)
12	400M	(1, 155987582, 105723490, 76332612, 3135442, 168381052, 75720068, 119726492, 175006849, 58051076, 117200356, 197634364)

```
            temp2 = temp1/(double)n;
            x[i] = modf(temp2, &intpart);;
          }
          temp += f1(x); // add function evaluation to temp
          tid += blockDim.x * gridDim.x;
    }

    // Reduction for summation of shared cache array elements into element
    // cache[0] for this block, cf. Sanders & Kandrot book p. 79
    cache[cacheIndex] = temp;

    // synchronize threads before reduction
    __syncthreads();

    // threadsPerBlock must be a power of 2
    int rp = blockDim.x/2;
    while (rp != 0) {
        if (cacheIndex < rp)
            cache[cacheIndex] += cache[cacheIndex + rp];
        __syncthreads(); // synchronize threads after this stage of reduction
        rp /= 2;
    }

    if (cacheIndex == 0) // store sum over block in q[blockIdx.x]
        q[blockIdx.x] = cache[0];
}
```

IOP Conf. Series: Journal of Physics: Conf. Series **1136** (2018) 012002 doi:10.1088/1742-6596/1136/1/012002

Appendix C. CUDA device functions for sample integrands

Appendix C.1. Device function for [3ls](e) integrand with tanh *transformation*

```
#include <math.h>

const int dim = 7;
const double C = 3.0;
const double half = 0.5;
const double emach = 5.0e-16;  // machine accuracy

__device__ double f1( double x[] ) {
  // Integrand for [31s](e) with tanh transformation
  // Integration result will be multiplied with pow(C*half,7.0) in main()

double x0,x1,x2,x3,x4,x5,x6,x7,x1sq,x4sq,x5sq,x6sq;
  double x12sq,x13sq,x12,x13,x1235,x0124,x4567,x02457;
  double arg[7],phip,dphi,deni,u2i,ai,t,tmp,dd,f0;
  int i;

  f0 = 0;
  dphi = 1.0;
  phip = 1.0;
  for(i = 0; i < dim; i++) {
     t = 2.0*x[i]-1.0;
     u2i = t*t;
     deni = 1.0-u2i;
     if(fabs(deni) <= emach) return(0);
     ai = C*t/deni;
     arg[i] = dphi*half*(tanh(ai)+1.0);
     tmp = deni*cosh(ai);
     phip = phip*dphi*(1.0+u2i)/tmp/tmp;
     dphi = dphi-arg[i];
  }
  if(phip == 0) return(0);
  x0 = arg[0];
  x1 = arg[1];
  x2 = arg[2];
  x3 = arg[3];
  x4 = arg[4];
  x5 = arg[5];
  x6 = arg[6];
  x7 = 1.0-x0-x1-x2-x3-x4-x5-x6;

  x1sq = x1*x1;
  x4sq = x4*x4;
  x5sq = x5*x5;
  x6sq = x6*x6;
  x12 = x1+x2;
  x13 = x1+x3;
  x1235 = x1+x2+x3+x5;
  x0124 = x0+x1+x2+x4;
  x4567 = x4+x5+x6+x7;
  x02457 = x0+x2+x4+x5+x7;
  x12sq = x12*x12;
  x13sq = x13*x13;

  dd = x02457*x1235*x4sq + x13sq*x4sq + x02457*x12sq*x4567
     - x0124*x02457*x1235*x4567 - x1sq*x1235*x4567 + 2.0*x1*x12*x13*x4567
     - x0124*x13sq*x4567 + 2.0*x02457*x12*x4*x5 + 2.0*x1*x13*x4*x5
     + x0124*x02457*x5sq + x1sq*x5sq + 2.0*x1*x1235*x4*x6 - 2.0*x12*x13*x4*x6
     + 2.0*x1*x12*x5*x6 - 2.0*x0124*x13*x5*x6 + x12sq*x6sq - x0124*x1235*x6sq;

  if(fabs(dd) > 0.0) f0 = 128.0*phip/dd/dd;
  return f0;
}
```

Appendix C.2. Device function for [2lb](b) integrand with Ψ_6 transformation

```
#include <math.h>

#define pi 3.14159265358979323846
#define pi60 188.49555921538759430760
#define con 335.54432

const int dim = 5;

__device__ double f1( double x[] ) {
  //Integrand for [2lb](b) with Psi_6 transformation

  double x1,x2,x3,x4,x5,x6,dd,f0;
  double arg[5],phip,pixi;
  int i;

  f0 = 0;
  phip = 1.0;
  for(i = 0; i < dim; i++) { // Sidi Psi_6 transformation
    pixi = pi*x[i];
    arg[i] = x[i]+(-45*sin(2*pixi)+9*sin(4*pixi)-sin(6*pixi))/pi60;
    phip = phip*pow(sin(pixi),6);
  }
  if(phip == 0) return(0);
  x1 = arg[0];                // Simplex-to-cube transformation
  x2 = (1.0-x1)*arg[1];
  x3 = (1.0-x1-x2)*arg[2];
  x4 = (1.0-x1-x2-x3)*arg[3];
  x5 = (1.0-x1-x2-x3-x4)*arg[4];
  x6 = 1.0-x1-x2-x3-x4-x5;

  phip = phip*(1.0-x1)*(1.0-x1-x2)*(1.0-x1-x2-x3)*(1.0-x1-x2-x3-x4);
  if(phip == 0) return(0);

  dd = (x1*x1+x5*x5+x1*x5)*(x2+x3+x4+x6)
       +(x2*x2+x3*x3+x6*x6+x2*x3+x2*x6+x3*x6)*(x1+x4+x5)
       +x4*x4*(x1+x2+x3+x5+x6)+3.0*x4*x5*x6+
       2.0*x1*x4*(x2+x3+x6)+2.0*x4*x5*(x2+x3);

  if(fabs(dd) > 0.0) f0 = con*phip/dd/dd;
  return f0;
}
```

References

[1] Aad G *et al.* 2012 *Phys. Lets. B* **716** 1–29
[2] Chatrchyan S *et al.* 2012 *Phys. Lets. B* **716** 30–61
[3] Butterworth J *et al.* 2014 Les Houches 2013: Physics at TeV colliders Standard Model Working Group report arXiv:1405.1067 [hep-ph], https://arxiv.org/abs/1405.1067
[4] Fujii K *et al.* 2017 Physics case for the 250 GeV stage of the International Linear Collider arXiv:1710.07621 [hep-ex], https://arxiv.org/abs/1710.07621
[5] de Doncker E, Shimizu Y, Fujimoto J and Yuasa F 2004 *Computer Physics Communications* **159** 145–156
[6] de Doncker E, Shimizu Y, Fujimoto J, Yuasa F, Cucos L and Van Voorst J 2004 *Nuclear Instruments and Methods in Physics Research A* **539** 269–273 hep-ph/0405098
[7] Yuasa F, de Doncker E, Fujimoto J, Hamaguchi N, Ishikawa T and Shimizu Y 2007 Precise numerical results of IR-vertex and box integration with extrapolation *PoS (ACAT07) 087* arXiv:0709.0777v2 [hep-ph]
[8] Yuasa F, Ishikawa T, Fujimoto J, Hamaguchi N, de Doncker E and Shimizu Y 2008 Numerical evaluation of Feynman integrals by a direct computation method *PoS (ACAT08) 122* arXiv:0904.2823 [hep-ph]
[9] de Doncker E, Fujimoto J, Kurihara Y, Hamaguchi N, Ishikawa T, Shimizu Y and Yuasa F 2010 Recursive box and vertex integrations for the one-loop hexagon reduction in the physical region *PoS (ACAT10) 073*
[10] de Doncker E, Fujimoto J, Hamaguchi N, Ishikawa T, Kurihara Y, Shimizu Y and Yuasa F 2010 *Springer Lecture Notes in Computer Science (LNCS)* **6017** 139–154
[11] de Doncker E, Fujimoto J, Hamaguchi N, Ishikawa T, Kurihara Y, Ljucovic M, Shimizu Y and Yuasa F 2010 Extrapolation algorithms for infrared divergent integrals arXiv:hep-ph/1110.3587, PoS (CPP2010)011

[12] Yuasa F, de Doncker E, Hamaguchi N, Ishikawa T, Kato K, Kurihara Y and Shimizu Y 2012 *Computer Physics Communications* **183** 2136–2144

[13] de Doncker E, Yuasa F and Kurihara Y 2012 *Journal of Physics: Conf. Ser.* **368**

[14] Kato K, de Doncker E, Ishikawa T, Kapenga J, Olagbemi O and Yuasa F 2016 *Journal of Physics: Conf. Series* **762** 012070, iopscience.iop.org/article/10.1088/1742-6596/762/1/012070

[15] de Doncker E, Yuasa F, Ishikawa T, Kapenga J and Olagbemi O 2016 *Procedia Computer Science* **80** 1428–1438 http://dx.doi.org/10.1016/j.procs.2016.05.462

[16] de Doncker E, Yuasa F, Kato K, Ishikawa T, Kapenga J and Olagbemi O 2018 *Computer Physics Communications* **224** 164–185 https://doi.org/10.1016/j.cpc.2017.11.001

[17] de Doncker E, Yuasa F, Kato K and Ishikawa T 2016 Automatic numerical integration and extrapolation for Feynman loop integrals *Workshop on Computational Particle Physics (CPP2016)* (Journal of Physics: Conf. Series (IoP JPCS))

[18] de Doncker E and Yuasa F 2017 *Procedia Computer Science* **108** 1773–1782 https://doi.org/10.1016/j.procs.2017.05.253

[19] de Doncker E and Kaugars K 2010 *Procedia Computer Science* **1** 117–124

[20] de Doncker E, Fujimoto J, Hamaguchi N, Ishikawa T, Kurihara Y, Shimizu Y and Yuasa F 2011 *Journal of Computational Science (JoCS)* **3** 102–112 doi:10.1016/j.jocs.2011.06.003

[21] de Doncker E and Yuasa F 2012 *Journal of Physics: Conf. Series (JPCS)* **402** doi:10.1088/1742-6596/402/1/012029

[22] de Doncker E, Yuasa F and Assaf R 2013 *Journal of Physics: Conf. Ser. (JPCS)* **454** doi:10.1088/1742-6596/454/1/012082

[23] de Doncker E and Yuasa F 2014 *Journal of Physics: Conf. Series (JPCS)* **523** doi:10.1088/1742-6596/523/1/012052

[24] de Doncker E, Yuasa F, Kato K, Ishikawa T and Olagbemi O 2015 *Journal of Physics: Conf. Series (JPCS)* **608** doi:10.1088/1742-6596/608/1/012071

[25] de Doncker E, Yuasa F, Kapenga J and Olagbemi O 2015 *Journal of Physics: Conf. Series (JPCS)* **640** doi:10.1088/1742-6596/640/1/012062

[26] de Doncker E, Yuasa F and Olagbemi F 2015 *Procedia Computer Science* **51** 1333–1342 doi:10.1016/j.procs.2015.05.318

[27] Sag T W and Szekeres G 1964 *Math. Comp.* **18** 245–253

[28] Sidi A 1993 *International Series of Numerical Mathematics* **112** 359–373

[29] Sidi A 2005 *Math. Comp.* **75** 327–343

[30] Robinson I and Hill M 2002 *Transactions on Mathematical Software* **28** 75–100 doi:10.1145/513001.513006

[31] Li T and Robinson 2010 *Transactions on Mathematical Software* **37** doi:10.1145/1824801.1824813

[32] Li T and Robinson 2010 plrint5d: A five-dimensional automatic cubature routine designed for a multi-core platform *ICCSA 2010, Lect. Notes in Computer Sc.* pp 44–59 https://link.springer.com/chapter/10.1007/978-3-642-12165-4_4

[33] Li Z, Wang J, Yan Q S and Zhao X 2016 *Chinese Physics C* **40** 033103, doi:0.1088/1674-1137/40/3/033103, arXiv:1508.02512v1 [hep-ph]

[34] de Doncker E, Almulihi A and Yuasa F High-speed evaluation of loop integrals using lattice rules In 18th Adv. Comp. and Anal. Tech. in Phys. Research (ACAT 2017, Seattle), Accepted

[35] Almulihi A and de Doncker E 2017 Accelerating high-dimensional integration using lattice rules on GPUs *Proc. 2017 Int. Conf. on Computational Science and Computational Intelligence (CSCI'17)* (CPS IEEE) Accepted

[36] Kawabata S 1995 *Computer Physics Communications* **88** 309–326

[37] Lepage G P 1978 *J. Comp. Phys.* **27** 192–203

[38] Hahn T 2005 *Computer Physics Communications* **168** 78–95 doi:10.1016/j.cpc.2005.01.010, arXiv:hep-ph/0404043v2i

[39] Korobov N M 1959 *Doklady Akademii Nauk SSSR* **124** 1207–1210 (Russ.)

[40] Korobov N M 1960 *Doklady Akademii Nauk SSSR* **132** 1009–1012 (Russ.). Eng. trans. Soviet Math. Doklady, 1, 696-700

[41] Korobov N M 1963 *Number-Theoretic Methods in Approximate Analysis* (Fizmatgiz, Moscow (Russ.))

[42] Hlawka E 1962 *Monatshefte für Mathematik* **66** 140–151

[43] Niederreiter H 1978 *Monatshefte für Mathematik* **86** 203–219

[44] Davis P J and Rabinowitz P 1984 *Methods of Numerical Integration* (Academic Press, New York)

[45] Niederreiter H 1988 Quasi-Monte-Carlo methods for multidimensional numerical integration *Numerical Integration III, International Series of Numerical Mathematics* 85 pp 157–171

[46] Sloan I and Joe S 1994 *Lattice Methods for Multiple Integration* (Oxford University Press, Oxford)

[47] Zaremba S K 1968 *Annales Polonici Mathematici* **21** 85–96

[48] Disney S A R and Sloan I H 1991 *Math. Comp.* **56** 257–266

[49] Sloan I H and Reztsov A 2002 *Math. Comp.* **71** 263–273

[50] Nuyens D and Cools R 2006 *Math. Comp.* **75** 903–920

[51] Nuyens D and Cools R 2006 *Journal of Complexity* **22** 4–28

[52] Sanders J and Kandrot E 2010 *CUDA by Example: An Introduction to General-Purpose GPU Programming* (Pearson)

[53] Iri M, Moriguti S and Takasawa Y 1970 *Kokyuroku of Res. Inst. for Math. Sci. Kyoto Univ.* **91** 82–118 (in Japanese); Engl. transl. in J. Comp. Appl. Math. 17 (1987), pp. 3-20

[54] Takahasi H and Mori M 1974 *Publications of the Research Institute for Mathematical Sciences* **9** 721–741

[55] Mori M 1978 *Publ. RIMS Kyoto Univ.* **14** 713–729

[56] Laporta S 2000 *Int. J. Mod. Phys. A* **15** 5087–5159 arXiv:hep-ph/0102033v1

IOP Conf. Series: Journal of Physics: Conf. Series **1136** (2018) 012003 doi:10.1088/1742-6596/1136/1/012003

Novel methods in track-based alignment to correct for time-dependent distortions of the ATLAS Inner Detector

Oscar Estrada Pastor[1], on behalf of the ATLAS Collaboration

[1]Instituto de Física Corpuscular (CSIC-UV), Parque Científico, Catedrático José Beltrán, 2 Paterna, Spain

E-mail: oscar.estrada@ific.uv.es

Abstract. ATLAS is a multipurpose experiment at the LHC proton-proton collider. Its physics goals require high resolution and unbiased measurement of all charged particle kinematic parameters. These critically depend on the layout and performance of the tracking system and the quality of its alignment. For the LHC Run-II, the system has been upgraded with the installation of a new pixel layer, the Insertable B-layer (IBL). The offline track alignment of the ATLAS tracking system has to deal with about 700,000 degrees of freedom defining its geometrical parameters, representing a considerable numerical challenge in terms of both CPU time and precision. An outline of the track based alignment approach and its implementation within the ATLAS software is presented. Special attention is paid to describe the techniques allowing to pinpoint and eliminate track parameter biases. During Run-II, the ATLAS Inner Detector Alignment framework has been adapted and upgraded to correct very short time scale movements of the sub-detectors. In particular, a mechanical distortion of the IBL staves up to 20 μm has been observed during data-taking. The techniques used to correct for this effect and to match the required Inner Detector performance are presented.

1. Introduction

The ATLAS detector [1] is a general purpose detector located at the Large Hadron Collider (LHC) at CERN. The first proton-proton collisions registered at the LHC took place in 2009 at a center of mass energy of 900 GeV. In 2010, the center of mass energy was set to 7 TeV and raised to 8 TeV in the final campaign of Run-I. A long shutdown stop (LS1) took place before Run-II started to raise the center of mass energy to 13 TeV and thus beating all the records. During this stop ATLAS upgraded the pixel detectors, adding the Insertable B-Layer (IBL), and thus becoming the closest detector to the beampipe.

The ATLAS detector is equipped with a central tracking system, the Inner Detector (ID), in order to reconstruct the trajectories of charged particles and to estimate their kinematic parameters. The accuracy of this process is limited by the finite resolution of the sensitive devices. Therefore misalignments of the active detector elements as well as geometrical distortions lead to a degradation of the resolution on reconstructed tracks and may lead to biases in the measured parameters.

Alignment is concerned with determining the actual geometry of the tracking system and its eventual changes in time. The aim is to compute the alignment parameters that result in the best description of all active detector modules. To do so, the ID has been aligned using a track-based technique [2], however any generic track-based alignment is exposed to a class of geometry deformations, the so called weak-modes, against which this procedure has very low or no sensitivity.

2. ATLAS Inner Detector

The ATLAS Inner Detector consists of three subdetectors, the Pixel detector (including the IBL), the Semiconductor Tracker (SCT), and the Transition Radiation Tracker (TRT), where all are embedded in a 2 T axial magnetic field. The ID has been designed to reconstruct the charged particles trajectories within a pseudorapidity range of $|\eta| < 2.5$ (see Figure 1 for a schematic view of the barrel region). The Pixel detector consists of 1744 silicon pixel modules arranged in three barrel layers and two end caps with three disks each. The expected hit resolution is 10 μm in $r\phi$ coordinates and 115 μm in the longitudinal coordinate. During LS1, the IBL was added as an additional layer to the Pixel detector, reducing the distance from the interaction point to the first tracking layer, as IBL is placed at 33.25 mm radius while the inner layer of the rest of the Pixel is placed at 50.5 mm, becoming the nearest sub-detector to the beampipe. The IBL is composed of 280 modules, mixing planar and 3D technology. The expected hit resolution is 8 μm in $r\phi$ coordinates and 40 μm in the longitudinal coordinate [3]. The SCT consists of 4088 silicon strip modules, arranged in four barrel layers and two end caps with nine wheels each. The intrinsic resolution is 17 μm in $r\phi$ coordinates and 580 μm in the longitudinal coordinate. The TRT is the outermost detector of the ID subdetectors and is made of 350848 straw tubes filled with Argon gas with a single hit resolution of 130 μm along $r\phi$.

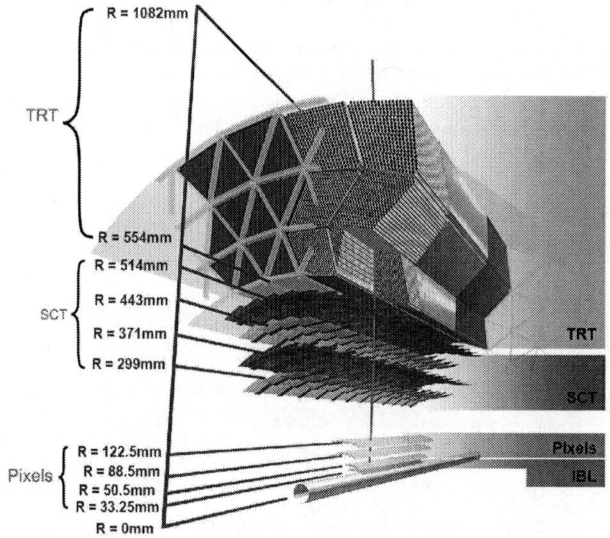

Figure 1. Schematic view of the ID barrel. The ID barrel is arranged in sub-detector layers. From innermost to outermost sub-detectors: IBL, Pixels, SCT and TRT.

3. Alignment procedure

The alignment procedure is based on the minimisation of the track-to-hit residuals for each module. Every sensor or mechanical assembly part is then considered as an alignable structure, which is treated as a rigid body with six degress of freedom (DoF) that uniquely define its position and orientation in space. These six DoF are three translations (T_x, T_y, T_z) of the centre

IOP Conf. Series: Journal of Physics: Conf. Series **1136** (2018) 012003 doi:10.1088/1742-6596/1136/1/012003

of the object and three rotations (R_x, R_y, R_z) around the Cartesian axes and will be referred to as the alignment parameters. Residuals are defined as the distance between the hit position as measured by the sensor and as computed in the track extrapolation. The minimisation of the χ^2 function, which is referred in equation (1), is used to find the alignment parameters that correct the module positions in order to describe accurately the real geometry of the detector:

$$\chi^2 = \sum_{tracks} \sum_{hits} \left(\frac{\vec{r}_h(\boldsymbol{\pi}, \mathbf{a})}{\sigma_h} \right)^2 = \sum_{tracks} \vec{r}(\boldsymbol{\pi}, \mathbf{a})^T \, V^{-1} \, \vec{r}(\boldsymbol{\pi}, \mathbf{a}) \tag{1}$$

where V is the covariance matrix of the detector measurements, $\mathbf{r}$ are the track-to-hit residuals depending on the alignment parameters, $\mathbf{a}$ the alignment parameters and $\boldsymbol{\pi}$ the track parameters.

One of the big successes of the alignment procedure is that it is performed in hierarchical levels starting from the biggest structure and finishing at the module level, mimicking the mechanical assembly. Starting with the largest physical structures at level 1, the detector subsystems are aligned separating the whole detector into end-caps and barrel regions in order to correct for collective movements. Level 2 treats individual barrel layers and end-cap disks as physical structures (barrel modules and end-cap wheels in case of the TRT). Level 3 corresponds to a silicon module and TRT wire level alignment. The levels are addressed consecutively during the alignment procedure. Table 1 shows the number of DoFs for each detector and level of alignment.

Table 1. Number of DoFs by detector and level of alignment. Assembly structures and silicon modules use 6 DoF. TRT wires use just 2 DoF.

Levels	Structures	DoFs
L1 (large structure)	8	48
L2 (*Si* layers/disks, TRT modules and wheels)	208	792
L3 (*Si* modules)	6112	36672
L3 (TRT wires)	350848	701696

4. Time-dependent deformations

During the LHC Run-I data taking campaign, the ATLAS Inner Detector was found to be stable for long periods of time, except for some "seismic" events (magnet ramp, power-cut, cooling failure,...), as Figure 2 shows.

However, in Run-II some sub-detectors have shown time-dependent deformations or movements, even within a single run, which required to introduce a new dynamic alignment scheme. Currently, two movements are known:

- Vertical movements of pixel sub-detector at the beginning of each run.
- IBL mechanical instabilities with temperature.

4.1. Pixel movements

Vertical movements of the pixel detector at the beginning of each run are registered due to change in the mass of the cooling liquid after the detector is switched on until an equilibrium between vapor and liquid is reached. Previously in Run-I no correction was applied, because only a small fraction of data was affected. During Run-II, an additional slow drift has also been observed. It takes places during all the run, in the opposite direction to the initial movement. The velocity

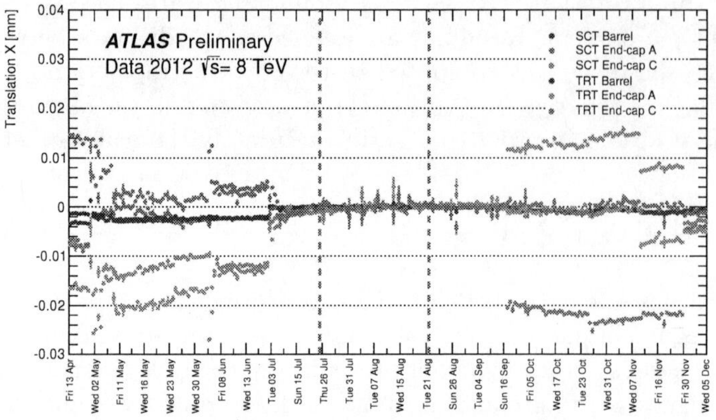

Figure 2. Evolution of the Level 1 alignment constant T_x during Run-I campaign

of this slow drift depends on the peak luminosity of LHC, which was increased during 2016. Figure 3 shows that Pixel quickly lifts 6 μm during the first hour of the run, followed by a gentle descent afterwards. A new dynamic alignment procedure was introduced in August 2016 which corrects for both these effects using time-dependent alignment. This required the introduction of an improved alignment database, in which the ID geometry database was restructured. Prompt alignment updates are computed currently within a run in 20-minute intervals and are running automatically inside the calibration loop process (updated within 24 hours after run is finished).

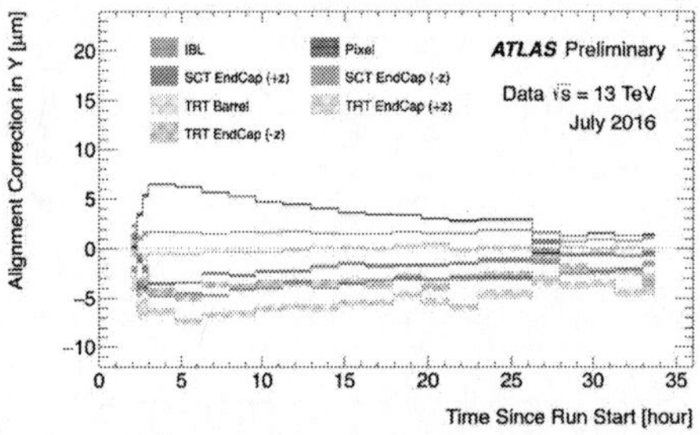

Figure 3. Correction in the Y vertical direction for the inner detector components as a function of time. Bands indicate statistical uncertainty. End-caps on the positive and negative side of the detector are shown separately. SCT barrel is kept fixed as a reference during the derivation of the alignment constants. While good stability was observed for all ID components in the X direction, the figure shows a fast movement of 6 μm for the Pixel sub-detector in the Y direction.

4.2. IBL mechanical instabilities

During collision data taking, the IBL staves were seen to bow even within a run. This was caused by an increased power consumption of the modules induced by irradiation, due to total ionization dose effect. Figure 4 (left) shows that the IBL staves bend depending on the different operating temperature conditions, while Figure 4 (right) shows that distortion magnitude changes in the same run, since the distortion varies with the integrated radiation dose and as a function of the LHC luminosity within a fill.

IOP Conf. Series: Journal of Physics: Conf. Series **1136** (2018) 012003 doi:10.1088/1742-6596/1136/1/012003

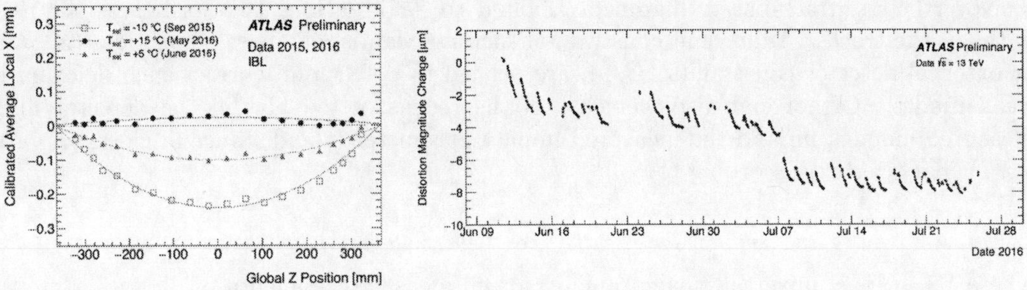

Figure 4. Left figure shows IBL local-X correction in the transverse plane averaged over all 14 IBL staves for 2015 data (red open squares), and for 2016 data using different temperature (+15, solid blue circles; +5, solid green triangles). No error bars associated with data are shown. Lines are presented just to highlight that trend. Right figure shows the relative bowing magnitude of the IBL staves, averaged over its 14 staves and computed with the default alignment (blue dots) and with respect to the aligned geometry (red open circles). The IBL operation temperature (T_{set}) for each period is also shown.

5. Run-II alignment results

The alignment performance is checked by evaluating the track-to-hit residuals. A comparison between residuals in data to those of perfectly aligned Monte Carlo simulation is also crucial. Figure 5 shows that the modules have been aligned at $\mathcal{O}(\mu\mathrm{m})$ level in both final 2016 data and initial 2017 data.

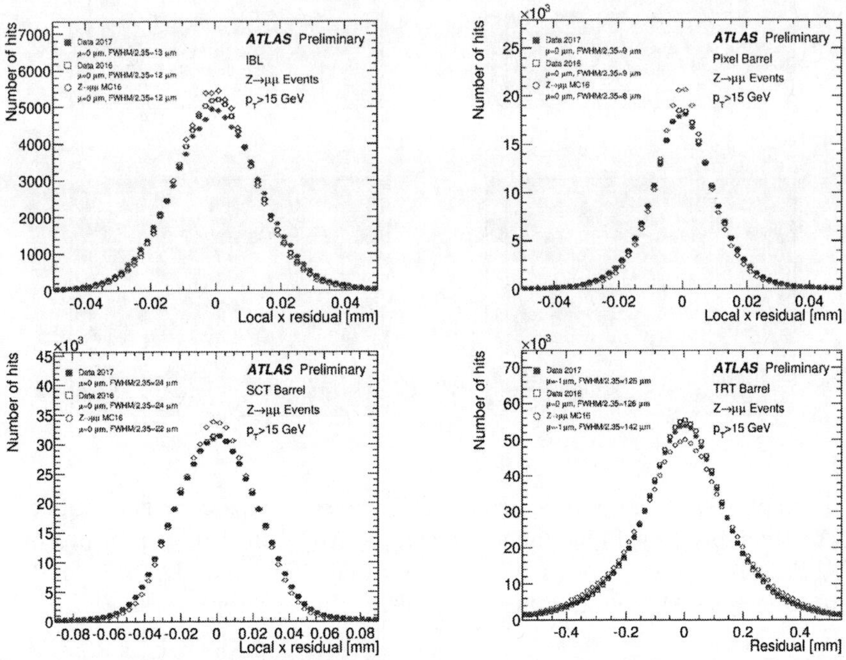

Figure 5. The residuals of the IBL, Pixel, SCT and TRT sub-detectors for 2017 data (gray squares) have been extracted with a sample of $Z \to \mu^+\mu^-$ events. Very similar residuals were found to the residuals obtained in the final 2016 reprocessed data (open squares) and a Monte Carlo simulated sample containing $Z \to \mu^+\mu^-$ events (open circles).

IOP Conf. Series: Journal of Physics: Conf. Series **1136** (2018) 012003 doi:10.1088/1742-6596/1136/1/012003

Nevertheless, track-based alignment is blind to deformations that preserves the helicoidal paths of the tracks. Additional constraints, such as beam spot, mass resonances (Z, J/ψ, K_s) or external detector constraints (E/p), are needed to detect and correct such deformations or weak modes. Orthogonal displacements of the reconstructed hits in the detector affect the measured momentum and the measured impact parameter (the distance of closest approach to the primary vertex) according to:

$$p_T^{reco} = p_T^{true}\left(1 + qp_T^{true}\delta_{sagitta}\right)^{-1} \tag{2}$$

where $\delta_{sagitta}$ is a universal bias parameter for all measured momenta.

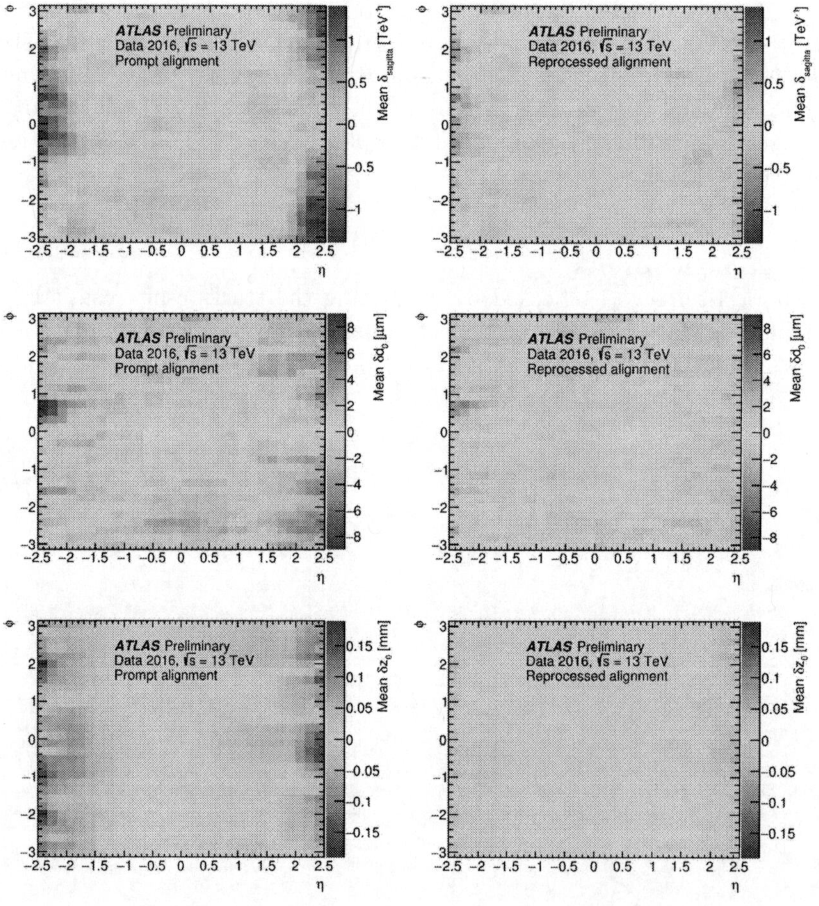

Figure 6. Prompt (left) and reprocessed (right) biases maps for the sagitta and impact parameters. For the reprocessed alignment the complete 30.4 fb^{-1} of proton-proton data taken after Technical Stop 1 are used, while for the prompt alignment only 20.3 fb^{-1} of data are used. Significant improvements in average biases are seen after the reprocessing with the updated alignment.

Momentum biases can be monitored using $Z \to \mu^+\mu^-$ events and an independent method based on electrons from E/p. While $Z \to \mu^+\mu^-$ events have been used to derive corrections for the impact parameters (i.e. d_0 and z_0) and sagitta biases, the E/p method is used as a cross-check for the sagitta biases, following the methods described in Ref. [4].

The first Inner Detector (ID) alignment constants for 2017 data-taking were derived using 82 pb^{-1} of proton-proton collision data delivered by the LHC in June 2017. The alignment was performed in several steps, considering different sub-detectors at increasing levels of detail, as explained in Section 3.

Figure 6 (left) shows weak modes maps using prompt alignment corrections (alignment constants derived shortly after the run has finished) and Figure 6 (right) shows corrections using reprocessed alignment (whose constants were derived after applying weak modes constraints and realigned). A clear improvement is observed for the sagitta and the impact parameters (i.e. d_0 and z_0) using reprocessed data.

6. Conclusions

The Inner Detector alignment framework has been successfully upgraded to cope with time-dependent deformations observed in Run-II. New techniques have been introduced to correct for IBL and Pixel rapid movements. In spite of the very challenging conditions, the alignment has been able to reach very precise results in a fully automatized process. Weak modes misalignments were identified and corrected for 2016 data. Initial 2017 data residuals look excellent already, being among the best that ATLAS recorded.

6.1. Acknowledgments

I would like to thank the ATLAS Tracking Combined Performance group for their help and support. Work supported by the Generalitat Valenciana and CSIC through the VALi+d PhD grant ACIF/2016/100 and by the Generalitat Valenciana through PrometeoII/2014/016.

References

[1] ATLAS Collaboration, The ATLAS Experiment at the CERN Large Hadron Collider 2008 *JINST* **3** S08003
[2] ATLAS Collaboration, Alignment Performance of the ATLAS Inner Detector Tracking System in 7 TeV proton-proton collisions at the LHC Tech. Rep. ATLAS-CONF-2010-067 URL `https://cds.cern.ch/record/1281342`
[3] ATLAS Collaboration, ATLAS Insertable B-Layer Technical Design Report Tech. Rep. CERN-LHCC-2010-013. ATLAS-TDR-19 URL `https://cds.cern.ch/record/1291633`
[4] ATLAS Collaboration, Study of alignment-related systematic effects on the ATLAS Inner Detector tracking Tech. Rep. ATLAS-CONF-2012-141 URL `https://cds.cern.ch/record/1483518`

IOP Conf. Series: Journal of Physics: Conf. Series **1136** (2018) 012004 doi:10.1088/1742-6596/1136/1/012004

Computational studying energy and spectral parameters of hadronic (pionic) atoms with account of the strong pion-nuclear interaction

Yu V Dubrovskaya, L A Vitavetskaya, I N Serga, D A Mironenko, R E Serga and A N Bystryantseva

Odessa State Environmental University, L'vovskaya str. 15, Odessa, 65016, Ukraine

E-mail: dubrovskayayv@gmail.com

Abstract. A relativistic computational approach to studying the energy and spectral characteristics of hadronic (pionic) atoms is presented and based on the Klein-Gordon-Fock equation with optimized π^--N interaction optical potential and relativistic many-body perturbation theory with correct treating radiation, electron-screening, nuclear effects. It is presented an effective scheme to computing energy levels shifts and widths, provided by a strong π^--N interaction and the interaction of the pion with QED vacuum (radiation width). It is based on using the model optimized optical complex π^--N interaction potential and relativistic energy approach with complex relativistic interparticle interaction potential. Data on the energy and spectral parameters for the pionic atom of ^{181}Ta are listed and compared with available experimental and alternative theoretical results.

1. Introduction

Studying the energy, spectral, radiative characteristics for exotic (hadronic, kaonic, pionic) atomic systems attracts a great interest as these atoms enable to probe different aspects of atomic and nuclear structure [1-10]. The knowledge of the corresponding properties for hadronic atomic systems can be used as a powerful tool for the study of particles and fundamental (for example, strong) interaction properties. It is worth to note that one of the most sensitive tests for the chiral symmetry breaking scenario in the modern nuclear physics is provided by investigation of the exotic hadronic atomic systems. These facts explain intense theoretical and experimental works on exotic atoms in different laboratories (CERN, Japan, Berkley, Virginia etc). Nowadays the transition energies in pionic (kaonic, muonic etc) atomic systems can be measured with an unprecedented precision. It allows to study the strong interaction at low energies measuring the energy and natural width of the ground level with a precision of few meV. There are a few theoretical approaches to studying spectra and spectral properties of the exotic (pionic, kaonic, muonic etc) atomic systems (see details in Refs. [1-55]). The most difficult aspects of the theoretical modelling are connected with the accurate treatment of the pion-nuclear strong interaction as the electromagnetic part of the problem can be in principle reasonably accounted for. In Refs. [2] it has been presented an effective relativistic approach to studying the energy and spectral characteristics of heavy hadronic (pionic) atoms. It is based on the Klein-Gordon-Fock equation with optimized π^--N interaction optical potential and relativistic many-body perturbation theory with correct accounting for the radiation, electron-screening, nuclear effects effects [31,32]. It has been developed an effective theory for calculating energy levels shifts and widths, provided by a strong π^--N interaction ("strong" width) and the interaction of the pion with

IOP Conf. Series: Journal of Physics: Conf. Series **1136** (2018) 012004 doi:10.1088/1742-6596/1136/1/012004

QED vacuum (radiation width) within the model optimized optical complex π^--N interaction potential and relativistic energy approach based on the Gell-Mann and Low formalism with complex relativistic interparticle interaction potential [56-68]. Here we apply it to computing the energy and spectral parameters of the pionic ^{181}Ta atom. The obtained theoretical data are listed and compared with available experimental and other theoretical results. All calculations are performed with using the modified blocks [2] of the PC computational code "Superatom-ISAN" (see details in Refs. [60-88]).

2. The theoretical method

2.1. The Klein-Gordon-Fock equation and electromagnetic interaction potentials

The relativistic electron wave functions are determined from solution of the Klein-Gordon-Fock equation (pion is the Boson with spin 0, mass: m_{π^-}=139.57018 MeV, r_{π^-}=0.672±0.08 fm) with a general potential (the latter includes an electric and polarization potentials of a nucleus plus the strong pion-nuclear interaction potential):

$$\{\frac{1}{c^2}[E+eV_0(r)]^2+\hbar^2\nabla^2-m^2c^2\}\varphi(x)=0\,. \tag{1}$$

Here E is the total energy of the system (sum of the mass energy mc^2 and binding energy ε_0). The central potential V_0 is the sum of the electric potential of a nucleus, vacuum-polarization potential and the strong interaction potential. As usually (see [2,31,32], the nuclear potential for the spherically symmetric density (as the first approximation) $\rho(r|R)$ can be presented as follows:

$$V_{nucl}(r|R)=-(1/r)\int_0^r dr'r'^2\rho(r'|R)+\int_r^\infty dr'r'\rho(r'|R)\,. \tag{2}$$

Probably nowadays the most consistent approach to determination of $\rho(r|R)$ is in using the relativistic mean-field (RMF) model, which has been designed as a renormalizable meson-field theory for nuclear matter and finite nuclei. However, quite physically reasonable acceptable approximation is given by the known Gauss or Fermi nuclear models. In particular, the density can be approximated by the Fermi function:

$$\rho(r)=\rho_0/\{1+\exp[(r-c)/a)]\}\,, \tag{3}$$

where the parameter a=0.523 fm, the parameter c is chosen in such a way that it is true the following condition for average-squared radius: $\langle r^2\rangle^{1/2}$=(0.836·A$^{1/3}$+0.5700)fm. The effective computational algorithm for definition of the potential $V_{nucl}(r|R)$ is based on the known Ivanov-Ivanova et al [56] method of differential equations (see Refs. [2-4,31,32]). It allows to determine the nuclear potential by solution of the differential equations system (the Fermi model):

$$V'nucl(r,R)=(1/r^2)\int_0^r dr'r'^2\rho(r',R)\equiv(1/r^2)y(r,R),\quad y'(r,R)=r^2\rho(r,R), \tag{4}$$

$$\rho'(r)=(\rho_0/a)\exp[(r-c)/a]\{1+\exp[(r-c)/a)]\}^2 \tag{5}$$

with the corresponding boundary conditions.

The next important theoretical step is a consistent and accurate accounting for the QED effects in computing the energy parameters of pionic atoms. To take into account the radiation (QED) corrections, namely, the important effect of the vacuum polarization one could use the procedure, which is in details described in the Refs. [3,4]. The important radiation contributions are given by the standard Uehling-Serber term and the Källen-Sabry and Wichmann-Kroll corrections of higher orders (such as $[\alpha\,(Z\alpha)]^n$ (n=2,..), $\alpha^2(\alpha Z)$, $\alpha\,(Z\alpha)^n$ (n=3) etc; α being the fine structure constant). In order to take into consideration an effect of the vacuum polarization in the first perturbation theory order one should use the generalized Uehling-Serber potential with modification to account for the high-order radiative corrections. It can be written as follows:

$$U(r) = -\frac{2\alpha}{3\pi r}\int_1^\infty dt\,\exp(-2rt/\alpha Z)\left(1+1/2t^2\right)\frac{\sqrt{t^2-1}}{t^2} \equiv -\frac{2\alpha}{3\pi r}C(g), \tag{6}$$

where $g=r/(\alpha Z)$. More correct and consistent approach is presented in Refs. [42,43,52-62]. An accounting of the nuclear finite size effect modifies the potential (6) as follows:

$$U^{FS}(r) = -\frac{2\alpha^2}{3\pi}\int d^3r'\int_m^\infty dt\exp\left(-2t|r-r'|/\alpha Z\right)\times\left(1+\frac{1}{2t^2}\right)\frac{\sqrt{t^2-1}}{t^2}\frac{\rho(r')}{|r-r'|}, \tag{7}$$

According to Refs. [3], the modified Uehling-Serber potential can be determined as a quadrature (7) and approximated with high precision by a simple analytical function. The use of new approximation of the Uehling potential permits one to decrease the calculation errors for this term down to 0.5 – 1%. A method for calculation of the self-energy part of the Lamb shift is based on an idea by Ivanov-Ivanova [58], which generalizes the known hydrogen-like method by Mohr and radiation model potential method by Flambaum-Ginges (look details in Refs. [46-50]).

2.2. Strong pion-nuclear interactions in pionic atomic system

The most difficult aspect of the theoretical study of pionic atomic systems is connected with a consistent and adequate account for the strong pion-nuclear interaction. Naturally, the most fundamental and exact microscopic theory of the strong interactions is provided by the modern quantum chromodynamics. Usually, one can consider the regimes of relatively low and high energies (asymptotic freedom) in physics of pionic systems. In a case of the low energies so called coupling constant increases to the order 1 and, therefore, this perturbation methods fail to describe the interaction of strongly interacting hadrons. At present time there are different approaches to treating pion-nuclear interaction even at relatively low energies (look details in Refs. [1,2,10,23-28,33-44]). From the computational viewpoint a more simplified approach to treatment of the strong interaction in the pionic atomic system is provided by the optical potential model (c.g. [1,33]). The most popular version of this method is based on using the known generalized Ericson-Ericson potential:

$$V_{\pi^- N} = V_{opt}(r) = -\frac{4\pi}{2m}\left\{q(r)\nabla\frac{\alpha(r)}{1+4/3\pi\xi\alpha(r)}\nabla\right\},$$

$$q(r) = \left(1+\frac{m_\pi}{m_N}\right)\{b_0\rho(r)+b_1[\rho_n(r)-\rho_p(r)]\} + \left(1+\frac{m_\pi}{2m_N}\right)\{B_0\rho^2(r)+B_1\rho(r)\delta\rho(r)\} \tag{8}$$

$$\alpha(r) = \left(1+\frac{m_\pi}{m_N}\right)^{-1}\{c_0\rho(r)+c_1[\rho_n(r)-\rho_p(r)]\} + \left(1+\frac{m_\pi}{2m_N}\right)^{-1}\{C_0\rho^2(r)+C_1\rho(r)\delta\rho(r)\}$$

where $\rho_{p,n}(r)$ are distribution of density of the protons and neutrons, respectively, ξ is a parameter such that $\xi=0$ corresponds to the case of “no correlation” while $\xi=1$ in the case of accounting of anticorrelations between nucleons); the isoscalar and isovector parameters b_0, c_0, B_0, b_1,c_1, C_0 B_1, C_1 are corresponding to the s-wave and p-wave (repulsive and attracting potential member) scattering length respectively in the combined spin-isospin space with taking into account for absorption of pions (with different channels for p-p pair $B_{0(pp)}$ and *p-n* pair $B_{0(pn)}$), the Lorentz-Lorentz effect in the p-wave interaction and isospin and spin dependence of an amplitude $\pi^- N$ scattering: $(b_0\rho(r) \to b_0\rho(r) + b_1\{\rho_p(r) - \rho_n(r)\}$ [2]).

2.3 *Complex energy of pionic atomic system*

The energy of a multielectron hadronic atom can be represented as follows:

$$E \approx E_{KG} + E_{FS} + E_{VP} + E_N; \tag{9}$$

Here E_{KG} -is the energy of a pion in a nucleus (Z, A) with the point-like charge (dominative contribution in (9)), E_{FS} is the contribution due to the nucleus finite size effect, E_{VP} is the radiation correction due to the vacuum-polarization effect, E_N is the energy shift due to the strong interaction V_N. The strong pion-nucleus interaction contribution into energy can be directly found from the solution of the Klein-Gordon-Fock equation with the corresponding pion-nucleon potential, for example, in the optical potential approximation (8). Since the corresponding optical potential contains complex parameters, the relevant energy eigen-values of the Klein-Gordon-Fock equation for the definite state $(i = nl)$ are complex valued too:

$$E_i = \mathrm{Re}\, E_i + i\, \mathrm{Im}\, E_i = \mathrm{Re}\, E_i - (i/2)\Gamma_i, \tag{10}$$

where Γ_i is a width of pionic energy level, which is determined by the imaginary part of (10). The total width of any level is determined as by the strong pion-nuclear interaction contribution Γ_i^s (pion absorption) as by the electromagnetic contribution Γ_i^{rad}. The latter is determined by a probability of the electromagnetic radiation transition (including the Auger process probability Γ_i^A) on the lower level. In order to calculate electromagnetic interaction characteristics (for example, probabilities of transitions in spectrum of the pionic atom) we have implemented the known (in a theory of usual multi-electron atom) relativistic energy formalism [56-68]. In this approach (an energy of any excited state being a complex quantity) a shift of the total energy level is usually represented as:

$$\Delta E_i = \mathrm{Re}\,\Delta E_i + i\, \mathrm{Im}\,\Delta E_i = \mathrm{Re}\,\Delta E_i - (i/2)\Gamma_i^{rad}, \tag{11}$$

where Γ_i^{rad} is a radiation width, and the corresponding radiative transition probability in the usual atomic system $P \sim \Gamma_i^{rad}$. In order to compute the latter we use the generalized relativistic energy approach. It is worth to remind that an initial general energy formalism combined with an empirical model potential method in a theory of atoms and multicharged ions has been developed by Ivanov-Ivanova et al [57-62]; further more general ab initio gauge-invariant version of relativistic energy approach has been presented by Glushkov-Ivanov [65-67]. The latter has been used in this work.

3. The results and conclusions

In Table 1 we list the values of the optical potential parameters, which have been used in our and other calculations [34-44,2]. It is worth to explain the used classification and abbreviations of the corresponding sets of optical potential parameters: Tauscher, ξ=0 – Tau1; Tauscher, ξ=1 – Tau2; Batty et al – Bat; Seki et al – Sek; Nagels – Nag; de Laat-Konijin et al – Laat, this theory - Odes [2,33-44]. Usually the potential parameters were initially obtained by calibration of the experimental data on pion-nuclear scattering for the light and medium nuclei. Further application of the model to the heavy systems and relatively low-lying states showed imperfections (in some cases) of these sets of the parameter values (the known pion anomalies). For example, experimental (low-energy scattering of pions; LAMPF) results for relatively low-lying states of pion at levels 3d, 2p, 1s (in such heavy atoms as Ta, Bi et al) have shown that the appropriate values of width due to strong interaction a factor two and more are lower than the values specified within the optical potential model using the earliest parameterization [1,33-35]. Under the parameterization of the optical potential authors [41,43] left without changing the parameters settings that are the most reliably identified, namely: ReB_0, ImB_0, c_0, c_1, ReC_0 , ImC_0. At the same time the parameters whose values differ most strongly in different sets, in particular, b_1 (plus parameters, which are usually not included so far in the basic optical potential parameterizations, i.e. ImB_1, ImC_1), should be optimized. This is achieved by receiving preliminary calculated relationships for the energy shifts and widths (for a number of states of the following systems: ^{20}Ne, ^{133}Cs, ^{208}Pb) upon the b_1, ImB_1, ImC_1 parameter values. As illustration In Table 2 we list the dependences of shifts and widths for the 4f, 3d levels due to the strong pion nuclear interaction upon the parameter ImB_1 value for the pionic ^{181}Ta.

Table 1. The values of the optical potential parameters, used in different calculations (see text).

	Tauscher ξ=0 [34]	Tauscher ξ=1 [34]	Batty ξ=1 [35]	Seki ξ=1 [36]	Nagels ξ=1[40]	Row78 ξ=1[37]	Laat ξ=1 [43,44]	Odes ξ=1 [2]	
b_0	-0.0296	-0.0293	-0.017	0.003	-0.013	-0.004	0.007	0.003	m_π^{-1}
b_1	-0.077	-0.078	-0.13	-0.143	-0.092	-0.094	-0.075	-0.094	m_π^{-1}
ReB_0	0	0	-0.0475	-0.15	-	-	-0.18	-0.15	m_π^{-4}
ImB_0	0.0436	0.0428	0.0475	0.046	-	-	0.058	0.046	m_π^{-4}
c_0	0.172	0.227	0.255	0.21	0.209	0.23	0.266	0.21	m_π^{-3}
c_1	0.22	0.18	0.17	0.18	0.177	0.17	0.40	0.18	m_π^{-3}
ReC_0	-	-	-	0.11	-	-	0.07	0.11	m_π^{-6}
ImC_1	-	-	-	-	-	-	-0.34	-0.25	m_π^{-6}

Table 2. The dependence of shifts and widths for the 4f, 3d levels due to the strong pion nuclear interaction upon the parameter ImB_1 value for ^{181}Ta

ImB_1	ε_0^{4f} (^{181}Ta)	Γ_0^{4f} (^{181}Ta)	ε_0^{3d} (^{181}Ta)	Γ_0^{3d} (^{181}Ta)
0.00	0.508	0.41	13.5	34.9
0.02	0.517	0.39	14.1	30.8
0.04	0.525	0.37	14.7	27.5
0.06	0.533	0.35	15.2	24.8
0.08	0.543	0.33	15.8	22.6
0.10	0.554	0.30	16.3	20.3
0.12	0.566	0.28	16.8	18.3
Exp.	0.56±0.04	0.31±0.05	16.2±1.3	20.1±1.5

In Table 3 we list our calculated (the relativistic Klein-Gordon-Fock theory combined with energy approach) data on the3d, 4f, 5g levels radiation widths for π^--Ta atom. There are also listed the analogous data by Laat et al [43,44], obtained using the relativistic Klein-Gordon-Fock model and Hartree-Fock approximation. In Table 4 we list theoretical and experimental data (in keV) for 4f, 3d levels shift and width due to strong π-N interaction [2,34-44]. In our parameterization of a strong potential the most reliably defined (B_0, c_0, c_1, C_0) parameters are remained unchanged. The parameters whose values differ greatly in different sets, in particular, b_1 (b_1 = -0.094) plus still not included ones ImB_1, ImC_1 have been optimized by computing dependency of the strong shift on the parameters b_1, ImB_1,ImC_1 for number of π^--atoms.

Table 3. The radiation widths of the 3d, 4f, 5g levels in some $\pi^- A$ systems

Nucleus	Γ^1_{rad}(5g) [43]	Γ^2_{rad} (5g) This work	Γ^1_{rad} (4f) [43,44]	Γ^2_{rad} (4f) This work	Γ^1_{rad} (3d) [43,44]	Γ^2_{rad} (3d) This work
^{181}Ta	25.7	23.5	90.9	88.6	369.9	366.1

Table 4. Theoretical and experimental data for the 4f, 3d levels shift and widths (keV) provided by the strong pion-nuclear interaction for the π^- -^{181}Ta atomic system (see text)

4f	Exp	H-like Func.	Tau1 ξ=0	Tau2 ξ=1	Bat ξ=1	Sek ξ=1	Laat ξ=1	Our ξ=1
ε_{4f}	0.56±0.04	0.47	0.57	0.54	0.53	0.47	0.60	0.55
Γ_{4f}	0.31±0.05	0.16	0.31	0.30	0.31	0.27	0.31	0.31
ε_{3d}	16.2±1.3	-	19.6	16.4	10.4	4.4	14.4	16.1
Γ_{3d}	20,1±1,5	-	40,5	37,5	33,4	26,2	27,6	20.2

In Table 5 data on the *4f-3d, 5g-4f* transition energies for pionic atoms ^{181}Ta are presented. There are also listed the measured values of the Berkley, CERN and Virginia laboratories and alternative data obtained on the basis of computing within alternative versions of the Klein-Gordon-Fock (KGF) theory with taking into account for a finite size of the nucleus in the model uniformly charged sphere and the standard Uehling-Serber radiation correction (see [1,2,4,23] and Refs. therein).

Table 5. Transition energies (keV) in the spectra of π-^{181}Ta atom (see text)

Trans.	Berkley E_{EXP}	CERN E_{EXP}	E_{KGF+EM} [1, 4]	E_{KGF-EM} [23]	E_N [4,23]	E_N, This work
5g-4f	453.1±0.4	453.90±0.20	453.06	453.78	453.52 453.62	453.71
4f-3d	-	1008.4±1.3	-	-	992.75	1008.80

To conclude, let us note that the key factors of the acceptable agreement between experimental and theoretical data for complex pionic atoms are provided by a correct treatment of the nuclear, relativistic, radiative and inter-electron correlation corrections. Using different schemes for accounting of these correlations explains a difference between calculational data, obtained within the Klein-Gordon-Fock approach. The further refinement of the obtained data can be reached by means of accounting for such effects as a spatial distribution of the magnetic moment inside a nucleus, nuclear-polarization corrections etc.

References

[1] Ericson T and Weise T 1988 *Pions and Nuclei* (Oxford : Clarendon)
[2] Serga I, Dubrovskaya Y, Kvasikova A et al 2012 *J. Phys.: Conf. Series* **397** 012013; Serga I et al 2017 *Photoelectr.* **26** 68
[3] Deslattes R, Kessler E, Indelicato P et al 2003 *Rev. Mod. Phys.* **75** 35
[4] Menshikov L I and Evseev M K 2001 *Phys. Uspekhi.* **171** 150
[5] Glushkov A V 2008 *Relativistic quantum theory. Quantum mechanics of atomic systems* (Odessa: Astroprint)
[6] Khetselius O Yu 2008 *Hyperfine structure of atomic spectra* (Odessa: Astroprint)
[7] Lauss B 2009 *Nucl. Phys. A.* **827C** 401; PSI experiment R-98.01, http://pihydro gen.psi.ch
[8] CERN DIRAC Collaboration, CERN-SPSLC-2011-001 SPSLC-P-284-ADD.-2011
[9] Leon M, Seki R 1974 *Phys.Rev.Lett.* **32** 132
[10] Batty C J, Eckhause M, Gall K et al 1989 *Phys. Rev.C.* **40** 2154

IOP Conf. Series: Journal of Physics: Conf. Series **1136** (2018) 012004 doi:10.1088/1742-6596/1136/1/012004

[11] Glushkov A V, Ambrosov S V, Loboda A V, Chernyakova Y G, Khetselius O Y and Svinarenko A A 2004 *Nucl. Phys. A* **734** E21
[12] Glushkov A V 1992 *Russian Phys. Journ.* **35** (7) 641
[13] Glushkov A, Malinovskaya S, Prepelitsa G and Ignatenko V 2005 *J. Phys.: Conf. Series* **11** 199
[14] Glushkov A V, Malinovskaya S V, Khetselius O Yu, Loboda A V, Sukharev D E and Lovett L 2009 *Int. J. Quant. Chem.* **109** 1717
[15] Malinovskaya S V, Glushkov A V, Khetselius O Yu, Svinarenko A A, Mischenko E V and Florko T A 2009 *Int. J. Quant. Chem.* **109** 3325
[16] Glushkov A V, Malinovskaya S V, Chernyakova Y G, Svinarenko A A 2004 *Int. J. Quant. Chem.* **99** 889
[17] Glushkov A V, Ambrosov S, Ignatenko A, Korchevsky D A 2004 *Int. J. Quant.Chem.* **99** 936
[18] Glushkov A V 1990 *J. Struct. Chem.* **31**(4) 529
[19] Glushkov A V, Mansarliysky V F, Khetselius O Yu, Ignatenko A V, Smirnov A V and Prepelitsa G P 2017 *J. Phys.: Conf. Series* **810** 012034
[20] Glushkov A V 2012 *J. Phys.: Conf. Series* **397** 012011
[21] Glushkov A V 2014 *J. Phys.: Conf. Series* **548** 012020
[22] Khetselius O Yu 2012 *J. Phys.: Conf. Series* **397** 012012
[23] Olaniyi B, Shor A, Cheng S, Dugan G and Wu C S 1982 *Nucl. Phys. A*. **403** 572
[24] Schlesser S, Le Bigot E-Q, Indelicato P and Pachucki K 2011 *Phys.Rev. C* **84** 015211
[25] Indelicato P 1996 *Phys.Scripta T* **65** 57
[26] Indelicato P and Trassinelli M 2005 *arXiv:physics*. 0510126v1
[27] Santos J, Parente F, Boucard S, Indelicato P and Desclaux J 2005 *Phys. Rev. A* **71** 032501
[28] Glushkov A V 2005 *AIP Conference Proc.* **796** 206
[29] Glushkov A V, Khetselius O Yu and Malinovskaya S V 2008 *Europ. Phys. J. ST* **160** 195
[30] Glushkov A V, Khetselius O Yu and Malinovskaya S V 2008 *Molec. Phys.* **106** 1257
[31] Glushkov A V, Khetselius O Yu and Svinarenko A A 2012 *Advances in the Theory of Quantum Systems in Chemistry and Physics (Series Progress in Theoretical Chemistry and Physics* vol 22) eds P Hoggan, E Brändas, J Maruani, G Delgado-Barrio and P Piecuch (Dordrecht: Springer) pp 51-68
[32] Glushkov A V, Khetselius O Yu and Lovett L 2009 *Advances in the Theory of Atomic and Molecular Systems* (*Series Progress in Theoretical Chemistry and Physics* vol 20) eds P Piecuch, J Maruani, G Delgado-Barrio and S Wilson (Dordrecht: Springer) pp125-152
[33] Erikcson M, Ericson T and Krell M 1969 *Phys. Rev. Lett.* **22** 1189
[34] Tauscher L 1971 Proc.of Intern. Sem. "π-Meson Nucleus Interaction" (Strasbourg: CNRS) p 45
[35] Batty C J, Biagi S F, Friedman E and Hoath S D 1978 *Phys. Rev. Lett.* **440** 931; Batty C J, Friedman E and Gal A 1983 *Nucl. Phys.* A **402** 411
[36] Seki R, Masutani K and Jazaki K 1983 *Phys. Rev. C* **27** 1817
[37] Rowe G, Salamon M and Landau R H 1978 *Phys. Rev. C* **18** 584
[38] Anagnostopoulos D, Biri S, Boisbourdain V et al 2003 *Nucl. Inst. Meth. B* **205** 9
[39] Anagnostopoulos D, Gotta D, Indelicato P and Simons L 2003 *arXiv:physics*. 0312090v1
[40] Nagels M M, de Swart J, Nielsen H et al 1976 *Nucl. Phys. B* **109** 1
[41] Taal A, D'Achard van Enschut J, Berkhput J et al 1985 *Phys. Lett. B* **156** 296
[42] Taal A, David P, Hanscheid H, Koch J H, de Laat C T et al 1990 *Nucl. Phys. A* **511** 573
[43] de Laat C T, Taal A, Konijn J et al 1991 *Nucl. Phys. A* **523** 453
[44] de Laat C T, Taal A, Duinker W et al 1987 *Phys. Lett. B* **189** 7
[45] Strauch T 2009 *High-Precision Measurement of Strong-Interaction Effects in Pionic D* (Julich)
[46] Mohr P J 1983 *Atom. Dat. Nucl. Dat. Tabl.* **24** 453; 1993 *Phys. Scripta* **46** 44
[47] Khetselius O Yu 2009 *Int. J. Quant. Chem.* **109** 3330
[48] Khetselius O Yu 2009 *Phys. Scripta* **T135** 014023
[49] Khetselius O Yu 2008 *AIP Conference Proceedings* **1058** 363
[50] Flambaum V V and Ginges J S 2005 *Phys. Rev. A* **72** 052115

[51] Khetselius O Yu 2012 *Quantum Systems in Chemistry and Physics (Series Progress in Theoretical Chemistry and Physics* vol 26) eds K Nishikawa, J Maruani, E Brändas, G Delgado-Barrio and P Piecuch (Dordrecht: Springer) pp 217-229
[52] Khetselius O Yu, Florko T A, Svinarenko A A and Tkach T B 2013 *Phys. Scripta* **T153** 014037
[53] Khetselius O Yu 2010 *AIP Conf. Proc.* **1290** 29
[54] Glushkov A V, Loboda A V 2007 *J. Applied. Spectr.* **74**(2) 305
[55] Safranova M S, Johnson W, Safranova U I and Cowan T 2006 *Phys. Rev. A* **74** 022504
[56] Ivanova E P and Grant I P 1998 *J. Phys.B: At. Mol. Opt. Phys.* **31** 2871
[57] Ivanova E P and Ivanov L N 1979 *Atom. Data. Nucl. Data. Tabl.* **24** 95
[58] Ivanov L N, Ivanova E P and Aglitsky E V 1988 *Phys. Rep.* **166** 315
[59] Ivanov L N, Ivanova E P and Knight L 1993 *Phys. Rev. A* **48** 4365
[60] Ivanova E P, Ivanov L N, Glushkov A V and Kramida A E 1985 *Phys. Scripta* **32** 513
[61] Ivanova E P and Glushkov A V 1986 *J. Quant. Spectr. Rad. Transfer.* **36** 127
[62] Glushkov A V, Ivanov L N and Ivanova E P 1986 *Autoionization Phenomena in Atoms*, (Moscow: Moscow University Press) p 58-160
[63] Glushkov A V, Malinovskaya S V, Filatov V V 1989 *Sov. Phys. Journ.* **32**(12) 1010
[64] Glushkov A V 1991 *Sov. Phys. Journ.* 34(4) 307 (Ga)
[65] Glushkov A V 1990 *Soviet Phys. J.* **33**(1) 1
[66] Glushkov A V 1991 *Opt. Spectrosc.* **70** 555
[67] Glushkov A V, Ivanov L N 1992 *Phys. Lett. A.* **170** 33
[68] Glushkov A and Ivanov L N 1993 *J. Phys. B: At. Mol. Opt. Phys.* **26** L379
[69] Malinovskaya S V and Glushkov A V 1992 *Russian Phys. Journ.* **35** 999
[70] Glushkov A V, Malinovskaya S, Prepelitsa G, Ignatenko V 2005 *J. Phys.: Conf. Series* **11** 199
[71] Malinovskaya S V, Glushkov A V, Khetselius O Yu, Loboda A V, Lopatkin Yu M, Nikola L V, Svinarenko A A and Perelygina T B 2011 *Int. J. Quant. Chem.* **111** 288
[72] Glushkov A V, Ambrosov S V, Loboda A V, Gurnitskaya E P and Prepelitsa G P 2005 *Int. J. Quant.Chem.* **104** 562
[73] Glushkov A V, Khetselius O Yu and Svinarenko A A 2013 *Phys. Scripta* **T153** 014029
[74] Glushkov A V 1992 *J. Appl. Spectrosc.* **56**(1) 5
[75] Svinarenko A A 2014 *J. Phys.: Conf. Series* **548** 012039
[76] Glushkov A V 1990 *Sov. Phys. Journ.* **33**(9) 754
[77] Glushkov A V, Loboda A, Gurnitskaya E and Svinarenko A 2009 *Phys. Scripta* **T135** 014022
[78] Glushkov A V, Butenko Yu V, Serbov N G, Ambrosov S V, Orlova V E , Orlov S V, Balan A K and Dormostuchenko G M 1996 *J. Appl. Spectrosc.* **63** 28
[79] Glushkov A V, Efimov V A, Gopchenko E D, Dan'kov S V, Polishchyuk V N and Goloshchak O P 1998 *Optics and Spectrosc.* **84**(5) 670
[80] Florko T, Ambrosov S, Svinarenko A and Tkach T 2012 *J. Phys.: Conf. Series* **397** 012037
[81] Khetselius O Yu 2009 *J. Phys.: Conf. Series* **194** 022009
[82] Svinarenko A A, Nikola L, Prepelitsa P, Tkach T, Mischenko E 2010 *AIP Conf. Proc.* **1290** 94
[83] Svinarenko A A, Khetselius O Yu, Buyadzhi V V, Florko T A, Zaichko P A and Ponomarenko E L 2014 *J. Phys.: Conf. Series* **548** 012048
[84] Svinarenko A, Ignatenko A, Ternovsky V, Nikola V, Seredenko S and Tkach T 2014 *J. Phys.: Conf. Series* **548** 012047
[85] Glushkov A V, Khetselius O Yu, Lopatkin Yu M, Florko T A, Kovalenko O A and Mansarliysky V F 2014 *J. Phys.: Conf. Series* **548** 012026
[86] Buyadzhi V, Zaichko P, Antoshkina O, Kulakli T, Prepelitsa G, Ternovsky V and Mansarliysky V 2017 *J. Phys.: Conf. Series* **905** 012003
[87] Buyadzhi V V, Zaichko P A, Gurskaya M Y, Kuznetsova A A, Ponomarenko E L and Ternovsky V B 2017 *J. Phys.: Conf. Series* **810** 012047
[88] Glushkov A V, Khetselius O, Gurnitskaya E, Loboda A and Mischenko E 2009 *AIP Conf. Proc.* **1102** 172

IOP Conf. Series: Journal of Physics: Conf. Series **1136** (2018) 012005 doi:10.1088/1742-6596/1136/1/012005

Electron localization properties in high pressure hydrogen at the liquid-liquid phase transition by Coupled Electron-Ion Monte Carlo

Carlo Pierleoni

Department of Physical and Chemical Sciences, University of L'Aquila, Italy
Maison de la Simulation, CEA, CNRS, Univ. Paris-Sud, UVSQ, Université Paris-Saclay, 91191 Gif-sur-Yvette, France

Giovanni Rillo

Department of Physics, Sapienza University of Rome, Italy

David M.Ceperley

Department of Physics, University of Illinois Urbana-Champaign, Illinois, USA

Markus Holzmann

LPMMC, UMR 5493 of CNRS, Université Grenoble Alpes, France
Institut Laue-Langevin, BP 156, F-38042 Grenoble Cedex 9, France

Abstract. We analyze in detail the electronic properties of high pressure hydrogen around the liquid-liquid phase transition based on Coupled Electron-Ion Monte Carlo calculations. Computing the off-diagonal single particle density matrix and the momentum distribution we discuss localization properties of the electrons. The abrupt changes of these distributions indicate a metal to insulator transition occurring together with the structural transition from the atomic to molecular fluid. We further discuss the electron-proton and electron-electron pair correlation functions, which also change abruptly at the transition.

1. Introduction

The search of metallization in hydrogen has been and still is one of the major driving forces in high pressure research. In 1935, Wigner and Huntington discussed the processes of metallization of low temperature solid molecular hydrogen upon increasing pressure either by a continuous process or through a structural phase transition to a monoatomic phase. They predicted that a body centered cubic monoatomic hydrogen would become favorable over the molecular form above 25GPa. Despite almost a century of intense research and impressive progress in both experimental and theoretical methods, this question still remains open. Static pressure experiments by the Diamond Anvil Cell technique in solid hydrogen have reported the record pressure of $\sim$500GPa [1, 2] and the emergence of metallic or semi-metallic behaviors, but the mechanism of metallization is still unclear. There has been intense interest in this system during the past year with the first claim of experimental observation of metallic solid hydrogen, however followed by many criticisms [1, 3, 4, 5, 6]. On the theoretical side, *ab initio* DFT calculations using the most common exchange-correlation approximations, have not been found to be accurate enough to address the ordering of the phases of hydrogen near metallization and molecular dissociation [7, 8]. Only recently, more fundamental methods based on Quantum Monte Carlo approaches, together with adequate computer resources, have become available and have started to be used to investigate potential phases of crystalline hydrogen [9, 10]. We expect that in the near future those methods will contribute to clarify the scenario of solid hydrogen metallization and provide fundamental information for interpreting the experimental results.

The situation seems slightly easier in the higher temperature fluid, at least concerning theoretical methods. Experimentally, a discontinuous change in the reflectivity and the resistivity of fluid hydrogen has been observed, both with shock wave and static compression techniques [11, 12, 13, 14]. However, the location of this discontinuity and its nature is still under debate, results from these two different methods differing by as much $\sim$150GPa. Theoretical *ab initio* calculations also predict a discontinuity in the Equation of State, a first order phase transition between an insulating molecular fluid and a dissociated metallic fluid. Again, different approximations predict different locations of the transition line [15] with the most accurate predictions based on Quantum Monte Carlo (QMC)[15, 16, 17] being rather close to the results from the static compression method, although at slightly higher pressures [15].

These simulations also allow the characterization of the nature of the phase transition as concerns both nuclear and electronic properties. We have shown that at the transition, molecules suddenly dissociate and the electronic localization changes, accompanied by the emergence of a non-zero static electronic conductivity (the latter obtained at the DFT level) [15]. While we have recently extended and integrated the description of the change of nuclear structure at the transition[16], we have only given a rapid account of the electronic properties at the transition in the original publication [15]. In the present paper we provide a more complete account of the changes in the electronic properties leading to a coherent theoretical picture of the liquid-liquid phase transition in hydrogen.

The paper is organized as follows. In the next section, after a brief recap of the basic ingredients of the Coupled Electron-Ion method, we present a detailed account of the theory and practical implementation of single electron density matrix calculations and the electronic momentum distribution used to characterize the electronic localization properties. We then discuss the results of ref. [15] in more detail and show original results for the density-density, electron-proton and electron-electron, pair distribution functions.

2. Method

2.1. The Coupled-Electron-Ion Monte Carlo Method

The Coupled Electron-Ion Monte Carlo method is an *ab initio* method for finite temperature nuclei coupled with ground state electrons based on the Born-Oppenheimer approximation. It is described in detail in refs. [18, 19]. More recent implementations and technical details employed in the present calculations are provided in the Supplementary Information of ref. [15] together with a detailed discussion of the various systematic biases introduced by numerical approximations. In the following, we only discuss the methodological issues concerning the calculation of the reduced single particle density matrix and momentum distribution which have never been addressed in detail before, in particular the connection with twisted boundary conditions to reduce finite size effects.

2.2. The reduced single particle density matrix and momentum distribution

For any given many-body state of the electrons, $\Psi(\mathbf{R}_N)$, $\mathbf{R} \equiv (\mathbf{r}_1, \mathbf{r}_2, \ldots, \mathbf{r}_N)$, all observables involving only one particle can be calculated from the knowledge of the reduced single particle

IOP Conf. Series: Journal of Physics: Conf. Series **1136** (2018) 012005 doi:10.1088/1742-6596/1136/1/012005

density matrix

$$\rho^{(1)}(\mathbf{r},\mathbf{r}') = N\frac{\int d\mathbf{r}_2\cdots d\mathbf{r}_N \Psi^*(\mathbf{r},\mathbf{r}_2,\ldots,\mathbf{r}_N)\Psi(\mathbf{r}',\mathbf{r}_2,\ldots,\mathbf{r}_N)}{\int d\mathbf{R}_N|\Psi(\mathbf{R}_N)|^2} \tag{1}$$

The real space density distribution of the electrons is given by its diagonal elements, $\rho(\mathbf{r}) \equiv \rho^{(1)}(\mathbf{r},\mathbf{r})$. From a Monte Carlo sampling of the probability $\Pi(\mathbf{R}_N) \propto |\Psi(\mathbf{R}_N)|^2$ we can calculate the off-diagonal elements via reweighting

$$\rho^{(1)}(\mathbf{r},\mathbf{r}') = N\left\langle \delta(\mathbf{r}-\mathbf{r}_1)\frac{\Psi(\mathbf{r}',\mathbf{r}_2,\ldots,\mathbf{r}_N)}{\Psi(\mathbf{r},\mathbf{r}_2,\ldots,\mathbf{r}_N)}\right\rangle \tag{2}$$

where $\langle\cdot\rangle \equiv \int d\mathbf{R}_N\Pi(\mathbf{R}_N)[\cdot]$. Whereas the reduced single particle density matrix is rather directly calculated within variational Monte Carlo [20], the sampling must be adapted to obtain unbiased estimators within reptation Monte Carlo [21, 22].

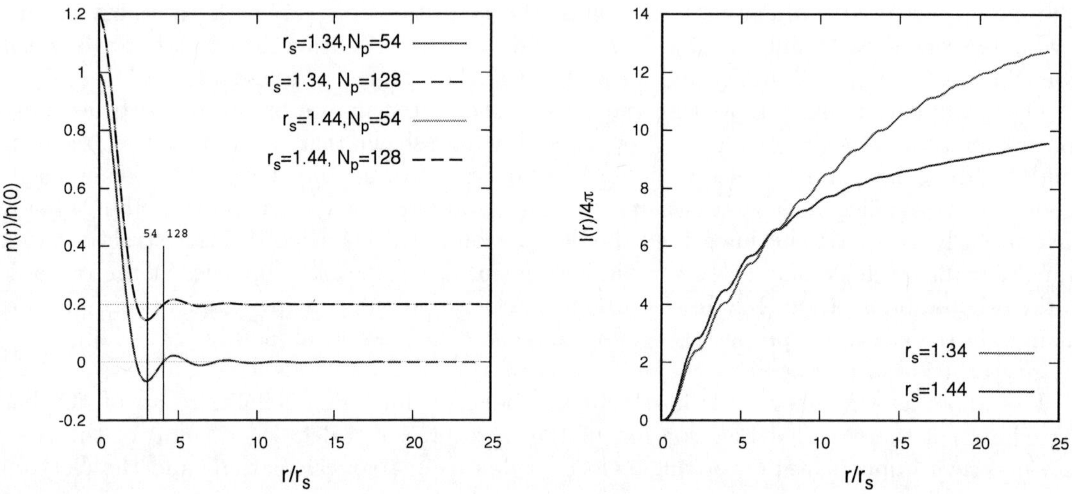

Figure 1. Left panel: The off-diagonal single particle density matrix, $n(r)$, at $T = 1200$K and two different densities $r_s = 1.34$ in the atomic and $r_s = 1.44$ in the molecular phase (with an offset of 0.2) of the liquid-liquid phase transition. Vertical lines indicate $L/2$, where L is the length of the simulation cell, for systems containing $N_p = 54$ and $N_p = 128$ protons. The use of twisted boundary conditions gives access to $n(r)$ at distances $r >> L/2$, without noticeable bias. Right panel: Convergence of the integral of the absolute value of the off-diagonal single particle density matrix, $I(r) = 4\pi \int_0^r dr' r'^2 |n(r')|/n(0)$, with respect to the distance r at $T = 1200$K. Its behavior at large distance provides a measure of electron localization in real space. Divergence occurs for delocalized, extended wave functions, typical for metallic systems, whereas the integral should asymptotically converge when electrons are localized (insulators). Although the region of convergence of the integral of the molecular fluid is not fully reached (despite the extended range accessed by the use of twisted boundary conditions), the atomic and molecular fluid are clearly distinguished.

Diagonalizing the reduced single particle density matrix we obtain the natural orbitals. For homogeneous systems, these are plane waves and $\rho^{(1)}(\mathbf{r},\mathbf{r}')$ depends only on the distance $|\mathbf{r}-\mathbf{r}'|$.

Since averaging over the proton configuration restores translational invariance in the fluid phase of hydrogen, we introduce the volume–averaged off-diagonal density matrix

$$n(r) = \frac{1}{V}\int d\mathbf{r}'\rho^{(1)}(\mathbf{r}+\mathbf{r}',\mathbf{r}') \tag{3}$$

whose Fourier transform gives the momentum distribution

$$n_k = \int d\mathbf{r}e^{-i\mathbf{k}\cdot\mathbf{r}}n(r) \tag{4}$$

These are normalized such that $n(0) = V^{-1}\sum_{\mathbf{k}} n_k$ gives the average electron density $\rho = 3/(4\pi r_s^3)a_B^{-3}$, usually given in terms of the dimensionless parameter r_s, where a_B is the Bohr radius.

For a normal Fermi liquid such as the homogeneous disorder-free electron gas, the momentum distribution is characterized by a discontinuity of the momentum distribution at the Fermi wave vector [23, 22], leading to oscillations in $n(r)$ whose envelop decays algebraically as r^{-3} at large distances.

2.3. Twist averaged boundary conditions

Using twisted boundary conditions for the wave functions [24], we have

$$\Psi_\theta(\mathbf{r}_1 + L\hat{\mathbf{x}}_\alpha, \mathbf{r}_2, \ldots) = e^{i\theta_\alpha}\Psi_\theta(\mathbf{r}_1, \mathbf{r}_2, \ldots) \tag{5}$$

where $\mathbf{x}_\alpha$ is the unit vector in the α direction ($\alpha = x, y, z$), and θ_α is the twist angle in this direction, $-\pi \leq \theta_\alpha < \pi$. For a general twist angle, we then have for the reduced density matrix

$$\rho_\theta^{(1)}(\mathbf{r} + L\mathbf{x}_\alpha, \mathbf{r}') = e^{i\theta_\alpha}\rho_\theta^{(1)}(\mathbf{r}, \mathbf{r}') \tag{6}$$

Averaging over N_θ different twist angles, we then get for the twist-averaged off-diagonal density matrix,

$$n_{\text{TABC}}(\mathbf{r} + L\hat{\mathbf{x}}_\alpha) = \frac{1}{N_\theta}\sum_{i=1}^{N_\theta} e^{i\theta_\alpha} n_{\theta_i}(\mathbf{r}) \tag{7}$$

Whereas $n_{\theta_i}(\mathbf{r})$ is only defined within the simulation box, the twist averaged density matrix can thus be extended to distances much larger than L by using a fine grid for the twist angles.

In Fig. 1, we show the twist averaged density matrix which smoothly extends the behavior of the off-diagonal matrix elements to distances much larger than $L/2$, the typical limit for simulations using strictly periodic boundary conditions. Comparison of two different system sizes does not show any noticeable bias. However, we note that slowly decaying finite-size effects due to two-particle correlations as described in Refs [25, 22] are not taken into account by twist averaged boundary conditions and have to be addressed separately.

Using a linear grid of twist angles, $\theta_\alpha = 2\pi n_\alpha/M$ with integer n_α, the twist averaged density matrix, $n_{\text{TABC}}(\mathbf{r})$, is a periodic function in a box of extension ML with a discrete Fourier transform on the reciprocal lattice $\mathbf{k} = (n_x, n_y, n_z)2\pi/ML$. The corresponding momentum distribution writes

$$n_{\mathbf{k}}^{\text{TABC}} = \int_{M^3V} d\mathbf{r}e^{-i\mathbf{k}\cdot\mathbf{r}}n_{\text{TABC}}(\mathbf{r}) = \frac{1}{N_\theta}\sum_{i=1}^{N_\theta}\sum_{\mathbf{m}}\int_V d\mathbf{r}e^{i\sum_\alpha[\theta_\alpha m_\alpha - k_\alpha(\mathbf{r}_\alpha + m_\alpha L)]}n_{\phi_i}(\mathbf{r}) \tag{8}$$

The summation over the integer values $m_\alpha = 0, \ldots M-1$ gives a periodic delta function which selects a unique twist angle θ_α modulus 2π. We can therefore write $\mathbf{k} = \mathbf{k}_0 + \mathbf{G}$ where

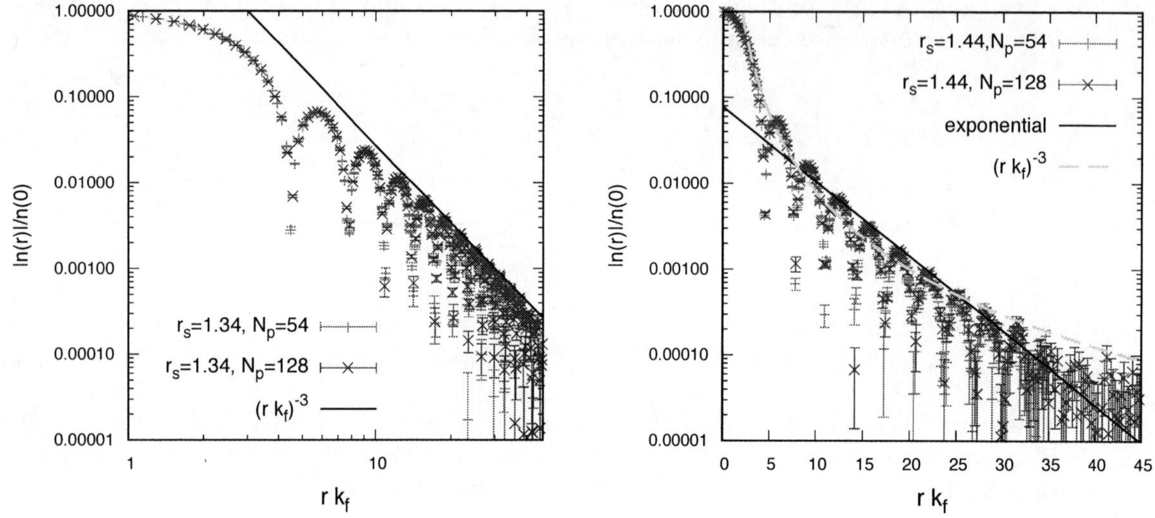

Figure 2. Absolute value of the off-diagonal part of the reduced single particle density matrix at $T = 1200$K in a logarithmic scale. For the atomic liquid at $r_s = 1.34$ (left panel), the decay of the envelope asymptotically decays as $(k_F r)^{-3}$, whereas it is closer to an exponential behavior for the molecular liquid at $r_s = 1.44$ (right panel).

$\mathbf{G} = (n_x, n_y, n_z)2\pi/L$ is on the reciprocal lattice of the original system of extension L, and write

$$n_{\mathbf{k}}^{\mathrm{TABC}} = \int_V d\mathbf{r} e^{-i(\mathbf{k}_0+\mathbf{G})\cdot\mathbf{r}} n_\theta(\mathbf{r}) \tag{9}$$

Therefore, the natural extension of the momentum distribution from periodic to twisted boundary conditions is given by

$$n_{\mathbf{G}+\mathbf{k}_0} = \int_V d\mathbf{r} e^{-i(\mathbf{k}_0+\mathbf{G})\cdot\mathbf{r}} n_\theta(\mathbf{r}) \tag{10}$$

where $k_0 = \theta/L$ shifts the grid in reciprocal space with respect to simulations using periodic boundary conditions. Twist averaged boundary conditions therefore allow us to extend the momentum distribution to an arbitrarily dense grid in reciprocal space. To impose a sharp Fermi surface, simulations should be done at fixed chemical potential using grand-canonical twist averaging where the number of electrons may vary with twist angle [26, 27].

3. Results

In the following we analyze the electronic properties of Coupled-Electron Ion Monte Carlo calculations [15] along five isotherms $T = 600K, 900K, 1200K, 1500K, 3000K$. For all temperatures $T \leq 1500K$, a weak first-order transition from a molecular to an atomic liquid, together with an insulator to metal transition, occurs upon increasing density [15]. In a recent paper, we have analyzed in details the structural properties of the protons, confirming the abrupt character of the molecular dissociation at the transition by employing a cluster analysis of the nuclear configurations visited during the CEIMC runs [16]. In the following, we analyze the electronic properties averaged over the nuclear configurations in more detail, and show that the change in the structural properties is accompanied with a sharp change of the electronic single particle density matrix as well as the electron-proton and electron-electron pair correlation

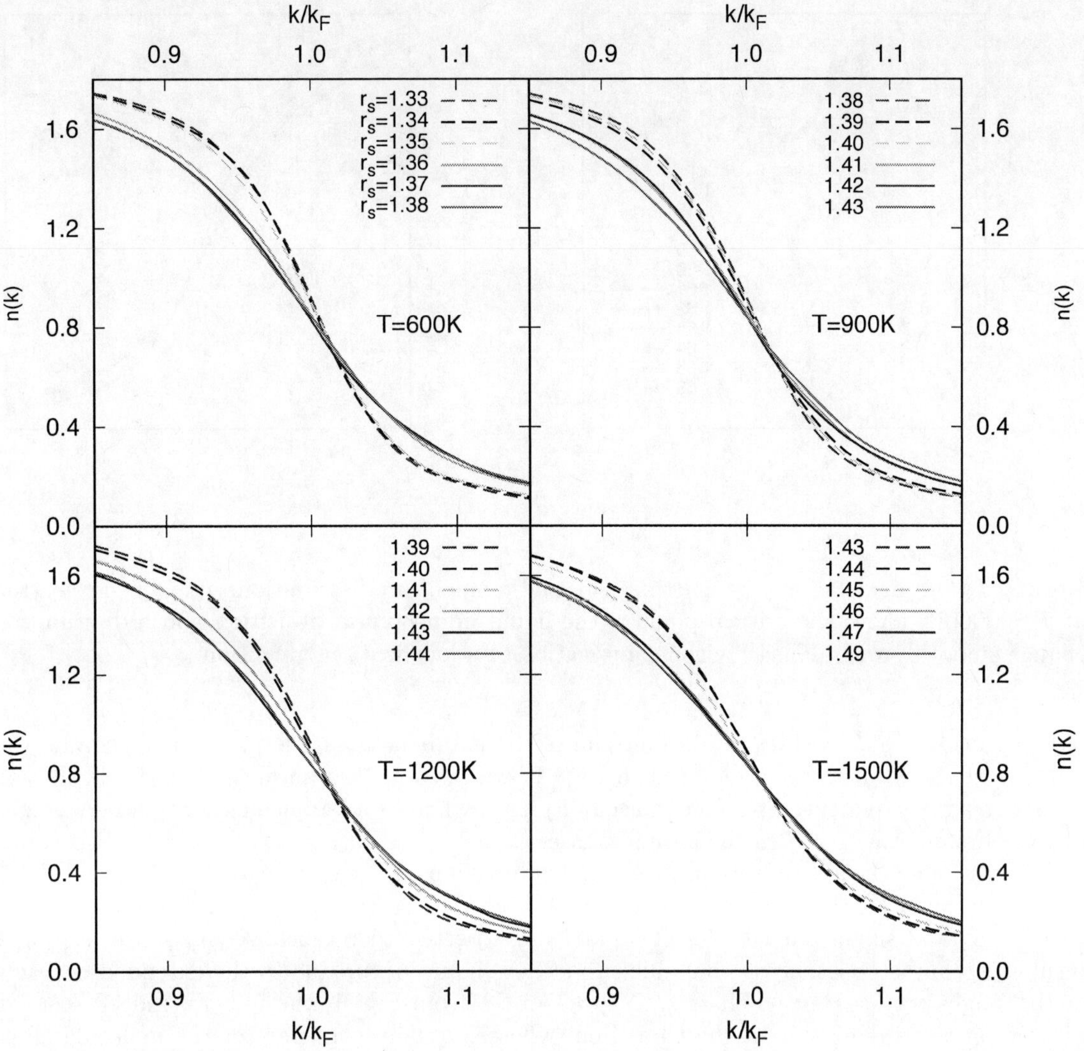

Figure 3. Momentum distribution, $n(k)$, of the electrons at different temperatures and densities. The distributions of the atomic fluid (dashed lines) are clearly distinguished from those of the molecular ones (full lines), except at the coexisting densities.

functions. All the following calculations of the off-diagonal density matrix are done at the VMC level using twisted boundary conditions with fixed number of electrons (canonical twist averaging). These results may still suffer from size effects due to two-particle correlations [25, 22] which decays slowly with increasing system size. However, these effects will not affect our main result showing a clear qualitative change in the off-diagonal single particle density matrix from the molecular to the atomic fluid.

3.1. The off-diagonal density matrix, electron localization, and momentum distribution

In Figure 1 and 2, we show the off-diagonal elements of the reduced single particle density matrix, $n(r)$, for the atomic ($r_s = 1.34$) and molecular ($r_s = 1.44$) liquid on both sides of the

IOP Conf. Series: Journal of Physics: Conf. Series **1136** (2018) 012005 doi:10.1088/1742-6596/1136/1/012005

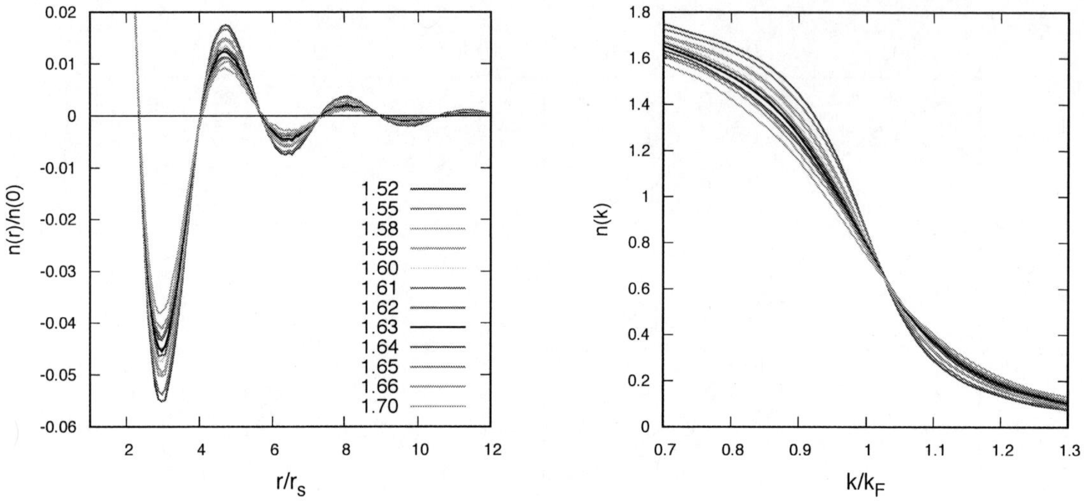

Figure 4. Off-diagonal density matrix, $n(r)$, and momentum distribution, $n(k)$, of the electrons at $T = 3000$K, above the critical point of the liquid-liquid phase transition. Both distributions change smoothly with density without indication of a localization transition.

phase transition at $T = 1200K$. At first glance the two behaviors are qualitatively similar, but because of the logarithmic scale used in Fig. 2, we see that the amplitude of the oscillations decays roughly exponentially in the molecular phase, whereas it is much closer to a Fermi-liquid like algebraic decay $\sim r^{-3}$ in the atomic phase.

Integration of the absolute value of $n(\mathbf{r})$ introduces a measure of electron localization [28, 15]. For any decay faster than r^{-3}, the integral $I(r) = 4\pi \int_0^r dr' r'^2 |n(r')|/n(0)$ converges with respect to $r \to \infty$ and single particle density matrix is expected to be localized in space leading to insulating character, otherwise metallic properties should occur due to the extended character of the single particle density matrix. As we can see from Fig. 1, the integral indeed seems to diverge on the atomic side of the transition, whereas it may converge for the molecular fluid. However, despite the extended region reached by the use of twisted boundary conditions, the integral has not yet obtained its asymptotic value.

The drastic change of the electronic single particle properties at the phase transition is even more evident in the momentum distribution, $n(k)$, shown in Fig. 3 for the four isotherms below the critical temperature. Along each isotherm, we report results for 6 densities (expressed in terms of r_s), around the transition point as done in ref [16]. In terms of k/k_F, where $k_F = (9\pi/4)^{1/3}(r_s a_B)^{-1}$ is the Fermi vector corresponding to the average electronic density, the distribution in Fourier space changes very smoothly with density or temperature within a single phase. However, the distribution changes rapidly at the transition, that of atomic liquid is substantially different from that of the molecular one[1].

Although the momentum distribution at k_F is much steeper for the atomic fluid, there is not a discontinuity at the Fermi surface that should be expected for a normal Fermi fluid. However, in our case, even in the metallic state, the electrons collide with protons, which can be considered as

[1] At T=1200K $n(k)$'s for the two coexisting densities are almost superimposed. Approaching the critical temperature the system spontaneously switch from one phase to the other along the trajectory resulting in an average behavior for long sampling. We believe this is the reason for this superimposition. At T=1500K, the generated trajectories are rather shorter.

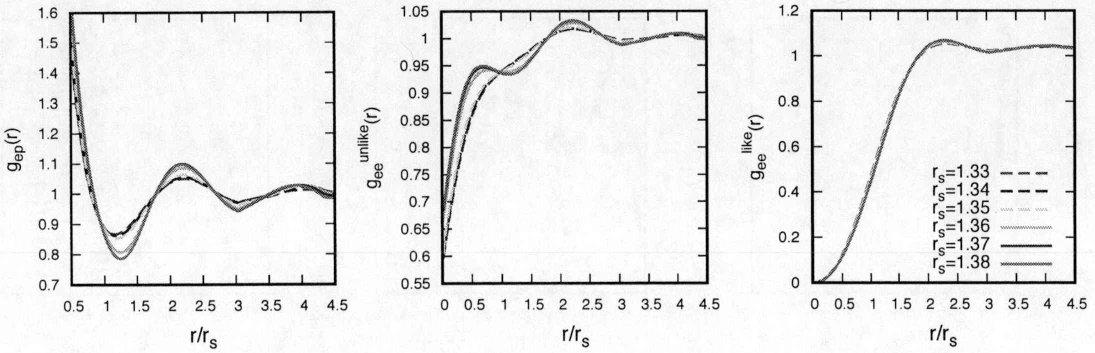

Figure 5. Electron-proton and electron-electron pair correlation with unlike and like spin, $g_{ep}(r)$, $g_{ee}^{\text{unlike}}(r)$, and $g_{ee}^{\text{like}}(r)$, respectively, at $T = 600K$. The distributions abruptly change at the transition from the molecular to the atomic fluid.

a static external potential within the Born-Oppenheimer approximation. In contrast to electrons moving in a periodic potential, even in the non-interacting electron approximation, the mean-free path (and life-time) of the electrons in the disorder potential created by the protons becomes finite and the discontinuity of the momentum distribution at the Fermi surface gets smoothed out.

At a higher temperature, $T = 3000K$, shown in Fig. 4, the off-diagonal density matrix as well as the momentum distribution changes continuously with density without any sign of a phase transition in the electronic properties, and we do not expect the occurrence of the metal to insulator transition at this temperature in agreement with the absence of a phase transition in the the structural analysis of the nuclear positions and in the equation of state [15, 16]. Therefore, the transition in the electronic properties, from a localized (insulating) to an extended (metallic) state coincides, within our precision, with the structural transition from the molecular to the atomic liquid at all temperatures.

3.2. The density-density correlation functions

The electron-proton and electron-electron correlation functions at low temperature are shown in Fig.5. Similar to the off-diagonal single particle properties discussed above, there is an abrupt change of the distributions at the liquid-liquid phase transition, more pronounced for unlike than for like spins in the electron-electron correlations. However, whereas the off-diagonal density matrix properties enter into transition matrix elements, e.g. needed for transport calculations, density-density correlation functions reflect structural properties. The changes in the electron-proton and unlike electron-electron correlation functions at the transition therefore have a rather direct interpretation in terms of the change from atomic to molecular bonds. Above the critical temperature, the pair correlation functions change smoothly with density (see Fig.6) again signalling the absence of a first order phase transition.

4. Conclusions

In this paper, we have presented the methodology for calculating off-diagonal density matrices using twisted boundary conditions and analyzed the changes in the electronic properties of hydrogen at the liquid-liquid phase transition. Our analysis shows an abrupt change in all electronic properties, indicating an insulator to metal transition which occurs simultaneously with the structural transition from the molecular to atomic fluid. At higher temperatures,

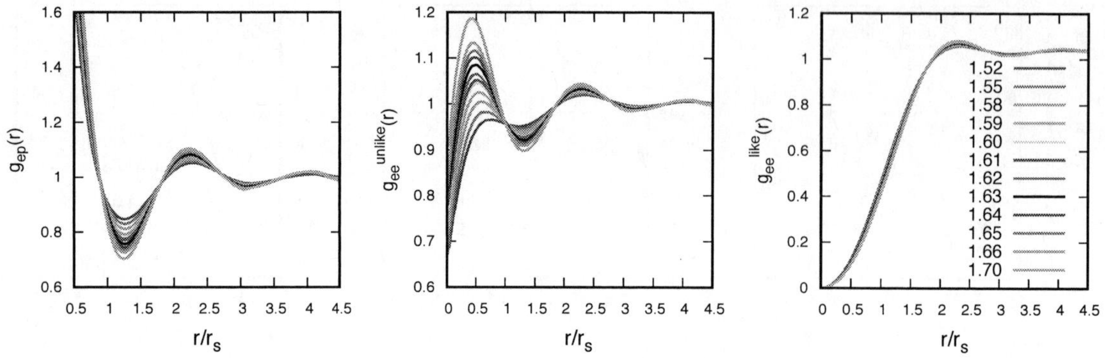

Figure 6. Electron-proton and spin-resolved electron-electron correlation functions, $g_{ep}(r)$, $g_{ee}^{\mathrm{unlike}}(r)$, and $g_{ee}^{\mathrm{like}}(r)$, respectively, at $T = 3000K$. Here, above the critical temperature of the structural transition, the distribution changes smoothly with density.

$T = 3000K$, above the critical point of the first-order transition, electronic properties change smoothly with density.

4.1. Acknowledgments

C.P. was supported by the Agence Nationale de la Recherche (ANR) France, under the program "Accueil de Chercheurs de Haut Niveau 2015" project: HyLightExtreme. D.M.C. was supported by DOE Grant NA DE-NA0001789 and by the Fondation NanoSciences (Grenoble). Computer time was provided by PRACE Projects 2013091918 and 2016143296 and by an allocation of the Blue Waters sustained petascale computing project, supported by the National Science Foundation (Award OCI 07- 25070) and the State of Illinois.

References

[1] Dias R P and Silvera I F 2017 *Science* **355** 715–718
[2] Eremets M, Drozdov A P, Kong P and Wang H 2017 URL `https://arxiv.org/abs/1708.05217`
[3] Goncharov A F and Struzhkin V V 2017 URL `https://arxiv.org/pdf/1702.04246.pdf`
[4] Loubeyre P, Occelli F and Dumas P 2017 URL `https://arxiv.org/pdf/1702.07192.pdf`
[5] Liu X D, Dalladay-Simpson P, Howie R T, Li B and Gregoryanz E 2017 URL `https://arxiv.org/pdf/1704.07601.pdf`
[6] Silvera I F and Dias R 2017 URL `https://arxiv.org/pdf/1703.03064.pdf`
[7] Clay R C, McMinis J, McMahon J M, Pierleoni C, Ceperley D M and Morales M A 2014 *Phys. Rev. B* **89** 184106
[8] Azadi S and Foulkes W M C 2013 *Phys Rev. B* **88** 014115
[9] Drummond N D, Monserrat B, Lloyd-Williams J H, Ríos P L, Pickard C J and Needs R J 2015 *Nature Communications* **6** 7794
[10] Rillo G, Morales M A, Ceperley D M and Pierleoni C 2018 *J. Chem. Phys.* **148** 102314
[11] Weir S T, Mitchell A C and Nellis W J 1996 *Phys. Rev. Letts.* **76** 1860–1863
[12] Knudson M D, Desjarlais M P, Becker A, Lemke R W, Cochrane K R, Savage M E, Bliss D E, Mattsson T R and Redmer R 2015 *Science* **348** 1455–1460
[13] Dzyabura V, Zaghoo M and Silvera I F 2013 *Proc. Nat. Acad. Sc.* **110** 8040–8044
[14] Zaghoo M, Salamat A and Silvera I F 2016 *Phys. Rev. B* **93** 155128
[15] Pierleoni C, Morales M A, Rillo G, Holzmann M and Ceperley D M 2016 *Proc. Natl. Acad. Sci.* **113** 4954–4957
[16] Pierleoni C, Holzmann M and Ceperley D 2018 *Contribution to Plasma Physics* **58** 99–106
[17] Mazzola G, Helled R and Sorella S 2018 *Phys. Rev. Lett.* **120** 025701
[18] Pierleoni C and Ceperley D 2006 The coupled electron-ion monte carlo method *Lecture Notes in Physics* vol 703 (Springer, Berlin) pp 641–683
[19] McMahon J M, Morales M A, Pierleoni C and Ceperley D M 2012 *Rev. Mod. Phys.* **84** 1607–1653

IOP Conf. Series: Journal of Physics: Conf. Series **1136** (2018) 012005 doi:10.1088/1742-6596/1136/1/012005

[20] McMillan W 1965 *Phys. Rev.* **138** 442–451
[21] Moroni S and Boninsegni M 2004 *J. Low Temp. Phys.* **136** 129
[22] Holzmann M, Bernu B, Pierleoni C, McMinis J, Ceperley D M, Olevano V and Delle Site L 2011 *Phys. Rev. Letts.* **107** 1–5
[23] Huotari S, Soininen J A, Pylkkänen T, Hämäläinen K, Issolah A, Titov A, McMinis J, Kim J, Esler K, Ceperley D M, Holzmann M and Olevano V 2010 *Phys. Rev. Lett.* **105** 086403
[24] Lin C, Zong F H and Ceperley D M 2001 *Physical review. E* **64** 016702
[25] Holzmann M, Bernu B, Olevano V, Martin R M and Ceperley D M 2009 *Phys. Rev. B* **79** 2–5
[26] Chiesa S, Ceperley D M, Martin R M and Holzmann M 2006 *Phys. Rev. Letts.* **97** 6–9
[27] Holzmann M, Clay R, Morales M A, Tubmann N M, Ceperley D M and Pierleoni C 2016 *Phys. Rev. B* **94** 035126
[28] Mazzola G and Sorella S 2015 *Phys. Rev. Letts.* **114** 1–5

The extended Hubbard model with attractive interactions

E G C P van Loon and M I Katsnelson

Radboud University, Institute for Molecules and Materials, NL-6525 AJ Nijmegen, The Netherlands

E-mail: evloon@science.ru.nl

Abstract. We discuss the phase diagram of the extended Hubbard model with both attractive and repulsive local and nonlocal interactions. The extended dynamical mean-field theory (EDMFT) and the dual boson method (DB) are compared. The latter contains additional nonlocal correlation effects that are not incorporated in EDMFT. We find that EDMFT and DB give almost identical results in the attractive V regime, where phase separation occurs. This is quite a difference with the previously studied repulsive V regime, where EDMFT and DB give very different phase boundaries for the checkerboard order phase, especially at small U.

The Hubbard model [1, 2, 3, 4] is the prototypical example of a system with strongly correlated electrons. It features competition between the kinetic and potential energy that leads to a plethora of interesting physics. In the Hubbard model, the potential energy is given by a purely local interaction U. In recent years, extensions of the Hubbard model which also include nonlocal interactions have also gained attention, often in the contex of two-dimensional systems such as surface systems [5], graphene [6], transition metal dichalcogenides [7] and ultracold fermions in optical lattices [8, 9, 10].

The extended Hubbard model is usually considered with repulsive interactions $V > 0$ and $U > 0$. This choice is motivated by physical considerations, since the Coulomb interactions between electrons is repulsive. In this work, we will also look at attractive interactions. Physically, a Hubbard model with attractive interactions is realizable in ultracold fermion experiments [11] and also in the condensed matter context [12, 13, 14]. Our motivation for studying attractive interactions here, however, comes from theoretical considerations. Strong nearest-neighbor repulsion leads to a charge-ordered state. The phase transition line from the Fermi liquid to this ordered phase has been determined in several computational approaches [15, 16, 17, 18, 19, 20, 21, 22, 23, 24, 25, 26, 27] and currently serves as the main way to compare these methods. In the attractive $V < 0$ regime, a second transition occurs, now to a phase separation into a high and a low density state [23, 28]. Our purpose is to use this transition as a new benchmark for computational approaches to the extended Hubbard model.

After the success of Dynamical Mean-Field Theory (DMFT) [29, 30] in the Hubbard model, Extended Dynamical Mean-Field Theory (EDMFT) [31, 32, 33, 34, 35, 36] was developed to incorporate nonlocal interactions. Just like DMFT, EDMFT includes local correlation effects via an effective impurity model. The idea of EDMFT is to encode the nonlocal interaction in a retarded interaction on the impurity level. Recent analysis, however, has shown that the predictions for the charge order transition in EDMFT do not match more advanced

IOP Conf. Series: Journal of Physics: Conf. Series **1136** (2018) 012006 doi:10.1088/1742-6596/1136/1/012006

theories very well [22, 37, 25, 26]. This should not come as a great surprise, the charge susceptibility intrinsically has a rich momentum structure, even in weakly interacting systems, and encapsulating this structure in a local quantity is difficult. To address this issue, new theories have been developed that add corrections with momentum dependence to (E)DMFT, either diagrammatically or via a cluster approach.

For the square lattice Hubbard model at half-filling, cluster-DMFT and the dual boson approach have given very similar results for the transition line between the Fermi liquid and the checkerboard charge density wave [25]. EDMFT overestimates the V needed to stabilize the charge-ordered phase. In EDMFT+GW, the Fock exchange is apparently crucial to get an accurate prediction of the phase boundary [26]. With this situation in mind, we now turn our attention to attractive local ($U < 0$) and nonlocal interactions ($V < 0$).

1. Model

The extended Hubbard model is given by the Hamiltonian

$$H = -t \sum_{\substack{\sigma \\ \langle jk \rangle}} c^{\dagger}_{k\sigma} c_{j\sigma} + \frac{1}{2} U \sum_{j} n_j n_j + \frac{1}{2} V \sum_{\langle jk \rangle} n_k n_j. \tag{1}$$

Here, $c^{\dagger}_{k\sigma}$ is the creation operator of an electron on site k with spin σ, n_j is the density on site j and $\langle jk \rangle$ denotes that j and k are nearest-neighbors. The parameter t is the hopping amplitude, and we use $t = 1$ to set the energy scale. The potential energy is given by the local Hubbard interaction U and the nonlocal Coulomb interaction V. We consider this model on the square lattice at half-filling, i.e., one electron per site.

The square lattice is highly symmetric, and this has big implications for the phase transitions that occur. A well known example is the Ising model on the square lattice, where the ferromagnetic transition for $J > 0$ and the antiferromagnetic transition for $J < 0$ show essentially the same behavior. The bipartite nature of the square lattice is responsible for this. Similar things happen in the Hubbard model [38]. In particular, in the Hubbard model (that is, with $V = 0$) on the square lattice the transformation $U \to -U$ interchanges charge and spin excitations and keeps the overall physics the same. This is called the particle-hole symmetry of the model.

The nonlocal interaction V breaks this symmetry since it explicitly couples to charge excitations and not to spin excitations. Note that the potential energy still has a symmetry related to the bipartite lattice: $V \to -V$ interchanges checkerboard order and phase separation, in analogy with the antiferromagnetism-ferromagnetism symmetry of the Ising model. So for $t = 0$, there would be a symmetry under $V \to -V$. It is the combination of kinetic and potential energy that finally destroys the symmetries. Of course, the combination of kinetic and potential energy is exactly what makes the Hubbard model interesting.

2. Method

Our main method is the dual boson (DB) [39] approach. We use the same methodology as in Ref. [22] and give a short overview here. The calculation is a two stage process. We start by performing a self-consistent EDMFT calculation, and then include the nonlocal correlation corrections of DB on top of the EDMFT results.

The EDMFT self-consistency procedure consists of the iterative solution to find an effective single-site problem, the impurity model, that gives a local self-energy and polarization so that the local properties of the impurity model and the lattice model correspond.

The impurity model is defined by a hybridization function Δ_ν and a screened interaction U_ω.

In the action formalism, the impurity model is given by

$$S^{\text{imp}} = -\sum_{\nu\sigma} c^*\nu\sigma \left[i\nu + \mu - \Delta_\nu\right] c_{\nu\sigma} + \sum_{\omega} U_\omega n_\omega n_\omega, \tag{2}$$

where ν and ω are fermionic and bosonic Matsubara frequencies respectively. The impurity model can be solved exactly using a continuous time quantum Monte Carlo solver [40, 41], we use the implementation of Ref. [42], based on the ALPS [43] libraries. This impurity solver uses improved estimators [44, 45] to accurately determine the impurity correlation functions $g_\nu = -\langle cc^*\rangle_\nu^{\text{imp}}$, $\chi_\omega = -\langle nn\rangle_\omega^{\text{imp}}$, $-\lambda_{\nu\omega} = \langle cc^*n\rangle^{\text{imp}}$, $\gamma_{\nu\nu'\omega} = -\langle cc^*cc^*\rangle^{\text{imp}}$, where $\langle x\rangle^{\text{imp}}$ denotes the connected, amputated expectation value of x in the impurity model [22]. Computationally, this is the most difficult part of the calculations.

The hybridization function and retarded interaction are determined self-consistently based on two self-consistency conditions

$$\begin{aligned} \frac{1}{N}\sum_{\mathbf{k}} \tilde{G}_{\mathbf{k}\nu} &= 0, \\ \frac{1}{N}\sum_{\mathbf{q}} \tilde{X}_{\mathbf{q}\omega} &= 0, \end{aligned} \tag{3}$$

where $\mathbf{k}$ and $\mathbf{q}$ are momenta and

$$\tilde{G}_{\mathbf{k}\nu} = \left[g_\nu^{-1} + \Delta_\nu - t_{\mathbf{k}}\right]^{-1} - g_\nu \tag{4}$$

$$\tilde{X}_{\mathbf{q}\omega} = \left[\chi_\omega^{-1} + U_\omega - V_{\mathbf{q}}\right]^{-1} - \chi_\omega \tag{5}$$

are the so-called dual Green's function and dual susceptibility. The first self-consistency condition ensures that the hybridization Δ of the impurity model describes the hopping (in momentum space) $t_{\mathbf{k}}$ of the lattice, the second that the retarded interaction U_ω describes the interactions U and V of the lattice (again given in momentum space).

When the self-consistency conditions (3) are satisfied, we calculate a single-shot of DB, using the ladder approach for the dual polarization operator, $\tilde{\Pi}$. It is given by [22]

$$\tilde{\Pi}_{\mathbf{q}\omega} = \frac{T}{N}\sum_{\mathbf{k}\nu} \lambda_{\nu+\omega,-\omega} \tilde{G}_{\mathbf{k}\nu} \tilde{G}_{\mathbf{k}+\mathbf{q}\nu+\omega} \Lambda_{\mathbf{q}\nu\omega} \tag{6}$$

$$\Lambda_{\mathbf{q}\nu\omega} = \lambda_{\nu\omega} - \frac{T}{N}\sum_{\mathbf{k}'\nu'} \Gamma_{\mathbf{q}\nu\nu'\omega} \tilde{G}_{\mathbf{k}'\nu'} \tilde{G}_{\mathbf{k}'+\mathbf{q}\nu'+\omega} \lambda_{\nu'\omega} \tag{7}$$

$$\Gamma_{\mathbf{q}\nu\nu'\omega} = \gamma_{\nu\nu'\omega} - \frac{T}{N}\sum_{\mathbf{k}''\nu''} \gamma_{\nu\nu''\omega} \tilde{G}_{\mathbf{k}''\nu''} \tilde{G}_{\mathbf{k}''+\mathbf{q}\nu''+\omega} \Gamma_{\mathbf{q}\nu''\nu'\omega}. \tag{8}$$

The temperature T and the total number of sites in the lattice N act to normalize the frequency and momentum sums. These are called ladder equations, expansion of the recursive definition of Γ leads to a ladder with an increasing number of rungs γ, and rails $\tilde{G}$.

From the dual polarization, we determine the physical susceptibility $\chi_{\omega,q} = \langle nn\rangle_{\omega,q}$ according to the DB formula for the susceptibility

$$\chi_{\omega,q}^{-1} = [\chi_\omega(1 + \chi_\omega \tilde{\Pi}_{\omega,q})]^{-1} + U_\omega - V_q. \tag{9}$$

This formula illustrates the difference between DB and EDMFT, since EDMFT corresponds to $\tilde{\Pi} = 0$. The combination $1 + \chi_{\omega=0}\tilde{\Pi}_{\omega=0,q}$ can be interpreted as the dual correction to the

EDMFT result. We also note that the susceptibility in this DB scheme satisfies the charge conservation law [46] at $q = 0$, i.e. $\chi_{\omega\neq 0, q=0} = 0$.

It is also possible to go further in the dual theory, on the one hand by including further diagrammatic corrections to the polarization and to the nonlocal self-energy, on the other hand by taking into account feedback of the nonlocal correlation effects onto the impurity model [47]. These are usually called inner and outer self-consistency, respectively. Here, we choose this relatively simple scheme since it allows for a direct comparison between EDMFT and DB and because it has been applied successfully to the repulsive ($U > 0$, $V > 0$) situation already [22].

Charge-ordered phases are indicated by a divergence of the zero-frequency susceptibility at some specific momentum q. The checkerboard order that occurs at $V > 0$ is visible at $q = (\pi, \pi)$. On the other hand, at $V < 0$, phase separation takes place, which is visible in a diverging compressibility, i.e. the susceptibility at $q = 0$. All our calculations are performed in the Fermi liquid phase where all susceptibilities stay finite, the phase boundary is determined by extrapolating $\chi^{-1}_{\omega=0,q}$ as a function of V and finding the value of V where it crosses zero.

We study the two dimensional square lattice extended Hubbard model at half-filling and at fixed temperature $\beta t = 10$. In all our calculations, the system is a Fermi liquid at $V = 0$. At sufficiently large U and sufficiently low temperature, the system will turn insulating (For recent study of the metal-insulator transition in the square lattice Hubbard model, see [48, 49]), but we do not consider that regime here. We also restrict our attention to the charge sector.

3. Results

In Fig. 1 we show the phase diagram of the extended Hubbard model according to the EDMFT and DB methods, at $\beta t = 10$. The left panel shows the inverse susceptibility. For fixed U, a larger value of V increases the checkerboard susceptibility and usually decreases the uniform (phase separation) susceptibility. An exception to this behavior is found at $U = 8$ and large values of V close to the checkerboard order. Here, the uniform susceptibility actually increases as a function of V. A comparison at fixed V shows that larger U decreases the magnitude of the charge susceptibility.

The phase boundaries in the right panel are obtained from the inverse susceptibility. The $U > 0$, $V > 0$ results are similar to those in the earlier literature [21, 22, 37, 25], as expected. Looking at $U < 0$, $V > 0$, the charge-ordered phases occurs at decreasing values of V as U decreases. The difference between the EDMFT and DB phase boundary stay approximately constant in the negative U region. At sufficiently large negative U, the checkerboard order appears even for $V = 0$. By particle-hole symmetry [38], going from U to $-U$ interchanges charge and spin fluctuations, so this corresponds to the diverging antiferromagnetic spin susceptibility at positive U in DMFT. The present DB approach reduces to DMFT for $V = 0$, so this was to be expected. However, this phase transition at $V = 0$ violates the Mermin-Wagner theorem and is an artifact of the mean-field spirit of the approximation. This shows that this is a regime where nonlocal corrections over very long length scales should be crucial.

Then, turning our attention to $V < 0$, we see that EDMFT produces a somewhat remarkable result. According to EDMFT, the phase separation occurs at exactly the same absolute value of V as the checkerboard order, i.e. $V_c^{ps} = -V_c^{co}$. This is caused by the fact that the local properties of EDMFT are invariant (This assumes a certain amount of symmetry of the lattice, which is satisfied by the square lattice and hypercubic lattices in general, see [47] for more details.) under the transformation $V \to -V$. Since EDMFT takes momentum dependence into account on a very rough level, this invariance means that $\chi_q \to \chi_{q+(\pi,\pi)}$ under the same transformation, interchanging the checkerboard susceptibility and the compressibility. As a result, the divergence of these two susceptibilities occur at the same absolute value of V. The issue is that EDMFT largely decouples the kinetic and potential energy, only linking them up in the impurity model, and the impurity model does not know about the lattice structure. The

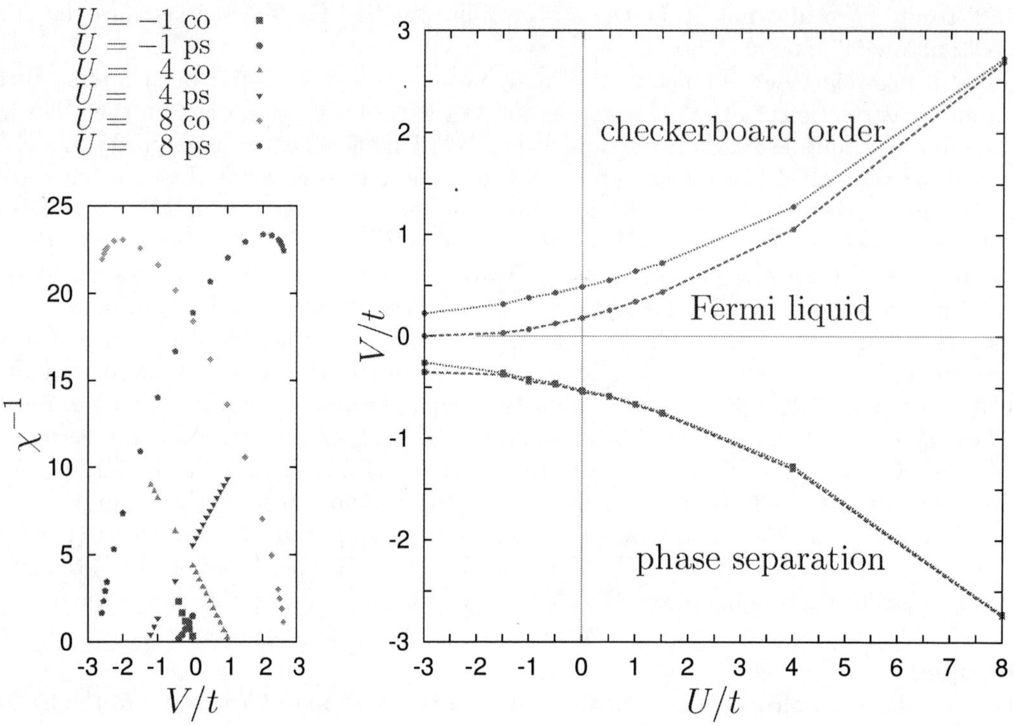

Figure 1. Left: Inverse susceptibility in DB as a function of V. The phase boundary is determined by extrapolating to $\chi^{-1} = 0$. For every value of U, both the checkerboard order (co) and phase separation (ps) susceptibilities are shown, corresponding to $q = (0, 0)$ and $q = (\pi, \pi)$. Right: Phase diagram at $\beta = 10$. The red dashed lines are the DB results and the brown dotted line corresponds to the simpler EDMFT approximation.

$V \to -V$ symmetry should be broken by the *interplay* of kinetic and potential energy, and EDMFT lacks the momentum structure of this interplay.

This invariance of the EDMFT approximation is not a true symmetry of the system, since repulsive and attractive interactions are physically quite different. Indeed, we find that DB gives different phase transition lines for repulsive and attractive V.

Unlike for repulsive V, however, the DB and EDMFT phase boundaries agree very well (but not exactly) for $V < 0$ along the entire studied range of U. This shows that the phase separation and the checkerboard charge-density wave are physically different. The former can be described accurately in a single-site, EDMFT, picture. From this, we conclude that nonlocal corrections to the compressibility are not very important. Earlier work on the compressibility in DB [50], also found relatively small corrections to the compressibility at half-filling.

Figure 2 shows the dual correction factor, Eq. 9 across the Brillouin Zone. At small $U = 1$, the nonlocal correction factor deviates strongly from unity, especially close to the checkerboard point (π, π). In the center of the Brillouin Zone, the correction factor is approximately unity and corrections to EDMFT are small. This explains why the EDMFT and DB phase separation happens at approximately the same V. The nonlocal correction factor is actually smaller than unity around $(0, \pi)$, indicating that the nonlocal corrections decrease the susceptibility compared to EDMFT. For larger $U = 8$, only small deviations from unity occur since charge fluctuations are strongly suppressed by the on-site interaction. The momentum structure has changed, now the mimimum occurs at $(0, 0)$ instead of at $(0, \pi)$. The small deviations from unity explain why

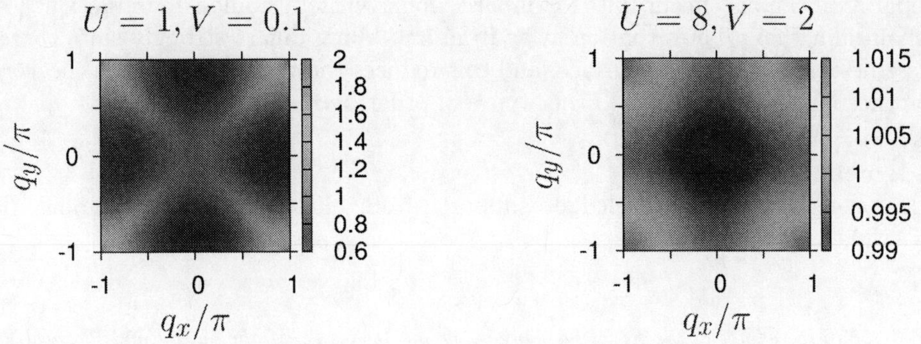

Figure 2. Nonlocal susceptibility correction factor $1+\chi\tilde{\Pi}$ at $\omega = 0$ as a function of momentum.

the EDMFT and DB phase boundaries are so close at large U.

The $V \to -V$ symmetry of EDMFT does affect the DB calculations, since they use EDMFT as a starting point. In particular, all impurity quantities that enter the DB calculation are the same at V and $-V$, and even a quantity like $\tilde{\Pi}$ is the same, since the nonlocal interaction V does not enter it directly. Taking into account feedback of the nonlocal DB correlations, contained in $\tilde{\Pi}$, onto the impurity model, the so-called outer self-consistency would break this partial symmetry [47]. In this work, only single-shot DB calculations were performed.

Another layer of feedback that was not considered in this work is that from the nonlocal contributions to the self-energy onto $\tilde{\Pi}$. The dual self-energy $\tilde{\Sigma}$ depends on $\tilde{X}^0$ which does depend on V explicitly, so this quantity is not symmetric under $V \to -V$. As a result, in DB calculations with inner self-consistency [22], where a dual self-energy $\tilde{\Sigma}$ is calculated and used to renormalize the dual Green's function $\tilde{G}$ that enters (9), the quantity $\tilde{\Pi}$ will lose the $V \to -V$ symmetry due to this feedback.

4. Discussion

Several schemes based on single-particle quantities only have been suggested to extend EDMFT [51, 52, 21, 5, 37, 26]. The advantage of these approaches compared to DB is that they do not require the (expensive) computation of the two-particle vertex of the impurity model. The approaches are mostly based on combining GW diagrams with a self-consistent impurity problem.

There is discussion in the literature on how to best combine these ideas, and this ambiguity is especially important since the schemes give qualitatively different numerical results [37]. Our findings here, and in particular the fact that the EDMFT and DB are very similar, indicate that the negative V region could be a good test bed for these approaches: They are expected to interpolate between EDMFT and DB, adding only some nonlocal correlation effects. But here EDMFT and DB give near identical result, so we expect EDMFT+GW schemes to also give very similar results. Extensions of DMFT that give significant deviations should be under serious scrutiny for double counting issues.

Checkerboard order becomes very favored for $U < 0$. In fact, DB overestimates this tendency and shows a transition at $V = 0$ already. This transition is related by particle-hole symmetry to a antiferromagnetic transition at $U > 0$ which is forbidden by the Mermin-Wagner theorem. This makes this a very challenging regime even for single-shot DB calculations. The Mermin-Wagner theorem can be restored by enforcing appropriate local constraints [53, 54]. This shows that the $U < 0$, $V > 0$ sector is an interesting place to compare computational approaches.

Phase separation can also occur in the Hubbard model without nonlocal interactions ($V = 0$) away from half-filling [55]. The situation away from half-filling differs strongly since there is no longer perfect nesting of the Fermi surface, and this reduces the tendency towards checkerboard order and antiferromagnetism so that other types of order become competitive.

Acknowledgments
E.G.C.P.v.L. and M.I.K. acknowledge support from ERC Advanced Grant 338957 FEMTO/NANO.

[1] Hubbard J 1963 *Proceedings of the Royal Society of London. Series A. Mathematical and Physical Sciences* **276** 238–257 URL http://rspa.royalsocietypublishing.org/content/276/1365/238.abstract
[2] Gutzwiller M C 1963 *Phys. Rev. Lett.* **10**(5) 159–162 URL http://link.aps.org/doi/10.1103/PhysRevLett.10.159
[3] Kanamori J 1963 *Progress of Theoretical Physics* **30** 275–289
[4] Hubbard J 1964 *Proceedings of the Royal Society of London. Series A. Mathematical and Physical Sciences* **281** 401–419
[5] Hansmann P, Ayral T, Vaugier L, Werner P and Biermann S 2013 *Phys. Rev. Lett.* **110**(16) 166401 URL http://link.aps.org/doi/10.1103/PhysRevLett.110.166401
[6] Schüler M, Rösner M, Wehling T O, Lichtenstein A I and Katsnelson M I 2013 *Phys. Rev. Lett.* **111**(3) 036601 URL http://link.aps.org/doi/10.1103/PhysRevLett.111.036601
[7] van Loon E G C P, Rösner M, Schönhoff G, Katsnelson M I and Wehling T O 2017 *ArXiv e-prints* (*Preprint* 1707.05640)
[8] Dutta O, Sowiński T and Lewenstein M 2013 *Phys. Rev. A* **87**(2) 023619 URL http://link.aps.org/doi/10.1103/PhysRevA.87.023619
[9] Dutta O, Gajda M, Hauke P, Lewenstein M, Lühmann D S, Malomed B A, Sowiński T and Zakrzewski J 2015 *Reports on Progress in Physics* **78** 066001 URL http://stacks.iop.org/0034-4885/78/i=6/a=066001
[10] van Loon E G C P, Katsnelson M I and Lemeshko M 2015 *Phys. Rev. B* **92**(8) 081106 URL http://link.aps.org/doi/10.1103/PhysRevB.92.081106
[11] Esslinger T 2010 *Annual Review of Condensed Matter Physics* **1** 129–152 URL https://doi.org/10.1146/annurev-conmatphys-070909-104059
[12] van der Marel D and Sawatzky G A 1988 *Phys. Rev. B* **37**(18) 10674–10684 URL https://link.aps.org/doi/10.1103/PhysRevB.37.10674
[13] Micnas R, Ranninger J and Robaszkiewicz S 1990 *Rev. Mod. Phys.* **62**(1) 113–171 URL https://link.aps.org/doi/10.1103/RevModPhys.62.113
[14] Strand H U R 2014 *Phys. Rev. B* **90**(15) 155108 URL https://link.aps.org/doi/10.1103/PhysRevB.90.155108
[15] Micnas R, Robaszkiewicz S and Chao K A 1984 *Phys. Rev. B* **29**(5) 2784–2789 URL https://link.aps.org/doi/10.1103/PhysRevB.29.2784
[16] Zhang Y and Callaway J 1989 *Phys. Rev. B* **39**(13) 9397–9404 URL http://link.aps.org/doi/10.1103/PhysRevB.39.9397
[17] Bursill R J and Thompson C J 1993 *Journal of Physics A: Mathematical and General* **26** 4497 URL http://stacks.iop.org/0305-4470/26/i=18/a=017
[18] Avella A and Mancini F 2004 *The European Physical Journal B - Condensed Matter and Complex Systems* **41** 149–162 ISSN 1434-6036 URL https://doi.org/10.1140/epjb/e2004-00304-9
[19] Avella A and Mancini F 2006 *Journal of Physics and Chemistry of Solids* **67** 142 – 145 ISSN 0022-3697 spectroscopies in Novel Superconductors 2004 URL http://www.sciencedirect.com/science/article/pii/S0022369705004646
[20] Amaricci A, Camjayi A, Haule K, Kotliar G, Tanasković D and Dobrosavljević V 2010 *Phys. Rev. B* **82**(15) 155102 URL https://link.aps.org/doi/10.1103/PhysRevB.82.155102
[21] Ayral T, Biermann S and Werner P 2013 *Phys. Rev. B* **87**(12) 125149 URL http://link.aps.org/doi/10.1103/PhysRevB.87.125149
[22] van Loon E G C P, Lichtenstein A I, Katsnelson M I, Parcollet O and Hafermann H 2014 *Phys. Rev. B* **90**(23) 235135 URL http://link.aps.org/doi/10.1103/PhysRevB.90.235135
[23] Lhoutellier G, Frésard R and Oleś A M 2015 *Phys. Rev. B* **91**(22) 224410 URL http://link.aps.org/doi/10.1103/PhysRevB.91.224410

[24] Stepanov E A, Huber A, van Loon E G C P, Lichtenstein A I and Katsnelson M I 2016 *Phys. Rev. B* **94**(20) 205110 URL https://link.aps.org/doi/10.1103/PhysRevB.94.205110
[25] Terletska H, Chen T and Gull E 2017 *Phys. Rev. B* **95**(11) 115149 URL https://link.aps.org/doi/10.1103/PhysRevB.95.115149
[26] Ayral T, Biermann S, Werner P and Boehnke L 2017 *Phys. Rev. B* **95**(24) 245130 URL https://link.aps.org/doi/10.1103/PhysRevB.95.245130
[27] Kapcia K J, Robaszkiewicz S, Capone M and Amaricci A 2017 *Phys. Rev. B* **95**(12) 125112 URL https://link.aps.org/doi/10.1103/PhysRevB.95.125112
[28] Frésard R, Steffen K and Kopp T 2016 *Journal of Physics: Conference Series* **702** 012003 URL http://stacks.iop.org/1742-6596/702/i=1/a=012003
[29] Metzner W and Vollhardt D 1989 *Phys. Rev. Lett.* **62**(3) 324–327 URL http://link.aps.org/doi/10.1103/PhysRevLett.62.324
[30] Georges A, Kotliar G, Krauth W and Rozenberg M J 1996 *Rev. Mod. Phys.* **68** 13 URL http://dx.doi.org/10.1103/RevModPhys.68.13
[31] Sengupta A M and Georges A 1995 *Phys. Rev. B* **52**(14) 10295–10302 URL https://link.aps.org/doi/10.1103/PhysRevB.52.10295
[32] Si Q and Smith J L 1996 *Phys. Rev. Lett.* **77**(16) 3391–3394 URL http://link.aps.org/doi/10.1103/PhysRevLett.77.3391
[33] Kajueter H 1996 Ph.D. thesis Rutgers University
[34] Smith J L and Si Q 2000 *Phys. Rev. B* **61**(8) 5184–5193 URL http://link.aps.org/doi/10.1103/PhysRevB.61.5184
[35] Chitra R and Kotliar G 2000 *Phys. Rev. Lett.* **84**(16) 3678–3681 URL http://link.aps.org/doi/10.1103/PhysRevLett.84.3678
[36] Chitra R and Kotliar G 2001 *Phys. Rev. B* **63**(11) 115110 URL http://link.aps.org/doi/10.1103/PhysRevB.63.115110
[37] Stepanov E A, Huber A, van Loon E G C P, Lichtenstein A I and Katsnelson M I 2016 *Phys. Rev. B* **94**(20) 205110 URL http://link.aps.org/doi/10.1103/PhysRevB.94.205110
[38] Mielke A 2015 *Many-Body Physics: From Kondo to Hubbard* (*Modeling and Simulation* vol 5) ed Pavarini E, Koch E and Coleman P (Jülich: Forschungszentrum Jülich GmbH Zentralbibliothek, Verlag) ISBN 987-3-95806-074-6 URL https://juser.fz-juelich.de/record/205123
[39] Rubtsov A N, Katsnelson M I and Lichtenstein A I 2012 *Annals of Physics* **327** 1320 ISSN 0003-4916 URL http://www.sciencedirect.com/science/article/pii/S0003491612000164
[40] Rubtsov A N, Savkin V V and Lichtenstein A I 2005 *Phys. Rev. B* **72**(3) 035122 URL http://link.aps.org/doi/10.1103/PhysRevB.72.035122
[41] Gull E, Millis A J, Lichtenstein A I, Rubtsov A N, Troyer M and Werner P 2011 *Rev. Mod. Phys.* **83** 349–404
[42] Hafermann H, Werner P and Gull E 2013 *Computer Physics Communications* **184** 1280 – 1286 ISSN 0010-4655 URL http://www.sciencedirect.com/science/article/pii/S0010465512004092
[43] Bauer B, Carr L D, Evertz H G, Feiguin A, Freire J, Fuchs S, Gamper L, Gukelberger J, Gull E, Guertler S, Hehn A, Igarashi R, Isakov S V, Koop D, Ma P N, Mates P, Matsuo H, Parcollet O, Pawłowski G, Picon J D, Pollet L, Santos E, Scarola V W, Schollwöck U, Silva C, Surer B, Todo S, Trebst S, Troyer M, Wall M L, Werner P and Wessel S 2011 *Journal of Statistical Mechanics: Theory and Experiment* **2011** P05001
[44] Hafermann H, Patton K R and Werner P 2012 *Phys. Rev. B* **85**(20) 205106 URL http://link.aps.org/doi/10.1103/PhysRevB.85.205106
[45] Hafermann H 2014 *Phys. Rev. B* **89**(23) 235128 URL http://link.aps.org/doi/10.1103/PhysRevB.89.235128
[46] Hafermann H, van Loon E G C P, Katsnelson M I, Lichtenstein A I and Parcollet O 2014 *Phys. Rev. B* **90**(23) 235105 URL http://link.aps.org/doi/10.1103/PhysRevB.90.235105
[47] Stepanov E A, van Loon E G C P, Katanin A A, Lichtenstein A I, Katsnelson M I and Rubtsov A N 2016 *Phys. Rev. B* **93**(4) 045107 URL http://link.aps.org/doi/10.1103/PhysRevB.93.045107
[48] Schäfer T, Geles F, Rost D, Rohringer G, Arrigoni E, Held K, Blümer N, Aichhorn M and Toschi A 2015 *Phys. Rev. B* **91**(12) 125109 URL http://link.aps.org/doi/10.1103/PhysRevB.91.125109
[49] van Loon E G C P, Hafermann H and Katsnelson M I 2018 *Phys. Rev. B* **97**(8) 085125 URL https://link.aps.org/doi/10.1103/PhysRevB.97.085125
[50] van Loon E G C P, Hafermann H, Lichtenstein A I and Katsnelson M I 2015 *Phys. Rev. B* **92**(8) 085106 URL http://link.aps.org/doi/10.1103/PhysRevB.92.085106
[51] Sun P and Kotliar G 2002 *Phys. Rev. B* **66**(8) 085120 URL http://link.aps.org/doi/10.1103/PhysRevB.66.085120
[52] Ayral T, Werner P and Biermann S 2012 *Phys. Rev. Lett.* **109**(22) 226401 URL http://link.aps.org/doi/10.1103/PhysRevLett.109.226401

[53] Vilk Y M and Tremblay A M 1997 *J. Phys. I France* **7** 1309–1368 URL http://dx.doi.org/10.1051/jp1:1997135

[54] Krien F, van Loon E G C P, Hafermann H, Otsuki J, Katsnelson M I and Lichtenstein A I 2017 *Phys. Rev. B* **96**(7) 075155 URL https://link.aps.org/doi/10.1103/PhysRevB.96.075155

[55] Otsuki J, Hafermann H and Lichtenstein A I 2014 *Phys. Rev. B* **90**(23) 235132 URL http://link.aps.org/doi/10.1103/PhysRevB.90.235132

IOP Conf. Series: Journal of Physics: Conf. Series **1136** (2018) 012007 doi:10.1088/1742-6596/1136/1/012007

Solving Quantum Spin Glasses with Off-Diagonal Expansion Quantum Monte Carlo

Itay Hen[1,2] **and Tameem Albash**[1,2]

[1] Information Sciences Institute, University of Southern California, Marina del Rey, California 90292, USA

[2] Department of Physics and Astronomy and Center for Quantum Information Science & Technology, University of Southern California, Los Angeles, California 90089, USA

E-mail: itayhen@isi.edu

Abstract. We study the performance of the Off-Diagonal Expansion quantum Monte Carlo algorithm on quantum spin glasses, specifically, 3-regular random antiferromagnets augmented with a transverse magnetic field. We demonstrate the advantages, sometimes by orders of magnitude, of the technique by comparing its runtimes against existing state-of-the-art schemes such as Path Integral quantum Monte Carlo and Stochastic Series Expansion.

1. Introduction

Solving for the ground state of spin glasses—disordered, frustrated Ising spin systems—is a formidable challenge. Spin glasses are prototypical classically-hard (also known as NP-hard) optimization problems [1, 2], and they stymie all algorithms designed to solve them, so much so that specialized hardware has been built to simulate and study their thermal equilibrium properties [3, 4, 5]. When such Ising spins systems are augmented with a non-commuting (quantum) component, forming quantum spin glasses, they continue to pose a significant challenge to Quantum Monte Carlo (QMC) algorithms.

QMC algorithms are known to be notoriously inefficient in describing quantum spin glasses in 'almost classical' parameter regimes, i.e., inside the spin-glass phase, where updates resulting from thermal fluctuations are expected to be far more dominant than those resulting from quantum fluctuations [6, 7, 8]. Since QMC methods normally evolve via configuration updates that are based on quantum fluctuations, the acceptance rates of quantum updates, e.g., single spin (or cluster) flips in the dual Ising system, decrease dramatically in classical regimes, often causing QMC algorithms to dramatically slow down or 'freeze' (see, e.g., Ref [9]). In these regimes, these models are known to exhibit additional structure that is not only interesting from a scientific standpoint but is also of critical importance to adiabatic approaches to quantum computation like quantum annealing [10, 11, 12, 13, 14, 15, 16, 17, 18, 19], since the minima in the spin glass phase may be the relevant bottlenecks for the quantum algorithm as opposed to the critical point associated with the paramagnetic-spin glass phase transition. One reason that efficient classical thermal updates are typically hard to implement within the framework of QMC algorithms is that these algorithms do not normally converge to classical Monte Carlo algorithms in the limit where the model becomes classical.

Here, we analyze the performance of a recently developed algorithm, named the off-diagonal expansion (ODE) quantum Monte Carlo algorithm [20], devised specifically to have the algorithmic flexibility to simulate interacting many-body systems ranging from the fully-quantum to the fully-classical. The algorithm is based on a novel decomposition of the canonical quantum partition function into a sum of Boltzmann-like weights, and it converges to the usual

IOP Conf. Series: Journal of Physics: Conf. Series **1136** (2018) 012007 doi:10.1088/1742-6596/1136/1/012007

decomposition of the classical partition function in the limit where the Hamiltonian of the system becomes classical. Based on this unique decomposition, the ODE algorithm aims to improve the convergence rates of simulated systems for which existing techniques are often inefficient.

We study random 3-regular MAX2SAT instances augmented with a transverse magnetic field. This class of instances corresponds to a particular choice of the Ising Hamiltonian, whereby each spin is coupled antiferromagnetically (with strength 1) with exactly three other spins picked at random. This class of instances is known to exhibit a quantum spin glass phase transition and is notoriously difficult to simulate by standard QMC techniques [9, 21], making it suitable to illustrate the strengths of the ODE algorithm.

Within the ODE approach the quantum imaginary-time dimension of the algorithm is 'elastic', i.e., it can stretch or shrink dynamically depending on the strength of the quantum part of the system — the off-diagonal portion of the Hamiltonian. In addition, the method does not introduce Trotter-type errors as in discrete-time path-integral QMC, a source of errors that normally occurs from an insufficient discretization of the imaginary-time dimension (over-discretization tends to sharply reduce the acceptance rates of the QMC updates). Moreover, in the classical limit where off-diagonal terms vanish, the algorithm naturally reduces to a classical thermal simulation. As we illustrate, the above properties allow the method to naturally overcome certain inefficiencies typically encountered by other QMC algorithms.

2. Decomposition of the partition function

We start by providing a brief description of the decomposition of the quantum partition function on which the ODE algorithm relies on, and the reader is referred to Ref. [20] for more details and a full derivation.

The canonical quantum partition function of a system described by a Hamiltonian H is given by $Z = \mathrm{tr}\left[\mathrm{e}^{-\beta H}\right]$. The Hamiltonian is first written in the form

$$H = H_{\mathrm{c}} - \sum_j \Gamma_j V_j \ , \tag{1}$$

where H_{c} is the classical portion of the Hamiltonian, i.e., a diagonal operator in some known basis, which we refer to as the computational basis, and whose basis states will be denoted by $\{|z\rangle\}$. The $\{\Gamma_j\}$ are generally complex-valued parameters, and $\{V_j\}$ are off-diagonal operators satisfying $[V_j, H_{\mathrm{c}}] \neq 0$. In order for the decomposition of the partition function to be feasible, the off-diagonal operators are chosen to obey $V_j|z\rangle = |z'\rangle$ for every basis state $|z\rangle$, where $|z'\rangle \neq |z\rangle$ is also a basis state. We take the Γ_j parameters to be identical, namely that $\Gamma_j = \Gamma, \forall j$, and we study the quantum random 3-regular MAX2SAT model whose Hamiltonian is

$$H = \sum_{\langle i,j \rangle} \sigma_i^z \sigma_j^z - \Gamma \sum_j \sigma_j^x \ , \tag{2}$$

where $H_{\mathrm{c}} = \sum_{\langle i,j \rangle} \sigma_i^z \sigma_j^z$ is the classical Hamiltonian of 3-regular MAX2SAT whereby each spin is coupled antiferromagnetically (with strength 1) with exactly three other spins picked at random, and the off-diagonal operators are the spin-flip terms $V_j = \sigma_j^x$.

The partition function can be written as a sum over sequences $S_q = V_{i_1} V_{i_2} \ldots V_{i_q}$ of q off-diagonal operators, with an associated classical 'effective energies' $E_{(0,\ldots,q)}$ as

$$Z = \sum_{\{z\}} \sum_{q} \sum_{\{S_q\}} \frac{(\beta\Gamma)^q}{q!} \langle z|S_q|z\rangle \mathrm{e}^{-\beta E_{(0,\ldots,q)}} \ , \tag{3}$$

where $E_{(0,\ldots,q)}$ is the 'effective classical energy' of the configuration. The effective classical energy is calculated from the multiset of energies $[E_0, \ldots, E_q]$, where E_0 is the classical energy associated with $|z\rangle$ and E_i is the classical energy associated with the computational basis state

$|z'\rangle = \prod_{k=1}^{i} V_k|z\rangle$. To calculate $E_{(0,\ldots,q)}$, we can use the divided differences recursion relation (see Ref. [20]) to arrive at

$$E_{(0,\ldots,q)} = \bar{E} - \frac{1}{\beta}\log\frac{2q\sinh\beta\Delta E}{\beta(E_q - E_0)}\,, \tag{4}$$

where $2\bar{E} = E_{(1,\ldots,q)} + E_{(0,\ldots,q-1)}$ and $2\Delta E = E_{(1,\ldots,q)} - E_{(0,\ldots,q-1)}$. In the limiting case where all energies in the sequence are equal, the above relation neatly becomes $E_{(0,\ldots,q)} = E_{(0)} = E_0$. The initial condition for the above recursion is simply $e^{-\beta E_{(i)}} = e^{-\beta E_i}$.

Since by construction the term $\langle z|S_q|z\rangle$ evaluates to either 0 or to 1 (the operation $S_q|z\rangle$ returns a basis state $|z'\rangle$ and therefore $\langle z|S_q|z\rangle = \langle z|z'\rangle = \delta_{z,z'}$), the partition function can be neatly presented in its final form as a sum over only non-vanishing terms:

$$Z = \sum_{\{S_q:\langle z|S_q|z\rangle=1\}} \frac{(\beta\Gamma)^q}{q!} e^{-\beta E_{(0,\ldots,q)}}\,. \tag{5}$$

We interpret the terms in the sum in Eq. (5) as weights, i.e., $Z = \sum_{\{\mathcal{C}\}} W_{\mathcal{C}}$, where the set of configurations $\{\mathcal{C}\}$ is all the distinct pairs $\{|z\rangle, S_q\}$. Because of the form of $W_{\mathcal{C}}$,

$$W_{\mathcal{C}} = \frac{(\beta\Gamma)^q}{q!} e^{-\beta E_{(0,\ldots,q)}}\,, \tag{6}$$

we refer to it as a 'generalized Boltzmann weight' (or, GBW). We shall refer to $E_{(0,\ldots,q)}$ as the 'effective classical energy' of the configuration $\mathcal{C}$ and denote it at times for brevity simply by $E_{\mathcal{C}}$.

In order to interpret the $W_{\mathcal{C}}$ terms as actual weights, they must be non-negative for any simulated system that is not plagued by the sign problem [22]. As is evident from the above expression, even values of q always yield positive weights regardless of the sign of Γ for the Hamiltonian Eq. (2). Therefore, in order for the $\langle z|S_q|z\rangle$ terms to evaluate to one rather than to zero, off-diagonal operators must always be produced and annihilated in pairs, implying that the total sign of the weight, Eq. (6), is positive.

3. The ODE algorithm

We now describe the basic ingredients of our Off-Diagonal Expansion (ODE) algorithm, based on the above partition function decomposition as it applies to the transverse-field Ising model, Eq. (2). We first establish the computational complexity associated with implementing this new algorithm, discussing in detail generic updates as well as measurements. We then present some results that allow us to fully characterize and to quantify the advantages of the algorithm over state-of-the-art QMC methods, specifically path-integral QMC and SSE.

3.1. General description of the algorithm

An ODE configuration is a pair $\mathcal{C} = \{|z\rangle, S_q\}$ where $|z\rangle$ corresponds to a classical bit configuration and $S_q = V_{i_1}V_{i_2}\cdots V_{i_q}$ is a sequence of (possibly repeated) off-diagonal operators. As was discussed above, each configuration $\mathcal{C}$ induces a list of states $Z = \{|z_0\rangle = |z\rangle, |z_1\rangle, \ldots, |z_q\rangle = |z\rangle\}$, which in turn also generates a corresponding multiset of diagonal energies $M_{\mathcal{C}} = \{E_0, E_1, \ldots, E_q\}$ of not-necessarily-distinct values (recall that $E_i = \langle z_i|H_c|z_i\rangle$). For systems with discretized energy values, the multiset can be stored efficiently in a 'multiplicity table' $\{m_0, m_1, \ldots, m_j, \ldots\}$, where m_j is the multiplicity of the energy E_j in the multiset. Given $M_{\mathcal{C}}$, the evaluation of the effective classical energy $E_{\mathcal{C}}$ and the GBW $W_{\mathcal{C}}$ follow from the definition of the GBW, Eq. (6).

The initial configuration of the ODE algorithm is a random classical configuration $|z\rangle$ and the empty sequence $S_{q=0} = 1$. The weight of this initial configuration is

$$W_{\mathcal{C}_{\text{init}}} = e^{-\beta E(z)}\,, \tag{7}$$

i.e., the classical Boltzmann weight of the initial random state $|z\rangle$. Here the effective classical energy $E_{\mathcal{C}_{\text{init}}}$ is the classical energy of $|z\rangle$.

3.2. Updates

We next describe the basic update moves for the algorithm. We consider here only generic local updates but note that updates of the global type can be specifically tailored to the system in question. An update is considered local if it changes the multiset M_C by a finite (i.e., by a system-size independent) number of terms, e.g., $M_C \to M_C + \{E(z_i)\} - \{E(z_j)\}$. The basic updates are summarized in Fig. 1 and are discussed in detail below.

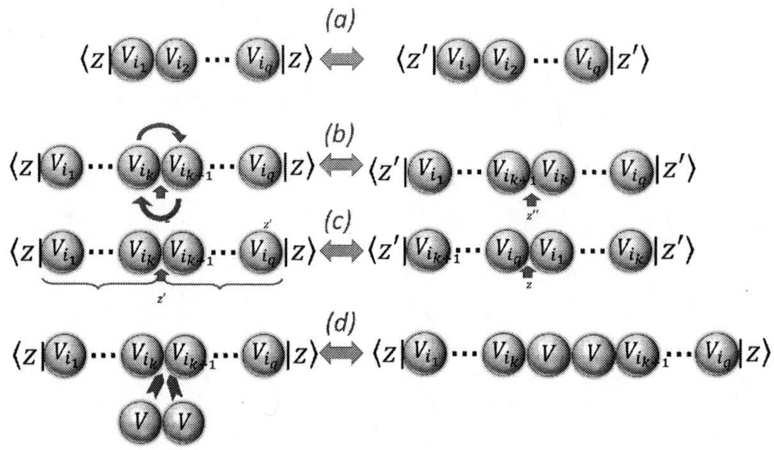

Figure 1. Basic update moves of the ODE algorithm. (a) Classical moves (e.g., a single bit flip), whereby only the initial state z is changed to z' leaving S_q unchanged. (b) Local swap, whereby two adjacent operators $V_{i_k} V_{i_{k+1}}$ are interchanged changing the state between them from z' to z''. (c) Block swap, whereby two partitions of the sequence are interchanged. This changes the initial state from z to z' as well as the ordering of the sequence. (d) Pair creation/annihilation, whereby a new pair of operators is inserted or deleted.

3.2.1. Classical moves Classical moves are any moves that involve a manipulation of the classical state $|z\rangle$ while leaving S_q unchanged [see Fig. 1(a)]. In a single bit-flip classical move, a spin from the classical bit-string state $|z\rangle$ of $\mathcal{C}$ is picked randomly and is flipped, generating a state $|z'\rangle$ and hence a new configuration $\mathcal{C}'$. Performing this change requires recalculating the energies associated with the sequence S_q leading to a new multiset $M_{\mathcal{C}'}$ and can become computationally intensive if q is large. Classical moves should therefore be attempted with low probabilities if q large. Simply enough, the acceptance probability for a classical move is

$$p = \min\left(1, \frac{W_{\mathcal{C}'}}{W_{\mathcal{C}}}\right) = \min\left(1, e^{-\beta\Delta E}\right), \tag{8}$$

where $\Delta E = E_{\mathcal{C}'} - E_{\mathcal{C}}$ is the difference between the effective classical energies of the proposed configuration $\mathcal{C}'$ and current configuration $\mathcal{C}$.

In the absence of a quantum part to the Hamiltonian ($\Gamma = 0$), not only are classical moves the only moves necessary, but they are also the only moves with a nonzero acceptance probability. In this case, the ODE algorithm automatically reduces to a classical Monte Carlo algorithm keeping the size of the imaginary-time dimension at zero ($q = 0$) for the duration of the simulation.

3.2.2. Local swap A local swap is the swapping of neighboring off-diagonal operators in the sequence S_q. A random pair of adjacent off-diagonal operators in the sequence is picked and swapped [as shown in Fig. 1(b)]. If the state between V_{i_k} and $V_{i_{k+1}}$ is $|z\rangle$ and is $|z'\rangle$ after the swap, then the swap involves adding an energy $E(z')$ and removing an energy $E(z)$ from the energy multiset [note that $E(z)$ and $E(z')$ may be the same]. The acceptance probability for the move is as in Eq. (8) with $M_{\mathcal{C}'} = M_{\mathcal{C}} + \{E(z')\} - \{E(z)\}$.

3.2.3. Block-swap A block swap [Fig. 1(c)] is a local update that involves a change of the classical state z. Here, a random position k in the sequence S_q is picked such that the sequence is split into two (non-empty) parts, $S_q = S_1 S_2$, with $S_1 = V_{i_1} \cdots V_{i_k}$ and $S_2 = V_{i_{k+1}} \cdots V_{i_q}$. The classical state $|z'\rangle$ at position k in the sequence is given by

$$\langle z'| = \langle z|S_1 = \langle z|V_{i_1} \ldots V_{i_k} \, , \tag{9}$$

where $|z\rangle$ is the classical state of the current configuration. The state $|z'\rangle$ has energy $E(z')$, and the state $|z\rangle$ has energy $E(z)$. We consider a new configuration defined by $\langle z'|S_2 S_1|z'\rangle$. The multiplicity table of this configuration differs from that of the current configuration by having one fewer $E(z)$ state and one additional $E(z')$ state. The weight of the new configuration is then proportional to $e^{-\beta M_{\mathcal{C}'}}$ where the multiset $M_{\mathcal{C}'} = M_{\mathcal{C}} + \{E(z')\} - \{E(z)\}$. The acceptance probability is as in Eq. (8) with the aforementioned $M_{\mathcal{C}'}$.

3.2.4. Creation/annihilation of off-diagonal operators The moves presented so far have left the size of S_q unchanged. The creation/annihilation move shown in Fig. 1(d) has the effect of changing the value of q by 2, i.e., $q \to q \pm 2$, which in the transverse-field Ising model corresponds to creating or destroying off-diagonal operators σ_j^x in pairs. We implement this via the insertion or deletion of two adjacent, identical operators. With probability $p_{\rm del}$ (e.g., $p_{\rm del} = 1/2$) we try to annihilate an adjacent pair, and with probability $1 - p_{\rm del}$ we try to insert a pair.

For pair insertion, we randomly pick an internal insertion point in the sequence (we denote this internal state by $|z'\rangle$) and a random V to insert. This adds two new energies $E(z')$ and $E(z'')$ to the multiset, where $|z''\rangle = V|z'\rangle$. The acceptance probability for pair creation is given by

$$p = \min\left(1, \frac{p_{\rm del}}{1 - p_{\rm del}} \frac{N\beta^2\Gamma^2}{(q+2)(q+3)} {\rm e}^{-\beta\Delta E}\right) \tag{10}$$

where as before $\Delta E = E_{\mathcal{C}'} - E_{\mathcal{C}}$ is the difference between the effective classical energies of the proposed configuration $\mathcal{C}'$ and current configuration $\mathcal{C}$ and $M_{\mathcal{C}'} = M_{\mathcal{C}} + \{E(z'), E(z'')\}$. For deletion, we randomly pick an internal point in the sequence. If the two operators to the side of the insertion point are not identical, no deletion is performed, and the move is rejected. If the two operators are identical, they are deleted and the relevant energies $E(z')$ and $E(z'')$ are removed from the multiplicity table. The probability of acceptance for the deletion move is

$$p = \min\left(1, \frac{1 - p_{\rm del}}{p_{\rm del}} \frac{q(q+1)}{\beta^2\Gamma^2 N} {\rm e}^{-\beta\Delta E}\right) , \tag{11}$$

where as before $\Delta E = E_{\mathcal{C}'} - E_{\mathcal{C}}$ and $M_{\mathcal{C}'} = M_{\mathcal{C}} - \{E(z'), E(z'')\}$.

The size of the imaginary time dimension q comes strictly from off-diagonal terms and shrinks or grows depending on the strength of the 'quantum component' of the model. This property is expected to be heavily utilized in order to overcome the freezing of QMC algorithms in almost classical regimes. In these regimes, q is small, and the algorithm reduces to being a classical thermal algorithm.[1]

3.3. Measurements

An integral part of any QMC algorithm is the acquisition of various properties of the model such as average energy, magnetization, specific heat and correlation functions. In the ODE algorithm (as in SSE), diagonal (classical) measurements are measured differently than off-diagonal ones.

[1] This is to be contrasted with the standard SSE formalism where one normally introduces an additional parameter L in order to fix the size of imaginary time dimension for more efficient weight calculations. The fixing of the size of imaginary time may adversely affect the convergence of the algorithm if it is chosen to be too large. Here, this parameter too is spurious.

3.3.1. Diagonal measurements Diagonal operators D obey $D|z\rangle = d(z)|z\rangle$ where $d(z)$ is a number that depends both on the operator and the state it acts on. Since $\langle z|DS_q|z\rangle = d(z)\langle z|S_q|z\rangle$, for any given configuration $\mathcal{C} = (|z\rangle, S_q)$, there is a contribution $d = d(z)$ to the diagonal operator thermal average $\langle D\rangle$. To improve statistics, we can also consider rotations in (the periodic) imaginary time. To do that, we may consider 'virtual' block-swap moves (see Sec. 3.2.3) that rotate S_q and as a result also change the classical configuration from $|z\rangle$ to $|z_i\rangle$. The contribution to the expectation value of a diagonal operator D thus becomes:

$$d = \frac{1}{\mathcal{Z}} \sum_{i=0}^{q-1} d(z_i) \mathrm{e}^{-\beta E_{\mathcal{C}_i}} . \tag{12}$$

where $E_{\mathcal{C}_i}$ is the effective classical energy associated with configuration $\mathcal{C}_i$ whose multiset is $M_{\mathcal{C}_i} = M_{\mathcal{C}} + \{E(z_i)\} - \{E(z)\}$ (recall that $z_0 \equiv z$, so $M_{\mathcal{C}_0} = M_{\mathcal{C}}$). The normalization factor $\mathcal{Z}$ above is the sum

$$\mathcal{Z} = \sum_{j=0}^{q-1} \mathrm{e}^{-\beta E_{\mathcal{C}_j}} = \sum_j m_j \mathrm{e}^{-\beta E_{\mathcal{C}_j}} \tag{13}$$

over all nonzero multiplicities m_j. In the case where $D = H_{\rm c}$ the above expression simplifies to:

$$d = \frac{1}{\mathcal{Z}} \sum_{i=0}^{q-1} E(z_i) \mathrm{e}^{-\beta E_{\mathcal{C}_i}} = \frac{1}{\mathcal{Z}} \sum_j m_j E(z_j) \mathrm{e}^{-\beta E_{\mathcal{C}_j}} . \tag{14}$$

3.3.2. Off-diagonal measurements We next consider the case of measuring the expectation value of an off-diagonal operator V_k, namely, $\langle V_k\rangle$. To do this, we interpret the instantaneous configuration as follows

$$W_{\mathcal{C}} = \frac{(\beta\Gamma)^q \mathrm{e}^{-\beta E_{\mathcal{C}}}}{q!} \langle z|S_q|z\rangle = \left(\frac{\beta\Gamma}{q\mathrm{e}^{-\beta\Delta E}}\right) \left[\frac{(\beta\Gamma)^{q-1} \mathrm{e}^{-\beta E_{\mathcal{C}'}}}{(q-1)!} \langle z|S_{q-1} V_{i_q}|z\rangle\right] , \tag{15}$$

where $\Delta E = E_{\mathcal{C}'} - E_{\mathcal{C}}$ and $\mathcal{C}'$ is the configuration associated with the multiset $M_{\mathcal{C}'} = M_{\mathcal{C}} - \{E(z)\}$. In the above form, we can reinterpret the weight $W_{\mathcal{C}}$ as contributing

$$v_k = \delta_{k,i_q} \frac{q\mathrm{e}^{-\beta\Delta E}}{\beta\Gamma} , \tag{16}$$

to $\langle V_k\rangle$.

As in the case of the diagonal measurements, one can take advantage of the periodicity in the imaginary time direction to improve statistics by rotating the sequence such that any of the elements of S_q becomes the last element of the sequence (see Sec. 3.2.3), weighted accordingly by the block-swap probability. By doing so, v_k becomes

$$v_k = \sum_i \frac{q}{\beta\Gamma} \frac{\mathrm{e}^{-\beta E_{\mathcal{C}_i}}}{\sum_{j=0}^{q-1} \mathrm{e}^{-\beta E_{\mathcal{C}_j}}} \frac{\mathrm{e}^{-\beta E_{\mathcal{C}'}}}{\mathrm{e}^{-\beta E_{\mathcal{C}_i}}} = \frac{q N_k}{\mathcal{Z}\beta\Gamma} \mathrm{e}^{-\beta E_{\mathcal{C}'}} \tag{17}$$

where $M_{\mathcal{C}_i} = M_{\mathcal{C}} + \{E(z_i)\} - \{E(z)\}$, the sum $\sum_i$ is over all rotated configurations $\mathcal{C}'$ whose S_q ends with V_k, and N_k is the number of times V_k appears in the sequence S_q.

3.4. Quantum-classical parallel tempering

As we reviewed in Sec. 2, the ODE partition function decomposition naturally reduces to the classical one when the strength of the off-diagonal terms in the Hamiltonian are sent to zero. As we show next, this allows us to naturally unify the classical Parallel Tempering (CPT) algorithm (also known as 'exchange Monte Carlo') [23, 24] and its quantum counterpart (QPT, see e.g., Refs [25, 26]). CPT is a refinement of the simulated annealing algorithm [27], whereby N_T

replicas of an N-spin system at inverse-temperatures $\beta_1 < \beta_2 < \ldots < \beta_{N_T}$ undergo Metropolis spin-flip updates independently of one another and in addition, replicas with neighboring temperatures regularly attempt to swap their temperatures with probabilities that satisfy detailed balance [28]. In this way, each replica performs a random-walk on the temperature axis, which generally allows for quicker equilibration of the system in comparison to other techniques. Analogously in QPT, temperature is replaced by a parameter Γ of the (quantum) Hamiltonian, e.g., the strength of the transverse magnetic field in the transverse Ising model, and each replica performs a random-walk on the Γ axis.

Here, we shall consider replicas along a curve in the β-Γ plane at points $\{(\beta_1, \Gamma_1), \ldots, (\beta_{N_T}, \Gamma_{N_T})\}$, then a parallel tempering swap probability between the i-th and $(i+1)$-th replica is given by:

$$P = \min\left(1, \frac{W_{\mathcal{C}_i}(\beta_{i+1}, \Gamma_{i+1}) W_{\mathcal{C}_{i+1}}(\beta_i, \Gamma_i)}{W_{\mathcal{C}_i}(\beta_i, \Gamma_i) W_{\mathcal{C}_{i+1}}(\beta_{i+1}, \Gamma_{i+1})}\right), \tag{18}$$

where the above weight ratio is conveniently simplified to:

$$\frac{W_{\mathcal{C}_i}(\beta_{i+1}, \Gamma_{i+1}) W_{\mathcal{C}_{i+1}}(\beta_i, \Gamma_i)}{W_{\mathcal{C}_i}(\beta_i, \Gamma_i) W_{\mathcal{C}_{i+1}}(\beta_{i+1}, \Gamma_{i+1})} = \left(\frac{\beta_i \Gamma_i}{\beta_{i+1} \Gamma_{i+1}}\right)^{q_{i+1} - q_i} \frac{\mathrm{e}^{-\beta_i (E'_{\mathcal{C}_{i+1}} - E_{\mathcal{C}_i})}}{\mathrm{e}^{-\beta_{i+1} (E_{\mathcal{C}_{i+1}} - E'_{\mathcal{C}_i})}}, \tag{19}$$

where $E_{\mathcal{C}_i}$ and $E_{\mathcal{C}_{i+1}}$ are the effective classical energies of configurations $\mathcal{C}_i$ and $\mathcal{C}_{i+1}$, respectively and $E'_{\mathcal{C}_i}$ and $E'_{\mathcal{C}_{i+1}}$ are the effective classical energies of these configurations when calculated with switched β and Γ.

4. Performance of ODE on quantum spin glasses

Having laid the groundwork for the ODE QMC algorithm, we present in this section some results that highlight some of its properties and advantages over existing QMC techniques, specifically a discrete-time Path Integral quantum Monte Carlo (PIQMC) algorithm with cluster-updates[2] and with the continuous-time SSE algorithm with cluster updates [32, 9].

To demonstrate the advantages of ODE over existing PIQMC and SSE, we carry out multiple runs of parallel tempering QMC simulations on random MAX2SAT instances augmented with a transverse field [see Eq. (2)] with various problem sizes N, a range of β (inverse temperature) and Γ (quantum strength) values.

Runtime comparisons of ODE against both PIQMC and SSE are summarized in Figs. 2–5. The figures depict the performance of ODE, PIQMC and SSE as it is reflected by the thermal average of x-magnetization $H_X = -\sum_{i=1}^{N} \sigma_i^x$, of the classical Hamiltonian H_c, and of the specific heat per spin $C = \beta^2 \left(\langle H^2 \rangle - \langle H \rangle^2\right)/N$ (other observables exhibit similar behavior) for increasing problem size and different values of β and Γ. Each data point in each of the panels is an average over 8 random instances.

The first three figures, Figs. 2–4, depict the results of PT simulations done along a fixed Γ line in the β-Γ plane, that is, different β values. As is clearly evident from the figures, while for large temperatures ($\beta \leq 2$) the performance of all three algorithms is comparable, with decreasing temperature and increasing problem size, ODE performs up to four or more orders of magnitude faster than the other two algorithms. Specifically, a large fraction of the instances did not finish running over the 24 hour window allocated for each of the simulations, whereas ODE had converged after a few seconds.

We note that while PIQMC can perform comparably to the ODE for the evaluation of the average H_c, PIQMC behaves poorly for the evaluation of the average of H_X. Similarly, while SSE performs better than PIQMC for evaluating the average of H_X, it performs worse for evaluating the average of H_c. From this perspective, we can conclude that the performance of ODE is also consistently better for different observables, which is reflected in its superior performance for calculating the average of the specific heat in Fig. 4.

[2] We use Wolff cluster updates [29] along the imaginary time direction [30, 31].

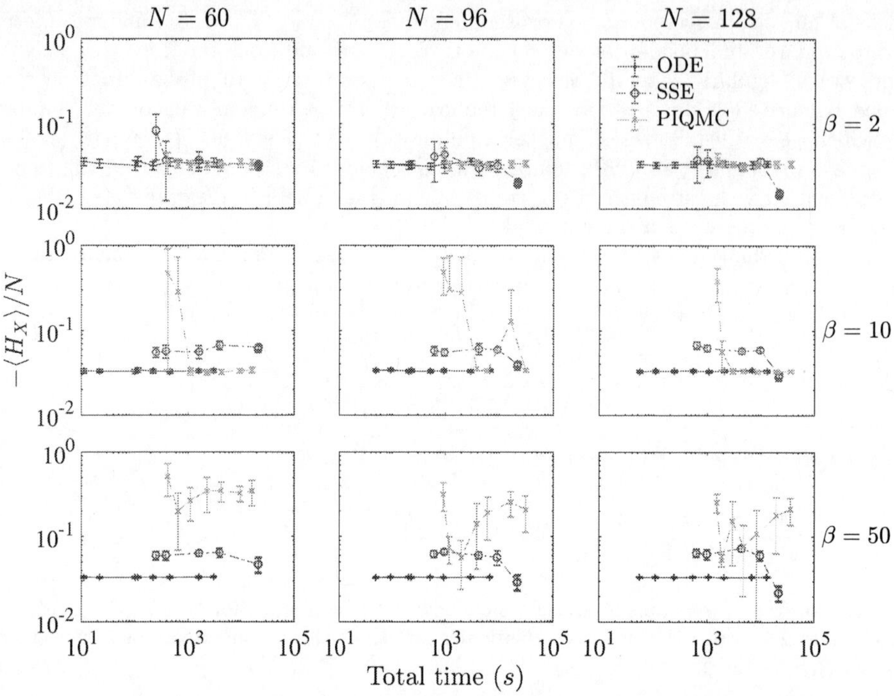

Figure 2. Thermal average of the off-diagonal Hamiltonian as obtained by ODE, PIQMC and SSE for a fixed $\Gamma = 0.1$. Results are obtained using PT simulations of different system sizes $N = 60, 96$ and 128, each with 11 replicas at inverse temperatures, $\beta \in \{0.1, 0.2, 0.5, 1, 2, 5, 10, 20, 30, 40, 50\}$. A subset of the results are shown for $\beta = 2, 10$ and 50. Each data point is the mean value of 10^3 bootstraps performed on the 500 measurements of the x-magnetization-per-spin taken in the course of simulations with different runtimes (horizontal axis), while the error bars correspond to the 2σ confidence interval generated by the bootstrap. For PIQMC we use 1280 Trotter slices. As is evident, ODE converges to the true value at times orders of magnitude faster than both SSE and PIQMC [33].

Figure 5 depicts the results of PT simulations done along a fixed β line in the β-Γ plane, that is, different Γ values. Interestingly, here all three algorithms seem to perform comparably, indicating that ODE is, as expected, at its optimum in near classical regimes.

To demonstrate the performance of the ODE algorithm away from the 'almost classical' regime, i.e., in regimes where the quantum strength parameter Γ is not small, we show in Fig. 6 some results obtained for $\Gamma = 0.4$ and 0.5 for different problem sizes. Here too, we find a regime where ODE converges faster to the true values by several orders of magnitude than PIQMC (not shown in this figure are SSE results, which we expect performs similarly to PIQMC in these regimes as is evident from the data presented in the preceding figures).

5. Conclusions

We have presented results for the performance of the ODE algorithm for a class of quantum spin glasses, 3-regular MAX2SAT with a transverse field, relative to PIQMC and SSE. Near the classical regime, when Γ the strength of the transverse is small, a clear advantage can be observed over PIQMC and SSE, with at times orders of magnitude faster performance. This advantage is expected because the ODE method has an elastic quantum dimension, which can shrink appropriately in the near classical limit.

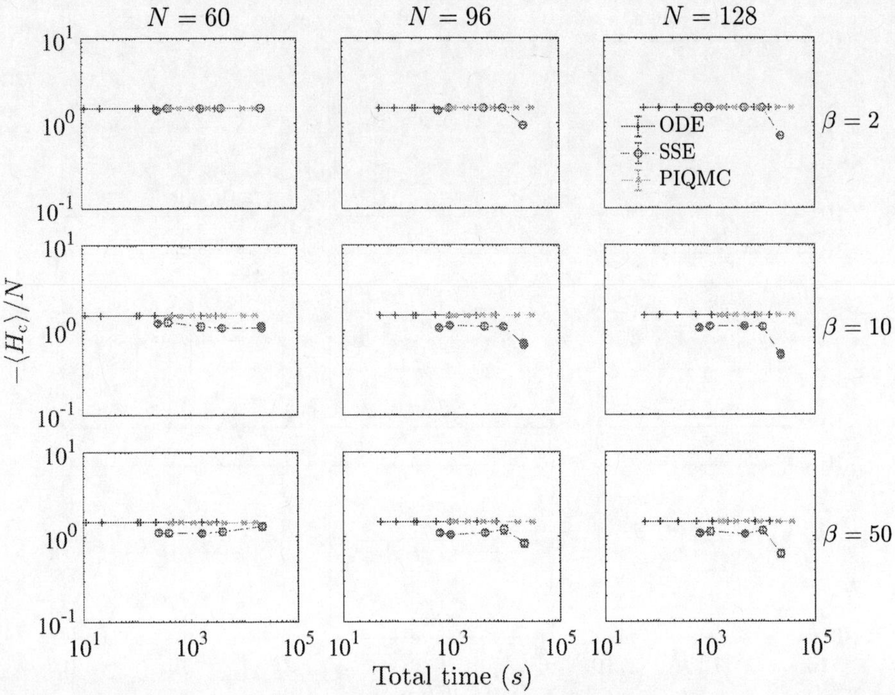

Figure 3. Thermal average of the classical Hamiltonian as obtained by ODE, PIQMC and SSE for a fixed $\Gamma = 0.1$. Results are obtained using PT simulations of different system sizes $N = 60, 96$ and 128, each with 11 replicas at inverse temperatures, $\beta \in \{0.1, 0.2, 0.5, 1, 2, 5, 10, 20, 30, 40, 50\}$. A subset of the results are shown for $\beta = 2, 10$ and 50. Each data point is the mean value of 10^3 bootstraps performed on the 500 measurements of the x-magnetization-per-spin taken in the course of simulations with different runtimes (horizontal axis), while the error bars correspond to the 2σ confidence interval generated by the bootstrap. For PIQMC we use 1280 Trotter slices. As is evident, ODE converges to the true value at times orders of magnitude faster than both SSE and PIQMC [33].

We have demonstrated how the algorithm applies to the transverse-field Ising model. It would be interesting to see how it performs with respect to existing techniques on other models considered difficult to simulate. Another aspect worth studying is the existence of additional updates that are more global in nature in order to further speed up convergence. These will more likely have to be specifically tailored to the system in question. Last, methods to facilitate the evaluation of the generalized Boltzmann weights are of significance as these scale in the worst case as the square of the imaginary time dimension. More efficient methods will serve to further increase the usefulness of the ODE algorithm. We leave the resolution of these questions for future work.

Acknowledgments
Computation for the work described here was supported by the University of Southern California's Center for High-Performance Computing (`http://hpcc.usc.edu`). The research is based upon work (partially) supported by the Office of the Director of National Intelligence (ODNI), Intelligence Advanced Research Projects Activity (IARPA), via the U.S. Army Research Office contract W911NF-17-C-0050. The views and conclusions contained herein are those of the authors and should not be interpreted as necessarily representing the official policies

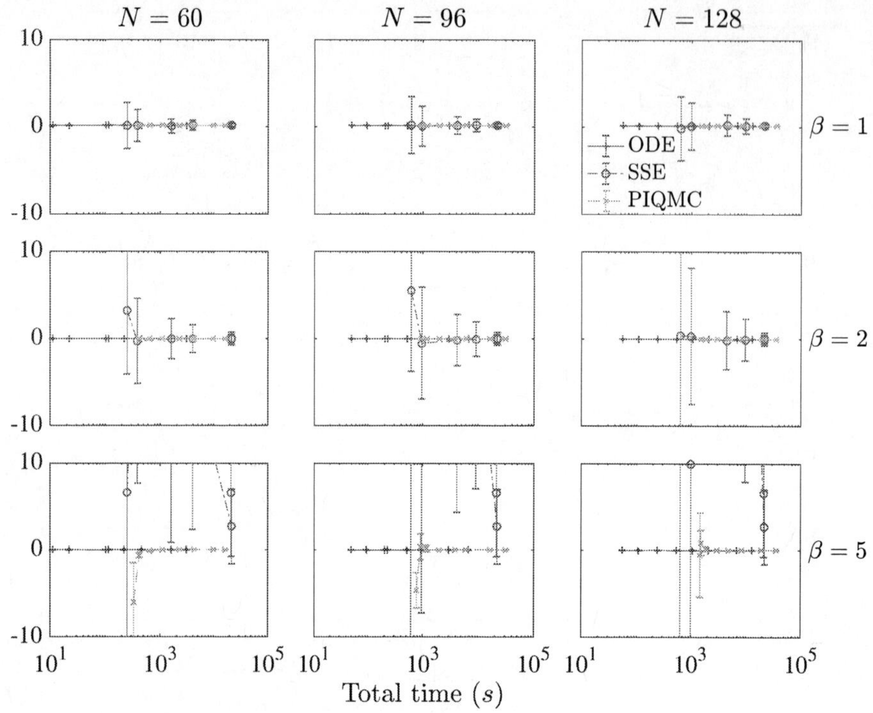

Figure 4. Thermal average of the specific heat per spin $C = \beta^2\left(\langle H^2\rangle - \langle H\rangle^2\right)/N$ as obtained by ODE, PIQMC and SSE for a fixed $\Gamma = 0.1$. Results are obtained using PT simulations of different system sizes $N = 60, 96$ and 128, each with 11 replicas at inverse temperatures, $\beta \in \{0.1, 0.2, 0.5, 1, 2, 5, 10, 20, 30, 40, 50\}$. A subset of the results are shown for $\beta = 1, 2$ and 5. Each data point is the mean value of 10^3 bootstraps performed on the 500 measurements of the x-magnetization-per-spin taken in the course of simulations with different runtimes (horizontal axis), while the error bars correspond to the 2σ confidence interval generated by the bootstrap. For PIQMC we use 1280 Trotter slices. As is evident, ODE converges to the true value at times orders of magnitude faster than both SSE and PIQMC [33].

or endorsements, either expressed or implied, of the ODNI, IARPA, or the U.S. Government. The U.S. Government is authorized to reproduce and distribute reprints for Governmental purposes notwithstanding any copyright annotation thereon.

References

[1] A. P. Young, ed., *Spin Glasses and Random Fields.* Singapore: World Scientific, 1998.

[2] F. Barahona, "On the computational complexity of Ising spin glass models," *J. Phys. A: Math. Gen*, vol. 15, no. 10, p. 3241, 1982.

[3] F. Belletti, M. Cotallo, A. Cruz, L. A. Fernandez, A. Gordillo, A. Maiorano, F. Mantovani, E. Marinari, V. Martin-Mayor, A. Muñoz Sudupe, D. Navarro, S. Perez-Gaviro, J. J. Ruiz-Lorenzo, S. F. Schifano, D. Sciretti, A. Tarancon, R. Tripiccione, and J. L. Velasco, "Simulating spin systems on IANUS, an FPGA-based computer," *Comp. Phys. Comm.*, vol. 178, pp. 208–216, 2008.

[4] F. Belletti, M. Guidetti, A. Maiorano, F. Mantovani, S. F. Schifano, R. Tripiccione, M. Cotallo, S. Perez-Gaviro, D. Sciretti, J. L. Velasco, A. Cruz, D. Navarro, A. Tarancon, L. A. Fernandez, V. Martin-Mayor, A. Muñoz-Sudupe, D. Yllanes, A. Gordillo-Guerrero, J. J. Ruiz-Lorenzo, E. Marinari, G. Parisi, M. Rossi, and G. Zanier, "Janus: An FPGA-based system for high-performance scientific computing," *Computing in Science and Engineering*, vol. 11, p. 48, 2009.

[5] M. Baity-Jesi, R. A. Baños, A. Cruz, L. A. Fernandez, J. M. Gil-Narvion, A. Gordillo-Guerrero, D. Iniguez, A. Maiorano, F. Mantovani, E. Marinari, V. Martin-Mayor, J. Monforte-Garcia, A. Muñoz Sudupe,

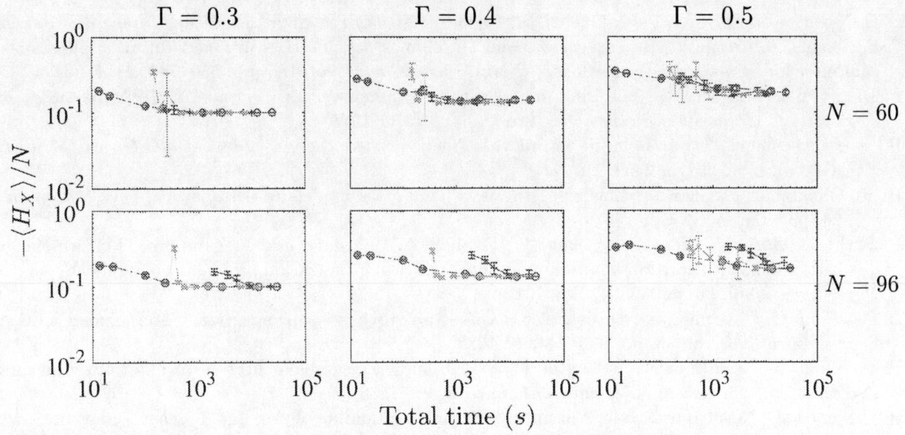

Figure 5. Thermal average of the off-diagonal Hamiltonian as obtained by ODE, PIQMC and SSE for a fixed $\beta = 10$. Results are obtained using PT simulations of different system sizes $N = 60$ and 96, each with 5 replicas at transverse field strengths, $\Gamma \in \{0.1, 0.2, 0.3, 0.4, 0.5\}$. A subset of the results are shown for $\Gamma = 0.3, 0.4$ and 0.5. Each data point is the mean value of 10^3 bootstraps performed on the 500 measurements of the x-magnetization-per-spin taken in the course of simulations with different runtimes (horizontal axis), while the error bars correspond to the 2σ confidence interval generated by the bootstrap. For PIQMC we use 1280 Trotter slices. Here, all three algorithms perform similarly[33].

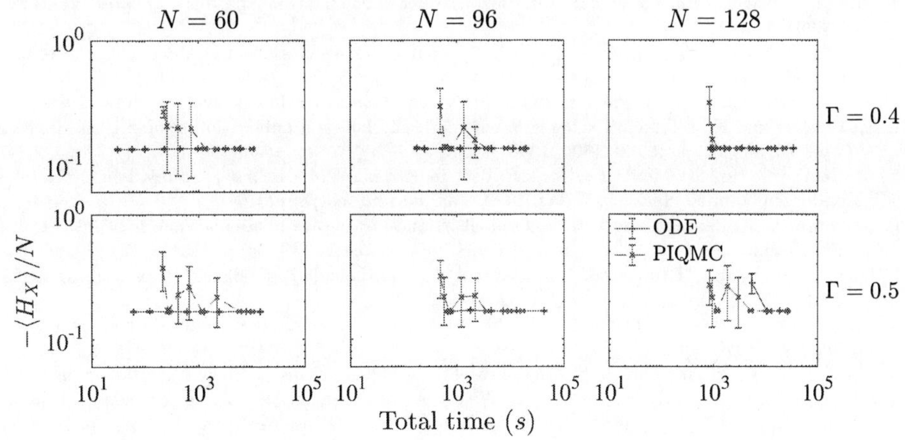

Figure 6. Thermal average of the off-diagonal Hamiltonian as obtained by ODE and PIQMC at large Γ values. Results are obtained using PT simulations of different system sizes $N = 60, 96$ and 128, each with 7 replicas at inverse temperatures, $\beta \in \{0.1, 0.2, 0.5, 1, 2, 5, 10\}$. Results are shown for $\beta = 10$. Each data point is the mean value of 10^3 bootstraps performed on the 500 measurements of the x-magnetization-per-spin taken in the course of simulations with different runtimes (horizontal axis), while the error bars correspond to the 2σ confidence interval generated by the bootstrap. For PIQMC we use 1280 Trotter slices. As is evident, even at large Γ values ODE can converge faster than PIQMC by several orders of magnitude [33].

D. Navarro, G. Parisi, S. Perez-Gaviro, M. Pivanti, F. Ricci-Tersenghi, J. J. Ruiz-Lorenzo, S. F. Schifano, B. Seoane, A. Tarancon, R. Tripiccione, and D. Yllanes, "Janus II: a new generation application-driven computer for spin-system simulations," *Comp. Phys. Comm*, vol. 185, pp. 550–559, 2014.

[6] P. Pfeuty and R. J. Elliott, "The ising model with a transverse field. ii. ground state properties," *Journal of Physics C: Solid State Physics*, vol. 4, no. 15, p. 2370, 1971.

[7] R. B. Stinchcombe, "Ising model in a transverse field. i. basic theory," *Journal of Physics C: Solid State Physics*, vol. 6, no. 15, p. 2459, 1973.

[8] B. K. Chakrabarti, "Critical behavior of the ising spin-glass models in a transverse field," *Phys. Rev. B*, vol. 24, pp. 4062–4064, Oct 1981.

[9] E. Farhi, D. Gosset, I. Hen, A. W. Sandvik, P. Shor, A. P. Young, and F. Zamponi, "Performance of the quantum adiabatic algorithm on random instances of two optimization problems on regular hypergraphs," *Phys. Rev. A*, vol. 86, p. 052334, Nov 2012.

[10] B. Apolloni, C. Carvalho, and D. de Falco, "Quantum stochastic optimization," *Stochastic Processes and their Applications*, vol. 33, pp. 233–244, 12 1989.

[11] B. Apolloni, N. Cesa-Bianchi, and D. de Falco, "A numerical implementation of quantum annealing," in *Proceedings of the Ascona/Locarno Conference*, p. 97, 1988.

[12] R. L. Somorjai, "Novel approach for computing the global minimum of proteins. 1. general concepts, methods, and approximations," *The Journal of Physical Chemistry*, vol. 95, pp. 4141–4146, 05 1991.

[13] P. Amara, D. Hsu, and J. E. Straub, "Global energy minimum searches using an approximate solution of the imaginary time schroedinger equation," *The Journal of Physical Chemistry*, vol. 97, pp. 6715–6721, 2015/01/03 1993.

[14] A. B. Finnila, M. A. Gomez, C. Sebenik, C. Stenson, and J. D. Doll, "Quantum annealing: A new method for minimizing multidimensional functions," *Chemical Physics Letters*, vol. 219, pp. 343–348, 3 1994.

[15] T. Kadowaki and H. Nishimori, "Quantum annealing in the transverse Ising model," *Phys. Rev. E*, vol. 58, p. 5355, Nov. 1998.

[16] J. Brooke, D. Bitko, T. F., Rosenbaum, and G. Aeppli, "Quantum annealing of a disordered magnet," *Science*, vol. 284, no. 5415, pp. 779–781, 1999.

[17] J. Brooke, T. F. Rosenbaum, and G. Aeppli, "Tunable quantum tunnelling of magnetic domain walls," *Nature*, vol. 413, pp. 610–613, Oct. 2001.

[18] E. Farhi, J. Goldstone, S. Gutmann, and M. Sipser, "Quantum computation by adiabatic evolution," *ArXiv*, 2000.

[19] E. Farhi, J. Goldstone, S. Gutmann, J. Lapan, A. Lundgren, and D. Preda, "A quantum adiabatic evolution algorithm applied to random instances of an NP-Complete problem," *Science*, vol. 292, pp. 472–475, Apr. 2001.

[20] T. Albash, G. Wagenbreth, and I. Hen, "Off-diagonal expansion quantum monte carlo," *Phys. Rev. E*, vol. 96, p. 063309, Dec 2017.

[21] C.-W. Liu, A. Polkovnikov, and A. W. Sandvik, "Quantum versus classical annealing: Insights from scaling theory and results for spin glasses on 3-regular graphs," *Phys. Rev. Lett.*, vol. 114, p. 147203, Apr 2015.

[22] M. N. . G. Barkema, *Monte Carlo Methods in Statistical Physics.* Oxford Uinversity Press, 1999.

[23] K. Hukushima and K. Nemoto, "Exchange monte carlo method and application to spin glass simulations," *J. Phys. Soc. Japan*, vol. 65, p. 1604, 1996.

[24] E. Marinari, "Optimized Monte Carlo methods," in *Advances in Computer Simulation* (J. Kertész and I. Kondor, eds.), p. 50, Springer-Verlag, 1998. (arXiv:cond-mat/9612010).

[25] P. Sengupta, A. W. Sandvik, and D. K. Campbell, "Bond-order-wave phase and quantum phase transitions in the one-dimensional extended hubbard model," *Phys. Rev. B*, vol. 65, p. 155113, Apr 2002.

[26] I. Hen and A. P. Young, "Exponential complexity of the quantum adiabatic algorithm for certain satisfiability problems," *Phys. Rev. E*, vol. 84, p. 061152, Dec 2011.

[27] S. Kirkpatrick, C. D. Gelatt, and M. P. Vecchi, "Optimization by simulated annealing," *Science*, vol. 220, pp. 671–680, May 1983.

[28] A. Sokal, "Monte carlo methods in statistical mechanics: Foundations and new algorithms," in *Functional Integration: Basics and Applications* (C. DeWitt-Morette, P. Cartier, and A. Folacci, eds.), Plenum, 1997.

[29] U. Wolff, "Collective Monte Carlo updating for spin systems," *Phys. Rev. Lett.*, vol. 62, pp. 361–364, Jan 1989.

[30] R. Martoňák, G. E. Santoro, and E. Tosatti, "Quantum annealing by the path-integral Monte Carlo method: The two-dimensional random Ising model," *Phys. Rev. B*, vol. 66, p. 094203, Sep 2002.

[31] S. Boixo, T. F. Ronnow, S. V. Isakov, Z. Wang, D. Wecker, D. A. Lidar, J. M. Martinis, and M. Troyer, "Evidence for quantum annealing with more than one hundred qubits," *Nat. Phys.*, vol. 10, pp. 218–224, 03 2014.

[32] A. W. Sandvik, "Stochastic series expansion method for quantum Ising models with arbitrary interactions," *Phys. Rev. E*, vol. 68, p. 056701, 2003.

[33] The simulations were performed on individual cores of Intel Xeon E5-2640 CPUs running at 2.60GHz.

IOP Conf. Series: Journal of Physics: Conf. Series **1136** (2018) 012008 doi:10.1088/1742-6596/1136/1/012008

Simulation of electron energy loss spectra with the *turboEELS* and *thermo_pw* codes

Oleksandr Motornyi[1], **Michèle Raynaud**[1], **Andrea Dal Corso**[2], **Nathalie Vast**[1]

[1] Laboratoire des Solides Irradiés, École Polytechnique - CEA DRF IRAMIS - CNRS UMR 7642, Paris-Saclay University, 91128 Palaiseau cédex, France

[2] Scuola Internazionale Superiore di Studi Avanzati (SISSA), Via Bonomea 265, IT-34136 Trieste, Italy, IOM-CNR Trieste, Italy

E-mail: oleksandr.motornyi@polytechnique.edu

Abstract.
For some materials like noble metals, electron energy loss spectra have a complex structure that makes them difficult to analyze without the help of *ab initio* calculations. Various theoretical approaches can be used for this purpose, among which the time-dependent density functional perturbation theory (TDDFPT) which has been widely used to study plasmons in a number of bulk and surface systems. In the present paper we present a comparison of the results and performance of two different numerical implementations of TDDFPT: the Sternheimer and Liouville-Lanczos methods. The former approach is implemented in the *thermo_pw* module and the latter one in the *turboEELS* code of the QUANTUM ESPRESSO package for electronic structure calculations. In the present paper a comparison is made for bulk bismuth, a semimetal, taking into account spin-orbit coupling, as well as for bulk gold, a noble metal. We show that for these two examples, both codes gives identical results and the *turboEELS* code has a better performance than the *thermo_pw* code, and point out in which cases the usage of *thermo_pw* alone or of both codes can be advantageous.

1. Introduction

Electron energy loss (EEL) spectroscopy [1] is a powerful experimental method that allows one to measure the change in kinetic energy -the energy loss- of electrons after their interaction with a sample. This technique can be used to obtain structural and chemical information about a specimen -from the high-energy loss region of the spectra- as well as information about valence electron excitations -from the low-energy loss region of the spectra-. In particular, the valence region of the spectra contains information about the band structure and dielectric properties of the material: plasmons, surface plasmons and interband transitions. For some materials like aluminum, valence EEL spectra has a simple structure containing one single, well-defined, plasmon peak for which the Drude model is a pertinent approximation. In materials where the behavior of valence electron is far from the (quasi) free electron model however, the EEL spectrum contains many peaks from single, to collective and to superposition of collective and single excitations, that make it difficult to analyze.

In order to perform a thorough analysis it is convenient to simulate EEL spectra computing the dielectric function. Indeed, in an EEL spectroscopy experiment, the inelastic scattering probability of an incoming electron by the electrons of the solid, given by the double-differential cross-section $d^2\sigma/(d\omega d\Omega)$ is measured, which is linked to the macroscopic dielectric function ϵ_M:

$$\frac{d^2\sigma}{d\omega d\Omega} \sim Im[\frac{1}{\epsilon_M(\mathbf{Q},\omega)}], \tag{1}$$

where $\mathbf{Q}$ is a momentum transfer, ω is an electron energy loss, and $d\Omega$ is the elemental solid-angle in which the scattering occurs.

Various *ab initio* methods that allow one to calculate spectral properties of solids have been developed in the past decades, addressing different aspects of the problem [2]. Time-dependent density functional theory (TDDFT) has been widely used to study plasmons in a number of bulk and surface systems [3, 4]. We note however that even for a material as simple as covalent silicon, one might need to account for many-body (excitonic) effects beyond TDDFT for instance to properly describe the shape of the plasmon peak [5].

In the present paper we will focus on the time-dependent density functional perturbation theory (TDDFPT) and linear response approach [6, 7] to the calculation of the dielectric function. Two implementations, the Sternheimer and Liouville-Lanczos methods, will be compared for bulk bismuth and gold. The paper is organized as follows. In Sec. 2, a brief overview and comparison of the theoretical methods: classic TDDFT approach, and the Liouville-Lanczos and Sternheimer approaches to TDDFPT, will be given. In Sec. 3, results obtained using different approaches and corresponding implementations will be shown along with the comparison of the performance of the codes. Finally, conclusions will be drawn in Sec. 4.

2. Comparison of theoretical approaches to linear response calculations

2.1. Dyson-like equation

The dielectric function in eq. (1) is determined by the macroscopic density response function (or susceptibility) χ_M [8]:

$$\epsilon_M(\mathbf{Q},\omega)^{-1} = 1 + v_G(\mathbf{Q})\chi_M(\mathbf{Q},\omega), \tag{2}$$

where $v_G(\mathbf{Q}) = 4\pi e^2/|\mathbf{Q}|^2$ is the Fourier transform of the Coulomb potential. χ_M is related to the microscopic density response function $\chi_{\mathbf{G},\mathbf{G}}(\mathbf{q},\omega)$:

$$\chi_M(\mathbf{Q},\omega) = \chi_{\mathbf{G},\mathbf{G}}(\mathbf{q},\omega), \tag{3}$$

where $\mathbf{Q} = \mathbf{q} + \mathbf{G}$, with $\mathbf{G}$ being a reciprocal lattice vector and $\mathbf{q}$ a wavevector in the first Brillouin zone. Within the traditional linear response approach of TDDFT, χ of the interacting many-body system is related to the density response function χ^0 of the non-interacting Kohn-Sham (KS) system through a Dyson-like screening equation:

$$\chi = \chi^0 + (v + f_{xc})\chi. \tag{4}$$

f_{xc} is the exchange-correlation kernel and χ^0 is given by

$$\chi^0_{\mathbf{G},\mathbf{G}'}(\mathbf{q},\omega) = \frac{1}{V}\sum_{\mathbf{k}}^{BZ}\sum_{n,n'} \frac{f_{n,\mathbf{k}} - f_{n',\mathbf{k}+\mathbf{q}}}{\hbar\omega + \varepsilon^0_{n,\mathbf{k}} - \varepsilon^0_{n',\mathbf{k}+\mathbf{q}} + i\eta} \langle\varphi^0_{n,\mathbf{k}}|e^{-i(\mathbf{q}+\mathbf{G})\mathbf{r}}|\varphi^0_{n',\mathbf{k}+\mathbf{q}}\rangle\langle\varphi^0_{n',\mathbf{k}+\mathbf{q}}|e^{i(\mathbf{q}+\mathbf{G})\mathbf{r}}|\varphi^0_{n,\mathbf{k}}\rangle, \tag{5}$$

where $\varphi^0_{n,\mathbf{k}}, \varepsilon^0_{n,\mathbf{k}}, f_{n,\mathbf{k}}$ are respectively the unperturbed single-particle wavefunctions, their eigenvalues and occupation numbers, V is the volume of the unit-cell and n, n' are indices that span all of the occupied and unoccupied bands. The accurate calculation of χ in a wide frequency range using eq. (4) and (5) will require a separate calculation for each value of the frequency and knowledge of all of the empty states up to the requested energy, which makes this task cumbersome.

2.2. Sternheimer equations

On the other hand, the density response-function can be defined through the first-order correction to the density $n(\mathbf{r},t) = n^0(\mathbf{r}) + n'(\mathbf{r},t)$ of the system under a weak external perturbation $V_{ext}(\mathbf{r},t) = V^0_{ext}(\mathbf{r}) + V'_{ext}(\mathbf{r},t)$:

$$n'(\mathbf{r},t) = \int_{-\infty}^{\infty} dt' \int d\mathbf{r}' \chi(\mathbf{r},\mathbf{r}',t-t') V'_{ext}(\mathbf{r}',t'). \tag{6}$$

In the case of EEL spectroscopy, the external perturbation is $V'_{ext}(\mathbf{r}',t') \sim e^{i(\mathbf{q}\cdot\mathbf{r}-\omega t)}$. One can perform a direct calculation of the χ using density functional perturbation theory (DFPT) techniques without explicit calculation of empty states [9] and without solving the Dyson-like equation (4). In this framework a system under external perturbation can be described by the TDDFPT KS equation:

$$i\hbar \frac{\partial \varphi_{n,\mathbf{k}}(\mathbf{r},t)}{\partial t} = \hat{H}^{KS}(\mathbf{r},t)\varphi_{n,\mathbf{k}}(\mathbf{r},t), \tag{7}$$

where $\varphi_{n,\mathbf{k}}(\mathbf{r},t) = \varphi^0_{n,\mathbf{k}}(\mathbf{r}) + \varphi'_{n,\mathbf{k}}(\mathbf{r},t)$ is the KS wavefunction decomposed into the unperturbed KS wave and its first-order variation. $\hat{H}^{KS}(\mathbf{r},t) = \hat{H}^0(\mathbf{r}) + V'(\mathbf{r},t)$ is the KS Hamiltonian which consists in the static unperturbed Hamiltonian $\hat{H}^0(\mathbf{r})$ and of the (linearized) perturbation part $V'(\mathbf{r},t)$. This equation and its complex-conjugate can be linearized and, taking into account only the first-order response, yield the so-called Sternheimer equation [9] which, in the frequency domain, reads:

$$(\hat{H}^0 - \varepsilon^0_{n,\mathbf{k}} - \hbar\omega)\varphi'_{n,\mathbf{k}}(\mathbf{r},\omega) + \hat{P}_c V'_{Hxc}(\mathbf{r},\omega)\varphi^0_{n,\mathbf{k}}(\mathbf{r}) = -\hat{P}_c V'_{ext}(\mathbf{r},\omega)\varphi^0_{n,\mathbf{k}}(\mathbf{r}), \tag{8}$$

$$(\hat{H}^0 - \varepsilon^0_{n,-\mathbf{k}} + \hbar\omega)\varphi'^*_{n,-\mathbf{k}}(\mathbf{r},-\omega) + \hat{P}_c V'^*_{Hxc}(\mathbf{r},-\omega)\varphi^0_{n,\mathbf{k}}(\mathbf{r}) = -\hat{P}_c V'^*_{ext}(\mathbf{r},-\omega)\varphi^0_{n,\mathbf{k}}(\mathbf{r}), \tag{9}$$

where $\hat{P}_c$ is a projector on empty states and the first order correction to the density $n'(\mathbf{r},t)$ reads:

$$n'(\mathbf{r},t) = \sum_{\mathbf{k}}^{BZ} \sum_{n} \varphi^{0*}_{n,\mathbf{k}}(\mathbf{r}) \left[\varphi'^*_{n,-\mathbf{k}}(\mathbf{r},-\omega) + \varphi'_{n,\mathbf{k}}(\mathbf{r},\omega)\right] \tag{10}$$

where n now only spans the occupied bands. For metals, partially occupied bands should be accounted for using the smearing approach [10, 11]. However, for the sake of simplicity it is ommited in the present work, and only the insulating case is considered. This set of equations can be solved self-consistently for each value of ω, allowing us to obtain the susceptibility χ of the system.

2.3. The Liouville-Lanczos equation

Equivalently, instead of eq. (7), one can use the quantum Liouville equation [12, 13, 14]:

$$i\hbar \frac{d\hat{\rho}(t)}{dt} = \left[\hat{H}_{KS}(t), \hat{\rho}(t)\right], \tag{11}$$

where $\hat{\rho}(t)$ is the reduced one-electron KS density-matrix whose kernel reads:

$$\rho(\mathbf{r},\mathbf{r}',t) = \sum \varphi_v(\mathbf{r},t)\varphi^*_v(\mathbf{r}',t), \tag{12}$$

and $n(\mathbf{r},t) = \rho(\mathbf{r},\mathbf{r},t)$. Equation (11) can be linearized in the same manner as eq. (7), leading to the linearized Liouvillian equation:

$$(\hbar\omega - \hat{\mathcal{L}}) \cdot \hat{\tilde{\rho}}'(\omega) = \left[\tilde{V}'_{ext}(\omega), \hat{\rho}^0\right]. \tag{13}$$

The action of the Liouvillian superoperator $\hat{\mathcal{L}}$ onto $\hat{\rho}$ is defined as:

$$\hat{\mathcal{L}} \cdot \hat{\rho}' = \left[\hat{H}^0, \hat{\rho}'\right] + \left[V'_{Hxc}[\hat{\rho}'], \hat{\rho}^0\right]. \tag{14}$$

Equation (13) can be solved using the Lanczos recursive scheme for all of the frequencies at once, resulting in the susceptibility χ.

2.4. Comparison of the methods

Each of three methods briefly described in Sec. 2.1 to 2.3 allows us to perform a first-principle calculation of the linear density response function $\chi(\mathbf{q}, \omega)$ and, thus, to obtain the dielectric and loss function of a given system, that can be directly compared to the experimental measurements (see eq. (1)). Each method has advantages and drawbacks, the understanding of which is necessary to choose the approach the most adapted to the target-problem.

Calculation of χ using a Dyson-like equation [15] is a powerful method that allows one to access and analyze individual electronic excitations in the material. However, this information comes for a relatively high cost: computation of χ^0 in eq. (5) requires the computation of corresponding empty states, and computation of χ in eq. (4) involves multiplication and inversion of large matrices, making this calculations demanding in terms of central processing unit (CPU) time and random access memory (RAM).

On the contrary, usage of DFPT techniques [9] in the Sternheimer approach allows one to avoid the computation of the (numerous) empty states, by introducing the projector on the empty states $\hat{P}_c$ in eq. (8) and (9). This reduces the RAM and CPU time usage. Nevertheless, the CPU time usage will remain high for the calculation over a wide frequency range, as the self-consistent solution of eq. (8) and (9) should be repeated for each value of the desired frequency.

The Lanczos recursion algorithm used to solve the linearized Liouvillian equation (13) allows us to avoid the explicit computation of the susceptibility for each frequency value, separating the calculation into two steps. The first step of the Lanczos recursion is time consuming. It is used to generate a tridiagonal matrix of coefficients which is used in the second step (post-processing) to calculate the frequency-dependent susceptibility χ. This scheme allows one to reduce CPU time usage significantly. However, there are two main drawbacks of the Liouville-Lanczos approach: firstly, it is possible to perform calculations only in the case of an adiabatic exchange and correlation functional V'_{Hxc} in eq. (14). In the case where V'_{Hxc} has an explicit frequency dependence, one will have to repeat the recursive procedure for each desired value of the frequency, increasing the CPU time tremendously, therefore making this procedure inefficient. Secondly, within both Liouville-Lanczos and Sternheimer approaches, it is not possible to retrieve the information from which orbitals a specific peak in a spectrum consists of, thus making an analysis for understanding the physical origins more demanding.

3. Comparison of the results and performance of two codes: *turboEELS* and *thermo_pw*

3.1. Computational details

Ground state properties have been calculated using the density functional theory (DFT) [16, 17] in the plane wave and pseudopotential method within the local density approximation [18] for bismuth and the generalized gradient approximation [19] for gold.

For bulk Bi, a 2-atom trigonal unit-cell was used, with a plane wave basis set limited with a cutoff energy of 60 Ry. A Monkhorst-Pack grid of 14^3 **k**-points has been used to sample the Brillouin zone [20]. For the calculation of EEL spectra, a small transferred momentum of $\mathbf{q} = 0.013\text{Å}^{-1}$ in the (111) trigonal direction was chosen. An imaginary frequency was introduced, so that $\hbar\omega$ in eq. (8) and (9) reads $\hbar\omega + i\eta$, with $\eta = 0.135eV$. Indeed, the Sternheimer equations are solved with this broadening. At variance, in the Liouville-Lanczos method the broadening is used only in the post-processing step, when the χ response function is built from the recursion coefficients. Calculations were performed using both scalar-relativistic - no spin-orbit coupling (SOC) and full-relativistic - with SOC normconserving pseudopotentials [21]. The number of iterations in the Lanczos recursion was 8000.

For bulk Au a 1-atom face-centered cubic unit-cell was used, with a plane wave basis set limited with a cutoff energy of 60 Ry. A Monkhorst-Pack grid of 32^3 **k**-points has been used to sample the Brillouin zone. For the calculation of EEL spectra, a small transferred momentum of $\mathbf{q} = 0.039\text{Å}^{-1}$ was used and an imaginary frequency $\eta = 0.135eV$ was added. Calculations were performed using scalar relativistic (without SOC) normconserving pseudopotential [22]. The number of iterations in the Lanczos recursion was 5000.

Convergence was carefully checked. In the *thermo_pw* the procedure has an explicit convergence threshold parameter on the variation of the density. This threshold was set to 10^{-12}. In the Liouville-Lanczos approach, the recursion chain should be cut after a certain number of iterations, meaning that convergence should be verified manually by performing additional iterations.

Both Sternheimer (*thermo_pw* code, unpublished [23]) and the Liouville-Lanczos (*turboEELS* code [14]) approaches are implemented in the Quantum ESPRESSO package for electronic structure calculations [24, 25]. Both codes support normconserving and ultrasolf pseudopotentials. Several levels of MPI parallelization are implemented: a plane-wave parallelization and a **k**-point parallelization.

3.2. Results
In figure 1, we report EEL spectra for bulk Bi obtained both with the *thermo_pw* code, and with *turboEELS*: results are in perfect agreement with each other and with previous calculations [3], both with and without SOC. We point out that the TDDFPT calculations were performed with, as input, data coming from the same (DFT) ground state self-consistent calculation.

Figure 2 shows the comparison of the EEL spectra for bulk Au obtained by *thermo_pw* and *turboEELS* codes.Obtained results are in agreement with previous ab initio calculations of the EELS for Au [26]. In the figure 2 one can see a severely damped plasmon peak around 2.5 eV and numerous interband transition at higher energies.

Figure 3 shows the relative difference between EEL spectra computed with the *thermo_pw* and *turboEELS* codes. Beyond 2 eV the agreement is perfect and the two codes agree within 1%. Below 2 eV, the loss function is tending to 0 and the magnitude of the relative difference becomes higher as the loss function does not tend to 0 in the same way. Indeed, oscillations that are intrinsic to the Lanczos, approach are magnified in the difference at low energy. As in the case for Bi, agreement is perfect, that indicates that both codes are well-suited to handle either semimetallic or metallic materials.

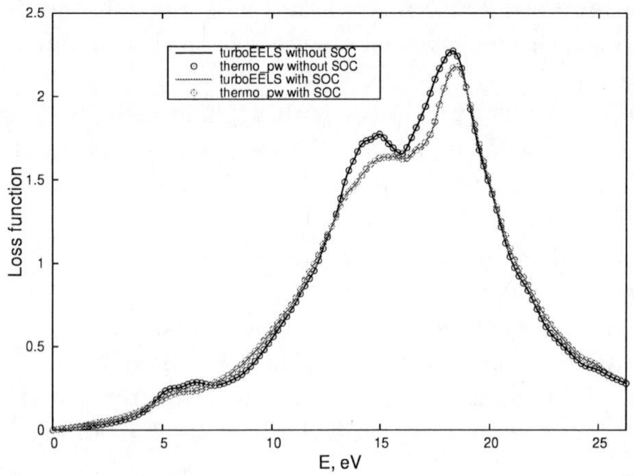

Figure 1: Bulk Bi. Loss function for a transferred momentum $\mathbf{q} = 0.013\text{Å}^{-1}$ in the z-direction obtained using: *turboEELS* code without (——) and with ········ spin-orbit coupling (SOC), and *thermo_pw* code without (O) and with (O) SOC.

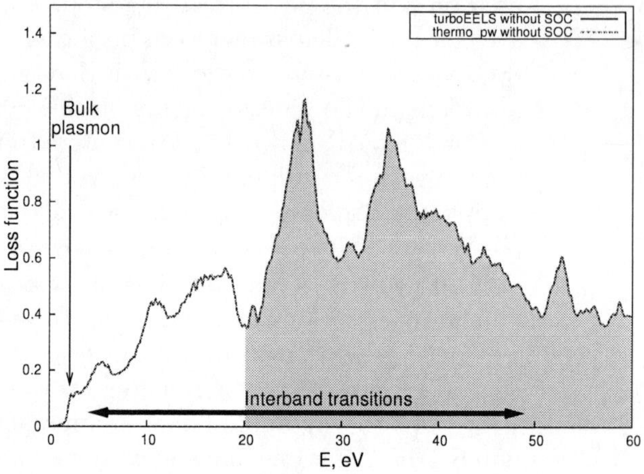

Figure 2: Bulk Au. Loss function for a vanishing transferred momentum is the x-direction, calculated with the *turboEELS* code (——) and *thermo_pw* code (- - - -). Shaded area marks the energy region where our pseudopotential is less accurate and arrows indicate the position of bulk plasmon and the region on interband transitions.

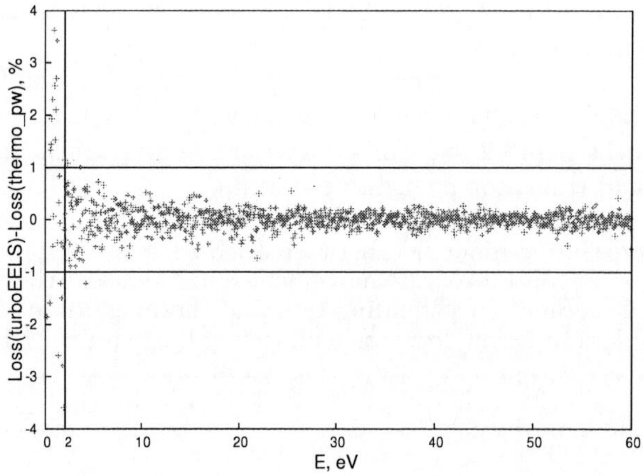

Figure 3: Bulk Au. Relative difference between the loss functions computed for a vanishing transferred momentum is the x-direction, calculated with the *turboEELS* and *thermo_pw* codes.

3.3. Performance

In figure 4, we show the comparison of the performance of both codes in terms of CPU time for bulk Bi without SOC. One can see that the *turboEELS* code is more efficient than the *thermo_pw* code on a wide frequency range. We note that calculations in a low frequency region (0-10 eV) are considerably faster, by a factor of 2-3, than in the middle and high frequency region (10-30 eV).

In table 1, we show the amount of time, number of CPU cores and number of frequencies used

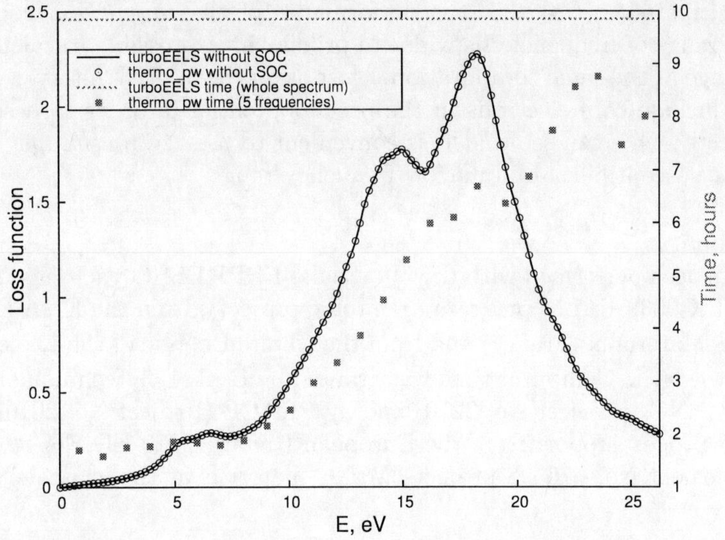

Figure 4: Bulk Bi. Performance comparison of *turboEELS* and *thermo_pw* codes without spin-orbit coupling. Loss functions obtained with *turboEELS* (——) and *thermo_pw* (○) codes are shown (left-hand side abscissa). Time required for *turboEELS* (- - - -) to compute the whole spectrum and for *thermo_pw* (■) to compute the loss function for five frequencies in the corresponding frequency range is shown (right-hand side abscissa).

to obtain curves in fig. 2 and 4. In the last column of table 1, the number of frequency is given, showing on average how many frequencies can be calculated with *thermo_pw* when consuming the same amount of CPU time needed for *turboEELS* to calculate the whole spectra. We point out that the *ratio* may however depend on the selected frequency range.

Performance comparison for the calculation with SOC is not shown in the present work. Taking SOC into account roughly increases the required CPU time by a factor of 4 for both codes, without influencing their relative performance.

Table 1: Performance comparison of the *turboEELS* and *thermo_pw* codes for bulk Bi and Au.

System	# of CPU cores	# of frequencies (ω)	Time (h)	average # of ω for *thermo_pw* to match *turboEELS* CPU time
Bi *turboEELS*	72	any	10	
Bi *thermo_pw*	72	130	118	11
Au *turboEELS*	240	any	0.75	
Au *thermo_pw*	240	600	16	28

4. Conclusions

In this work we have done a comparison between results and performance of the codes and corresponding methods that allow us to calculate EEL spectra. It has been found that results obtained using the Sternheimer equation and the Liouville-Lanczos approach are in perfect agreement with each other for semiconducting (not shown), semimetallic and metallic samples.

However, as anticipated in Sec. 2.4, the performance of the *turboEELS* code is significantly better when the range of frequencies is wide. In principle it is possible to calculate EEL spectra on a 0-250 eV range at the same computational cost as on the 0-25 eV interval, although it might be necessary to include core electrons in the pseudopotential in order to describe high-energy excitations correctly. One can see that it is convenient to use the *thermo_pw* code for a precise calculation of the susceptibility in a narrow frequency range.

Acknowledgement

Calculations have been performed with the QUANTUM ESPRESSO packages [25]. Support from the ANR-10-LABX-0039-PALM program (Femtonic project), from the DGA (France), from the program NEEDS-Matériaux (France) and from the Chaire Énergie of the École Polytechnique is gratefully acknowledged. Computer time was granted by École Polytechnique through the LLR-LSI project, by GENCI (Project No. 2210) and by PRACE (Project No. 2010PA3750). Andrea Dal Corso acknowledges support from the European Union H2020-EINFRA-2015-1 programme under grant agreement No. 676598 project "MaX - materials at the exascale".

References

[1] Egerton R F 1996 *Electron Energy-Loss Spectroscopy in the Electron Microscope* 2nd ed (New York and London: Plenum)
[2] Botti S, Schindlmayr A, Sole R D and Reining L 2007 *Rep. Prog. Phys.* **70** 357
[3] Timrov I, Markov M, Gorni T, Raynaud M, Motornyi O, Gebauer R, Baroni S and Vast N 2017 *Phys. Rev. B* **95** 094301
[4] See in particular Ref. 9-14 and 28-48 of Ref. [3]
[5] Olevano V and Reining L 2001 *Phys. Rev. Lett.* **86** 5962
[6] Runge E and Gross E 1984 *Phys. Rev. Lett.* **52** 997
[7] Gross E K U and Kohn W 1985 *Phys. Rev. Lett.* **55** 2850
[8] Car R, Tosatti E, Baroni S and Leelaprute S 1981 *Phys. Rev. B* **24** 985
[9] Baroni S, de Gironcoli S, Corso A D and Giannozzi P 2001 *Rev. Mod. Phys.* **73** 515
[10] de Gironcoli S 1995 *Phys. Rev. B* **51** 6773
[11] Methfessel M and Paxton A 1989 *Phys. Rev. B* **40** 3616
[12] Rocca D, Gebauer R, Saas Y and Baroni S 2008 *J. Chem. Phys.* **128** 154105
[13] Timrov I, Vast N, Gebauer R and Baroni S 2014 *Phys. Rev. B* **88** 064301
[14] Timrov I, Vast N, Gebauer R and Baroni S 2015 *Computer Physics Communications* **06/2015** dOI: 10.1016/j.cpc.2015.05.021
[15] Onida G, Reining L and Rubio A 2002 *Rev. Mod. Phys.* **74** 601
[16] Hohenberg P and Kohn W 1964 *Phys. Rev.* **136** B864
[17] Kohn W and Sham L 1965 *Phys. Rev.* **140** A1133
[18] Perdew J P and Zunger A 1981 *Phys. Rev. B* **23** 5048
[19] Perdew J, Burke K and Ernzerhof M 1996 *Phys. Rev. Lett.* **77** 3865
[20] Monkhorst H and Pack J 1976 *Phys. Rev. B* **13** 5188
[21] Markov M, Sjakste J, Fugallo G, Paulatto L, Lazzeri M, Mauri F and Vast N 2016 *Phys. Rev. B* **93** 064301
[22] Pseudopotential was generated using parameters provided in `http://bohr.inesc-mn.pt/~jlm/pseudo.html`
[23] Code can be obtained from `http://people.sissa.it/~dalcorso/thermo_pw_dist.html`
[24] Giannozzi P, Baroni S, Bonini N, Calandra M, Car R, Cavazzoni C, Ceresoli D, Chiarotti G, Cococcioni M, Dabo I, Dal Corso A, De Gironcoli S, Fabris S, Fratesi G, Gebauer R, Gerstmann U, Gougoussis C, Kokalj A, Lazzeri M, Martin-Samos L, Marzari N, Mauri F, Mazzarello R, Paolini S, Pasquarello A, Paulatto L, Sbraccia C, Scandolo S, Sclauzero G, Seitsonen A, Smogunov A, Umari P and Wentzcovitch R 2009 *J. Phys.: Condens. Matter* **21** 395502
[25] Giannozzi P, Andreussi O, Brumme T, Bunau O, Nardelli M B, Calandra M, Car R, Cavazzoni C, Ceresoli D, Cococcioni M, Colonna N, Carnimeo I, Corso A D, de Gironcoli S, Delugas P, Ferretti R A D J A, Floris A, Fratesi G, Fugallo G, Gebauer R, Gerstmann U, Giustino F, Gorni T, Jia J, Kawamura M, Ko H Y, Kokalj A, Küçükbenli E, Lazzeri M, Marsili M, Marzari N, Mauri F, Nguyen N L, Nguyen H V, de-la Roza A O, Paulatto L, Poncé S, Rocca D, Sabatini R, Santra B, Schlipf M, Seitsonen A P, Smogunov A, Timrov I, Thonhauser T, Umari P, Vast N, Wu X and Baroni S 2017 *J. Phys.: Condens. Matter* **29** 465901 URL `http://www.quantum-espresso.org`
[26] Alkauskas A, Schneider S D, Hébert C, Sagmeister S and Draxl C 2013 *Phys. Rev. B* **88** 195124

Time evolution of an unstable soliton solution to dust acoustic plasma with trapped electrons

Yuttakarn Rattanachai[1] and Sarun Phibanchon[2]

[1] Faculty of Sciences and Liberal Arts, Rajamangala University of Technology Isan
744 Saranarai Road, Muang, Nakorn Ratchasima, 30000, Thailand
[2] Faculty of Education, Burapha University
169 Longhardbangsaen road, Sansuk, Muang, Chonburi, 20131, Thailand

E-mail: sarunp@go.buu.ac.th

Abstract. The positive dust charged grains is considered as the immobile particles compared to ions and electrons. Using the assumption that some electrons can be trapped in the dust charge potential while others move freely in the plasma with external magnetic field. The reductive perturbation method shows that the weakly nonlinear waves equation gives a soliton solution. The spectral method is applied for the time evolution of the perturbed soliton solution to the sinusoidal function with the long-wavelength. This shows that the unstable plane soliton will be transformed into the higher dimensions.

1. Introduction

Plasma is the fourth state of the matter which provides the variety of the oscillations. It is also believed that our universe consists more than 60% of this state. [1]. However, in general, there are various of sizes and charges in plasmas as well as a neutral particles. These neutral particle can also be charged as both the negative and positive charges [2] called the dusty plasmas. There are many examples of dusty plasmas which found in planetary rings, cometary comae and tails as well as in asteroid zones and mesosphere and magnetosphere [2, 3]. As like the plasmas, dusty plasmas contains many modes of oscillations such as dust acoustic (DA) waves [4], where dust mass gives the inertia and the thermal pressures to the electrons and dust ion acoustic waves (DIA) [5]. The dusty plasma is also found in man-made laboratories such as a dust in fusion devices or during the processes of microelectronic fabrication. We have found this phenomenon in various physical systems, this is the main reason why some recently works are dedicated to the dusty plasmas. One remarkable phenomena that can be found in the weakly nonlinear ion-acoustic waves is the soliton [6]. There are many scientific works on this phenomenon [7] as well as DA and DIA [2, 3]. To obtain the soliton equation, the electron distribution function can be described as the Maxwellian function [6, 8] or non-thermal case [9, 10, 11, 12].

Soliton is not only found in plasmas but also in others continuous media such as fiber optics [13, 14], surface and internal water waves [15] or superfluid ^{4}He [16]. The main characters of soliton are traveling without change its structure and particle-like collisions.

The environment of the dusty plasma mainly contains the electrons, positive ions and obviously some dust particles. The interaction between charged particles and dust particles gives a charging of the dust grains. The size of dust particles is ranged from ten nanometers to hundreds of microns. They are quite big compare to the electrons and ions. To observe the dusty acoustic waves, we will only consider the dust charged grains.

In this work we propose that some positive charged dust grains [17] with some trapped electrons in their electric potentials. This situation can be provided the soliton equation for the weakly nonlinear wave equation. The transverse perturbation with a long-wavelength to the soliton solution gives the transformation of the perturbed plane soliton to the higher dimensional soliton.

2. Dusty plasma with trapped electrons

To obtain the weakly nonlinear equation for the dusty plasma with some trapped electrons [9, 18], we will apply the reductive perturbation method on the governed equations for the dust particles. These equations are mass continuity, momentum and Poisson equations in which are written as follows

$$\frac{\partial n}{\partial t} + \boldsymbol{\nabla} \cdot (n\mathbf{v}) = 0, \tag{1}$$

$$\frac{\partial \mathbf{v}}{\partial t} + (\mathbf{v} \cdot \boldsymbol{\nabla})\mathbf{v} = -\boldsymbol{\nabla}\phi + \mathbf{v} \times \Omega\hat{x}, \tag{2}$$

and

$$\boldsymbol{\nabla}^2\phi = \mu_e n_e - \mu_i n_i - n_d, \tag{3}$$

respectively, where n_i, n_e and n_d are ion, electron and dust particle density, respectively, $\Omega = B\sqrt{\epsilon_0/n_{d0}m_d}$, n_{d0} is an unperturbed dust particle density and m_d is a dust particle mass. The unperturbed number of electrons residing on the electrons and positive ions are μ_e and μ_i, respectively. All these quantities are dimensionless, they are normalized by the following characteristic quantities: n_d by the n_{d0}, v by the ion sound velocity $(k_BT_e/m_i)^{1/2}$, ϕ by (k_BT_e/e), t by the plasma frequency, $\omega_{pi} = (4\pi e^2 n_{e0}/m_i)^{1/2}$ and x by the electron Debye length, $\lambda_d = (k_BT_e/4\pi e^2 n_{e0})^{1/2}$, where k_B denotes the Boltzmann's constant. On the assumption that some electrons can be trapped into the dust charged potential, the electron distribution function can be considered as followed Schamel's works [9, 18],

$$n_e = e^{\phi}\,\mathrm{erfc}(\sqrt{\phi}) + \frac{1}{\sqrt{\beta}}\begin{cases} e^{\beta\phi}\,\mathrm{erf}(\sqrt{\beta\phi}) & ;\beta \geqslant 0 \\ \frac{2}{\sqrt{\pi}}W(\sqrt{-\beta\phi}) & ;\beta < 0 \end{cases} \tag{4}$$

where $W(x)$ is a Dawson function,

$$W(x) = e^{x^2}\int_0^x e^{-t^2}dt,$$

and $\beta = T_{ef}/T_{et}$ denotes the ratio of temperatures between free and trapped electrons. The stretch variables is introduced as follows,

$$\tau = \epsilon^{3/2}t, \quad \xi = \epsilon^{1/2}(x-t), \quad \chi = \epsilon^{1/2}y, \text{ and } \quad \mu = \epsilon^{1/2}z, \tag{5}$$

where ϵ is a small parameter to measure the weakness of the wave amplitude, $0 < \epsilon < 1$. The dependent variables (n, $\mathbf{v}$ and ϕ) are expanded as

$$\begin{aligned} n &= 1 + \epsilon n_1 + \epsilon^2 n_2 + \ldots \\ \phi &= \epsilon\phi_1 + \epsilon^2\phi_2 + \ldots \\ v_x &= \epsilon v_{x1} + \epsilon^2 v_{x2} + \ldots \\ v_y &= \epsilon^{3/2} v_{y1} + \epsilon^2 v_{y2} + \ldots \\ v_z &= \epsilon^{3/2} v_{z1} + \epsilon^2 v_{x2} + \ldots \end{aligned} \tag{6}$$

and consider (4) and n_i as

$$n_e(\phi) = 1 + \phi - \frac{4}{3}b\phi^{3/2} + \frac{\phi^2}{2} + O(\phi^{5/2}),$$
$$n_i(\phi) = 1 + \phi + \frac{\phi^2}{2} + O(\phi^3). \tag{7}$$

where $b = (1-\beta)/\sqrt{\pi}$. We substitute (5)-(7) into (1)-(3), the first order, ϵ, can be expressed as

$$-\frac{\partial n_1}{\partial \xi} + \frac{\partial v_{x1}}{\partial xi} = 0, \quad \frac{\partial v_{x1}}{\partial \xi} = \frac{\partial \phi_1}{\partial xi},$$
$$\frac{\partial \phi_1}{\partial \chi} = \Omega v_{z1}, \quad \frac{\partial \phi_1}{\partial \mu} = \Omega v_{y1}, \quad \phi_1 = n_1$$

and three-half order, $\epsilon^{3/2}$, can be shown as

$$-\frac{\partial v_{y1}}{\partial \chi} = -\frac{\partial v_{z1}}{\partial \mu}, \quad \frac{\partial v_{y1}}{\partial \xi} = -\Omega v_{z2}, \quad \frac{\partial v_{z1}}{\partial \xi} = \Omega v_{y2}.$$

We next consider the second order, ϵ^2, including $b = \epsilon^{1/2}\hat{b}$,

$$\frac{\partial n_1}{\partial \tau} - \frac{\partial n_2}{\partial \xi} + \frac{\partial}{\partial \xi}[v_{x2} + n_1 v_{x1}] + \frac{\partial v_{y2}}{\partial \chi} + \frac{\partial v_{z2}}{\partial \mu} = 0,$$
$$\frac{\partial v_{x1}}{\partial \tau} - \frac{\partial v_{x2}}{\partial \xi} + v_{x1}\frac{\partial v_{x1}}{\partial \xi} = -\frac{\partial \phi_2}{\partial \xi}$$
$$\frac{\partial^2 \phi_1}{\partial \xi^2} + \frac{\partial^2 \phi_1}{\partial \chi^2} + \frac{\partial^2 \phi_1}{\partial \mu^2} = \mu_e\left[\phi_2 - \frac{4b}{3}\phi_1^{3/2} + \frac{\phi_1^2}{2}\right] - \mu_i\left[\phi_2 + \frac{\phi_1^2}{2}\right] - n_2.$$

We can now rearrange all these equations and express in terms of the first order variables,

$$2\frac{\partial \phi_1}{\partial \tau} + 2\phi_1\frac{\partial \phi_1}{\partial \xi} + 2b\mu_e\phi_1^{1/2}\frac{\partial \phi_1}{\partial \xi} + \frac{\partial^3 \phi_1}{\partial \xi^3} + \left(1 + \frac{1}{\Omega^2}\right)\frac{\partial}{\partial \xi}\left[\frac{\partial^2 \phi_1}{\partial \chi^2} + \frac{\partial^2 \phi_1}{\partial \mu^2}\right] = 0.$$

This is a weakly nonlinear equation for the dusty plasma with trapped electrons. For more convenience, we introduce some new variables to eliminate all coefficients,

$$\phi_1 = \Gamma^2\phi, \quad \tau = \frac{t}{\Gamma^3\sqrt{2}}, \quad \xi = \frac{x}{\Gamma\sqrt{2}}, \quad \chi = \frac{\sqrt{\sigma}y}{\Gamma}, \quad \mu = \frac{\sqrt{\sigma}z}{\Gamma},$$

where $\sigma = (1 + 1/\Omega^2)/2$ and $\Gamma = b\mu_e$, the dimensionless form can be written as

$$\frac{\partial \phi}{\partial t} + \phi\frac{\partial \phi}{\partial x} + \phi^{1/2}\frac{\partial \phi}{\partial x} + \frac{\partial^3 \phi}{\partial x} + \frac{\partial}{\partial x}\left[\frac{\partial^2 \phi}{\partial y^2} + \frac{\partial^2 \phi}{\partial z^2}\right] = 0. \tag{8}$$

We will refer this equation as Schamel-Korteweg-de Vries-Zakharov-Kutznetsov (SKdVZK) equation, this equation also admits the soliton solution [19] for 1D case, namely,

$$\phi = \left(\frac{60\eta^2}{1 + \sqrt{1 + 75\eta^2}\cosh 2\eta(x - x_0)}\right)^2. \tag{9}$$

IOP Conf. Series: Journal of Physics: Conf. Series **1136** (2018) 012009 doi:10.1088/1742-6596/1136/1/012009

3. Time evolution of an unstable soliton

To study the unstable soliton, we will apply the long-wavelength function to the soliton solution in the perpendicular direction to the soliton direction [20, 21, 19]. This ansatz can be used as the initial condition for the time evolution,

$$\phi = \phi_0 + \epsilon u(x) e^{iky+\gamma t},$$

where γ determines the growth rate of soliton, ϵ denotes a small number, $\epsilon << k$, $u(x)$ can be any real functions and ϕ_0 is the unperturbed soliton, (9). After substitute this ansatz into (8) and collect to the 1-st order of ϵ, we then have

$$\frac{d}{dx}(Lu(x)) = -\gamma u(x), \tag{10}$$

where

$$L = \frac{d^2}{dx^2} + \phi_0 + \phi_0^{1/2} - (c - k^2).$$

c denotes the wave speed and ϕ_0 is the equation in (9). Moreover, there is another method called the variational method [22] for calculating the growth rate. This method is mostly used for the cubic nonlinear Schrödiger equation and its modification [23, 24]. The Lagrangian density will be constructed as

$$\mathcal{L} = \frac{\psi_x \psi_t}{2} - \frac{c\psi_x^2}{2} + \frac{4}{15}\psi_x^{5/2} + \frac{\psi_x^3}{6} + \frac{\psi_{xx}^2}{2} + \frac{\psi_{xy}^2}{2}, \tag{11}$$

where $\psi_x = \phi(x)$. To calculate the growth rate, we need to guess the trial function for (11). For these methods, the growth rate can be found all the possible values of k to make soliton unstable. To evaluate this result, we will apply the spectral method to SKdVZK equation. The basic idea is to apply a Fourier transformation, for the 2D case,

$$\Phi_{p,q} = F(\phi_{l,m}) = \sum_{l=0}^{N_x-1}\sum_{m=0}^{N_y-1} \phi_{l,m} e^{\mathrm{i}(\xi_p x_l + \chi_q y_m)},$$

where N_x and N_y are the number of mesh points in the x and y direction, respectively. $(x_l, y_m) = (lL_x/N_x, mL_y/N_y)$, in which L_x and L_y are the lengths of the domain in the x and y directions, respectively. $(\xi_p, \chi_q) = 2\pi(p/L_x, q/L_y)$ for $p = 0, \ldots, N_x - 1$ and $q = 0, \ldots, N_y - 1$. The SKdVZK after taking a Fourier transform is written as

$$\Phi_t + F(\phi\phi_x + \phi^{1/2}\phi_x) - \mathrm{i}(\xi^3 + \xi\chi^2)\Phi = 0.$$

For the temporal space, the Runge-Kutta method [25] is applied to the time derivative and the initial condition is set as

$$\phi_0 = \left(\frac{60\eta^2}{1+\sqrt{1+75\eta^2}\cosh 2\eta(x-x_0)}\right)^2 \left(1 + \epsilon\cos\left(\frac{2\pi y}{\lambda}\right)\right),$$

where x_0 denotes a starting point and the sinusoidal function is used for the long wave-length function. All parameters are set for time evolution is $L_x \in [-25, 25]$ and $L_y \in [-25, 25]$, $x_0 = 0$, $\eta = 0.2$, $dt = 0.0005$, $N_x = 128$ and $N_y = 128$. The periodic boundary conditions in x-direction has also been applied. The initial perturbed plane soliton is shown in Figure 1(1). Figure 1(2)-(4) show the evolution of the unstable soliton, and become the higher solitons as shown in Figure 1(5)-(6), this 2D soliton moves faster than the unperturbed plane soliton. There are also others two solitons found because there are possible space that more solitons can be existed.

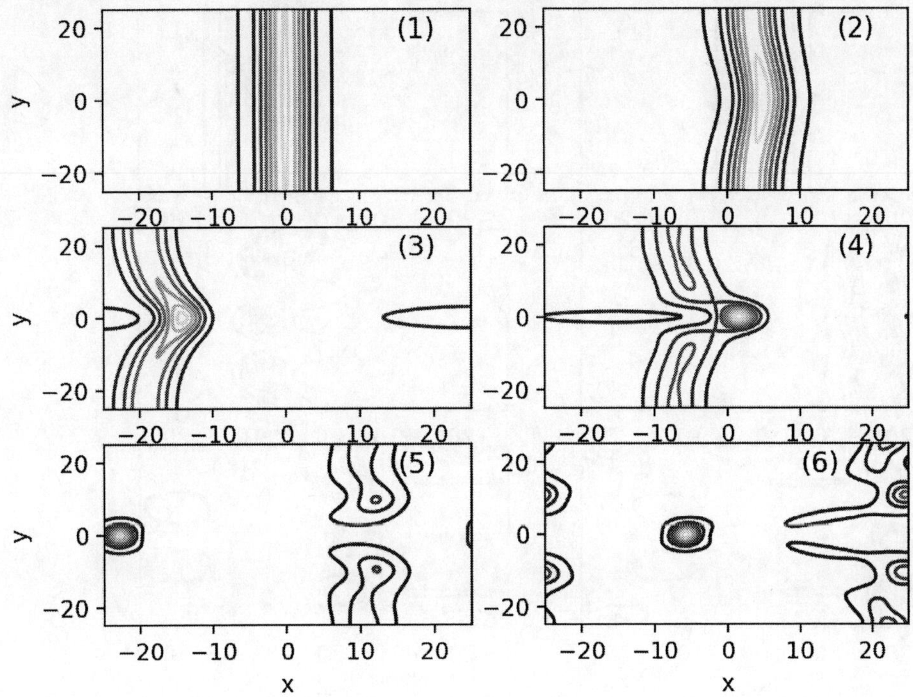

Figure 1. Time evolution of an unstable soliton with $\eta = 0.2$ with (1) t=0, (2) t= 85, (3) t= 130, (4) t= 150, (5) t = 180 and (6) t = 200.

To obtain more 2D solitons, we double the period of the sinusoidal function and the transformation of unstable soliton is shown in Figure 2. For this case, 3 solitons emerge simultaneously.

4. Conclusion

The reductive perturbation method can be used to obtain the soliton equation for the weakly nonlinear wave equation. The long wavelength perturbation method is used to determine the unstable growth rate of the soliton. The unstable dust ion-acoustic soliton will be transformed into the higher dimensional solitons. The number of higher solitons depends on the length of perpendicular direction. The higher soliton dimension might be useful for study some behaviors such as their collisions [26].

5. Acknowledgments

SP wishes to acknowledge support from Faculty of Education, Burapha University. YR would like to thank the Higher Education Commission for the support.

6. References

[1] Chen F 2016 *Introduction to Plasma Physics and Controlled Fusion, 3rd Ed.* (Spinger)

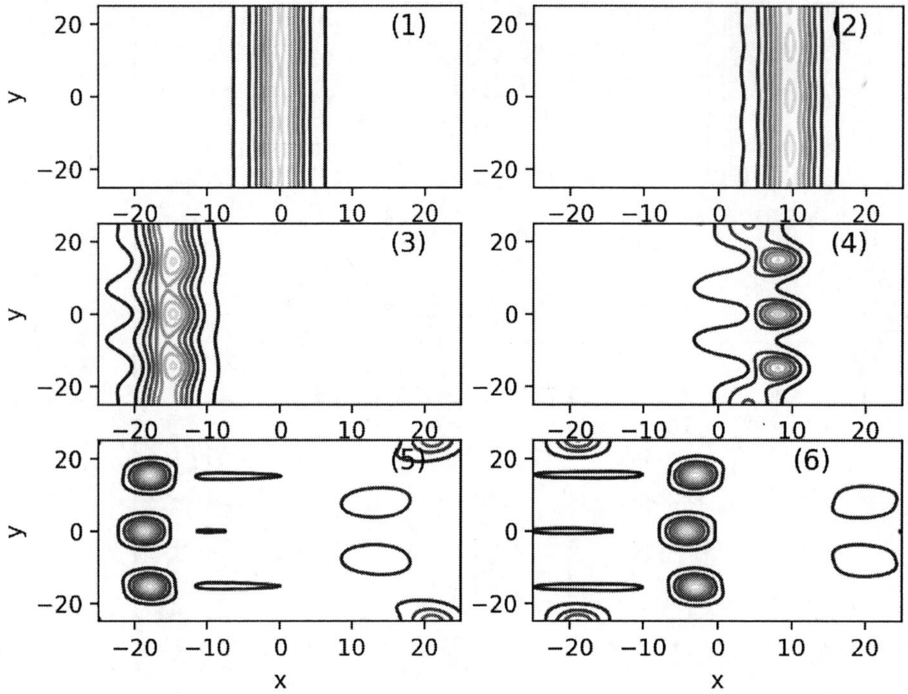

Figure 2. Time evolution of an unstable soliton with $\eta = 0.2$ with (1) t=0, (2) t= 15, (3) t= 55, (4) t= 88, (5) t = 120 and (6) t = 140.

[2] Shukla P and Mamun A 2002 *Introduction to dusty plasma physics* (IOP)
[3] Verheest F 2000 *Waves in Dusty Space Plasmas* (Kluwer Academics)
[4] Rao N, Shukla P and Yu M 1990 *Planet. Space Sci* **38** 543
[5] Barkan A, D'Angelo N and Merlino R 1996 *Planetary and Space Science* **44** 239–242
[6] Washimi H and Taniuti T 1966 *Phys. Rev. Lett.* **17** 996–8
[7] Infeld E and Rowlands G 2000 *Nonlinear Waves, Solitons and Chaos* 2nd ed (Cambridge: Cambridge University Press)
[8] El-Wakil S A, Zahran M A, El-Shewy E K and Mowafy A E 2006 *Physica Scripta* **74** 503–509
[9] Schamel H 1972 *Plasma Phys.* **14** 905–24
[10] Cairns R A, Mamun A A, Bingham R and Shukla P K 1996 *Phys. Scr.* **T63** 80–6
[11] Anowar M and Mamun A 2009 *IEEE Transactions on Plasma Science* **37** 1638–1645
[12] Misra A and Wang Y 2015 *Communications in Nonlinear Science and Numerical Simulation* **22** 1360–1369
[13] Agrawal G 2001 *Nonlinear fiber optics* 3rd ed (Academic Press)
[14] Mollenauer L F and Gordon J P 2006 *Solitons in Optical Fibers: Fundamentals and Applications* (ElsevierAcademic Press)
[15] Khusnutdinova K R, Stepanyants Y A and Tranter M R 2018 *Physics of Fluids* **30** 022104
[16] Ancilotto F, Levy D, Pimentel J and Eloranta J 2018 *Physical Review Letters* **120**
[17] Baluku T K, Hellberg M A and Mace R L 2008 *Physics of Plasmas* **15** 033701
[18] Schamel H 1973 *J. Plasma Phys.* **9** 377–87
[19] Allen M A, Phibanchon S and Rowlands G 2007 *JPP* **73** 215–29
[20] Rowlands G 1969 *J. Plasma Phys.* **3** 567–76
[21] Allen M A and Rowlands G 1993 *J. Plasma Phys.* **50** 413–24
[22] Bettinson D C and Rowlands G 1998 *J. Plasma Phys.* **59** 543–54
[23] Anderson D, Lisak M and Berntson A 2001 *Pramana J. Phys.* **57** 917–36

[24] Phibanchon S and Rattanachai Y 2013 The soliton instability to the quintic nonlinear Schrödinger equation *Burapha University International Conference* pp 621–627
[25] Press W H, Teukolsky S A, Vetterling W T and Flannery B P 1992 *Numerical Recipes in C* 2nd ed (Cambridge: Cambridge University Press)
[26] Phibanchon S and Allen M A 2007 Time evolution of perturbed solitons of modified Kadomtsev-Petviashvili equations *The 2007 International Conference on Computational Science and its Applications (ICCSA 2007)* (Los Alamitos, CA, USA: IEEE Computer Society) pp 20–3

IOP Conf. Series: Journal of Physics: Conf. Series **1136** (2018) 012010 doi:10.1088/1742-6596/1136/1/012010

Advanced computational approach to studying Rydberg and autoionization resonances in spectra of lanthanides: Ytterbium

V B Ternovsky, A A Kuznetsova, E V Ternovsky, D A Mironenko, A V Smirnov

Odessa State Environmental University, L'vovskaya str.15, Odessa-16, 65016, Ukraine

E-mail: ternovskyvb@gmail.com

Abstract. We present the results of computing the Rydberg autoionization resonances in spectra of the ytterbium atom, in particular, data on the energies and widths of the autoionization resonances $4f^{13}$ $[^2F_{7/2}]6s^2np[5/2]_2$, $4f^{13}$ $[^2F_{7/2}]6s^2nf[5/2]_2$, which are arisen in result of excitation of the 4f-shell electrons. For some values of the principal quantum number the presented data are compared with the corresponding available experimental data by Jong-Hoon Yi et al, which are obtained by method of three-photon laser polarization spectroscopy. As the method of computing we apply the method of a relativistic many-body perturbation theory with the Dirac-Kohn-Sham zeroth approximation combined with the generalized relativistic energy approach. This approach takes effectively into account the important exchange-correlation corrections as the corrections of the perturbation theory second and higher orders (for example, the polarization interaction, quasiparticles screening corrections) and effect of the non-Coulomb grouping levels in the Rydberg spectra etc.

1. Introduction

The intensive experimental and theoretical investigations in the field of astrophysical analysis, laboratory, thermonuclear plasma diagnostics, fusion research etc stimulate a great traditional interest to accurate computing the spectra, radiative and autoionization decay characteristics of the Rydberg atomic (atoms in the highly excited states with large values of the principal quantum number $n >> 1$) systems (see, for example [1-10]). It is also provided by a great number of different important applications in laser physics and quantum electronics. For example, it is worth to mention new possible lasing generation schemes in the short-wave range with using the Rydberg systems, different topics of atomic and molecular optics and spectroscopy, quantum informatics and computing etc. At last, it is also important to remind about such unique and interesting nonlinear optical phenomena such as the Bose-condensate in vapors of the Rydberg alkali-metal atoms, the Rydberg matter, fountains of the cold Rydberg atoms, atomic (nuclear) clocks etc. At present time the Rydberg atoms and ions of relatively light elements are intensively and quite adequately studied, however, the situation became more dramatic for the Rydberg atoms and ions of the heavy and superheavy elements.

Over the last several decades, a large number of theoretical studies of atomic structure has been performed using the traditional comprehensive methods such as the well-known Hartree-Fock (HF) and Dirac-Fock (DF) methods in different versions, methods of the many-body perturbation theory (PT) with the HF and DF zeroth approximations, density functional theory, R-matrix, coupled-cluster theories, and more simplified approaches such as the quantum defect and Coulomb approximations, pseudo- and model potential methods, the classical and quasiclassical models and many others (see, for example, [6-21] and Refs. therein). These methods have been successfully used in computing the spectroscopic properties of different light and middle Rydberg atoms. In a case of the heavy Rydberg atoms in a free state or in an external electromagnetic field a modern level of description of the Rydberg atoms is not sufficiently satisfactory. A precise accounting for the relativistic and exchange-correlation effects, including an effect of the non-Coulomb grouping levels in the Rydberg spectra (the effect, which, as a rule, is not correctly treated within simplified Coulomb and quantum defect models) is of a great interest and importance [6,8-10].

This paper is devoted to theoretical studying the Rydberg autoionization resonances in spectra of heavy elements, namely, the ytterbium atom. The ytterbium atom as the other lanthanides elements is intensively studied [22-36], nevertheless, for example, the most accurate data on spectral characteristics of the Rydberg states are absent hitherto.

As the method of computing the autoionization resonances spectra we apply the method of formally exact relativistic many-body perturbation theory with the Dirac-Kohn-Sham (DKS) zeroth approximation combined with a generalized energy approach (version [37-45]). This approach allows to take effectively into account the exchange-correlation corrections of the perturbation theory second order and higher orders (polarization interaction, quasiparticles screening, etc.). The wave function zeroth basis is found from the Dirac equation with ab initio potential, which includes the optimized Dirac-Kohn-Sham potential. The correlation corrections of the PT high orders are taken into account within the Green functions method with using the Feynman diagram's technique. These corrections correspond to the effects of electron screening, particle-hole interactions, mass operator iterations etc. It is important to note that the method is constructed on the base of the same ideas as the well-known formalism of the PT with the model potential zeroth approximation by Ivanov-Ivanova et al [33-35,46-50]. However, there are two key differences; namely, the PT zeroth approximation [20,64-70] is the Dirac-Kohn-Sham one. In order to calculate the radiative transition and autoionization resonances parameters a new version of relativistic energy approach is used. It generalizes the known model relativistic energy approach in a case of the multielectron atom that has been proposed by Ivanov-Ivanova *et al* [34-36, 50-52]. A generalized gauge-invariant relativistic energy approach in a case of the multielectron atomic systems has been developed by Glushkov-Ivanov-Ivanova et al [52-60].

We present the results of the test computing energies and widths of the excited (autoionization) states for the Yb 7s6p configuration and compare these data with the available experimental (compilation) and theoretical data, obtained within different theoretical approaches such as multiconfiguration DF method, the relativistic many-body PT with the model potential zeroth approximation by Ivanov-Ivanova et al [33,34,49,50]. For the first time we list our theoretical data on the energies and widths of the autoionization resonances $4f^{13}$ $[^2F_{7/2}]6s^2np[5/2]_2$, $4f^{13}$ $[^2F_{7/2}]6s^2nf[5/2]_2$ and compare these data with the corresponding available experimental data by Jong-hoon Yi et al [31,32].

2. Relativistic energy approach and optimized relativistic many-body perturbation theory

The method is based on the convenient field procedure, which includes computing the energy shifts ΔE of the degenerate electron states. More exactly, speech is about constructing secular matrix M (with using the Gell-Mann and Low adiabatic formula for ΔE), which is already complex in the relativistic theory, and its further diagonalization. It allows to compute the energies and decay probabilities of a non-degenerate excited state for a complex atomic system [46-50]. The secular matrix elements can be further expanded into a PT series on the interparticle interaction. Further the standard Feynman diagrammatic technique is usually used. Generally speaking, the secular matrix M can be represented as follows:

$$M = M^{(0)} + M^{(1)} + M^{(2)} + M^{(3)}, \tag{1}$$

where $M^{(0)}$ is the contribution of the vacuum diagrams of all PT orders (this contribution determines only the general levels spectrum shift); $M^{(1)}$, $M^{(2)}$, $M^{(3)}$ are contributions of the 1-, 2- and 3-

quasiparticle (QP) diagrams respectively. The matrix $M^{(1)}$ can be presented as a sum of the independent one-QP contributions. Substituting these quantities into (1) one could have summarized all the one-QP diagrams contributions. In the empirical methods here the experimental values of one-electron energies are usually used, however, the necessary experimental quantities (especially for the rare-earth and other elements) are not often available. The detailed procedure for computing $\mathrm{Re}\,M^{(2)}$ is presented, for example, in Refs. [9,33-36,46-50].

The relativistic Dirac Hamiltonian of an atomic multielectron system is as follows (the atomic units are used) [9]:

$$H = \sum_i \{\alpha c p_i - \beta c^2 - Z/r_i\} + \sum_{i>j} \exp(i|\omega|r_{ij})(1-\alpha_i\alpha_j)/r_{ij} \tag{2}$$

where Z is a charge of nucleus, α_i ,α_j are the Dirac matrices, ω_{ij} is the transition frequency, c – the velocity of light.

The interelectron interaction potential (second term in (3)) takes into account the retarding effect and magnetic Breit interaction in the lowest order on parameter of the fine structure constant α. In the PT zeroth approximation we used the following mean-field potential:

$$V_{MF} = V^{DKS}(r) = [V^{D}_{Coul}(r) + V_X(r) + V_C(r|a)] \tag{3}$$

with the standard Coulomb V^{D}_{Coul}, generalized exchange V_X and correlation Vc potentials [15-18]. In particular, the Kohn-Sham exchange potential looks as [15,16]:

$$V_X[\rho(r),r] = V_X^{KS}(r)\cdot\{\frac{3}{2}\ln\frac{[\beta+(\beta^2+1)^{1/2}]}{\beta(\beta^2+1)^{1/2}} - \frac{1}{2}\}, \tag{4}$$

where $\beta = [3\pi^2\rho(r)]^{1/3}/c$. The corresponding generalized Lundqvist-Gunnarsson correlation potential is as follows [17,18]:

$$V_C[\rho(r),r] = -0.0333\cdot b\cdot\ln[1+18.3768\cdot\rho(r)^{1/3}], \tag{5}$$

where b is the optimization parameter [53].

The most complicated problem of the relativistic PT computing the rare-earth elements spectra is in an accurate, precise accounting for the multi-electron exchange-correlation effects (including polarization and screening effects, a continuum pressure etc), which can be treated as the effects of the PT second and higher orders. Using the standard Feynman diagrammatic technique one should analyze two kinds of diagrams (the polarization and ladder ones), which describe the polarization and screening exchange-correlation effects. The detailed description of the polarization diagrams and the corresponding analytical expressions for matrix elements of the polarization QPs interaction (through the polarizable core) potential is presented, for example, in Refs. [42-44,61-66]. An effective approach to accounting for the multi-electron polarization contributions is described earlier and based on using the effective relativistic two-quasiparticle polarizable operator [61] (generalization of the nonrelativistic operator [50]), which is included into the PT first order matrix elements.

It is important to remind that in the QED theory, the photon propagator D(12) plays the role of the interelectron interaction (look Eq.(1)) Naturally the analytical form of D(12) depends on the gauge, in which the electrodynamical potentials are written. The gauge dependence of the radiative amplitudes in the approximate calculations is a well known fact and is in details investigated by Grant, Armstrong, Aymar and Luc-Koenig, Ivanov-Ivanova-Glushkov et al (see [2-5,17,18,34-36,52-60] and Refs. therein). Grant has investigated the gauge connection with the limiting non-relativistic form of the transition operator and formulated the conditions for approximate functions of the states, in which the amplitudes of the photo processes are gauge invariant. Glushkov-Ivanov developed a new procedure of the accurate treating the PT lowest order multielectron effects, in particular, the gauge dependent radiative contribution for the certain class of the photon propagator gauge [52,53]. The corresponding contribution is considered to be the typical representative of the important multielectron exchange-correlation effects, whose minimization is a reasonable criteria in the searching for the optimal PT one-QP basis.

Below we will be interested by studying the spectra of the autoionization resonances in the ytterbium atom and calculating their energies and widths. Usually the excited states of the ytterbium atom can be treated as the states with the 2-QP (also 3-QP) above the electron core [Xe]$4f^{14}$. Within the energy approach [33,34,50-53] the radiative and autoionization widths are determined by the square of an electron interaction matrix element having the form:

$$V_{1234}^{\omega} = [(j_1)(j_2)(j_3)(j_4)]^{1/2} \sum_{\lambda\mu} (-1)^{\mu} \begin{pmatrix} j_1 j_3 & \lambda \\ m_1 - m_3 & \mu \end{pmatrix} \times Q_{\lambda}(1234), \tag{6}$$

$$Q_{\lambda} = Q_{\lambda}^{\rm Qul} + Q_{\lambda}^{\rm Br},$$

where j_i is the total single electron momentums, m_i – the projections; Q^{Qul} is the Coulomb part of interaction, Q^{Br} - the Breit part. For example, the Coulomb part $Q_{\lambda}^{\rm Qul}$ can be expressed in terms of the known radial integrals R_{λ} and angular coefficients S_{λ} [46-50]:

$$Q_{\lambda}^{\rm Qul} = \{R_l(1243)S_{\lambda}(1243) + R_{\lambda}(\tilde{1}2 4\tilde{3})S_{\lambda}(\tilde{1}24\tilde{3}) + R_{\lambda}(1\tilde{2}\tilde{4}3)S_{\lambda}(1\tilde{2}\tilde{4}3) + R_{\lambda}(\tilde{1}\tilde{2}\tilde{4}\tilde{3})S_{\lambda}(\tilde{1}\tilde{2}\tilde{4}\tilde{3})\} \tag{7}$$

The tilde in (7) designates that the large component of radial part of the one-QP Dirac function f must be replaced by the small one g. Below is given the example of the radial integral [4,62]:

$$R_{\lambda}(1243) = \iint dr_1 r_1^2 r_2^2 f_1(r_1) f_3(r_1) f_2(r_2) f_4(r_2) Z_{\lambda}^{(1)}(r_<) Z_{\lambda}^{(1)}(r_>). \tag{8}$$

where the function Z is connected with the standard Bessel functions as:

$$Z_{\lambda}^{(1)} = \left[\frac{2}{|\omega_{13}|\alpha Z}\right]^{\lambda+1/2} \frac{J_{\lambda+1/2}(\alpha|\omega_{13}|r)}{r^{\lambda}\Gamma(\lambda+3/2)}. \tag{9}$$

The angular coefficients are defined by standard way [50]. The Breit part of Q is defined in the similar way as above, but the contribution of our interest is a real part. According to the Ivanova-Ivanov et al method [34,35,46, 50,51], calculation of the integrals of the type (9) $R_{\lambda}(1243)$ is reduced to the solution of a system of differential equations:

$$\left.\begin{aligned} y_1' &= f_1 f_3 Z_{\lambda}^{(1)}(\alpha|\omega|r) r^{2+\lambda}, \\ y_2' &= f_2 f_4 Z_{\lambda}^{(1)}(\alpha|\omega|r) r^{2+\lambda}, \\ y_3' &= [y_1 f_2 f_4 + y_2 f_1 f_3] Z_{\lambda}^{(2)}(\alpha|\omega|r) r^{1-\lambda}. \end{aligned}\right\} \tag{10}$$

In addition it is easily to show that

$$y_3(\infty) = \mathbf{Re} R_{\lambda}(1243), \quad y_1(\infty) = X_{\lambda}(13). \tag{11}$$

The system of differential equations includes also equations for functions $f/r^{|æ|-1}$, $g/r^{|æ|-1}$, $Z_\lambda^{(1)}, Z_\lambda^{(2)}$. The formulas for the autoionization (Auger) decay probability include the radial integrals $R_\alpha(\alpha k\gamma\beta)$, where one of the functions describes electron in the continuum state. When calculating this integral, the correctly normalized function should be used. It is important also to note that the calculation is carried out in the *jj*-coupling scheme representation. The transition to the intermediate coupling scheme has been realized by diagonalization of the secular matrix. Indeed, only $\mathrm{Re}M$ should be diagonalized. The imaginary part is converted by means of the matrix of eigenvectors $\{C_{mk}\}$, obtained by diagonalization of $\mathrm{Re}M : \mathrm{Im}M_{mk} = \sum_{ij} C^*_{mi} M_{ij} C_{jk}$; M_{ij} are the matrix elements in the *jj*-coupling scheme, and M_{mk} in the intermediate coupling scheme representation. This procedure is correct to terms of the order of $\mathrm{Im}M/\mathrm{Re}M$ [26]. In order to reach the corresponding optimization of the relativistic wave functions, we have used the procedure [53], which derives an undoubted profit in the routine many-body calculations as it provides the way of the refinement of the atomic characteristics calculations, based on the "first principles". All computing was performed with using the modified PC code "Superatom-ISAN" (version 93). Other details can be found in Refs. [9,37-50].

3. Some results and conclusions

Below we present the results for some Rydberg autoionization resonances in the spectrum of Yb. As illustration, in table 1 we present the test data of our computing energies (E4) and widths (Γ4) of the excited (autoionization) states of the YbI 7s6p configuration (accounted from the ground state: $4f^{14}6s^2$ $^1S_0$ Yb). For comparison in table 1 we also list the experimental (Letokhov et al) and theoretical data [9,22,27-29,33,34]: E1,Γ1- PT with the model potential zeroth approximation by Ivanov et al [33]; E2,Γ2- the optimized PT with the DF zeroth approximation [28]; E3- multiconfiguration DF method of Ref. [27] (the classification in [27] differ from Ref. [28,33]).

Table 1. Energies E (10 cm^{-1}) and widths Γ(cm^{-1}) of YbI 7*s*6*p* configuration states (see text)

Term	Theory E3	Theory E1	Theory Γ1	Theory Γ2	Theory E4	Theory Γ4	Exp. E_{exp}	Exp. Γ_{exp}
$^3P_0^0$	6123	5980	0.7	1.25	5940	1.20	5913	1.1
$^3P_1^0$	6208	6000	3.0	1.10	6025	1.06	6042	0.95
$^3P_2^0$	6242	6260	0.7	1.51	6258	1.54	6252	1.6
$^1P_1^0$	6422	6360	1.8	2.48	6370	2.50	6366	2.6

An analysis of the presented data shows physically reasonable agreement between the values of energies E1-E4 and E_{exp}, however, the theoretical and experimental values of the widths significantly differ. The reason of difference is explained by insufficiently exact estimate of the radial integrals, using the non-optimized bases and some other additional calculation approximations [9,17,18,28-34]. Besides, one should take into account for the important multi-body exchange-correlation effects.

In tables 2,3 we list our data on the energies (E_{th}) and widths (Γ_{th}) of the autoionization resonances $4f^{13}$ $[^2F_{7/2}]6s^2np[5/2]_2$, $4f^{13}$ $[^2F_{7/2}]6s^2nf[5/2]_2$ which are arisen in result of excitation of the 4f-shell electrons. For some values of the principal quantum number the presented data are compared with the corresponding available experimental data by Jong-hoon Yi et al, which are obtained by method of three-photon laser polarization spectroscopy [31,32]. Let us note that the obtained results are in a physically reasonable agreement with experimental data. Secondly, an accurate account for the exchange –correlation effects, including an effect of the non-Coulomb grouping levels in the Rydberg spectra plays a principally critical role. Moreover, as the detailed analysis shows, the difference between theoretical and experimental values on widths can reach a few times. It is important to note that the widths of the studied Rydberg resonances are very little.

Table 2. Energies (10 cm^{-1}) and widths (cm^{-1}) of the Yb autoionization resonances $4f^{13}$ $[^2F_{7/2}]6s^2np[5/2]_2$ (see text)

N	E_{exp}	Γ_{exp}	E_{th}	Γ_{th}
12	7012	1.5	7012	1.7
13	7048	0.4	7043	0.5
15	7091	1.2	7092	1.4
20	7143	0.6	7143	0.7
25	7161	1.3	7161	1.5
26	7163	0.6	7163	0.8
30	7170	0.5	7169	0.7
31	7171	0.4	7171	0.5
33	7173	0.6	7173	0.6
34	-	-	7174	0.3
35	-	-	7175	0.5
46	-	-	7180	0.4

Table 3. Energies (cm^{-1}) and widths (cm^{-1}) of the Yb autoionization resonances $4f^{13}$ $[^2F_{7/2}]6s^2nf[5/2]_2$ (see text)

n	E_{exp}	Γ_{exp}	E_{th}	Γ_{th}
12	7096	0.5	7096	0.7
13	7110	0.4	7111	0.5
15	7131	1.4	7131	1.6
20	7156	0.8	7156	0.9
25	7167	0.5	7167	0.8
26	7169	0.5	7169	0.7
30	7173	0.4	7173	0.5
31	7174	0.5	7174	0.6
33	-	-	7175	0.4
34	-	-	7176	0.8
35	-	-	7177	0.5
46	-	-	7181	0.3

This may be connected with a complex energetics of the studied atom [24-26,48], It provides an unusual physics of autoionization resonances in spectra of the lanthanides atoms and their decay mechanisms, especially in comparison with the conventional standards of spectroscopy of the inert gases, alkali and alkaline-earth atoms (the typical resonance widths are of tens and hundreds of cm^{-1}). Speech is about the possible unusual features in behaviour of the autoionization resonances of the lanthanides atoms in a sufficiently weak dc electric (or laser) field that can be detected by a laser spectroscopy methods [9,34,78,79]. One should consider different types of autoionization decay for the lanthanides (and possibly actinides) atoms, in particular, the well known classical Beutler-Fano decay channel and additional the re-orientation type decay channel, predicted by Ivanov et al [33-36,79]. Resonances of the opposite parity are mixed with each other in a sufficiently weak DC electric (laser) field. In case of the heavy lanthanides atoms it results in new effect of the giant broadening some resonances with the little width. For the first time, for Tm, the possibility of such an effect has been predicted by Glushkov-Ivanova-Ivanov [52,79]. These interesting spectral features of the Rydberg autoionization resonances can be used in the effective selective photo-ionization schemes of separating heavy isotopes and some other important applications [1-4,67-80].

Acknowledgements

The authors would like to thank the organizers of the CCP-2017 (Paris) for invitation to make contributions. The useful comments of the anonymous referees are very much acknowledged too.

References

[1] Firstenberg O, Adams C S and Hofferberth S 2016 *J. Phys. B: At. Mol. Opt. Phys.* 49 152003
[2] Pritchard J D, Weatherill K J and Adams C S 2013 *Annual Review of Cold Atoms and Molecules* vol 1 ed K W Madison et al pp 301–50
[3] Saffman M, Walker T G and Mølmer K 2010 *Rev. Mod. Phys.* 82 2313
[4] Ivanova E P 2011 *Phys. Rev. A.* 84 043829
[5] Ivanova E P and Grant I P 1998 *J. Phys.B: At. Mol. Opt. Phys.* 31 2871
[6] Grant I P 2008 *Relativistic Quantum Theory of Atoms and Molecules* (Oxford, University Press)
[7] Cowan R D 1981 *The Theory of Atomic Structure Spectra* (Berkeley: University of California Press)
[8] Shavitt I and Bartlett R J 2009 *Many-body Methods in Chemistry and Physics* (Cambridge: Cambridge University Press)
[9] Glushkov A V 2008 *Relativistic Quantum Theory. Quantum mechanics of atomic systems* (Odessa: Astroprint)
[10] Khetselius O Yu 2008 *Hyperfine Structure of Atomic Spectra* (Odessa: Astroprint)
[11] Khetselius O Yu 2008 *AIP Conf. Proc.* **1058** 363
[12] Khetselius O Yu 2012 *J. Phys.: Conf. Ser.* **397** 012012
[13] Safronova U I and Safronova M S 2009 *Phys. Rev.* A **79** 022512.
[14] Fischer C F 2000 *Comput. Phys. Commun.* **128** 635
[15] Kohn J W and Sham L J 1964 *Phys. Rev. A.* **140** 1133
[16] Gross E G and Kohn W 2005 *Exchange-Correlation Functionals in Density Functional Theory* (New York: Plenum)
[17] Glushkov A V 2006 *Relativistic and correlation effects in spectra of atomic systems* (Odessa: Astroprint).
[18] Glushkov A V, Khetselius O Y and Lovett L 2009 *Advances in the Theory of Atomic and Molecular Systems. Series: Progress in Theoretical Chemistry and Physics* (Dordrecht: Springer) **20** pp 125-152
[19] Malinovskaya S V, Glushkov A V, Dubrovskaya Y V, Vitavetskaya L A 2006 *Recent Advances in the Theory of Chemical and Physical Systems, Series: Progress in Theoretical Chemistry and Physics* (Springer) **15** pp 301-307
[20] Glushkov A V, Malinovskaya S V, Gurnitskaya E P, Khetselius O Yu and Dubrovskaya Yu V 2006 *J. Phys.: Conf. Ser.* **35** 425
[21] Glushkov A V, Malinovskaya S V, Prepelitsa G P and Ignatenko V M 2005 *J. Phys.: Conf. Ser.* **11** 199
[22] Martin W C, Zalubas R and Hagan L 1978 Atomic Energy levels – The rare-earth elements
[23] Wyart J-F and Camus P 1979 *Phys. Scripta.* **20** 43
[24] Camus P, Debarre A and Morillon C 1980 *J. Phys. B: At. Mol. Phys.* **13** 1073
[25] Aymar M, Debarre A, Robaux O 1980 *J. Phys. B: At. Mol. Opt. Phys.* **13** 1089
[26] Baumann M, Braun M, Gaiser A, Liening H 1985 *J. Phys. B: At. Mol. Opt. Phys.* **18** L601
[27] Karacoban B and Özdemir L 2011 *Acta Phys. Pol.*A **119** 342
[28] Glushkov A V, Khetselius O Yu and Svinarenko A A 2013 *Phys. Scr.* **T153** 014029
[29] Svinarenko A A 2014 *J. Phys.: Conf. Ser.* **548** 012039
[30] Fang D-W, Xie W-J, Zhang Y, Hu X, Liu Y Y 2001 *J. Quant. Spectr. Rad. Transfer.* **69** 469
[31] Jong-Hoon Yi, Lee J, Kong H J 1995 *Phys. Rev. A.* **51** 3053
[32] Jong-Hoon Yi, Park H, Lee J 2001 *J.of Korean Phys.Soc.* **39** 916
[33] Bekov G I, Vidolova-Angelova E, Ivanov L N, Letokhov V S and Mishin V I 1981 *JETP.* **80** 866

[34] Ivanov L N and Letokhov V S 1985 *Com. Mod. Phys.* D **4** 169
[35] Vidolova-Angelova E, Ivanov L N, Ivanova E P and Angelov D A 1986 J. *Phys. B: At. Mol. Opt. Phys.* **19** 2053
[36] Vidolova-Angelova E and Ivanov L N 1991 *J. Phys. B: At. Mol. Opt. Phys*. **24** 4147
[37] Glushkov A V 1991 *Opt. Spectrosc.* **70** 555
[38] Glushkov A V, Ambrosov S V, Loboda A, Chernyakova Y G, Khetselius O Y, Svinarenko A A 2004 *Nucl. Phys. A* **734** E21
[39] Glushkov A V, Ambrosov S V, Loboda A V, Gurnitskaya E P and Prepelitsa G P 2005 *Int. J. Quant.Chem*. **104** 562
[40] Glushkov A V, Lovett L, Khetselius O Yu, Gurnitskaya E P, Dubrovskaya Yu V and Loboda A V 2009 *Int. Journ. Modern Phys. A* **24** 611
[41] Glushkov A V, Ambrosov S V, Loboda A V, Gurnitskaya E P and Khetselius O Yu 2006 *Recent Advances in the Theory of Chemical and Physical Systems, Series: Progress in Theoretical Chemistry and Physics* (Springer) **15** pp 285-299
[42] Glushkov A V, Malinovskaya S V, Khetselius O Yu, Loboda A V, Sukharev D E and Lovett L 2009 *Int. J. Quant. Chem.* **109** 1717
[43] Khetselius O Yu 2009 *Phys. Scripta.* **T135** 014023
[44] Khetselius O Yu 2009 *Int. Journ. Quant. Chem*. **109** 3330
[45] Malinovskaya S V, Glushkov A V, Khetselius O Yu, Svinarenko A A, Mischenko E V and Florko T A 2009 *Int. J. Quant. Chem.* **109** 3325
[46] Ivanova E P and Ivanov L N 1979 *Atom. Data. Nucl. Data. Tabl*. **24** 95
[47] Driker M N, Ivanova E P, Ivanov L N and Shestakov A F 1982 *J. Quant. Spectr. Rad. Transfer*. **28** 531
[48] Ivanova E P and Glushkov A V 1986 *J. Quant. Spectr. Rad. Transfer*. **36** 127
[49] Ivanov L N, Ivanova E P and Aglitsky E V 1988 *Phys. Rep.* **166** 315
[50] Ivanova E P, Ivanov L N, Glushkov A V and Kramida A E 1985 *Phys. Scr.* **32** 513
[51] Ivanov L N, Ivanova E P and Knight L 1993 *Phys. Rev. A* **48** 4365
[52] Glushkov A V, Ivanov L N and Ivanova E P 1986 *Autoionization Phenomena in Atoms*, (Moscow: Moscow State University) p 58
[53] Glushkov A V, Ivanov L N 1992 *Phys. Lett. A.* **170** 33
[54] Glushkov A V 2012 *Quantum Systems in Chemistry and Physics, Series: Progress in Theoretical Chemistry and Physics* (Dordrecht: Springer) **26** pp 231-252
[55] Glushkov A V 1992 *JETP. Lett.* **55** 97
[56] Glushkov A V 2005 *AIP Conference Proc.* **796** 206
[57] Glushkov A, Loboda A, Gurnitskaya E and Svinarenko A 2010 *Phys. Scripta.* **T135** 014022
[58] Glushkov A V 2012 *J. Phys.: Conf. Ser*. **397** 012011
[59] Khetselius O Y 2015 *Frontiers in Quantum Methods and Applications in Chemistry and Physics. Series: Progress in Theoretical Chemistry and Physics* (Cham: Springer) 29 pp 55-76
[60] Glushkov A V 2014 *J. Phys.: Conf. Ser*. **548** 012020
[61] Glushkov A V 1990 *Soviet Phys. J.* **33**(1) 1
[62] Glushkov A V 1990 *J. Struct. Chem.* **31**(4) 529
[63] Glushkov A V, Butenko Yu V, Serbov N G, Ambrosov S V, Orlova V E , Orlov S V, Balan A K and Dormostuchenko G M 1996 *J. Appl. Spectrosc.* **63** 28
[64] Glushkov A V, Malinovskii A V, Efimov V A, Kivganov A F, Khokhlov V N, Vitavetskaya L A, Borovskaya G A 1998 *J. Struct. Chem*. **39**(2) 179
[65] Glushkov A V, Efimov V A, Gopchenko E D, Dan'kov S V, Polishchyuk V N, Goloshchak O P 1998 *Opt. and Spectr.* **84**(5) 670
[66] Glushkov A V, Efimov V A, Gopchenko E D, Dan'kov S V, Polishchyuk V N, Goloshchak O P 1998 *Optika and Spektr.* **84**(5) 747
[67] Glushkov A V, Khetselius O Yu and Malinovskaya S V 2008 *Europ. Phys. Journ* ST **160** 195

[68] Glushkov A V, Khetselius O Yu and Malinovskaya S V 2008 *Molec. Phys.* **106** 1257
[69] Glushkov A V, Ambrosov S V, Ignatenko A V and Korchevsky D A 2004 *Int. J. Quant.Chem.* **99** 936
[70] Glushkov A V, Malinovskaya S V, Chernyakova Y G and Svinarenko A A 2004 *Int. J. Quant.Chem.* **99** 889
[71] Malinovskaya S V, Dubrovskaya Yu V, Vitavetskaya L A 2005 *AIP Conference Proc.* **796** 201
[72] Khetselius O Y 2012 *Quantum Systems in Chemistry and Physics. Ser.: Progress in Theoretical Chem. and Phys.* (Dordrecht: Springer) **26** pp 217-229
[73] Glushkov A V, Khetselius O Yu, Lopatkin Yu M, Florko T A, Kovalenko O A, Mansarliysky V F 2014 *J. Phys.: Conf. Ser.* **548** 012026
[74] Glushkov A V, Kondratenko P A, Buyadgi V V, Kvasikova A S, Shakhman A N, Sakun T N 2014 *J. Phys.: Conf. Ser.* **548** 012025
[75] Svinarenko A A, Khetselius O Yu, Buyadzhi V V, Florko T A, Zaichko P A, Ponomarenko E L 2014 *J. Phys.: Conf. Ser.* **548** 012048
[76] Letokhov V S 1986 *Phys.-Uspekhi.* **29** 70
[77] Letokhov V S and Ivanov L N 1975 *Quantum Electr.* **2** 585; 1975 *JETP* **68** 1748
[78] Glushkov A and Ivanov L N 1993 *J. Phys. B: At. Mol. Opt. Phys.* **26** L379
[79] Letokhov V S, Ivanov L N and Glushkov A V 1992 *Preprint of Institute for Spectroscopy of Russian Academy of Sciences* (Troitsk: ISAN) 92-5-AS
[80] Glushkov A V 2013 *Advances in Quantum Methods and Applications in Chemistry, Physics, and Biology. Series: Progress in Theoretical Chemistry and Physics* (Cham: Springer) **27** pp 161-177

IOP Conf. Series: Journal of Physics: Conf. Series **1136** (2018) 012011 doi:10.1088/1742-6596/1136/1/012011

Advanced computational approach in electron-collisional spectroscopy of atoms and multicharged ions in plasmas

V V Buyadzhi, E V Ternovsky, Yu G Chernyakova, T B Tkach, A A Buyadzhi, P A Zaichko

Odessa State Environmental University, L'vovskaya str.15, Odessa-16, 65016, Ukraine

E-mail: buyadzhivv@gmail.com

Abstract. An advanced computational approach to determination of the electron-collisional strengths and cross-sections for atomic ions in the Debye plasmas is presented and used to calculate effective collision strengths of the Kr^{26+} Ne-like ion excitation states for temperature $T=5\times10^6K$ and density $n_e==10^{14}cm^{-3}$. The obtained results are compared with the R-matrix data by Griffin et al and other theoretical estimates. The approach is based on the generalized relativistic energy formalism and relativistic many-body perturbation theory with the Debye shielding model Hamiltonian for electron-nuclear and electron-electron systems. The optimized one-electron representation in the perturbation theory zeroth approximation is constructed by means of the correct treating the gauge dependent multielectron contribution of the lowest perturbation theory corrections to the radiation widths of atomic levels.

1. Introduction

The properties of laboratory and astrophysical plasmas have drawn considerable attention over the last decades [1-6]. It is known that multicharged ions play an important role in the diagnostics of a wide variety of plasmas. Similar interest is also stimulated by importance of this information for correct determination of the characteristics for plasma in thermonuclear (tokamak) reactors, searching new mediums for X-ray range lasers. The electron-ion collisions play a major role in the energy balance of plasmas. For this reason, modelers and diagnosticians require absolute cross sections for these processes. The cross sections for electron-impact excitation of ions are needed to interpret spectroscopic measurements and for simulations of plasmas using collisional-radiative models. Such well-known atomic methods such as multi-configuration Dirac–Fock, relativistic distorted-wave method, R-matrix and others approaches [4-18] have been intensively applied to problems considered. At present time a considerable interest has been encapsulated to studying elementary atomic processes in plasmas environments (for example, see [1-54] and Refs. therein) because of the plasmas screening effect on the plasmas-embedded atomic systems. In many papers the calculations of various atomic and ionic systems embedded in the Debye plasmas have been performed [4-10, 16-20]; it is well-known that the Debye model is justified only in the limit of high temperature and low density. However, a development of the advanced computational quantum-mechanical models for the further accurate computing oscillator strengths, electron-collisional strengths and cross-sections for the atomic ions in plasmas, including the Debye plasmas, remained very actual and difficult problem.

In this paper we develop and apply an advanced computational approach to determination of the electron-collisional strengths and cross-sections for the atomic ions in the Debye plasmas. The approach is based on the generalized relativistic energy formalism and relativistic many-body perturbation theory (PT) with the Debye shielding model Hamiltonian for electron-nuclear and electron-electron systems [17-25]. The PT formalism is constructed on the base of the same fundamental points as the well-known PT approach with the model potential zeroth approximation by Ivanov-Ivanova et al [1-3,26-32]. However there are a few fundamental differences. For example, in our case the PT zeroth approximation [18-20] is in fact the Dirac-Debye-Hückel one. In order to compute the radiative and collisional parameters an advanced version of a relativistic energy approach is used [47,55-64]. It is important to remind that a model relativistic energy approach in a case of a multielectron atom has been developed by Ivanov-Ivanova *et al* [26-30]. A generalized gauge-invariant version of relativistic energy approach in a case of the multielectron atomic systems has been developed by Glushkov-Ivanov-Ivanova (see Refs. [30,33-36]). Earlier this approach has been successfully applied to many actual problems of modern atomic, nuclear, mesonuclear, molecular optics and spectroscopy etc (look Refs. [37-52,55-68]). The method [33] is used to construct the optimized one-electron representation in the PT zeroth approximation by means of the correct treating the gauge dependent multielectron contribution of the lowest PT corrections to the radiation widths of atomic levels. The presented approach has been used to calculate effective collision strengths of the Kr^{26+} Ne-like ion excitation states for temperature $T=5\times10^{6}$K and density n_e==$10^{14}cm^{-3}$. The obtained results are compared with the R-matrix data by Griffin et al and model potential data [5,20]. All computing was performed with using the modified PC code "Superatom-ISAN" (version 93). Other details can be found in Refs. [17-22].

2. Generalized energy approach in scattering theory

In the theory of the non-relativistic atom a convenient field procedure is known for calculating the energy shifts ΔE of degenerate states. This procedure is connected with the secular matrix M diagonalization [26-32]. In constructing M, the Gell-Mann and Low adiabatic formula for ΔE is used. A similar approach, using the Gell-Mann and Low formula with the QED scattering matrix, is applicable in the relativistic atom theory. The secular matrix elements are already complex in the PT second order and their imaginary parts are connected with the radiation decay possibility. The total energy shift of the state is presented in the form:

$$\Delta E = \mathrm{Re}\Delta E + \mathrm{i}\ \mathrm{Im}\Delta E, \qquad \mathrm{Im}\ \Delta E = -\Gamma/2, \tag{1}$$

where Γ is interpreted as the level width. The whole calculation of the energies and decay probabilities of a non-degenerate excited state is reduced to calculation and diagonalization of the complex matrix M and definition of matrix of the coefficients with eigen-state vectors $B^{IK}_{ie,iv}$ [27-31]. To calculate all necessary matrix elements one must use the basis's of the 1QP relativistic functions. In ref. [33] it has been proposed "ab initio" optimization principle for construction of cited basis's. There is used the minimization of the gauge dependent multielectron contribution of the lowest QED PT corrections to the radiation widths of atomic levels. In the fourth order of QED PT there appear diagrams, whose contribution into the ImδE accounts for the polarization effects. This contribution describes collective effects and it is dependent upon the electromagnetic potentials gauge (the gauge non-invariant contribution ΔE_{ninv}). The minimization of the functional $Im\ \Delta E_{ninv}$ leads to the integral differential equation, that is numerically solved [33].

The Dirac-Debye shielding model Hamiltonian for electron-nuclear and electron-electron subsystems is as follows (atomic units are used) [18-20]:

IOP Conf. Series: Journal of Physics: Conf. Series **1136** (2018) 012011 doi:10.1088/1742-6596/1136/1/012011

$$H=\sum_i[\alpha cp-\beta mc^2-Z\exp(-\mu r_i)/r_i]+\sum_{i>j}\frac{(1-\alpha_i\alpha_j)}{r_{ij}}\exp(-\mu r_{ij}), \tag{2}$$

where c is the velocity of light and Z is a charge of the atomic ion nucleus, ω_{ij} is the transition frequency; α_i , α_j are the Dirac matrices. The plasmas environment effect is modelled by the shielding parameter μ , which describes a shape of the long-rang potential. The parameter μ is connected with the plasmas parameters such as temperature T and the charge density n as follows:

$$\mu\sim\sqrt{e^2 n/k_BT}\,. \tag{3a}$$

Here e is the electron charge and k_B is the Boltzman constant. The density n is given as a sum of the electron density N_e and the ion density N_k of the k-th ion species with the nuclear charge q_k :

$$n=N_e+\sum_k q_k^2 N_k\,. \tag{3b}$$

It is very useful to remind the simple estimates for the shielding parameter. For example, under typical laser plasmas conditions of $T\sim$ 1keV and $n\sim 10^{22}$ cm^{-3} the parameter μ is of the order of 0.1 in atomic units; in the EBIT plasmas $T\sim$ 0.05keV, $n\sim10^{18}$ cm^{-3} and $\mu\sim10^{-3}$. We are interested in studying the spectral parameters of ions in plasmas with the temperature $T\sim$ 0.1-1keV (10^6-10^7K) and $n\sim10^{14}$-10^{26} cm^{-3} ($\mu\sim10^{-5}$-10^0). It should be noted that indeed the Debye screening for the atomic electrons in the Coulomb field of nuclear charge is well understood due to the presence of the surrounding plasma electrons with high mobility. On the other hand, the contribution due to the Debye screening between electrons would be orders of magnitude smaller. Majority of the previous works on the spectroscopy study have considered the screening effect only in the electron-nucleus potential where the electron-electron interaction potential is truncated at its first term of the standard exponential expansion for its dominant contribution [5-10,18,19]. However, as the authors [8] note, it is also important to take into account the screening in the electron- electron interactions for large plasma strengths to achieve more realistic results in the search for stability of the atomic structure in the plasma environment.

In the PT zeroth approximation we use the mean-field potential, which includes the Yukawa-type potential (insist of the pure Coulomb one) plus exchange Kohn-Sham potential [54,55] and additionally the modified Lundqvist-Gunnarsson correlation potential (with the optimization parameter b) as in Refs. [18,19,55-58]. The most complicated problem of the relativistic PT computing the radiative and collisional characteristics of the multielectron atomic systems is in an accurate, precise accounting for the exchange-correlation effects (including polarization and screening effects, a continuum pressure etc) as the effects of the PT second and higher orders . Using the standard Feynman diagrammatic technique one should consider two kinds of diagrams (the polarization and ladder ones), which describe the polarization and screening exchange-correlation effects. The polarization diagrams take into account the quasiparticle (external electrons or vacancies) interaction through the polarizable core, and the ladder diagrams account for the immediate quasiparticle interaction. The detailed description of the polarization diagrams and the corresponding analytical expressions for matrix elements of the polarization quasiparticles interaction (through the polarizable core) potential are presented in Refs. [6-8,31,59-65].

An effective approach to accounting of the polarization diagrams contributions is in adding the effective two-quasiparticle polarizable operator into the PT first order matrix elements. In Ref. [31] the corresponding non-relativistic polarization functional has been derived. More correct relativistic expression has been presented in the Refs. [55, 59] and used in our work. According to Ref. [59], the polarization potential is as follows:

$$V_{pol}^{d}(r_1r_2)=X\left\{\int\frac{dr'\left(\rho_c^{(0)}(r')\right)^{1/3}\theta(r')}{|r_1-r'|\cdot|r'-r_2|}-\int\frac{dr'\left(\rho_c^{(0)}(r')\right)^{1/3}\theta(r')}{|r_1-r'|}\int\frac{dr''\left(\rho_c^{(0)}(r'')\right)^{1/3}\theta(r'')}{|r''-r_2|}\Big/\left\langle\left(\rho_c^{(0)}\right)^{1/3}\right\rangle\right\},$$

$$\left\langle\left(\rho_c^{(0)}\right)^{1/3}\right\rangle=\int dr\left(\rho_c^{(0)}(r)\right)^{1/3}\theta(r),$$

$$\theta(r)=\left\{1+\left[3\pi^2\cdot\rho_c^{(0)}(r)\right]^{2/3}/c^2\right\}^{1/2} \tag{4}$$

where X is the numerical coefficient.

The justification of the energy approach in the scattering problem is in details described in Refs. [18-20,27-30]. The scattered part of energy shift *Im ΔE* appears first in the second order of the atomic perturbation theory. For example, the collisional de-excitation cross section is defined as follows:

$$\sigma(IK\to 0)=2\pi\sum_{j_{in},j_{sc}}(2j_{sc}+1)\{\sum_{j_{ie},j_{iv}}<0\mid j_{in},j_{sc}\mid j_{ie},j_{iv},J_i>B_{ie,iv}^{IK}\}^2 \tag{5}$$

$$<0\mid j_{in},j_{sc}\mid j_{ie},j_{iv},J_i>=\sqrt{(2j_{ie}+1)(2j_{iv}+1)}(-1)^{j_{ie}+1/2}\times\sum_{\lambda}(-1)^{\lambda+J_i}\times$$

$$\times\{\delta_{\lambda,J_i}/(2J_i+1)Q_\lambda(sc,ie;iv,in)+\begin{bmatrix}j_{in}\cdots j_{sc}\cdots J_i\\ j_{ie}\cdots j_{iv}\cdots\lambda\end{bmatrix}Q_\lambda(ie;in;iv,sc)\}, \tag{6}$$

where $Q_\lambda=Q_\lambda^{\mathrm{Qul}}+Q_\lambda^{\mathrm{Br}}$ is the sum of the Coulomb and Breit matrix elements. For example, the Coulomb part Q_λ^{Qul} contains the radial R_λ and angular S_λ integrals as follows see details in Refs.[17-19, 27-31]):

$$\begin{aligned}Q_\lambda^{\mathrm{Coul\text{-}Yuk}}=\{&R_\lambda(1243)S_\lambda(1243)+R_\lambda(\tilde{1}24\tilde{3})S_\lambda(\tilde{1}24\tilde{3})+\\&+R_\lambda(1\tilde{2}\tilde{4}3)S_\lambda(1\tilde{2}\tilde{4}3)+R_\lambda(\tilde{1}\tilde{2}\tilde{4}\tilde{3})S_\lambda(\tilde{1}\tilde{2}\tilde{4}\tilde{3})\}\end{aligned} \tag{7}$$

Here the tilde designates that the large radial Dirac component f must be replaced by the small Dirac component g, and instead of l_i, $l_i'=l_i-1$ should be taken for $j_i<l_i$ and $l_i'=l_i+1$ for $j_i>l_i$. The plasmas shielding model is incorporated and based on introducing the Yukawa - type electron-nuclear attraction and electron-electron repulsion potentials. The relativistic wave functions are calculated by solution of the Dirac-Kohn-Sham equation with the optimized one-parameter exchange-correlation potential. The calibration of the single model potential parameter has been performed on the basis of the special ab initio procedure within [33] (see also [30,55]).

3. Results and conclusions

In table 1 we present the theoretical data on the effective collision strengths of the Kr^{26+} Ne-like ion excitation states for the temperature $T=5\cdot10^6$ K and the electron density $n_e=10^{14}$ cm^{-3}. The R-matrix data by DF Griffin et al (RM) [5] and model potential (MP) data [20] are listed for comparison too. It should be noted that strong compensation of different PT terms is a characteristic feature of the states with vacancies in the core. This is one of the main reasons for the fact that the accuracy of conventional a priori calculations of such states does not always satisfy the requirements arising in many applications. Summation over $j_{im}j_{sc}$ in (18) spreads over the range 1/2-23/2. For some levels the corrections due the correlation effects change the results by a factor of 2-3,5.

Table 1. The effective collision strengths of the Kr^{26+} Ne-like ion excitation states for the temperature $T=5\cdot10^6$ K and electron density $n_e=10^{14}$ cm^{-3} (see text).

Term	RM	MP	Our paper
$2p^53s\ (3/2,1/2)_2$	8.29(-3)	8.13(-3)	8.17(-3)
$2p^53s\ (3/2,1/2)_1$	9.36(-3)	9.19(-3)	9.23(-3)
$2p^53p\ (3/2,1/2)_1$	3.49(-3)	3.38(-3)	3.41(-3)
$2p^53p\ (3/2,1/2)_2$	4.30(-3)	4.18(-3)	4.24(-3)
$2p^53s\ (1/2,1/2)_0$	1.32(-3)	1.21(-3)	1.26(-3)
$2p^53s\ (1/2,1/2)_1$	7.69(-3)	7.56(-3)	7.60(-3)
$2p^53p\ (3/2,3/2)_3$	4.03(-3)	3.89(-3)	3.94(-3)
$2p^53p\ (3/2,3/2)_1$	3.14(-3)	3.01(-3)	3.06(-3)
$2p^53p\ (3/2,3/2)_2$	3.36(-3)	3.12(-3)	3.16(-3)
$2p^53p\ (3/2,3/2)_0$	8.67(-3)	8.49(-3)	8.55(-3)
$2p^53p\ (1/2,1/2)_1$	2.69(-3)	2.54(-3)	2.59(-3)
$2p^53p\ (1/2,3/2)_1$	2.80(-3)	2.72(-3)	2.75(-3)
$2p^53p\ (1/2,3/2)_2$	3.27(-3)	3.16(-3)	3.21(-3)
$2p^53d\ (3/2,3/2)_0$	1.24(-3)	1.13(-3)	1.18(-3)
$2p^53p\ (1/2.1/2)_0$	1.71(-2)	1.58(-2)	1.63(-2)
$2p^53d\ (3/2,3/2)_1$	3.45(-3)	3.31(-3)	3.36(-3)
$2p^53d\ (3/2,3/2)_3$	3.80(-3)	3.67(-3)	3.74(-3)
$2p^53d\ (3/2,5/2)_2$	4.13(-3)	3.96(-3)	4.01(-3)

Using of the shielding approach and an account for the highly-lying excited states is quantitatively important for the adequate description of the collision strengths.

To conclude, let us note that in any case our calculation encourages us to believe that using relativistic energy approach with the optimal Dirac-Kohn-Sham one-electron PT basis and shielding model block is quite consistent and effective one from the point of view of the theory correctness and results exactness. This fact was surely confirmed by calculations of the oscillator strengths, radiative widths in atoms and multicharged ions (c.f. [3,18-30,66-70]).

Acknowledgements

The authors would like to thank the organizers of the CCP-2017 (Paris) for invitation to make contributions. The useful comments of the anonymous referees are very much acknowledged too.

References

[1] Ivanova E P and Grant I P 1998 *J. Phys.B: At. Mol. Opt. Phys.* **31** 2871
[2] Ivanova E P 2011 *Phys. Rev. A.* **84** 043829
[3] Ivanova E P and Zinoviev N A 2001 *Phys. Lett. A* **274** 239
[4] Nakamura N, Kavanagh A P, Watanabe H, Sakaue H A, Li Y, Kato D, Curell F J and Ohtani S 2007 *Journ. Phys. Conf. Ser.* **88** 012066
[5] Griffin D C, Balance C P, Mitnik D M, Berengut J C 2008 *J. Phys. B* **41** 21520
[6] Bielińska-Wąż D, Karwowski J, Saha B and Mukherjee P K 2004 *Phys. Rev. E* **69** 016404
[7] Kar S and Ho Y K 2006 *Phys. Rev. A* **73** 032502
[8] Madhulita Das M, Sahoo B K and Pal S 2014 *J. Phys. B: At. Mol. Opt. Phys.* **47** 175701
[9] Li Y, Wu J, Hou Y and Yuan J 2008 J. *Phys. B: At. Mol. Opt. Phys.* **41** 145002
[10] Chen Z B, Dong C Z, Jiang J and Xie L Y 2015 *J. Phys. B: At. Mol. Opt. Phys.* **48** 144030
[11] Grant I P 1984 *Int. J. Quant. Chem.* **25** 23

[12] Bartschat K, Hudson E T, Scott M P, Burke P G, and Burke V M 1996 *J. Phys. B: At. Mol. Opt. Phys.* **29** 115
[13] Descouvemont P, Tursunov E M and Baye D 2006 *Nucl. Phys. A* **765** 370
[14] Dyall K G, Grant I P, Johnson C T, Parpia F A and Plummer E P 1989 *Comput. Phys. Commun.* **55** 425
[15] Zhang H, Sampson D H, Clark R E H and Mann J B 1987 *At.Data Nucl. Data Tabl.* **37** 17
[16] Khetselius O Yu 2008 *Hyperfine structure of atomic Spectra* (Odessa: Astroprint)
[17] Glushkov A V 2008 *Relativistic Quantum theory. Quantum mechanics of atomic systems* (Odessa: Astroprint)
[18] Glushkov A V, Khetselius O Yu, Loboda A V, Ignatenko A, Svinarenko A , Korchevsky D, Lovett L 2008 *AIP Conf. Proc.* **1058** 175
[19] Glushkov A V, Malinovskaya S V, Loboda A V, Shpinareva I M, Gurnitskaya E P, Korchevsky D A 2005 *J. Phys.: Conf. Ser.* **11** 188
[20] Glushkov A V 1990 *Sov. Phys. Journ.* **33**(9) 754
[21] Glushkov A V, Kondratenko P A, Lepikh Y I, Fedchuk A P, Lovett L, Svinarenko A A 2009 *Int. Journ. Quant. Chem.* **109**(14) 3473
[22] Malinovskaya S V, Glushkov A V, Khetselius O Yu, Svinarenko A A, Mischenko E V and Florko T A 2009 *Int. J. Quant. Chem.* **109** 3325
[23] Glushkov A V, Khetselius O Yu and Svinarenko A A 2013 *Phys. Scr.* **T153** 014029
[24] Glushkov A V, Loboda A, Gurnitskaya E and Svinarenko A 2009 *Phys. Scr.* **T135** 014022
[25] Glushkov A V, Khetselius O Yu, Lopatkin Yu M, Florko T A, Kovalenko O A, Mansarliysky V F 2014 *J. Phys.: Conf. Ser.* **548** 012026
[26] Ivanova E P and Ivanov L N 1979 *Atom. Data. Nucl. Data. Tabl.* **24** 95
[27] Ivanov L N, Ivanova E P and Knight L 1993 *Phys. Rev. A* **48** 4365
[28] Driker M N, Ivanova E P, Ivanov L N and Shestakov A F 1982 *J. Quant. Spectr. Rad. Transfer.* **28** 531
[29] Ivanov L N, Ivanova E P and Aglitsky E V 1988 *Phys. Rep.* **166** 315
[30] Glushkov A V, Ivanov L N and Ivanova E P 1986 *Autoionization Phenomena in Atoms*, (Moscow: Moscow State University) p 58
[31] Ivanova E P, Ivanov L N, Glushkov A V and Kramida A E 1985 *Phys. Scr.* **32** 513
[32] Ivanova E P and Glushkov A V 1986 *J. Quant. Spectr. Rad. Transfer.* **36** 127
[33] Glushkov A V, Ivanov L N 1992 *Phys. Lett. A.* **170** 33
[34] Glushkov A V 1992 *JETP. Lett.* **55** 97
[35] Glushkov A and Ivanov L N 1993 *J. Phys. B: At. Mol. Opt. Phys.* **26** L379
[36] Glushkov A V 2012 Quantum Systems in Chemistry and Physics, Series: Progress in Theoretical Chemistry and Physics (Dordrecht: Springer) **26** pp 231-252
[37] Glushkov A V 2012 *J. Phys.: Conf. Ser.* **397** 012011
[38] Glushkov A V 2014 *J. Phys.: Conf. Ser.* **548** 012020
[39] Khetselius O Yu 2009 *Phys. Scripta.* **T135** 014023
[40] Khetselius O Yu 2010 *AIP Conf. Proc.* **1290** 29
[41] Khetselius O Yu 2008 *AIP Conf. Proc.* **1058** 363
[42] Glushkov A V, Ambrosov S V, Ignatenko A V, Korchevsky D A 2004 *Int. J. Quant.Chem.* **99** 936
[43] Glushkov A V, Malinovskaya S V, Khetselius O Yu, Loboda A V, Sukharev D E and Lovett L 2009 *Int. J. Quant. Chem.* **109** 1717
[44] Glushkov A V, Malinovskaya S V, Prepelitsa G P and Ignatenko V M 2005 *J. Phys.: Conf. Ser.* **11** 199
[45] Glushkov A V, Khetselius O Yu and Malinovskaya S V 2008 *Europ. Phys. J. ST* **160** 195
[46] Glushkov A V, Khetselius O Yu and Malinovskaya S V 2008 *Molec. Phys.* **106** 1257
[47] Malinovskaya S V, Glushkov A V, Dubrovskaya Y V, Vitavetskaya L A 2006 *Recent Advances in the Theory of Chemical and Physical Systems, Ser.: Progress in Theoretical*

Chem. and Phys. (Dordrecht: Springer) **15** pp 301-307
[48] Glushkov A V 2012 *Quantum Systems in Chemistry and Physics. Ser.: Progress in Theoretical Chem. and Phys.* (Dordrecht: Springer) **26** pp 231-252
[49] Glushkov A V, Khetselius O Y and Lovett L 2009 *Advances in the Theory of Atomic and Molecular Systems. Ser.: Progress in Theoretical Chem. and Phys.* (Dordrecht: Springer) **20** pp 125-152
[50] Glushkov A V, Lovett L, Khetselius O Yu, Gurnitskaya E P, Dubrovskaya Yu V and Loboda A V 2009 *Int. Journ. Modern Phys. A* **24** 611
[51] Glushkov A V, Malinovskaya S V, Gurnitskaya E P, Khetselius O Yu and Dubrovskaya Yu V 2006 *J. Phys.: Conf. Ser.* **35** 425
[52] Khetselius O Y 2012 *Quantum Systems in Chemistry and Physics. Ser.: Progress in Theoretical Chem. and Phys.* (Dordrecht: Springer) **26** pp 217-229
[53] Kohn J W and Sham L J 1964 *Phys. Rev. A.* **140** 1133
[54] Gross E G and Kohn W 2005 *Exchange-Correlation Functionals in Density Functional Theory* (New York: Plenum)
[55] Glushkov A V 2006 *Relativistic and correlation effects in spectra of atomic systems* (Odessa: Astroprint).
[56] Khetselius O Y 2012 *J. Phys.: Conf. Ser.* **397** 012012
[57] Glushkov A V, Mansarliysky V F, Khetselius O Yu, Ignatenko A V, Smirnov A V, Prepelitsa G P 2017 *J. Phys.: Conf. Ser.* **810** 012034
[58] Glushkov A V, Ambrosov S V, Loboda A V, Gurnitskaya E P and Prepelitsa G P 2005 *Int. Journ. Quant.Chem.* **104** 562
[59] Glushkov A V 1990 *Soviet Phys. J.* **33**(1) 1
[60] Glushkov A V 1990 *J. Struct. Chem.* **31**(4) 529
[61] Glushkov A V 2005 *AIP Conference Proc.* **796** 206
[62] Glushkov A V and Loboda A V 2007 *J. Applied. Spectr.* **74**(2) 305
[63] Malinovskaya S V, Dubrovskaya Yu V, Vitavetskaya L A 2005 *AIP Conference Proc.* **796** 201
[64] Glushkov A V, Malinovskii A V, Efimov V A, Kivganov A F, Khokhlov V N, Vitavetskaya L A, Borovskaya G A 1998 *J. Struct. Chem.* **39**(2) 179
[65] Glushkov A V, Efimov V A, Gopchenko E D, Dan'kov S V, Polishchyuk V N, Goloshchak O P 1998 *Opt. and Spectr.* **84**(5) 670
[66] Khetselius O Yu 2009 *Int. J. Quant. Chem.* **109** 3330
[67] Glushkov A V, Khetselius O Yu, Svinarenko A A, Prepelitsa G P, Mischenko E V 2010 *AIP Conference Proc.* **1290** 263
[68] Glushkov A V, Khetselius O Yu, Svinarenko A A, Prepelitsa G P, Shakhman A N 2010 *AIP Conference Proc.* **1290** 269
[69] Svinarenko A A 2014 *J. Phys.: Conf. Ser.* **548** 012039
[70] Khetselius O Yu, Glushkov A V, Gurskaya M Yu, Kuznetsova A A, Dubrovskaya Yu V, Serga I N, Vitavetskaya L A 2017 *J. Phys.: Conf. Ser.* **905** 012029

IOP Conf. Series: Journal of Physics: Conf. Series **1136** (2018) 012012 doi:10.1088/1742-6596/1136/1/012012

Modified operator perturbation theory computational approach to atomic systems in a DC electric field: Stark resonances

A V Glushkov, A A Kuznetsova, M Yu Gurskaya, A V Ignatenko and A O Makarova

Odessa State Environmental University, L'vovskaya str.15, Odessa-16, 65016, Ukraine

E-mail: glushkovav@gmail.com

Abstract. New computational realization of the modified operator perturbation theory method to computing the Stark resonances energies and widths for the non-hydrogenic (non-H) atomic systems in a DC electric field is presented. The method is based on the operator form of the perturbation theory of the Schrödinger equation for the non-H atomic systems and includes the physically reasonable distorted-waves approximation in the frame of the formally exact quantum-mechanical procedure. The Stark resonances energies and widths in the lithium atom are calculated and compared with results of calculations on the basis of the method of complex eigenvalue Schrödinger equation by Themelis-Nicolaides, the complex absorbing potential method by Sahoo-Ho and the B-spline-based coordinate rotation method approach by Hui-Yan Meng et al.

1. Introduction

The Stark effect is one of the best known problems in quantum mechanics, but at the same time one of the most difficult (outside the weak-field region) [1-8]. At last years it attracts a great interest especially in the multielectron atoms that is stimulated by a whole range of interesting phenomena to be studied such as quasi-discrete state mixing, a zoo of Landau- Zener anticrossings, autoionization in the multielectron atoms, the effects of potential barriers (shape resonances), new kinds of resonances above threshold etc [11-56]. An external electric field shifts and broadens the bound state atomic levels. The standard quantum-mechanical approach relates the complex eigenenergies (EE) $E = E_r + i\Gamma/2$ and complex eigenfunctions (EF) to the shape resonances. The field effects drastically increase upon going from one excited level to another. The highest levels overlap forming a "new continuum" with lowered boundary.

The calculation difficulties inherent to the standard quantum mechanical approach are well known. Here one should mention the well-known Dyson phenomenon. The Wentzel-Kramers-Brillouin (WKB) approximation overcomes these difficulties for the states lying far from the " new continuum" boundary. Some modifications of the WKB method (see review in Ref. [58]) are introduced in Stebbings and Dunning, Kondratovich and Ostrovsky, Popov et al; Ivanov-Letokhov [34,35] have fulfilled the first estimations of the effectiviness of the selective ionization of the Rydberg atom using a DC electric and laser fields within the quasiclassical model. Different calculational procedures are used in the Pade and then Borel summation of the divergent Rayleigh-Schrödinger perturbation theory

(PT) series (Franceschini et al 1985, Popov et al 1990) and in the sufficiently exact numerical solution of the difference equations following from expansion of the wave function over finite basis (Benassi ans Grecchi 1980, Maquet et al 1983, Kolosov 1987, Telnov 1989, Anokhin-Ivanov 1994), complex-coordinate method, quantum defect approximation etc [11-55]. Hehenberger, McIntosh and E. Brändas [11] have applied the Weyl's theory to the Stark effect in the hydrogen atom. They have shown that one of the interesting features of Weyl's theory is that it requires a complex parameter and complex solutions to the differential equations making it a powerful tool for the treatment of resonance states. Rittby, Elander and Brändas [12] have applied the Weyl's theory and the complex-rotation method to phenomena associated with a continuum spectrum. Brändas and Froelich have shown that a complex scale transformation of the time–dependent Schrödinger equation leads to a symmetric EE value problem containing both bound states and resonance (complex) EE values as solutions. It is worth to note that application of the complex coordinate method to a resonance problem has been justified in [20-23]. Themelis and Nicolaides [41-44] adopted an *ab initio* theory to compute the complex energy of multielectron atomic states. Their approach is based on the state-specific construction of a non-Hermitian matrix according to the form of the decaying-state EF which emerges from the complex eigenvalue Schrodinger equation (CESE) theory. Sahoo and Ho [45,46] carried out the calculation the Stark resonances energies and widths in the lithium atom on the base of the complex absorbing potential (CAP) method. Jianguo Rao et al and Hui-Yan Meng et al [38-40] have presented the B-spline-based coordinate rotation method plus the model potential approach and applied it to investigate the complex energies of low-lying resonances of the hydrogen and lithium atoms in an electric field. In Refs.[56-58] it has been presented a consistent uniform quantum approach to the solution of the non-stationary state problems including the DC (Direct Current) strong-field Stark effect and also scattering problem It is based on the operator form of the perturbation theory (OPT) for the Schrödinger equation.

In this work we present new computational realization of the modified OPT method for the non-H atomic systems and apply it to studying the Stark resonances parameters for the lithium atom in a DC electric field. The Stark resonances parameters energies and widths are calculated and compared with the data of calculations on the basis of the alternative sophisticated complex eigenvalue approaches [40,42,45].

2. Modified operator perturbation theory approach to multielectron atom in an electric field

As usually [56,57], the Schrödinger equation for the electron function taking into account the uniform electric field and field of the nucleus (Coulomb units are used: for length, 1 unit is h^2/Ze^2m; for energy 1 unit is mZ^2e^4/h^2) is:

$$[-(1 - N/Z) / r + V_m(r) + \varepsilon z - 1/2\Delta - E]\, \psi = 0, \tag{1}$$

where E is the electron energy, Z is the nucleus charge, N is the number of electrons in the atomic core (for the hydrogen atom: Z=1, N=0), V_m is a model potential that describes interaction with the electron shells for multi-electron atom (for the hydrogen atom V_m=0). Firstly, we only deal with the Coulomb part of the electron-atomic residue interaction. The non-Coulomb part, as well as relativistic effects, can be approximately accounted for next step. The separation of variables in the parabolic coordinates ($\xi = r + z,\ \ \eta = r - z,\ \ \varphi = \tan^{-1}(y/x)$):

$$\psi(\zeta,\eta,\varphi) = f(\zeta)\, g(\eta) (\zeta \cdot \eta)^{|m|/2} \exp(im\varphi)/(2\pi)^{1/2} \tag{2}$$

transforms it to the system of two equations for the functions f, g:

$$f'' + \frac{|m|+1}{t} f' + [1/2E + (\beta_1 - N/Z) / t - 1/4\varepsilon(t)\, t\,] f = 0, \tag{3}$$

$$g'' + \frac{|m|+1}{t} g' + [1/2E+\beta_2 / t + 1/4\varepsilon(t)\ t\,]\, g = 0, \tag{4}$$

coupled through the constraint on the separation constants:

$$\beta_1 + \beta_2 = 1 \tag{5}$$

Let us underline that here and below variable t denotes the argument common for the whole differential equations system (4). For the uniform electric field $\varepsilon(t) = \varepsilon$. Potential energy in equation (4) has the barrier. Two turning points for the classical motion along the η axis, t_1 and t_2, at a given energy E are the solutions of the quadratic equation ($\beta = \beta_1$, $E = E_0$):

$$t_2=\{[\,E^2_0 - 4\varepsilon(1-\beta)]^{1/2} - E_0\}/\varepsilon, \tag{6}$$

$$t_1=\{-[E^2_0 - 4\varepsilon(1-\beta)]^{1/2} - E_0\}/\varepsilon, \quad t_1< t_2 \tag{7}$$

To simplify the calculational procedure, the uniform electric field ε in (3) and (4) should be substituted by the function [56]:

$$\varepsilon(t) = \frac{1}{t}\,\varepsilon\left[(t-\tau)\frac{\tau^4}{\tau^4+t^4}+\tau\right] \tag{8}$$

with sufficiently large τ (τ=1.5t_2). The motivation of a choice of the $\varepsilon(t)$ and some physical features of electron motion along the η-axis are presented in Refs. [56-58]. Here we only underline that the function $\varepsilon(t)$ practically coincides with the constant ε in the inner barrier motion region, i.e. $t<t_2$ and disappears at $t>t_2$. It is important that the final results do not depend on the parameter τ. It is carefully checked in the numerical calculation. The scattering states energy spectrum now spreads over the range $(-\varepsilon\tau/2,+\infty)$, compared with $(-\infty,+\infty)$ in the uniform field. In contrast to the case of a free atom in scattering states in the presence of the uniform electric field remain quantified at any energy E, i.e. only definite values of β_1 are possible. The latter are determined by the confinement condition for the motion along the η-axis. The same is true in our case, but only for $E \subset \left(-\frac{1}{2}\varepsilon\tau, +\frac{1}{2}\varepsilon\tau\right)$. Ultimately, such a procedure provides construction of realistic functions of the bound and scattering states. In a certain sense, this completely corresponds to the advantages of the distorted-wave approximation known in scattering theory [61,62].

The total Hamiltonian H (ς, ν, φ) does not possess the bound stationary states. According to OPT [37,40]), one has to define the zero order Hamiltonian H_0, so that its spectrum reproduces qualitatively that of the initial one. To calculate the width Γ of the concrete quasistationary state in the lowest PT order one needs only two zeroth–order EF of H_0: bound state function $\Psi_{Eb}(\varepsilon, \eta, \varphi)$ and scattering state function $\Psi_{Es}(\varepsilon, \eta, \varphi)$ with the same EE. It can be solved a more general problem: a construction of the bound state function along with its complete orthogonal complementary of scattering functions Ψ_E with $E \subset \left(-\frac{1}{2}\varepsilon\tau, +\infty\right)$. First, one has to define the EE of the expected bound state. It is the well-known problem of states quantification in the case of the penetrable barrier [2]. The system (3) and (4) with the total Hamiltonian H is solved under the conditions:

$$f(\mathrm{t})\rightarrow 0 \text{ at } t \Rightarrow \infty, \tag{9}$$

$$\partial x(\beta, E) / \partial E = 0$$

with

$$x(\beta, E) = \lim_{t\Rightarrow\infty} [\, g^2(t) + \{g'(t) / \mathrm{k}\}^2]\, t^{|m|+1}. \tag{10}$$

These two conditions quantify the bound energy E and separation constant β_1. Further one should solve the system of the ordinary differential equations (3) and (4) with probe pairs of E, β_1. The corresponding EF:

$$\psi_{Eb}(\zeta, \eta, \varphi) = f_{Eb}(\zeta)\, g_{Eb}(\eta)(\zeta\,\eta)^{|m|/2} \exp(im\varphi)(2\pi)^{--1/2}, \tag{11}$$

where $f_{Eb}(t)$ is the solution of (3) (with the just determined E, β_1) at $t \subset (0,\infty)$ and $g_{Eb}(t)$ is the solution of (4) (with the same E, β_1) at $t < t_2$ (inside barrier) and $g(t) = 0$ otherwise.

These bound state EE, eigenvalue β_1 and EF for the zero-order Hamiltonian H_0 coincide with those for the total Hamiltonian H at $\varepsilon \Rightarrow 0$, where all the states can be classified due to the quantum numbers n, n_1, n_2, m (principal, parabolic, azimuthal) connected with E, β_1, m by the well-known expressions. The scattering state functions:

$$\psi_{E's}(\zeta, \eta, \varphi) = f_{E's}(\zeta)\, g_{E''s}(\eta)\, (\zeta\,\eta)^{|m|/2} \exp(im\varphi)(2\pi)^{-1/2} \tag{12}$$

is orthogonal to the above defined bound state function and to each other. In addition, these functions must describe the motion of the ejected electron, i.e. $g_{E's}$ must satisfy the equation (4) asymptotically. Following the OPT ideology [56], we choose the next form of $g_{E's}$:

$$g_{E's}(t) = g_1(t) - z_2' g_2(t) \tag{13}$$

with $f_{E's}$ and $g_1(t)$ satisfying the differential equations (3) and (4). The function $g_2(t)$ satisfies the non-homogeneous differential equation, which differs from (4) only by the right-hand term, disappearing at $t \Rightarrow \infty$. The total equation system, determining the scattering function, reads

$$f''_{E's} + \frac{|m|+1}{t} f'_{E's} + [1/2E' + (\beta_1' - N/Z)/t - 1/4\, \varepsilon(t)t] f_{E's} = 0,$$

$$g_1'' + \frac{|m|+1}{t} g_1' + [1/2E' + \beta_2'/t + 1/4\varepsilon(t)t] g_1 = 0, \tag{14}$$

$$g_2'' + \frac{|m|+1}{t} g_2' + [1/2E + \beta_2'/t + 1/4\varepsilon(t)t] g_2 = 2g_{Eb},$$

($\beta_1' + \beta_2' = 1$). At the given E', the only quantum parameter β_1' is determined by the natural boundary condition: $f_{E's} \Rightarrow 0$ at $t \Rightarrow \infty$. Of course: $\beta_1' = \beta_1$, $f_{E's} = f_{Eb}$ at $E' = E$; only this case is needed in the particular problem we deal with here. The coefficient z_2' ensures the orthogonality condition:

$$\langle \Psi_{Eb} | \Psi_{E's} \rangle = 0. \tag{15}$$

The imaginary part of state energy in the lowest PT order is as follows:

$$\mathrm{Im}E = \Gamma/2 = \pi|<\Psi_{Eb}|H|\Psi_{Es}>|^2 \quad (16)$$

with the total Hamiltonian H. The state functions Ψ_{Eb} and Ψ_{Es} are assumed to be normalized to 1 and by the $\delta(k-k')$ condition, accordingly. The matrix elements $\langle\Psi_{Eb}|H|\Psi_{E's}\rangle$ entering the high- order PT corrections can be determined in the same way. They can be expressed through the set of one-dimensional integrals, described in details in Refs. [56-58].

In contrast to the hydrogen atom, the non-relativistic Schrödinger equation for an electron moving in the field of the atomic core in many-electron atom (in particular, an alkali element) and a uniform external electric field does not allow separation of variables in the parabolic coordinates ξ, η, φ [2].One of the ways this problem could be related to the use of effective potentials, chosen in such a way (for example, in the Miller-Green approximation [51]) that to achieve the separation of variables in the Schrödinger equation. Here the model potential approach [59,60] or the quantum defect approximation [57] can be used. One may introduce the ion core charge z^* for the multielectron atom. According to standard quantum defect theory, the relation between quantum defect value μ_1, electron energy E and principal quantum number n is: $\mu_1 = m - z^*(-2E)^{-1/2}$. The quantum defect in the parabolic coordinates $\delta(n_1n_2m)$ is connected to the quantum defect value of the free $(\varepsilon = 0)$ atom by the following relation [57]:

$$\delta(n_1n_2m) = (1/n)\sum_{l=m}^{n-1}(2l+1)(C^{JM}_{J,M-mlm})^2\mu_l, \quad (17)$$

$$J = (n-1)/2, M = (n_1 - n_2 + m)/2.$$

Using the quantum defect approximation allows to modify the OPT method for the non-H atoms. In general form speech is about solving the problem associated with the use of a screened Coulomb approximation and further taking into account the non-Coulomb potential of the atomic residue in the calculation of mixing levels of an atom in an external field. The details of this procedure are presented in Ref. [57]. Further let us underline that the whole computational procedure at known resonance energy E and separation parameter β_1 has been reduced to the solution of one system of the ordinary differential equations. This master system includes the differential equations for the state functions f_{Eb}, g_{Eb}, f_{Es}, g_{Es}, as well as the equations for the integrals to compute (16) etc. The procedure is sufficiently simple and realized as the numerical code with using the fourth-order Runge–Kutta method of solving the differential equations (the atomic code "Superatom-ISAN") [3,56-59].

3. Some results and conclusions

We have applied the developed computational approach to calculating the complex energy eigenvalues representing the shifted and broadened 2s state of lithium atom as a function of electric field strength. Sahoo and Ho [45] performed the calculation on the basis of a complex absorbing potential (CAP) method. Themelis and Nicolaides [42] adopted *ab initio* theory to compute the complex energy of multielectron atomic states. Their approach is based on the state-specific construction of a non-Hermitian matrix according to the form of the decaying-state eigenfunction which emerges from the complex eigenvalue Schrodinger equation (CESE) theory. Meng et al [40] has elaborated the B-spline based coordinate rotation (B-CR) approach. In Table 1 we present our data on the eigenvalues EE (in atomic units: a.u.) representing the shifted and broadened 2s state of lithium atom as a function of electric field strength (in a.u.).

IOP Conf. Series: Journal of Physics: Conf. Series **1136** (2018) 012012 doi:10.1088/1742-6596/1136/1/012012

Table 1. Complex eigenvalues representing the shifted and broadened 2s state of lithium atom as a function of the field strength, calculated by different methods (see text).

Li 2s	B-CR [40]	B-CR [40]	CAP [45]	CAP [45]	CESE [42]	Semiclassical [42]	This work	This work
ε (a.u.)	E_r (a.u.)	Γ/2 (a.u)	E_r (a.u.)	Γ/2 (a.u)	Γ/2 (a.u)	Γ/2 (a.u)	E_r (a.u.)	Γ/2 (a.u)
0.0050	−0.20009	–	−0.20019	7.20[−9]	–	4.65[−11]	-0.20012	–
0.0055	−0.20052	3.80[−10]*	−0.20062	7.45[−9]	5.91[−10]	7.77[−10]	−0.20055	5.45[−10]
0.0060	−0.20100	3.74[−9]*	−0.20109	8.85[−9]	5.46[−9]	7.99[−9]	−0.20103	4.94[−9]
0.0065	−0.20151	2.49[−8]*	−0.20161	3.14[−8]	3.52[−8]	5.66[−8]	−0.20154	3.16[−8]
0.0070	−0.20207	1.24[−7]*	−0.20217	1.61[−7]	1.69[−7]	3.00[−7]	−0.20211	1.52[−7]
0.0075	−0.20268	4.87[−7]	−0.20277	4.83[−7]	6.49[−7]	1.26[−6]	−0.20272	5.08[−7]
0.0080	−0.20333	1.57[−6]	−0.20342	1.48[−6]	2.06[−6]	4.37[−6]	−0.20336	1.73[−6]
0.0090	−0.20477	1.05[−5]	−0.20486	1.05[−5]	1.33[−5]	3.41[−5]	−0.20480	1.21[−5]
0.0100	−0.20642	4.50[−5]	−0.20651	4.77[−5]	5.50[−5]	1.72[−4]	−0.20645	4.81[−5]
0.0125	−0.21147	4.76[−4]	−0.21155	4.68[−4]	5.46[−4]	2.95[−3]	−0.21149	4.96[−4]
0.0150	−0.21749	1.78[−3]	−0.21754	1.76[−3]	1.97[−3]	1.82[−2]	−0.21751	1.83[−3]
0.0175	−0.22393	4.03[−3]	−0.22397	4.06[−3]	4.35[−3]	6.35[−2]	−0.22394	4.24[−3]

For comparison the analogous results, obtained on the basis of the CAP, CESE, B-CR methods [40,42,45] are presented. Analysis of the data shows that the positions (energies) of the Stark resonances in the present calculation are in a physically reasonable agreement with theoretical data obtained by other, in particular, CESE and B-CR methods. However, the results for the width of resonance differ more significantly from each other. For example, the CAP calculation for the width of the 2s state at strength F< 0.0060 a.u. gives systematically larger values than obtained by the CESE, B-CR and our methods. The resonance width values are higher than the corresponding B-CR data and correspondingly a little less than the values, obtained within the CESE method for all strengths of the electric field under consideration. Concerning the widths of resonances it should paid to attention on convergence aspect for the CAP and CESE method. As it has been underlined in [40], in the case of a weak electric field (naturally the widths of resonances became very small), the methods have difficulties in obtaining a stable value of a width. In order to obtain the well-converged results, it is necessary to use larger basis size. Naturally, in a limit of a weak electric field the well-known quasiclassical WKB approximation and standard PT [1,2] calculation will be more appropriate. One of the advantages of the B-CR method is possibility to apply in the case of increasing field strengths without a significant computational effort growth, however, the convergence of the width Γ to obtain reliable complex eigenvalues should be carefully carried out. In the CAP method, there is no systematic way of choosing a scaling factor in an quite artificial complex potential, which is added to the original atomic Hamiltonian. One of the serious advantages of the modified OPT method is that an increasing a field strength does not lead to an increase of computational effort and there is no a convergence problem. The approach provides not only resonance state function definition but also the construction of the complex EE state function along with its complete orthogonal complementary of the scattering functions. In Refs. [63,64] the operator PT ideology has been used to consider a problem of resonances in the heavy ions collisions and AC Stark effect. The OPT approach can be used in calculating the Stark resonance parameters in a case of the strong electric fields and it is of a great interest for many modern atomic and molecular physics applications.

Acknowledgements

The authors would like to thank Prof. R. Spezia and Prof. R. Vuilleumier and colleagues for invitation to make contributions to CCP-2017 (Paris, France). The useful comments of the anonymous referees are very much acknowledged too.

IOP Conf. Series: Journal of Physics: Conf. Series **1136** (2018) 012012 doi:10.1088/1742-6596/1136/1/012012

References

[1] Bethe H A and Salpeter E E 1957 *Quantum mechanics of One- and Two-Electron Atoms* (Berlin: Springer).

[2] Landau L D and Lifshitz E M 1965 *Quantum Mechanics* (Oxford: Pergamon).

[3] Glushkov A V 2008 *Relativistic Quantum theory. Relativistic quantum mechanics of atomic systems* (Odessa: Astroprint)

[4] Zimmerman M L, Littman M G, Kash M M and Kleppner D 1979 *Phys. Rev.* A **20** 2251

[5] Fabre C, Goy P and Haroche S 1977 *J. Phys. B: At. Mol. Phys.* **10** L183

[6] Harmin D A 1982 *Phys. Rev. Lett.* **49** 128

[7] Svinarenko A A, Khetselius O Yu, Buyadzhi V V, Florko T A, Zaichko P A and Ponomarenko E L 2014 *J. Phys.: Conf. Ser.* **548** 012048

[8] Khetselius O Yu 2012 *J. Phys.: Conf. Ser.* **397** 012012

[9] Buyadzhi V V, Zaichko P A, Antoshkina O A, Kulakli T A, Prepelitsa G P, Ternovsky V B and Mansarliysky V F 2017 *J. Phys.: Conf. Ser.* **905** 012003

[10] Florko T A, Ambrosov S V, Svinarenko A A and Tkach T B 2012 *J. Phys.: Conf. Ser.* **397** 012037

[11] Hehenberger M, McIntosh H V and Brändas E 1974 *Phys. Rev. A* **10** 1494

[12] Rittby M, Elander N and Brändas E 1981 *Phys. Rev. A* **24** 1636

[13] Froelich P, Davidson E R, Brändas E 1983 *Phys. Rev. A* **28** 2641

[14] Lipkin N, Moiseyev N and Brändas E 1989 *Phys. Rev. A* **40** 549

[15] Moiseyev N 1998 *Phys. Rep.* **302** 211

[16] Reinhardt W P 1982 *Int. J. Quant.Chem.* **21** 133

[17] Reinhardt W P 1982 *Annu. Rev. Phys. Chem.* **33** 223

[18] Cerjan C, Hedges R, Holt C, Reinhardt W P, Scheibner K and Wendoloski J J 1978 *Int. J. Quant. Chem.* **14** 393

[19] Maquet A, Chu S-I and Reinhardt W P 1983 *Phys. Rev. A* **27** 2946

[20] Simon B 1979 *Phys. Lett. A* **71**, 211

[21] Herbst I W and Simon B 1981 *Commun.Math, Phys.* **80**, 181

[22] Aquilar J and Combes J M 1971 *Commun. Math. Phys.* **22** 269

[23] Balslev E and Combes J M 1971 *Commun. Math. Phys.* **22** 280

[24] Luc-Koenig E and Bachelier A 1980 *J. Phys.B: At. Mol. Phys.* **13** 1743

[25] Franceschini V, Grecchi V and Silverstone H 1985 *Phys. Rev. A* **32** 1338

[26] Benassi L and Grecchi V 1980 *J. Phys. B: At. Mol. Opt. Phys.*. **13** 911

[27] Anokhin S B and Ivanov M V 1984 *Opt. Spectr.* **59** 499

[28] Telnov D A 1989 *J. Phys. B.: At.Mol. Opt.Phys*. **22** 1399

[29] Kolosov V V 1987 *J. Phys. B: At. Mol. Opt. Phys.* **20** 2359

[30] Filho O, Fonseca A, Nazareno H and Guimarães P 1990 Phys. Rev. A **42** 4008

[31] Kondratovich V and Ostrovsky V 1984 *J. Phys. B: At. Mol. Opt. Phys.* **17** 2011

[32] Popov V S 1993 *Phys. Lett. A* **173** 63

[33] Popov V S, Mur V D, Sergeev A V and Weinberg V M 1990 *Phys. Lett. A* **149** 418

[34] Ivanov L N and Letokhov V S 1975 *JETP* **68** 1748

[35] Letokhov V S, Ivanov L N and Glushkov A V *Laser separation of heavy lanthanides and actinides isotopes: Autoionization resonances and decay in electric field* 1992 *Preprint of Institute for Spectroscopy of Russian Academy of Sciences* (Troitsk: ISAN) 92-5-AS

[36] Khetselius O Yu 2008 *AIP Conf. Proc.* **1058** 363

[37] Wangt K, Hot T-S and Chu S-I 2006 *Phys. Rev. A* **73** 023403

[38] Rao J, Liu W and Li B 1994 *Phys. Rev. A* **50** 1916

[39] Rao J and Li B 1995 *Phys. Rev. A* **51**, 4526 (1995).

[40] Meng H-Y, Zhang Y-X, Kang S, Shi T-Y and Zhan M-S 2008 *J. Phys. B: At. Mol. Opt. Phys.* **41** 155003

[41] Themelis S I and Nicolaides C A 2000 *J. Phys. B: At. Mol. Opt. Phys.* **33** 5561

[42] Themelis S I and Nicolaides C A 2001 *J. Phys. B: At. Mol. Opt. Phys.* **34** 2905
[43] Mercouris T and Nikolaides C A 2003 *Phys. Rev. A* **67** 063403
[44] Nicolaides C A 2017 *Adv. Quant. Chem.* **74** 149
[45] Sahoo S and Ho Y K 2000 *J. Phys. B: At. Mol. Opt. Phys.* **33** 5151
[46] Ho Y-K 1983 *Phys.Rep.* **99** 3
[47] Ivanov I A and Ho Y-K 2004 *Phys.Rev. A* **69** 023407
[48] Khetselius O Yu 2009 *Int. J. Quant. Chem.* **109** 3330
[49] Khetselius O Yu 2009 *Phys.Scr.* **T135** 014023
[50] Khetselius O Yu 2010 *AIP Conf. Proc.* **1290** 29
[51] Miller K J and Green A E S 1974 *J.Chem.Phys*. **60** 2617
[52] Svinarenko A A 2014 *J. Phys.: Conf. Ser*. **548** 012039
[53] Grutter M, Zehnder O, Softley T P and Merkt F, 2008 *J. Phys. B: At. Mol. Opt. Phys*. 41 115001
[54] Merkt F, Osterwalder A, Seiler R, Signorell R, Palm H, Schmutz H and Gunzinger R 1998 *J. Phys. B: At. Mol. Opt. Phys.* **31** 1705
[55] Stambulchik E and Maron I *2008 J. Phys. B: At. Mol. Opt. Phys.* **41** 095703
[56] Glushkov A V and Ivanov L N 1993 *J. Phys. B: At. Mol. Opt. Phys.* **26** L379
[57] Glushkov A V, Ambrosov S V, Ignatenko A V, Korchevsky D A 2004 *Int. J. Quant. Chem.* **99** 936
[58] Glushkov A V 2013 *Advances in Quantum Methods and Applications in Chemistry, Physics, and Biology. Series: Progress in Theoretical Chemistry and Physics, Eds.* Hotokka M, Maruani J, Brändas E, Delgado-Barrio G (Cham: Springer) **27** pp 161-178
[59] Ivanova E P , Ivanov L N, Kramida A E and Glushkov A V 1985 *Phys.Scr.* **32** 512
[60] Khetselius O Yu, Florko T A, Svinarenko A A and Tkach T B 2013 *Phys. Scr.* **T153** 014037
[61] Madison D H and Al-Hagan 2010 *J. At. Mol. Opt. Phys.(Hindawi)* **2010** 367180
[62] Whelan T and Walters H R J 1990 *J. Phys. B: At. Mol. Opt. Phys.* **23** 2989
[63] Glushkov A V 2005 *AIP Conf. Proc*. **796** 206
[64] Glushkov A V 2014 *J. Phys.: Conf. Ser*. **548** 012020

IOP Conf. Series: Journal of Physics: Conf. Series **1136** (2018) 012013 doi:10.1088/1742-6596/1136/1/012013

Multilevel Lattice Boltzmann-Particle Dynamics simulations at the Physics-Biology interface

Massimo Bernaschi[1], **Simone Melchionna**[2],**Philippe Derreumaux**[3], **Fabio Sterpone**[3], **Sauro Succi**[1,4]

[1] IAC-CNR, via dei Taurini 19, 00185, Rome, Italy
[2] ISC-CNR, Dipartimento di Fisica, Universitá Sapienza, P.le A. Moro, 5, 00185, Rome, Italy
[3] LBT, IBPC, CNRS UPR9080, Univ. Paris Diderot, Sorbonne Paris Cité, 13 rue Pierre et Marie Curie, 75005, Paris, France
[4] IACS-SEAS, Harvard University, Oxford Street 33, 02432, Cambridge, USA

E-mail: succi@iac.cnr.it

Abstract. We discuss current multilevel Lattice Boltzmann-Particle Dynamics (LBPD) simulations of complex states of flowing matter at the interface between physics and biology. We also offer a few considerations on prospective Exascale simulations in the direction of computational physiology.

1. Introduction

The quantitative description of flowing matter traditionally relies upon two basic pillars: continuum fluid mechanics and atomistic/molecular dynamics. The former describes matter as a collection of space-filling (continuum) fields, say density, pressure and flow, whereas the latter explicitly acknowledges the discrete nature of the microscopic constituents of matter, i.e., atoms and molecules. By definition, they deal with very different ranges of space and time scales, which we denote generically as macro and micro, respectively. As computing power and algorithms grow faster and stronger, the distinction between the macro and microlevels becomes increasingly fuzzier, as the corresponding range of operation start to develop a sizeable overlap. More precisely, current upper-end molecular dynamics simulations can simulate micrometric regions over milliseconds [1, 2, 3, 4], which is sufficient to generate a sizeable overlap with continuum fluid dynamics; a cubic micron of water contains about thirty billion molecules, more than enough for the continuum description to apply. Under such conditions, it is only natural to approach the quantitative simulation of flowing matter by multiscale strategies capable of accomodating both micro and microlevels within a unified computational harness.

Such strategy, multiscale modelling for short, experienced a tremendous boost over the last two decades across virtually all walks of science and engineering. Particularly noteworthy, though, is the case of computational biology, which offers a pletora of instances whereby the concurrent interaction of several scales is key to a proper understanding of the phenomena in point. Perhaps the most spectacular case being the molecular mechanics/quantum-mechanics simulation of proteins, which made it to the Nobel recognition in 2013 [5, 6].

In this paper we discuss another instance of multiscale modelling that we label LBPD, for Lattice Boltzmann Particle Dynamics.

The LB part takes care of the continuum sector, typically the flow solvent, while the PD is in charge of molecular motion. The distinctive trait of LBPD is that *both* micro and macro descriptions are inherently based on *mesoscale* (coarse-grained) representation of both solvent and molecular components. In other words, LBPD is strongly rooted within the two pillars of kinetic theory: the Boltzmann equation and stochastic particle dynamics, as we are going to briefly detail in the sequel.

1.1. Lattice Boltzmann

The basic object of Boltzmann kinetic theory is the distribution function $f(\vec{r}, \vec{v}; t)$ measuring the probability density of finding a molecule at position $\vec{r}$ and time t with molecular velocity $\vec{v}$.

This makes a pretty demanding computational object, which obeys a complicated non-linear integro-differential equation in 6 dimensional space, plus time. Therefore, notwithstanding impressive advances, solving the Boltzmann equation remains a very challenging task to this day.

The basic idea of the Lattice Boltzmann method is to constrain the velocity degrees of freedom to a discrete lattice with sufficient symmetry to protect the conservation laws which secure the emergence of standard hydrodynamic behavior in the macroscopic limit.

The classic LB scheme in single-relaxation time (BGK) form reads as follows [7, 8]:

$$f_i(\vec{r} + \vec{c}_i; t + \Delta t) - f_i(\vec{r}; t) = -\omega \Delta t (f_i(\vec{r}; t) - f_i^{eq}(\vec{r}; t)) + S_i(\vec{r}; t)\Delta t, \; i = 0, b \qquad (1)$$

where f_i is the discrete Boltzmann distribution associated with the discrete velocity $\vec{c}_i$, $i = 0, b$ running over the discrete lattice, to be detailed shortly. In the above, ω is a relaxation parameter controlling the fluid viscosity and f_i^{eq} is the lattice local equilibrium, basically the local Maxwell-Boltzmann distribution truncated to the second order in the Mach number. The truncation is not a luxury, but a necessity dictated by Galilean invariance. The point is that the mass-momentum-momentum flux conservations in discrete velocity space (latin subscripts run over spatial dimensions x, y, z and p denotes the fluid pressure),

$$\sum_i f_i^{eq}\{1, c_{ia}, c_{ia}c_{ib}\} = \{\rho, \rho u_a, \rho u_a u_b + P\delta_{ab}\} \qquad (2)$$

can not be fulfilled for an *arbitrary* flow field u_a, if the local Maxwell-Boltzmann distribution is retained with the plain replacement $\vec{v} = \vec{c}_i$.

Finally, S_i is a generic source of mass/momentum/energy, describing the coupling of the fluid with the surrounding environment. This term is crucial for the extension of LB beyond Navier-Stokes hydrodynamics.

The above constraints, though, can be matched perturbatively by expanding the local Maxwellian to second order in the Mach number, which is sufficient to recover the (isothermal) Navier-Stokes equations, since these equations are quadratic in the fluid velocity. Full Galilean invariance for an arbitrary flow field u_a implies instead an *infinite* series in the Mach number, corresponding to the full expansion of the local Maxwellian in Hermite polynomials.

The actual expression of the discrete local equilibria reads as follows:

$$f_i^{eq} = w_i \rho (1 + u_i + \frac{1}{2} q_i) \qquad (3)$$

where $u_i = c_{ia}u_a/c_s^2$ and $q_i = (c_{ia}c_{ib} - c_s^2\delta_{ab})u_a u_b/c_s^4$ represent the dipole and quadrupole contributions, respectively. Hereafter, repeated indices are summed upon. The lattice equilibria are polynomial truncations of their continuum Maxwell-Boltzmann counterparts, which means that they are non-negative definite only in a finite range of fluid velocities, typically of the

IOP Conf. Series: Journal of Physics: Conf. Series **1136** (2018) 012013 doi:10.1088/1742-6596/1136/1/012013

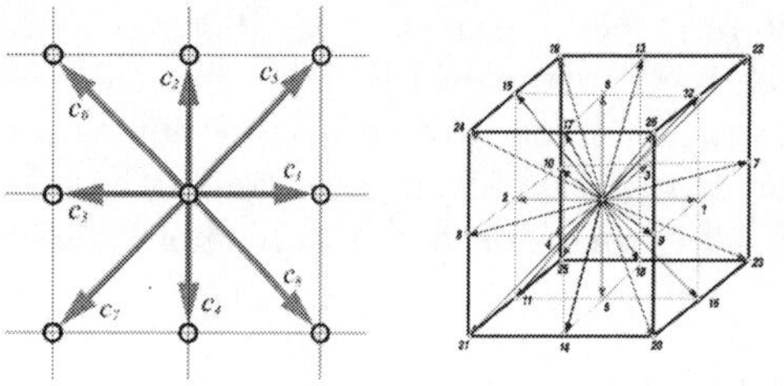

Figure 1. Examples of standard 2d and 3d LB lattices, with 9 (left) and 27 (right picture) discrete speeds, respectively. The 3d LB model with 19 speed model is the union of green and blue arrows in the right picture.

order of $u/c_s \sim 0.3$. This configures the LB method as an appropriate description of quasi-incompressible, low Mach-number flows.

In the above, w_i is a set of lattice-dependent weights, normalized to unity, which represent the lattice analogue of the global Maxwell-Boltzmann distribution. Finally $c_s^2 = \sum_i w_i c_{ia}^2$ is the lattice sound speed.

The corresponding discrete velocities are subsets of the D3Q27 mother lattice with 27 discrete speeds in three spatial dimensions. D2Q9 and D3Q27 are the direct tensor product of the elementary one-dimensional D1Q3 stencil, $c_{ix} = \{-1, 0, +1\}$, in two and three spatial dimensions, respectively (see Fig. 1).

The expansion (3) implies that hydrodynamic LB flows are bound to be quasi-incompressible, i.e., Mach number well below unity. Likewise, third order kinetic moments, describing energy and heat flux, are not correctly reproduced, since these terms require sixth-order isotropic lattice tensors. Such constraints can not be met by lattices confined to the first Brillouin region described by D3Q27: to this purpose, higher order lattices extending beyond the first Brillouin cell are necessary.

Based on (2) and (3), it can be shown that in order for the LB to recover the NS equations in the zero-Knudsen hydrodynamic limit, isotropic tensors only up to order four need to be reproduced exactly in the lattice.

The main computational merits of LB as a fluid solver have been amply documented in a vast literature, hence here we shall simply remind its main highlight: regardless of the spacetime complexity of the emerging macrocell physics, the LB information always move along straight lines, the molecular streamlines defined by the condition

$$\Delta\vec{r}_i - \vec{c}_i\Delta t = 0$$

This is the source of major computational benefits, including an outstanding amenability to parallel computing, a key aspect for leading edge LBPD simulations at the physics/biology interface.

It is also implies limitations towards complex geometries, as typically handled by body-fitted coordinates or finite volumes or elements, an interesting topic which is beyond the scope of the present paper.

1.2. Particle Dynamics

Most biological phenomena depend crucially on the interaction of biological bodies with the surrounding solvent, i.e. water. Biological bodies can range from small molecules, such as organic molecules or peptides, to large ones such as proteins, nucleic acids, up to various type of cells. Such wide spectrum is echoed by a corresponding variety of computational representations, a topic which makes a vigorous branch of modern research on mesoscale methods [9]. For brevity, here we shall refer to the biological body as to a collection of $p = 1, N$ "particles" following a suitable (stochastic) dynamics. i.e.

$$\frac{d\vec{r}_p}{dt} = \vec{v}_p \tag{4}$$

$$\frac{d\vec{v}_p}{dt} = \frac{\vec{F}_{pp} + \vec{F}_{fp}}{m_p} \tag{5}$$

$$\tag{6}$$

The key, of course, is in the specific expression of the force term, describing both particle-particle (pp) and fluid-particle interactions (fp). Consistently with the mesoscale nature of LBPD, both are the result of some form of coarse-graining. The pp-part is in charge of reflecting the effect of suitable force fields, while the fp-part typically consists of a deterministic term (systematic drag) and a stochastic one (thermal fluctuations), linked via the Fluctuation Dissipation theorem. From the practical standpoint, a key feature of the LBPD procedure is the cross-talking between the LB fluid and the PD dynamics. Indeed, it is clear that the particles react on the fluid through a particle-fluid force $\vec{F}_{pf}$ which is equal and opposite to $\vec{F}_{fp}$.

This is where technical details take the lead; a suitable description of these technical matters goes beyond the scope of this paper. Here, we just mention that the LBPD coupling typically involves G2P (Grid-to-Particle) interpolations to transfer the flow field into the actual particle position, and the conjugate P2G (Particle-to-Grid) interpolation to transfer the particle forces into the grid locations. The choice of the interpolation kernels can range from point-like Dirac deltas to smoother versions based on higher order finite-support shape functions, pretty much in the spirit of smoothed particle hydrodynamics [10]. An accurate and efficient implementation of these "handshaking" interfaces is key to the success of the LBPD scheme.

1.3. Remarks on LBPD

The LBPD procedure falls within the time-honoured tradition of hybrid particle-in-cell methods, in use since long for particle-field simulations [11, 12]. Nevertheless, LBPD presents some specific features, which we briefly discuss in the following. The mesoscopic representation of macromolecules and surrounding solvent allows to account not only for nanometric-scale hydrodynamics, but in principle, also for accurate macromolecular interactions. *The* key question in this respect is what degree of molecular realism can be regarded as sufficient for the purpose of a realistic biological representation. The answer is problem-specific: representing a lipid chain, a protein, or a DNA chain may require different degrees of chemical specificity, depending on the problem at hand. It is also true, however, that one can employ such coarse-grained potentials in a rather flexible way, as long as the mesoscopic properties, let's say at the nanometer / nanosecond scale, are correctly reproduced. Such scale, fortunately, is precisely where kinetic modelling, particularly for the liquid state of the biological solutions, and the force fields match in accuracy. At larger scales, say microns and above, such level of detail may become irrelevant, therefore a fair representation of thermodynamics, possibly via an equation of state and an accurate representation of fluid mechanics, fills most needs. The latter marks a major point for LBPD, specially in connection with massively parallel implementations in complex geometries. This blue-sky scenario, however, should cope also with long-range forces, such as the electrostatics ones. Fortunately, under typical cellular conditions electrostatic interactions are strongly screened above nanometric separations.

As discussed above, Boltzmann kinetic theory describes the aqueous solvent as a continuum in a probabilistic sense. On the other hand, the LB mesh spacing Δx is defined as a representation (coarse-graining) of the collective kinetic behaviour of a group of solvent molecules. In order to observe hydrodynamic behavior down to the mesh spacing distances, the mean-free-path generally falls below Δx. In the liquid state, the molecular mean free path extends over a few angstroms, and a sub-nanometric lattice spacing can be supported in LB, allowing for the hydrodynamic behaviour to emerge at a distance larger than Δx. Therefore, the lattice grid element must not be viewed as a volumetric entity that contains a fixed number of particles, but instead a numerical support for the probabilistic description of the averaged single particle trajectories. Horbach and Succi [13] have shown that this strategy is effective for the simulation of nano fluids and the obtained results agree very closely with molecular dynamics simulations of simple fluids.

In order to improve physico-chemical realism, with the incorporation of the relevant force-fields, LB models should also be enriched with water-like features, such as directional interactions, so as to make the LB-water solvent more realistic. Preliminary efforts in this direction have been developed in the past, but their thorough validation remains largely open [14].

Many applications naturally suggest themselves; thermal stability of proteins, the onset of neurodegenerative diseases due to peptidic aggregation, diffusion of proteins and trafficking in cellular crowding, being just some examples in point in which a tight coupling between realistic hydrodynamics and chemical specificity is essential.

In passing, we note that LBPD resonates well with the modern focus of in-cell and computational biology. While the past success of the fully atomistic approaches for simulating proteins and DNA was strongly linked to the development of the experimental structural biology [5], and flourished around the central paradigm of the structure/function relationship, the future boost of multi-scale modelling will represent the novel microscope to support system biology where biological function is declined in term of protein/protein multiple interactions, cascade effects and long range transport of information.

1.4. The OPEP coarse-grained model

We pause here to describe shortly the implicit solvent coarse-grained (CG) model for protein OPEP (Optimized Potential for Efficient peptide structure Prediction) [15] that has been integrated in the LBPD scheme [16] and that allows to treat biomolecules at a reasonable chemical accuracy but at a cheaper cost than traditional all-atoms molecular mechanics models. The OPEP force field reserves an atomistic description to the amino-acid backbone including hydrogens, but reduces all side-chain atoms to a single CG bead. The Hamiltonian of the force field is given by bonded and non bonded terms. In the former class we find standard potentials, e.g. harmonic, for bonds, angles and dihedrals. Some extra special terms allow also to map correctly the Ramachandra plots for secondary structures. While the model does not assign explicit charges to the particles, some interactions of electrostatic origins are modelled via ad-hoc potentials. This is the case of the Hydrogen Bond propensity and its cooperative effects via two-body and four-body terms, a critical ingredient to reproduce the sampling of α-helices and β-strands secondary structures. Also ion-pairing between charged amino-acids is modelled with effective potentials of mean force. The non bonded interactions are optimised using standard Lennard-Jones or other short-range type of potentials. As an effect of its resolution and the parametrized potentials, OPEP, at variance with mainstream CG models as Martini, folds peptides and proteins, is a very good tool to sample protein fluctuations including long range rearrangements, and protein aggregation. Because of the atomistic resolution of the backbone, a relatively short time-step as compared to other CG model is needed to correctly integrate the equation of motion, $1.5 fs$ that can be raised to $4\,fs$ if backbone mass rescaling is adopted to reduce the high frequency of hydrogen motions. For many purposes the details of intramolecular flexibility is not necessary, OPEP intramolecular interactions can therefore be replaced by an effective elastic network thus making possible to access longer time-scale in the simulation because of the reduced number of degrees of freedom and a larger time-step ($10\,fs$)

[17, 18].

2. LBPD SIMULATIONS AT THE PHYSICS/BIOLOGY INTERFACE

Several different examples have already shown the potential of LB simulations in multiple areas across Physics and Biology. A selected set of examples will be briefly reviewed in this Section, to provide a taste for the breadth of applications that have been tackled in the recent past. The Physics-Biology interface is enormously rich and varied, ranging from nanometric macromolecular phenomena, to peptidic aggregation and biopolymer translocation, to cellular motion and active matter organization. However, the topic goes way beyond the scope of the present work. Nevertheless, it is important to appreciate the great flexibility of the method to cope with multiple flow conditions and, more importantly, to accomodate the proper degree of chemical specificity required by biological applications.

2.1. Biopolymer Translocation

The translocation of biopolymers, and in particular DNA strands in nanometric pores, provides a showcase of the synergistic hydrodynamic effects assisting or interfering with the translocation process. Such process mimics a genuinely biologic one, whereby viral penetration takes place via the injection of viral DNA into the host cell's cytoplasm. Computer simulations can provide an exhaustive access to the translocation process, both for technological innovations and for a better understanding of the migration of small biopolymers. At technological level, analysis of nanopores could provide a new path to DNA fast-sequencing, provided the physics of translocation is better understood [19]. The key to nanopore-based sequencing is the ability to translocate DNA chains through a nanoconfined environment, where the genetic information can be decoded by optical mapping, ionic or electronic detection.

Translocating a long DNA chain into extremely narrow pores results in large entropy loss caused by the confinement and the need to stretch the DNA macromolecule creating a free energy barrier, which reduces DNA capture rates and causes clogging at the nanochannel/pore entrance. Most of translocation is modulated by the competing effects taking place in the chambers and in the pore, giving rise to a genuine multiscale scenario [20, 21, 22, 23]. When facing such complex set-up, all-atoms Molecular Dynamics methods, or even coarse-grained representations of the translocating biopolymer, necessarily have to give up the explicit representation of the solvent, thus imposing severe limitations to the simulation accuracy. On the other hand, resorting to other simulation methods, based on a direct solution of the Navier-Stokes equations, or by using other mesoscopic numerical methods (Lagrangian or Eulerian based), face several challenges including the generation of consistent noise under confinement, the stability of the numerical method and the request for time averaging of a process involving a handful of molecules/atoms flowing in a narrow space.

One more level of complication arises from the inclusion of electrokinetics, that is, by describing the multi-component saline solution that flows together with DNA from chamber to chamber. To handle such scenario ionic currents and solvent flow are coupled via the exchange of momentum, that gives rise to electro-osmotic and electrophoretic phenomena. The self-consistent equations can still be casted in the LBPD framework in a natural way [24]. In Figure 2 we provide two examples of polymer translocation studied via LBPD.

2.2. Proteins in cell: motion, conformations and folding

Protein functionality often comes along with important conformational changes that mechanically transport biochemical signals. The extreme case is the folding process that allows a protein to acquire a well defined 3D structure ready to function. Reproducing protein folding during its complete temporal span is a major challenge in computational science, with plenty of implications for medicine and drug design. For isolated small and mid-size proteins placed

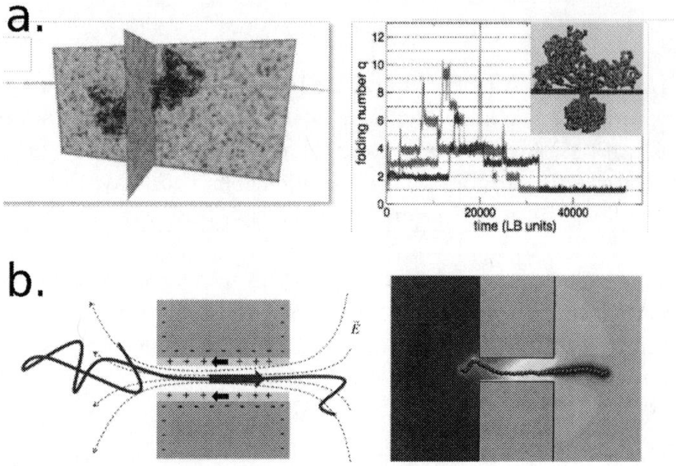

Figure 2. Panel a. Translocation of a neutral biopolymer in a narrow pore. Here, the translocating biopolymer is represented as a simple necklace of beads, by neglecting correlations stemming from the local molecular rigidity or backbone charge, but including thermal noise. Depending on the translocation conditions and pore size, the polymer threads as single-file (left-panel) or multi-file (right panel). Panel b. A charged biopolymer, translocating in a narrow pore. This model is representative of single-stranded DNA in which molecular rigidity is neglected. The charged macromolecule moves in an electrolyte solution, whereby a neutral solvent and counterions and coions migrate due to an externally applied electric field, giving rise to electro-osmotic flow that ultimately causes translocation [24].

in a water solution the problem can be considered computationally solved. All-atom Molecular Dynamics simulations performed at the millisecond timescale by the group of DE Shaw [25, 26] or calculations based on the enhanced sampling Replica Exchange strategy [27] demonstrated that molecular mechanics force fields developed along the years can fold successfully a protein in good agreement with experiments. Such efforts have allowed to explore the folding kinetics and paths, and test the reaction diffusion framework earlier introduced in conjunction with the funnel-theory. The matter changes drastically as soon as the focus is placed on a protein in its natural environment, that is the cell. Here, a protein experiences a strong crowded space where excluded volume effect, specific interactions and solvent mediated correlation significantly affect its dynamics and stability. The possibility to follow the "dance" of proteins in cellular conditions is considered the next challenge in molecular and computational biology, and is stimulating advances in spectroscopic single molecule techniques as well as ensemble methods. Computational biology is responding to the call too. To this purpose it is mandatory the use of simplified and cheap molecular force fields yet realistic enough to capture essential internal motions and intermolecular interactions. LB based approach can disclose the potentiality of a combined multi-scale approach including solvent mediated correlation in implicit solvent MD. In fact, in a cellular environment, along with the strong crowding condition, hydrodynamic interactions (HI) are expected to play a critical role by affecting the mobility of proteins, correlating at long range the cyclic conformational changes of molecular motors with the mobility of passive particles, selecting the paths of conformational changes – including folding events. In some special cases, the fluid environment plays a crucial role by triggering the activity of proteins, as in the case of catch-bonds that explicit their functionality under the presence of shear flow in cell adhesion or in blood coagulation. Several groups have started approaching the challenge, either by attempting to include HI in brownian dynamics[28], by using stochastic rotation dynamics combined with simple coarse-grained model [29] and with the incorporation of the implicit solvent coarse-grained model for protein OPEP within the LBPD harness[16, 17]. For instance, recent studies based on LBPD showed that crowding contributes to slow-down the dynamics of a target protein [15, 17] in very good agreement with Neutron Scattering and NMR experiments. Recent efforts were also directed to model catch-bonds proteins as the respond to Couette and Poiseuille flows [18].

2.3. Amyloid Aggregation

A special case where protein dynamics and conformational changes play a common role is the aggregation of misfolded soluble proteins into fibrils, the hallmark of several neurodegenerative diseases, such as Alzheimer, Parkinson, and Huntington. The aggregation of amyloid β proteins is a complex process, whereby amyloid fibrils extend up to hundred of nanometers, and the time scale of full growth exceeds hours *in vitro.* Understanding the mechanics of amyloid aggregation is crucial to the design of drugs able to prevent fibril formation and toxicity in the brain. Computer modelling is an essential tool to explore the aggregation process. Key thermodynamic insights have been obtained by using oversimplified molecular models [30], mimicking amyloid aggregation at microscopic level is still a challenge [31]. The all-atom approach is prohibitive and a bare implicit solvent model would lack the action of solvent mediated correlations. By deploying the LBPD methodology, it is shown that the effect of HIs is to decisively speed up amyloid aggregation [32]. As a matter of fact, solvent mediated interactions enhance the peptides mobility easing their encounter and collapse in an aggregated structure. Moreover, as the size of the aggregates increases, the latter acts as an active body, attracting isolated monomers which fuse together according to a coagulation model. This is an important, possibly the first, real-life application where macromolecular realism and solvent-mediated interactions are shown to provide a more complete picture for amyloid aggregation.

2.4. Towards Computational Physiology

Computational physiology deals with the dynamics of the main organs of the human body, such as the cardiovascular or the respiratory systems, which operate at macroscopic scales. LB hemodynamics has known a major burst of activity for the last decade, with several applications to coronary, carotid, and cerebral blood flow. These are macroscopic applications which fit within the continuum hydrodynamic picture. The challenge here, is to reach up to physiological scales (1-10 cm), without surrendering the essential microbiological features, the finite-size of red blood cells (8 micron) in the first place [33, 34, 35, 36, 37, 38] and possibly even their elasto-mechanical properties [39].

This is of major interest for many reasons; the granular nature of blood may have a significant impact on the recirculation patterns in the vicinity of natural geometrical irregularities, such as bifurcations, stenoses, aneurysms, or man-made ones, like stents and other medical devices. It is quite clear that a fully 4D (three-space dimensions plus time) real-time numerical and visual access at the blood dynamics from microns to the full-scale geometry, would disclose unprecedented opportunities for personalized precision-medicine. It is to be stressed here that statistical averaging is of little meaning in a physiological context, since each individual is a story of his own. As a result, the detailed access to the patient-specific 4D hemodynamic data across scales of motion, would offer a forward leap in the quality and accuracy of pre-emptive medicine [40, 41].

A direct extension is to simulate blood flows by accounting for the explicit presence of red blood cells (RBC) suspended in plasma. This is a typical case where the hydrodynamic medium hosts particles with finite-size, anisotropic shape, actually prolate ellipsoids that represent RBC cells to a first approximation. In hemodynamics, there are several situations, where the continuous nature of blood should be replaced by the two-component mixture of plasma

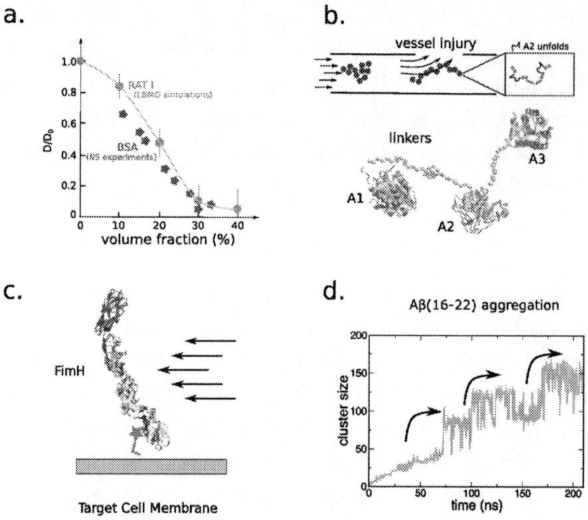

Figure 3. Panel **a**. Translational diffusion at the nanosecond timescale of the protein RAT I in solution at different concentrations and corresponding to a crowded occupied volume ϕ and computed via LBPD (blue circles) as compared to the experimental diffusion slowdown reported for a protein of similar size (BSA) and estimated by Neutron Scattering, details in Ref. [15]. Panel **b**. Modelling of a single domain of the von Willebrand factor protein. This elementary unit of the multi-domain chain is composed by the globular A1, A2 and A3 proteins linked by polypeditic chains. During blood coagulation, the vWf extends under the action of the shear and A2 partial unfolding contributes to further extend the chain. In the multiscale modelling of the vWf unit, A1 and A3 are represented by an elastic network, A2 is represented by the full flexible OPEP force field, while the linkers are modelled using a simple bead for each amino-acids and ad hoc ultra-coarse grained potentials (bond, angle, torsion and vdw). Panel **c**. Modelling of the FimH protein, representing the tip of E.coli pilum. This tip links E. coli to a target protein at the infected cell membrane and this binding is enhanced under shear flow. The behaviour under flow of the Fim proteins was recently investigated via LBPD and OPEP force field [18]. Panel **d**. Massive aggregation of amyloid peptide Aβ(16-22) followed in time for a system composed of 1000 peptides. After a first linear growth, the largest cluster increases size discontinuously by the fusion of preformed aggregates [32].

and cells. This is the case, for example, of stenotic vessels as caused by atherosclerosis, where common conditions are such that about 10 RBCs can pass in the lumen narrowing. In this condition, the granular nature of blood shows up, and non-trivial rheological properties emerge as a consequence of the concerted motion of plasma and RBCs. As we zoom into capillaries, with a lateral size of 100 microns and below, the motion of red blood cells reveals highly non-trivial effects and for capillaries with a diameter of a few microns, RBCs deform in order to squeeze in, in such a way that they manage to crowl into the micrometer-sized vessels. On the other hand, when inspecting larger-scale circulation, in the 100 - 500 microns range, it is generally sufficient to consider the cells as rigid bodies.

An all-embracing LBPD hemodynamic tool is still beyond reach, the various prominent codes being able to cover only three decades in space at best. Mesoparticle methods, such as Dissipative Particle Dynamics, can range from, say 0.1 to 1 microns [42], while extreme implementations of the MUPHY code [43] managed to cover about four decades, from centimetres to a few microns. Intermediate options are also available, like the multilevel code Hemocell, which is explicitly targeted to address the mechanical properties of RBCs so as to allow direct coupling between fluid flow and biological processes such as cellular adhesion [44]. While a super-code based on the modular coupling of the aforementioned approaches is in principle conceivable, it would be highly labor-intensive in practice. Most likely, each of these tools will keep growing on their own, by extending their range of operation and facilitating the mutual exchange of increasingly accurate information.

3. SIMULATIONS AT THE PHYSICS/CHEMISTRY/BIOLOGY INTERFACE: DREAMING AHEAD

Having briefly discussed the current LBPD applications, as a conclusive remark we next venture into some visionary, yet hopefully realistic, considerations for the future, mostly in connection with foreseeable progress in computer technology. From the conceptual viewpoint, these efforts fully inscribe within the very elegant multiscale integrative approach to Biology advocated in Denis Noble's books [45, 46].

The LB method is often quoted for its amenability to parallel processing. The performance of LB codes is best conveyed in terms of MLUPS, namely Million Lattice Updates Per Second; 1 MLUPS means updating one million sites by one step in time in one CPU second. Given that a LB code for Navier-Stokes hydrodynamics requires about 200 Flops per site and step, 1 MLUPS corresponds to approximately 200 Mflops/s. On current high-end laptops, a LB code features of the order of 5 MLUPS. On more complex applications, especially with LBPD interfaces, this figure lowers to about 1 MLUPS or below, due to the increased computational density, typically in the order of thousands Flops/site/step. At a 1 MLUPS rate, a 1Gflops/s computer updates one million sites/second, meaning that an LBPD simulation spanning, say, four decades in time (ten thousands time steps) would take just three hours.

We may realistically assume that the LBPD computational complexity scales linearly with the volume of the simulated four-dimensional spacetime, namely:

$$C \sim kV = kL^3T = kL^{3+\alpha}$$

where $k \sim 10^3$ is the computational density in Flops/site/step and $\alpha = 1$ for the ballistic regime and $\alpha = 2$ for the diffusive one. Even within the more demanding diffusive regime, a prospective Exaflop computer would ideally permit to span over four decades in space and over eight in time, *in matter of a day*!

With these advances in mind, several exciting scenarios open-up in front of the LBPD framework, such as the direct simulation of organelles, or highly complex physiological processes like haemostasis, to be detailed shortly. Before such dreams come true, though, many technical and conceptual issues have to be successfully addressed. To begin with, upscaling to such sizes requires *qualitatively* new parallel programming paradigms. In particular, as size increases, the cost of accessing memory kicks heavily in; for instance, the streaming step, which entails literally zero floating point operations, can currently account for a significant fraction (about 1/3) of the execution time on current leading edge LB simulations. Consequently, much care must be exercised to minimise the overheads of memory access, for instance by hierarchical organization of the data arrays in energy shells, i.e. discrete velocities with the same magnitude [47, 48].

The efficient parallelisation of LB codes requires additional skilful techniques, aimed at i) Overlapping communication and computation, ii) Distributing the workload evenly across the various processors, iii) Accessing data efficiently. The above items are only accrued in the case of multiscale LBPD, due to conflicting needs raised by grid-based data structures (LB) and grid-free particles (PD). Such techniques have indeed led to impressive performances; for instance, the MUPHY code has delivered up to 20 Pflops, with 90 percent parallel efficiency for the case of protein dynamics in crowded solutions, see Figure 4.a.

With Exascale machines in mind, another major issue to be considered is *fault-tolerance*, i.e. how to equip LBPD with feedback mechanisms capable of recovering from failures, as they

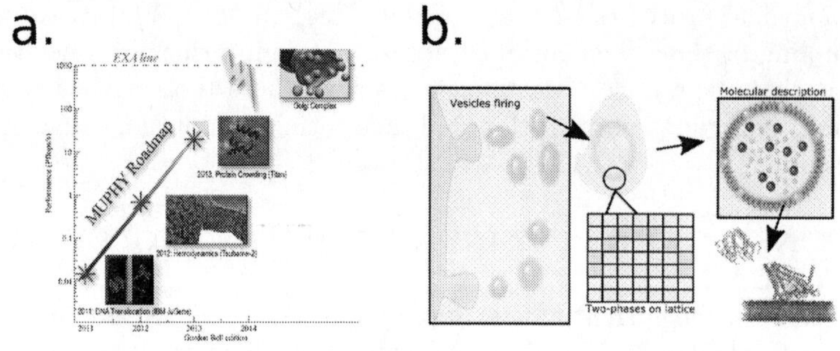

Figure 4. Panel a. The roadmap of the MUPHY code based on Gordon-Bell performance. Panel b. Multiscale representation of vesicles transporting proteins. Vesicle firing from a membrane where the vesicles are modelled as immiscible fluid (green) phase separating from the aqueous host (light blue background), the central panel zooms the vesicle bilayer with liquid water in the interior, the right panel depicts a molecular representation of the vesicle containers with an heterogenous protein suspension in the interior and membrane proteins embedded in the bilayer.

inevitably occur once millions and possibly billions of concurrent processes are in action. In this context, the new challenging applications would focus for instance on fundamental biological processes where the complex states of flowing matter across many scales of motion is essential, this includes for example the simulation of molecular transport in the cellular Golgi apparatus, the multi-scale investigation of haemostasis, neuronal firing and ions flow in neurones. Meeting the above challenges is all but a trivial task, but the pay-off might be immense.

4. Acknowledgements

The research leading to these results has received funding from the European Research Council under the Horizon 2020 Programme Grant Agreement n. 739964 ("COPMAT"). FS acknowledges funding from the ERC (FP7/2007-2013) Grant Agreement no.258748 and ANR-11-LABX-0011-01 (Dynamo).

References

[1] Salomon-Ferrer R, Götz A W, Poole D, Le Grand S and Walker R C 2013 *J. Chem. Theory Comput.* **9** 3878–3888
[2] Phillips J C, Braun R, Wang W, Gumbart J, Tajkhorshid E, Villa E, Chipot C, Skeel R D, Kale L and Schulten K 2005 *J. Comput. Chem.* **26** 1781–1802
[3] Abraham M J, Murtola T, Schulz R, Páll S, Smith J C, Hess B and Lindahl E 2015 *SoftwareX* **1** 19–25
[4] Lindorff-Larsen K, Piana S, Dror R and Shaw D 2011 *Science* **334** 517–520
[5] Levitt M 2001 *Nature Struct. Bio.* **8** 392–393
[6] Karplus M 2014 *Angew. Chem. Int. Ed. Engl.* **53** 9992
[7] Higuera F, Succi S and Benzi R 1989 *Europhys. Lett.* **9** 345
[8] Qian Y, d'Humières D and Lallemand P 1992 *Europhys. Lett.* **17** 479
[9] Dünweg B and Ladd A J 2009 *Advanced Computer Simulation Approaches for Soft Matter Sciences III* (Springer) pp 89–166
[10] Monaghan J J 1992 *Ann. Rev. Astron. and Astrophys.* **30** 543–574
[11] Eastwood J W 1991 *Comp. Phys. Comm.* **64** 252–266
[12] Hockney R W and Eastwood J W 1988 *Computer simulation using particles* (CRC Press)
[13] Horbach J and Succi S 2006 *Physical review letters* **96** 224503
[14] Succi S, Moradi N, Greiner A and Melchionna S 2014 *Front. Phys.* **2** 22
[15] Sterpone F, Melchionna S, Tuffery P, Pasquali S, Mousseau N, Cragnolini T, Chebaro Y, St-Pierre J F, Kalimeri M, Barducci A, Laurin Y, Tek A, Baaden M, Nguyen P H and Derreumaux P 2014 *Chem. Soc. Rev.* **43** 4871–4893

[16] Sterpone F, Derreumaux P and Melchionna S 2015 *J. Chem. Theory Comput.* **11** 1843–1853
[17] Chiricotto M, Sterpone F, Derreumaux P and Melchionna S 2016 *Philos. Trans. A* **374** 20160225
[18] Sterpone F, Melchionna S, Doutreligne S, Tran T, Nguyen P, Baaden M and Derreumaux P 2017 *Bioch. Bioph. Res. Comm.* **17** 31739
[19] Fyta M, Melchionna S and Succi S 2011 *J. Polym. Sci. B* **49** 985–1011
[20] Fyta M G, Melchionna S, Kaxiras E and Succi S 2006 *Multiscale Mod. Sim.* **5** 1156–1173
[21] Fyta M, Melchionna S, Succi S and Kaxiras E 2008 *Phys. Rev. E* **78** 036704
[22] Bernaschi M, Melchionna S, Succi S, Fyta M and Kaxiras E 2008 *Nano lett.* **8** 1115–1119
[23] Fyta M, Melchionna S, Kaxiras E and Succi S 2008 *Computing in Science & Engineering* **10**
[24] Datar A V, Fyta M, Marconi U M B and Melchionna S 2017 *Langmuir*
[25] Shaw D E, Maragakis P, Lindorff-Larsen K, Piana S, Dror R O, Eastwood M P, Bank J A, Jumper J M, Salmon J K, Shan Y *et al.* 2010 *Science* **330** 341–346
[26] Lindorff-Larsen K, Trbovic N, Maragakis P, Piana S and Shaw D E 2012 *J. Am. Chem. Soc.* **134** 3787–3791
[27] Pascheka D, Hempela S and García A E 2008 *Proc. Natl. Acad. Sci. USA* **105** 17754–17759
[28] Ando T, Chow E and Skolnick J 2013 *J. Chem. Phys.* **139** 121922
[29] Schofield J, Inder P and Kapral R 2012 *J. Chem. Phys.* **136** 205101
[30] Šarić A, Chebaro Y C, Knowles T P J and Frenkel D 2014 *Proc. Natl. Acad. Sci. USA* **111** 17869–17874
[31] Nasica-Labouze J, Nguyen P H, Sterpone F, Berthoumieu O, Buchete N V, Coté S, Simone A D, Doig A J, Faller P, Garcia A, Laio A, Li M, Melchionna S, Mousseau N, Mu Y, Paravastu A, Pasquali S, Rosenman D J, Strodel B, Tarus B, Viles J H, Zhang T, Wang C and Derreumaux P 2015 *Chem. Rev.* **115** 3518–3563
[32] Chiricotto M, Melchionna S, Derreumaux P and Sterpone F 2016 *J. Chem. Phys.* **145** 035102
[33] Pontrelli G, Halliday I, Melchionna S, Spencer T J and Succi S 2014 *Mathem. Comp. Mod. Dyn. Syst.* **20** 470–490
[34] Pontrelli G, Halliday I, Melchionna S, Spencer T J and Succi S 2012 *IFAC Proceedings Volumes* **45** 30–39
[35] Toschi F and Succi S 2005 *Europhys. Lett.* **69** 549
[36] Melchionna S 2011 *Macrom. Theory Sim.* **20** 548–561
[37] Pontrelli G, Halliday I, Spencer T J, König C S and Collins M W 2015 *Comp. Meth. Biomech. Biom. Eng.* **18** 351–361
[38] Matyka M, Koza Z and Mirosław Ł 2013 *Comp. Fluids* **73** 115–123
[39] Sloot P M and Hoekstra A G 2009 *Brief. Bioinf.* **11** 142–152
[40] Rybicki F J, Melchionna S, Mitsouras D, Coskun A U, Whitmore A G, Steigner M, Nallamshetty L, Welt F G, Bernaschi M, Borkin M *et al.* 2009 *The International Journal of Cardiovascular Imaging* **25** 289–299
[41] Omori T, Imai Y, Kikuchi K, Ishikawa T and Yamaguchi T 2015 *Annals of biomedical engineering* **43** 238–257
[42] Fedosov D A, Caswell B and Karniadakis G E 2010 *Biophysical journal* **98** 2215–2225
[43] Bernaschi M, Melchionna S, Succi S, Fyta M, Kaxiras E and Sircar J K 2009 *Comp. Phys. Comm.* **180** 1495–1502
[44] Artoli A M, Hoekstra A G and Sloot P M 2006 *J. Biomech.* **39** 873–884
[45] Noble D 2008 *The music of life: biology beyond genes* (Oxford University Press)
[46] Noble D 2016 *Dance to the Tune of Life: Biological Relativity* (Cambridge University Press)
[47] Shet A G, Siddharth K, Sorathiya S H, Deshpande A M, Sherlekar S D, Kaul B and Ansumali S 2013 *Inte. J. Mod Phys. C* **24** 1340011
[48] Shet A G, Sorathiya S H, Krithivasan S, Deshpande A M, Kaul B, Sherlekar S D and Ansumali S 2013 *Phys. Rev. E* **88** 013314

IOP Conf. Series: Journal of Physics: Conf. Series **1136** (2018) 012014 doi:10.1088/1742-6596/1136/1/012014

Nuclear quantum effects in molecular dynamics simulations

H Dammak[1], **M Hayoun**[2], **F Brieuc**[1], **G Geneste**[3]

[1] Laboratoire Structures Propriétés et Modélisation des Solides, CentraleSupélec, CNRS, Université Paris-Saclay, 91190 Gif-sur-Yvette, France
[2] Laboratoire des Solides Irradiés, Ecole Polytechnique, CNRS, CEA, Université Paris-Saclay, 91128 Palaiseau, France
[3] CEA, DAM, DIF, F-91297 Arpajon, France

hichem.dammak@centralesupelec.fr

Abstract. To take into account nuclear quantum effects on the dynamics of atoms, the path integral molecular dynamics (PIMD) method used since 1980s is based on the formalism developed by R. P. Feynman. However, the huge computation time required for the PIMD reduces its range of applicability. Another drawback is the requirement of additional techniques to access time correlation functions (ring polymer MD or centroid MD). We developed an alternative technique based on a quantum thermal bath (QTB) which reduces the computation time by a factor of ~20. The QTB approach consists in a classical Langevin dynamics in which the white noise random force is replaced by a Gaussian random force having the power spectral density given by the quantum fluctuation-dissipation theorem. The method has yielded satisfactory results for weakly anharmonic systems: the quantum harmonic oscillator, the heat capacity of a MgO crystal, and isotope effects in ^{7}LiH and ^{7}LiD. Unfortunately, the QTB is subject to the problem of zero-point energy leakage (ZPEL) in highly anharmonic systems, which is inherent in the use of classical mechanics. Indeed, a part of the energy of the high-frequency modes is transferred to the low-frequency modes leading to a wrong energy distribution. We have shown that in order to reduce or even eliminate ZPEL, it is sufficient to increase the value of the frictional coefficient. Another way to solve the ZPEL problem is to combine the QTB and PIMD techniques. It requires the modification of the power spectral density of the random force within the QTB. This combination can also be seen as a way to speed up the PIMD.

1. Introduction

Molecular dynamics (MD) simulation is frequently used to investigate and predict the properties of condensed matter in the classical limit. For a crystal, these calculations are valid for temperatures higher than the Debye temperature [1] (940 K for MgO). As shown in figure 1, the experimental heat capacity decreases when the temperature decreases and vanishes at $T = 0$ K, while MD simulation leads to a nearly constant value, corresponding to $3k_B$ per atom. The arrow in figure 1 points out this disagreement. Standard MD ignores the nuclear quantum effects (NQE), which are at the origin of the shape of the experimental heat capacity curve. This behavior is a direct consequence of the quantization of the energy of the vibration modes. Another example is the evolution as a function of temperature of the lattice parameter in lithium hydride (LiH) given in figure 2. At low temperature, the crystal rapidly reaches its ground state upon cooling, generating a freezing (saturation) of the lattice parameter whereas the standard MD predicts a linear behavior. The arrow in figure 2 emphasizes this discrepancy due to the quantization of the energy of the vibration modes and due to the total energy of the ground state, called zero-point energy (ZPE), which is larger than the potential energy minimum. In addition, the lattice parameter of lithium deuteride (LiD) behaves similarly but is shifted (see double arrow) with respect to the LiH one. This isotope effect, which is a particular case of NQE, is due to the change of the vibrational frequencies, in turn due to the difference in the atomic masses of H and D. In fact, the average energy of the quantum oscillator, at a given temperature, is frequency dependent. In the harmonic approximation, the average energy is given by the well-known expression:

$$\theta(\omega, T) = \hbar\omega\left[\frac{1}{2} + \frac{1}{\exp(\beta\hbar\omega)-1}\right] \tag{1}$$

where ω is the angular frequency of the oscillator, $\hbar$ is the reduced Planck constant and β the statistical temperature $(1/k_B T)$. Figure 3 shows the evolution of this energy as a function of temperature compared to the classical limit $(k_B T)$. One can note the freezing of the energy at low temperature $(T < \hbar\omega/k_B)$ corresponding to the ZPE, $\hbar\omega/2$. In contrast, in the classical description, the average energy does not depend on the frequency of the oscillator.

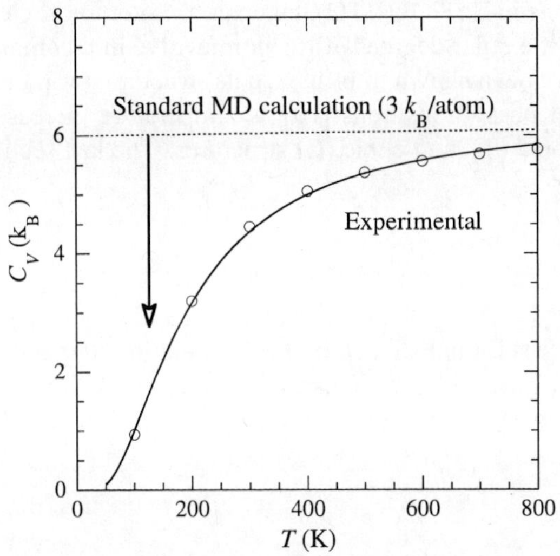

Figure 1. Heat capacity per molecule of the MgO crystal as a function of the temperature, T [2]. The full line is the Debye model [1].

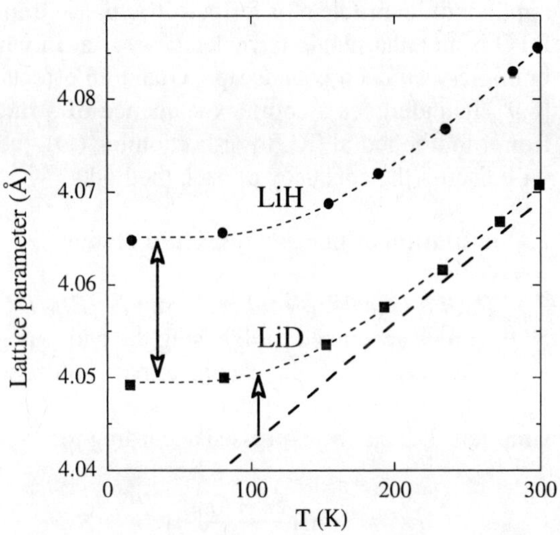

Figure 2. Experimental lattice parameters of LiH and LiD crystals as a function of the temperature, T [3]. The straight dashed line is the standard MD result.

These examples show that the quantum nature of nuclei can play a major role at low temperatures and/or in systems that contain light atoms. In this case, NQE cannot be neglected and must be taken into account especially in MD simulation, which is the focus of this paper. The historical method including NQE is based on Feynman's path integral [4] (PIMD or path integral Monte Carlo). It can provide exact results – under the assumption of distinguishable nuclei – but at the price of a high computational cost. An (approximate) alternative method includes the quantum fluctuations through a random force, whose power spectral density is related to $\theta(\omega, T)$ by the quantum mechanical fluctuation-dissipation theorem [5]. This is the quantum thermal bath (QTB) MD introduced in 2009 by Dammak *et al.* [6]. The two methods can be combined, as QTB-PIMD [7], to improve the efficiency of the PIMD.

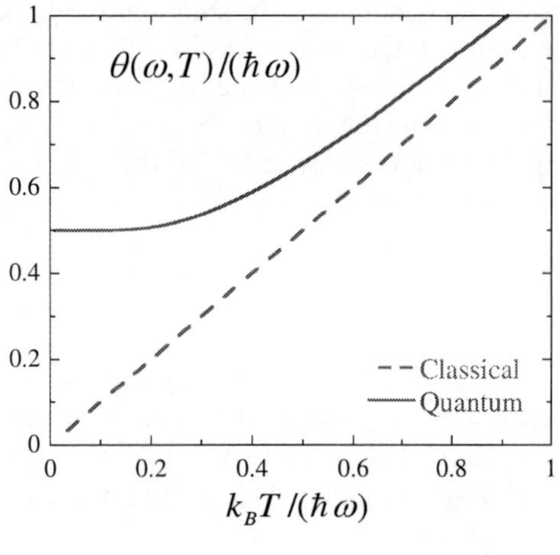

Figure 3. Normalized energy of the quantum and classical harmonic oscillators as a function of the reduced temperature.

In the following section, these three methods including NQE in MD simulation will be presented and applied to the problem of ferroelectric phase transitions in $BaTiO_3$ (BTO), taken as a prototypical case. BTO is an anharmonic ferroelectric crystal in which the polar degrees of freedom evolve in a complex multiple-well energy landscape. Quantum effects have been shown to influence its structural properties [8,9]. It undergoes a complex sequence of structural phase transitions [10] as temperature increases: from rhombohedral (R), to orthorhombic (O), tetragonal (T), and cubic (C) structures. The last section will discuss the adequacy of each method.

2. Simulation of nuclear quantum effects

2.1. Path Integral Molecular Dynamics (PIMD)

NQE can be accounted for by using the path-integral (PI) formalism [4]. In this formulation of quantum statistical mechanics, the canonical partition function Z is written as a discretized imaginary time path integral. For a quantum system containing N (discernible) particles of mass m (assumed as identical for simplicity), Z can be expressed according to:

$$Z = \lim_{P\to\infty} \left(\frac{2\pi\, m\, P\, k_B T}{h^2}\right)^{3NP/2} \times$$
$$\int_{\{\mathbf{r}^N\}^{(1)}} \cdots \int_{\{\mathbf{r}^N\}^{(P)}} \exp\left(-\beta V_{eff}\left(\{\mathbf{r}^N\}^{(1)}, \dots, \{\mathbf{r}^N\}^{(P)}\right)\right) \{d\mathbf{r}^N\}^{(1)} \dots \{d\mathbf{r}^N\}^{(P)} \quad (2)$$

The integral is over P (Trotter number) replicas of the system, labeled by the integer s, each replica being a set of N positions of the atoms $\{\mathbf{r}^N\}^{(s)} = \left(\mathbf{r}_1^{(s)}, \dots, \mathbf{r}_N^{(s)}\right)$. These replicas result from the discretization of the PI in imaginary time (imaginary time slices). The effective potential V_{eff} which depends on all atomic positions of all replicas is composed of two terms, the physical potential energy, V, computed in each replica (and averaged over them), and a harmonic coupling term, of angular frequency $\omega_P = \sqrt{P}/\beta\hbar$, between replicas:

$$V_{eff}\left(\{\mathbf{r}^N\}^{(1)}, \dots, \{\mathbf{r}^N\}^{(P)}\right) = \sum_{s=1}^{P} \left[\frac{1}{P} V\left(\{\mathbf{r}^N\}^{(s)}\right) + \sum_{i=1}^{N} \frac{1}{2} m\omega_P^2 \left(\mathbf{r}_i^{(s)} - \mathbf{r}_i^{(s+1)}\right)^2\right]. \quad (3)$$

Each particle i of the replica s is thus interacting through harmonic forces with the particles i of the replicas $(s+1)$ and $(s-1)$, forming a polymer ring that closes on itself by periodic boundary conditions, $\mathbf{r}_i^{(P+1)} = \mathbf{r}_i^{(1)}$.

In the limit where the Trotter number $P \rightarrow \infty$, this equivalent classical system has the same partition function as that of the quantum system. As a consequence, MD simulation can be applied to the classical equivalent to numerically estimate the static properties of the quantum system. In the microcanonical ensemble, the corresponding equation of motion of each particle i in each replica s writes

$$m\,\ddot{\mathbf{r}}_i^{(s)} = -\frac{1}{P}\nabla_{\mathbf{r}_i^{(s)}} V(\{\mathbf{r}^N\}^{(s)}) - m\omega_P^2\left(2\mathbf{r}_i^{(s)} - \mathbf{r}_i^{(s+1)} - \mathbf{r}_i^{(s-1)}\right). \tag{4}$$

In practical simulations, the Trotter number is finite, and must be chosen to converge the estimated properties. For instance, the average total energy of the system is given by the following estimator:

$$\langle E\rangle = \langle \textstyle\sum_{s=1}^P \sum_{i=1}^N \frac{\left(\mathbf{p}_i^{(s)}\right)^2}{2m} - \sum_{s=1}^P \sum_{i=1}^N \frac{1}{2} m\omega_P^2 \left(\mathbf{r}_i^{(s)} - \mathbf{r}_i^{(s+1)}\right)^2 \rangle + \langle \sum_{s=1}^P \frac{1}{P} V(\{\mathbf{r}^N\}^{(s)})\rangle. \tag{5}$$

where the first and second averages are the kinetic and the potential energies, respectively. This primitive estimator can be derived from $-\partial Ln(Z)/\partial\beta$.

In the canonical or isothermal-isobaric ensemble, Eq. (4) is modified by adding the forces due the thermostat and/or barostat. In this study of the ferroelectric phase-transitions in BTO, the extension of the Langevin method to the isothermal-isobaric ensemble which has been achieved by Quigley and Probert is used [11,12]. The ferroelectric properties of BTO were modeled by an effective Hamiltonian [13,14] derived from first-principles density-functional calculations. The degrees of freedom of this Hamiltonian are the local modes and the homogeneous strain tensor. The reduced local polar displacement $\mathbf{u}_i$ ($\mathbf{r}_i = a_0\mathbf{u}_i$) inside the cell i is related to the local dipolar moment $\mathbf{p}_i$ through the cell parameter a_0 (7.46 Bohr) and an effective charge Z^* (9.956 e): $\mathbf{p}_i = Z^* a_0 \mathbf{u}_i$. Using this model, a ferroelectric phase-transition temperature corresponds to the temperature at which the mean local mode (dipole) rotates and thus exhibits a change of the macroscopic polarization direction: [111], [101] and [001] for the R, O and T phases, respectively.

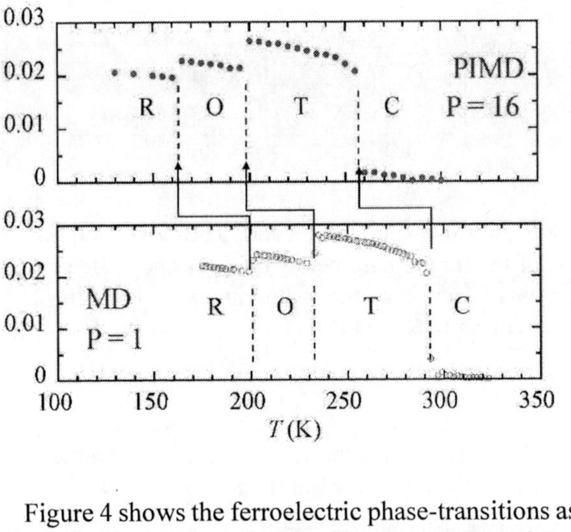

Figure 4. Evolution as a function of temperature of the average of the non-zero components of the reduced polarization (local modes) in $BaTiO_3$ as obtained by PIMD and standard MD. Vertical dashed lines show the three ferroelectric phase transitions from rhombohedral (R), to orthorhombic (O), to tetragonal (T), and to cubic (C) structures.

Figure 4 shows the ferroelectric phase-transitions as obtained by PIMD and standard MD. The PIMD calculation is converged with a Trotter number of $P = 16$ in the whole temperature range. The PIMD sequence of phase transitions, R-O-T-C, is identical to the one experimentally observed, but the values of the transition temperatures are different due to the model employed [15]. Hence, these PIMD results will be taken as a reference instead of the experimental ones, in order to test other techniques within the same model. Hence, the standard MD calculation ($P = 1$) overestimates the transition temperatures and the discrepancy is about 30-40 K. This shows an important NQE, which cannot be neglected.

2.2. Quantum Thermal Bath Molecular Dynamics (QTB-MD)

An alternative technique to include the NQE in MD simulations is the QTB method [6], which is based on a modification of the Langevin thermostat [16]. The component α of the random force, $R_{i\alpha}$, applied on the atom i, is not a *white noise* and its power spectral density, I_R, is derived from the quantum dissipation-fluctuation theorem [17], and is related to the Fourier transform of the autocorrelation function, $\langle R_{i\alpha}(t)R_{i\alpha}(t+\tau)\rangle$, according to the Wiener–Khinchin theorem:

$$\langle R_{i\alpha}(t)R_{i\alpha}(t+\tau)\rangle = \int_{-\infty}^{+\infty} I_R(\omega,T)\exp(-i\omega\tau)\frac{d\omega}{2\pi} \tag{6}$$

$$I_R(\omega,T) = 2m\,\gamma\,\theta(\omega,T) \tag{7}$$

where γ is the frictional coefficient. The equation of motion is thus:

$$m\,\ddot{\mathbf{r}}_i = -\nabla_{\mathbf{r}_i}V(\{\mathbf{r}^N\}) - m\gamma\dot{\mathbf{r}}_i + \mathbf{R}_i \tag{8}$$

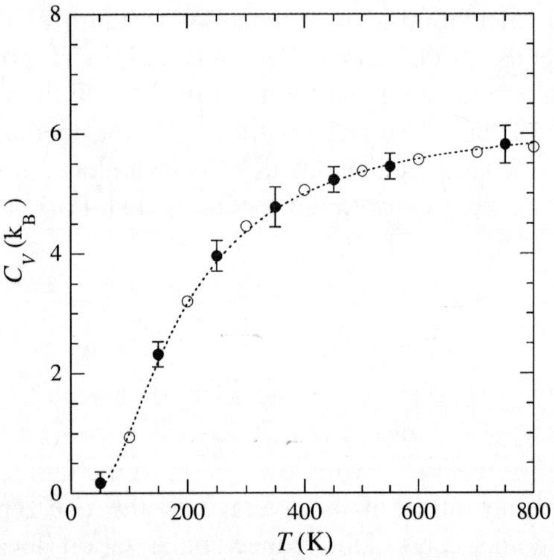

Figure 5. Heat capacity of the MgO crystal as computed by QTB-MD [6] (full circles) compared to experimental values (open circles). The dashed line is a guide for the eye.

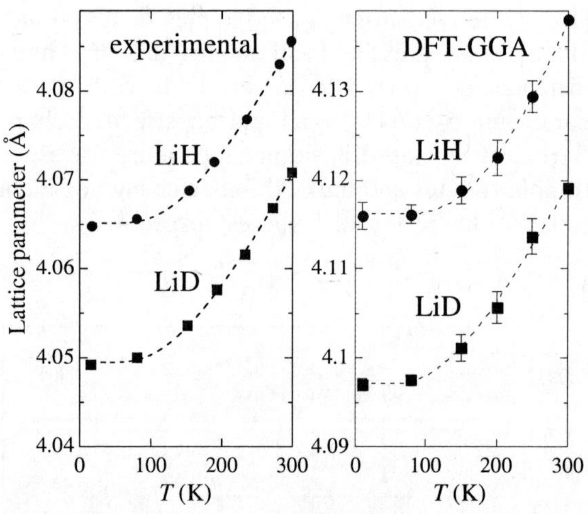

Figure 6. Experimental and QTB-MD computation of the lattice parameters of ^{7}LiH and ^{7}LiD crystals as a function of the temperature, *T*, by using DFT within the GGA [20].

In contrast to the Langevin thermostat, I_R is ω-dependent and the random force components are generated using the procedure detailed in References [18] and [19]. In summary, for n MD time steps, δt, the random force, $\tilde{R}_{i\alpha}$, is first generated in the Fourier space ($\omega_j = \frac{2\pi j}{n\,\delta t}$):

$$\tilde{R}_{i\alpha}(\omega_j) = (a_j + \mathrm{i}\,b_j)\sqrt{I_R(\omega_j)\,n\,\delta t\,/2} \tag{9}$$

where a_j and b_j are normally distributed random numbers, and i the imaginary number. $R_{i\alpha}(t_j = j\delta t)$ is then obtained by inverse Fourier transform.

QTB-MD provides the experimental behavior for the heat capacity in the MgO crystal, as shown in figure 5. The isotopic shift in the lattice parameter of ^{7}LiH and ^{7}LiD is also well reproduced. Figure 6 yields the experimental and computed lattice parameter at low temperature of both crystals. The discrepancies between experimental and QTB-MD data are due to the GGA functional used. Although exact only in the case of a purely harmonic systems, the QTB leads to satisfactory results in many anharmonic systems, such as MgO [6], LiH [20] and many others [21-24]. Unfortunately, the method can fail when dealing with anharmonic systems. Indeed, the QTB technique is subject to zero-point energy leakage (ZPEL), like any other method based on classical trajectories [25]. It consists in the transfer of energy from high-frequency vibrational modes to low-frequency vibrational modes, hence losing the distribution imposed through the power spectral density of equation (7). A case of ZPEL is given by the ferroelectric phase transitions in $BaTiO_3$. Figure 7a shows that only the T-C phase transition is observed when using usual values of the frictional coefficient – γ less than 0.5 THz. The T-C temperature is about 210 K instead of the PIMD reference value of 260 K. In addition, the low-temperature phases, R and O, are not observed even at $T = 0$ K. Fortunately, the ZPEL can be reduced or even suppressed by increasing the value of γ. As can be seen in figure 7a, the R-O-T-C sequence of transitions occurs from $\gamma = 1.2$ THz. The values of the transition temperatures converge toward the PIMD reference temperatures for higher values of γ. For $\gamma = 16$ THz, the QTB-MD simulation, with consecutive transition temperatures of 160 K, 190 K and 255 K, is in good agreement with the PIMD result (figure 7b). We conclude that using very high values of the frictional coefficient is an efficient solution in case of failure of the QTB due to ZPEL. The possible drawback of using such high values will be discussed in the last section.

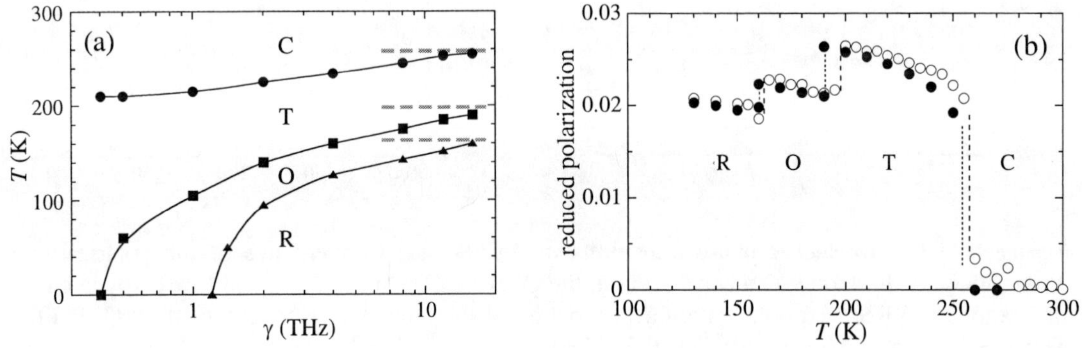

Figure 7. R-O, O-T, and T-C ferroelectric phase transitions in $BaTiO_3$. (a) Convergence of the phase-transition temperatures with the friction coefficient, γ, within the QTB-MD simulation. (b) Average of the non-zero components of the reduced polarization (local modes), as a function of the temperature, *T*, obtained by QTB-MD with γ = 16 THz (full circles), and by PIMD with a Trotter number *P* = 16 (open circles).

2.3. Combining QTB and PIMD

It is possible to combine the QTB and PIMD [26] in order *i*) to improve the convergence of the PIMD mostly at very low temperature and *ii*) to correct potential failures of the QTB-MD technique especially in the case of anharmonic systems. This combination requires the modification of the power spectral density of the random forces applied on each atom of each replicas. Indeed, for not converged Trotter number, quantum fluctuations are already partially included within the ring polymer of the PIMD. The

IOP Conf. Series: Journal of Physics: Conf. Series **1136** (2018) 012014 doi:10.1088/1742-6596/1136/1/012014

QTB random forces will thus only bring the missing part of the NQE which is dependent on the Trotter number. In practice, $\theta(\omega, T)$ of equation (7) is replaced by the adequate function $\kappa_P(\omega, T)$, which is solution of the following equation

$$\frac{1}{P}\sum_{k=0}^{P-1}\frac{\kappa_P(\omega_k,T)}{m\,\omega_k^2} = \frac{\theta(\omega,T)}{m\,\omega^2} \qquad (10)$$

where ω_k is the angular frequency of the normal modes of the ring polymer in the harmonic approximation:

$$\omega_k^2 = \frac{\omega^2}{P} + 4\omega_P^2 \sin^2\left(\frac{k\pi}{P}\right) \qquad (11)$$

The power spectral density is thus given by:

$$I_R(\omega, T) = 2m\,\gamma\,\kappa_P(\omega, T) \qquad (12)$$

Since the corresponding random forces are intended to be applied to the normal modes, the random forces applied on the atoms are obtained through an orthogonal transformation which can be found in Reference [7].

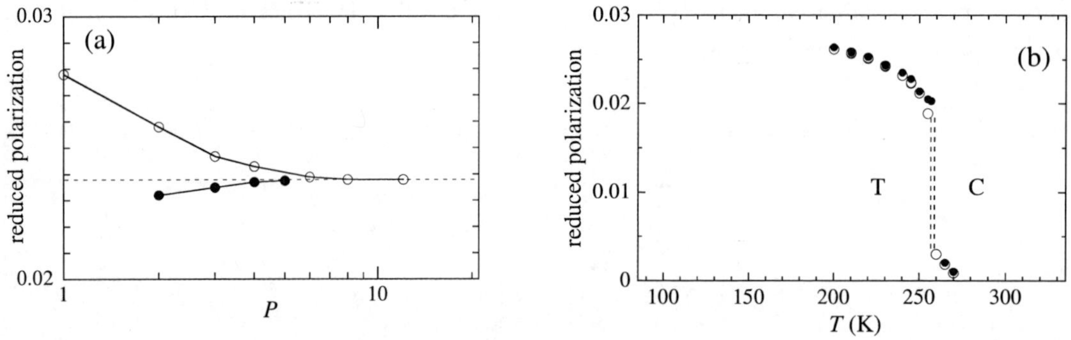

Figure 8. T-C ferroelectric phase transitions in $BaTiO_3$. (a) Convergence of the polarization at $T = 240$ K with the Trotter number, P, within the QTB-PIMD (full circles), and PIMD (open circles) simulations. (b) Reduced polarization as a function of the temperature, T, obtained by QTB-PIMD with $P = 2$ (open circles), and $P = 3$ (full circles).

Let us recall that for QTB-MD simulation, using usual values of the frictional coefficient, the T-C transition in BTO occurs at a temperature of ~210 K, which is much lower than the PIMD reference values (figure 7a). The QTB-PIMD method is applied to the BTO system, varying the Trotter number from 2 to 5 (and using a low value of $\gamma = 0.5$ THz). Figure 8a shows that convergence of the reduced polarization at 240 K in the tetragonal phase is reached for $P = 5$, whereas $P = 8$ is required using PIMD. Hence, the convergence is improved by the combination of the two methods. In figure 8b, QTB-PIMD method with $P = 2$ gives a transition temperature of $T = 257$ K, very close to that obtained by PIMD (P = 16, $T = 259$ K). It is worth noting that the failure of the QTB method is fixed solely by two replicas when combining QTB with PIMD. Concerning the R-O and O-T transitions, the transition temperatures are in agreement with the PIMD ones for $P = 3$. This means that the effects of the zero-point energy leakage associated with the use of the QTB have been suppressed by the combination. We conclude, as expected, that QTB-PIMD is more efficient than the standard PIMD, since less replicas are needed for convergence.

3. Discussion and Conclusion

The QTB method provides exact results in the case of purely harmonic systems, except for properties that could be affected by frictional forces present in any Langevin dynamics [27]. In addition, the main advantage of the method is that time correlation functions are directly accessible, as in standard MD. Moreover, the QTB is very simple to implement in an existing code.

For anharmonic systems, QTB-MD can fail due to ZPEL, which is the consequence of the coupling between vibrational modes. In this case, the resulting energy distribution does not match equation (1), it is intermediate between the quantum distribution and the classical homogeneous distribution. To suppress or reduce the ZPEL, it is necessary to increase the value of γ, which is the technical parameter of the thermal bath. Encouraging results have been obtained in the case of a Lennard-Jones aluminum crystal and phase transitions in BTO, as well as for simple models of anharmonic systems [19]. The disadvantage of a high value of γ is the broadening of vibrational peaks and a spurious high-frequency tail in the phonon density of states (DOS). Nevertheless, vibrational spectra are not always dramatically altered, this effect depends on the simulated system. It is therefore convenient to check the DOS when using a high value of γ within QTB-MD simulations.

Unfortunately, there exist systems for which increasing γ do not considerably reduces the ZPEL. The discrimination can be obtained through the strength of the anharmonicity. It can be evaluated with a dimensionless parameter, C, including a characteristic length, d, and a characteristic energy, V_0:

$$C = \frac{\hbar^2}{2mV_0d^2} \tag{13}$$

There exists a critical value of C, depending on the system, below which it can be considered as weakly anharmonic. For the Morse potential and the double-well model the critical values are about 0.01 and 0.1, respectively. Hence, weakly anharmonic systems can be successfully simulated by QTB-MD with an appropriate value of γ, whereas for strongly anharmonic systems QTB-MD should not be used. In the latter case, QTB-PIMD is an adequate method to take into account the NQE. Indeed, it allows a better convergence with the number of beads than PIMD. The disadvantage of both methods (PIMD and QTB-PIMD) is that time correlation functions are not directly accessible. The sequence of phase transitions in BTO has been successfully retrieved by QTB-MD with a high value of γ [19] and by QTB-PIMD [7]. In contrast, the values of the ferroelectric polarization in BTO close to the temperatures of the phase-transitions are better computed in QTB-PIMD.

For weakly anharmonic systems, some macroscopic properties are not affected by the ZPEL. In fact, the total energy is exact although the energy distribution is not correct. For instance, QTB-MD has been successfully used in numerous applications to highlight nuclear quantum effects or isotope effect on lattice parameter [6,20], on heat capacity of crystal [6] and carbon nanotube [22], and on binding energies in clusters [28].

References

[1] Ashcroft N W and Mermin N 1976 *Solid State Physics* (Philadephia, Saunders College Publishing)
[2] Anderson O L and Suzuki I 1983 *J. Geophys. Res.* **88** 3549
[3] Smith D K and Leider H R 1968 *J. Appl. Crystallogr.* **1** 246
[4] Tuckerman M E 2010 *Statistical Mechanics: Theory and Molecular simulation* (New York, Oxford University Press) 475-490
[5] Calen H B and Welton T A 1951 *Phys. Rev.* **83** 34
[6] Dammak H, Chalopin Y, Laroche M, Hayoun M and Greffet J J 2009 *Phys. Rev. Lett.* **103** 190601
[7] Brieuc F, Dammak H and Hayoun M 2016 *J. Chem. Theory Comput.* **12** 1351

[8] Geneste G, Dammak H, Hayoun M and Thiercelin M 2013 *Phys. Rev. B* **87** 014113
[9] Völkel G, Müller K A 2007 *Phys. Rev. B* **76** 094105
[10] Lines E and Glass A M 1979 *Principles and Applications of Ferroelectrics and Related Materials* (Oxford U K, Clarendon)
[11] Quigley D and Probert M I J 2004 *J. Chem. Phys.* **120** 11432
[12] Quigley D and Probert M I J 2005 *Comput. Phys. Commun.* **169** 322
[13] Zhong W and Vanderbilt D 1996 *Phys. Rev. B* **53** 5047
[14] Zhong W, Vanderbilt D and Rabe K M 1995 *Phys. Rev. B* **52** 6301
[15] In fact, the effective Hamiltonian has been derived on an ensemble of microscopic configurations computed within DFT-LDA. The discrepancies between DFT-LDA and experimental transition temperatures are mainly due to the Local Density Approximation.
[16] Bingelli N and Chelikowsky J R 1994 *Phys. Rev. B* **50** 11313
[17] Callen H B and Welton TA 1951 *Phys. Rev.* **83** 34
[18] Chalopin Y, Dammak H, Laroche M, Hayoun M and Greffet J J 2011 *Phys. Rev. B* **84** 224301
[19] Brieuc F, Bronstein Y, Dammak H, Depondt P, Finocchi F and Hayoun M 2016 *J. Chem. Theory Comput.* **12** 5688
[20] Dammak H, Antoshchenkova E, Hayoun M and Finocchi F 2012 *J. Phys.: Condens. Matter* **24** 435402
[21] Calvo F, Van-Oanh N, Parneix P and Falvo C 2012 *Phys. Chem. Chem. Phys.* **14** 10503
[22] Savin A V, Kosevich Y A and Cantarero A 2012 *Phys. Rev. B* **86** 064305
[23] Basire M, Borgis D and Vuilleumier R 2013 *Phys. Chem. Chem. Phys.* **15** 12591
[24] Bronstein Y, Depondt P, Finocchi F and Saitta A M 2014 *Phys. Rev. B* **89** 214101
[25] Ben-Nun M and Levine R D 1994 *J. Chem. Phys.* **101** 8768
[26] Brieuc F 2016 *Modelling and simulation of quantum effects in molecular dynamics: application to the study of proton conduction* (PhD Thesis, Université Paris-Saclay) https://www.theses.fr/2016SACLC060
[27] Pastorino C, Kreer T, Müller M and Binder K 2007 *Phys. Rev. E* **76** 026706
[28] Calvo F, Naumkin F Y, Wales D J 2012 *Chem. Phys. Lett.* **551** 38

IOP Conf. Series: Journal of Physics: Conf. Series **1136** (2018) 012015 doi:10.1088/1742-6596/1136/1/012015

Impact of packing fraction on diffusion-driven pattern formation in a two-dimensional system of rod-like particles

Yuri Yu Tarasevich[1], Valeri V Laptev[1,2], Valentiva V Chirkova[1], Nikolai I Lebovka[3,4]

[1]Astrakhan State University, 20A Tatishchev Street, Astrakhan, 414056, Russia
[2]Astrakhan State Technical University, 16 Tatishchev Street, Astrakhan, 414025, Russia
[3]F. D. Ovcharenko Institute of Biocolloidal Chemistry, NAS of Ukraine, 42 Boulevard Vernadskogo, 03142 Kiev, Ukraine
[4]Taras Shevchenko Kiev National University, Department of Physics, 64/13 Volodymyrska Street, 01601 Kiev, Ukraine

E-mail: tarasevich@asu.edu.ru

Abstract. Pattern formation occurring in a two-dimensional system of rod-like particles has been simulated using a lattice approach. The rod-like particles were modelled as linear k-mers of two mutually perpendicular orientations (k_x- and k_y-mers) on a square lattice with periodic boundary conditions (torus). Two different models of random sequential adsorption were used to produce an initial homogeneous and isotropic distribution of k-mers with different packing fraction values. By means of the Monte Carlo technique, translational diffusion of the k-mers was simulated as a random walk, while rotational diffusion was ignored, so, the k_x- and k_y-mers were considered as individual species. The system tended towards a well-organised nonequilibrium steady state in the form of diagonal stripes for relatively long k-mers ($k \geqslant 6$) and moderate packing fractions (in the interval $p_{down} < p < p_{up}$, where both the critical packing fractions p_{down} and p_{up} were dependent on k).

1. Introduction

A variety of complex nonequilibrium phase behaviours, orientational ordering, and self-organisation properties have been experimentally observed in vibrated systems composed of shape-anisotropic (particularly, rod-like) particles in two dimensions, see [1, 2, 3, 4, 5, 6, 7, 8]. These experiments demonstrated that, at a high packing fraction, the shape of the particles is important in their orientational ordering. Numerous examples of patterns and of phase behaviour in granular media together with appropriate references can be found in review [9].

In recent decades, much attention has been paid to the study of systems of linear k-mers (particles occupying k adjacent adsorption sites) deposited on 2D lattices. A linear k-mer represents the simplest model of a hard-core (completely rigid) rod-like particle with an aspect ratio of k. Computer simulations have been extensively applied to investigate percolation and jamming phenomena for the random sequential adsorption (RSA) [10] of k-mers (see, e.g., [11, 12, 13] and the references therein).

Recently, diffusion-driven pattern formation in a 2D system of k-mers has been studied by means of Monte Carlo (MC) simulations [14, 15]. The k-mers were orientated in two mutually

IOP Conf. Series: Journal of Physics: Conf. Series **1136** (2018) 012015 doi:10.1088/1742-6596/1136/1/012015

perpendicular directions. For those systems with equal numbers of species of two different kinds, i.e., k_x-mers and k_y-mers,

(i) nonequilibrium steady patterns in the form of stripe domains have been observed only when periodic boundary conditions (PBCs) apply in both directions;
(ii) this self-organisation is possible only in fairly dense systems;
(iii) these stripe domains have been observed only for relatively long k-mers, $k \geqslant 6$;
(iv) the relaxation time to reach a well-organised steady-state is drastically dependent on lattice size.

Furthermore, self-organisation can occur in systems with unequal numbers of k_x-mers and k_y-mers.

However, the effect of the packing fraction on such pattern formation does not appear to have been studied in detail. Particularly, jammed states have been produced using the RSA, as reported in [14, 15]. Note, that, while other mechanisms, e.g., RSA with diffusional relaxation can produce denser systems [16], it is not clear whether self-organisation is possible in such the dense systems.

The present conference paper is devoted to a detailed analysis of how the packing fraction affects pattern formation in a two-species diffusion system. For this, we have examined the effect of the packing fraction on such pattern formation using the above two kinds of RSA to produce initial homogeneous and isotropic states.

The rest of the paper is organised as follows. In section 2, the technical details of the simulations are described and all necessary quantities are introduced. Section 3 presents our principal findings. Section 4 summarises the main results.

2. Details of simulation

In our study, a lattice approach was used, the problem being simulated using a square lattice of size $L \times L$. All calculations were performed using only one lattice size $L = 256$. This size was chosen as the relaxation time increases drastically with increasing lattice size [14, 15]. We used toroidal boundary conditions, i.e., periodic boundary conditions (PBCs) along both the x and y axes.

The rod-like particles were presented as linear k-mers. Previous data had evidenced that self-organisation is not observed for short k-mers ($k < 6$) [14]]; accordingly in this work, the length of the k-mers was varied from 6 to 12. Longer particles were excluded from consideration due to an expected essential finite-size effect (the ratio L/k becomes too small for $k > 12$). Isotropic deposition of the k-mers was ensured, i.e., k-mers oriented along the x and y directions (k_x-mers and k_y-mers, respectively) were equiprobable in their deposition. This corresponded to the zero value of a mean order parameter of the system, defined as

$$s = \frac{|N_y - N_x|}{N_y + N_x}, \tag{1}$$

where N_x and N_y are the numbers of sites occupied by the k_x-mers and k_y-mers, respectively.

The diffusion of k-mers was simulated as a random walk. Only translational diffusion was taken into consideration. This is essentially the case for fairly dense systems in the jamming state, where rotational diffusion is impeded, especially for large values of k. Undoubtedly, for dilute systems, rotational diffusion can occur, however, this was ignored in our study.

There are two possibilities of random walk [17]. In the first case, the particles are "blind", i.e. a particle chooses one of the four possible directions and tries to move in this direction. If the attempt is unsuccessful, the particle does not try to find a different direction available for its movement. Such behaviour obeys detailed balance condition [18]. In the second case, a particle is "myopic", i.e., it consistently tries all four possible directions in a random order until its first successful attempt to move or until all possibilities have been tried. Note, that this type of kinetics can violate the detailed balance condition and drives the system to a nonequilibrium steady state [18]. Nevertheless, this approach is natural for "intellectual" particles, e.g., biological species that are looking for life resources, and for pedestrians. Thus, such particles can also be called "intellectual" because they choose one of the possible directions to move, if any exists, with equal probability.

An arbitrary k-mer was randomly chosen at each step and a translational shift by one lattice unit along either the longitudinal or the transverse axis of the k-mer was attempted. All four possible directions to shift the k-mer were attempted in a random order until a direction was found in which displacement of the particle is possible, or until all possible directions had been exhausted [17].

The time required for an attempted displacement of the total number of k-mers in the system, N, was taken as the MC time unit. Time counting was started from the value of $t_{MC} = 0$, being the initial moment (before diffusion), and the total duration of the simulation was 10^7 MC time units. There are two possibilities of random choice of the next particle [19]. The first possibility is to perform a random permutation of all N particles and then consistently to go through these randomly reordered particles. In this case, each particle is chosen once, and only once, at each MC step. The second possibility is a selection with return. In this case, at the current MC step, some particles may be chosen several times while others may end up not being chosen at all. Both algorithms lead to the same final states but the intermediate dynamics of the systems may differ [19]. The latter algorithm was used in [14, 15] and in the present research.

The RSA model [10] was used to produce an initial homogeneous distribution of linear k-mers. These k-mers were deposited randomly and sequentially, and their overlapping with previously placed particles was forbidden, i.e. excluded volume interaction was assumed. The packing fraction, p, was varied in the range $p \in [0.1, p_j]$, where p_j was the packing fraction at jamming. In the jammed state, no additional k-mer can be placed because the presented voids are too small or of inappropriate shape. To avoid confusion, here and below, we will use p_j for the packing fraction of the jammed state produced solely by standard RSA.

We additionally studied the effect of the packing fraction on self-organisation when using the extended RSA approach for obtaining packing fractions $p > p_j$. To produce such a dense state, we used a two-step algorithm.

At the first step, RSA was used to produce a jammed state.

At the second step, the particles were allowed to diffuse. Due to diffusion, some additional voids can occur. When a void is large enough, one additional particle was obligatorily deposited onto the lattice. The action resembles the purpose of shaking a full sugar bowl when we want to add more sugar to it. This additional deposition was stopped when the packing fraction reached the necessary value.

Figure 1 demonstrates how the packing fraction increases with time for one particular case. In the case of unlimited additional deposition, fairly dense packing fraction $p \approx 1.25p_j$ could be reached for $6 \leqslant k \leqslant 12$. It was found that additional particle deposition leads to suppression of self-organisation. Preliminary studies had shown that the initial structure is actually frozen, but becomes more and more dense during the additional deposition of particles.

After reaching the desired packing fraction, only diffusion of the particles was allowed. We will use the abbreviation RSADA for this process of extended RSA with diffusion and additional adsorption.

To characterise the temporal evolution of the system under consideration, several quantities were monitored at each given MC step.

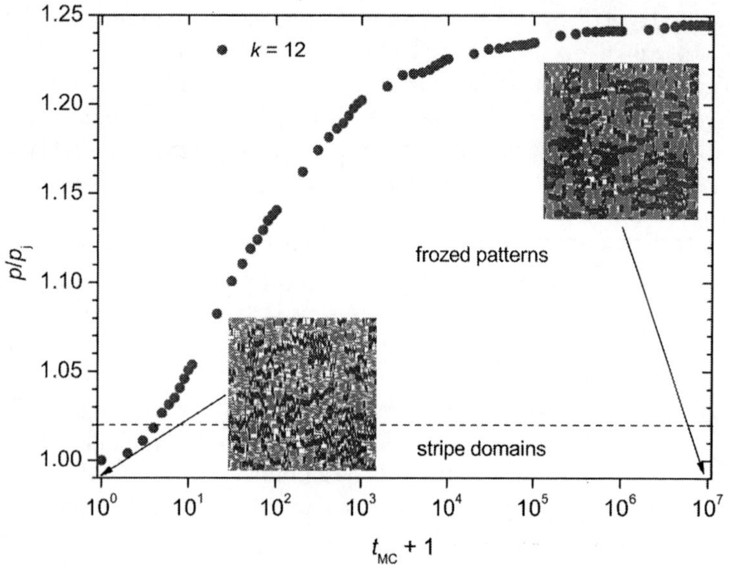

Figure 1. Growth of the packing fraction with time, $k = 12$. Insets demonstrate the initial jammed state produced using RSA alone ($p_j = 0.72$, $t_{MC} = 0$) and the final high-density state obtained using the extended RSA approach ($p = 0.89 \approx 1.25p_j$, $t_{MC} = 10^7$). Notice, that it would be possible to add more k-mers to the latter state even after 10^7 MC steps.

(i) The normalised number of clusters, n, i.e., the current value divided by the value at the initial state, $t_{MC} = 0$. In percolation theory, a cluster is a group of neighbouring occupied sites [20]. In our particular case, a cluster is a group of k-mers of the same kind, i.e., k_x-mers or k_y-mers, that are connected with one another. Two k-mers are considered to be connected when there is at least one pair of neighbouring sites that belong to different k-mers. The clusters built of k_x-mers and clusters built of k_y-mers were accounted separately and then averaged. We used the Hoshen–Kopelman algorithm [21] to count the clusters.

(ii) The local anisotropy, s, i.e., the order parameter s was calculated using (1) in a window of $l \times l$ sites and averaged over the entire set of windows. We used $l = 2^{-n}L$, $n = 1, 2, \dots, 5$. The principal analysis was performed for $l = L/4$. For this window size, we calculated ratios of each final value of s at $t_{MC} = 10^7$ to its initial value at $t_{MC} = 0$.

(iii) The fraction of interspecific contacts $n^*_{xy} = n_{xy}/(n_{xy} + n_x + n_y)$, where n_{xy} is the number of interspecific contacts between the different sorts of k-mers (i.e., k_x–k_y), n_x and n_y are the numbers of intraspecific contacts between k-mers of the same kind (i.e., k_x–k_x and k_y–k_y, respectively). For convenience of comparison, we additionally used the ratios of the final value of n^*_{xy} at $t_{MC} = 10^7$ to its initial value at $t_{MC} = 0$.

(iv) The shift ratio, R, i.e., the ratio of the number of shifts of the k-mers along the transverse axes to the number of shifts along their longitudinal axes during one MC step.

(v) The electrical conductivities, σ. We used the method described in [14] to transform the system under consideration into a random resistor network (RRN). The Frank–Lobb algorithm [22] was applied to calculate the electrical conductivity of such RRNs, (see figure A1 for details of the calculations).

All these quantities were averaged over 100 independent statistical runs for each pair (p, k).

3. Results

We examined the effect of the packing fraction, p, on pattern formation. Figure 2 presents examples of the final patterns ($t_{MC} = 10^7$) at different values of p for $k = 9$. Up to $p = 0.4$, no regular patterns occured despite visible rearrangement of the system (figure 2(a)–(d)). Then, stripe domains formed when the value of the packing fraction, p, was between $p_{down} = 0.5$ and the jamming concentration, p_j, (figure 2(e)–(h)). Slightly above p_j ($p \approx 1.02p_j$), stripe domains did not form within 10^7 MC steps (figure 2(i)), nevertheless, rearrangement of the system was still continuing. For larger values of p, the mobility of the k-mers was effectively constrained, namely, a k-mer could perform small longitudinal irregular oscillations with zero mean square displacement (figure 2(j)).

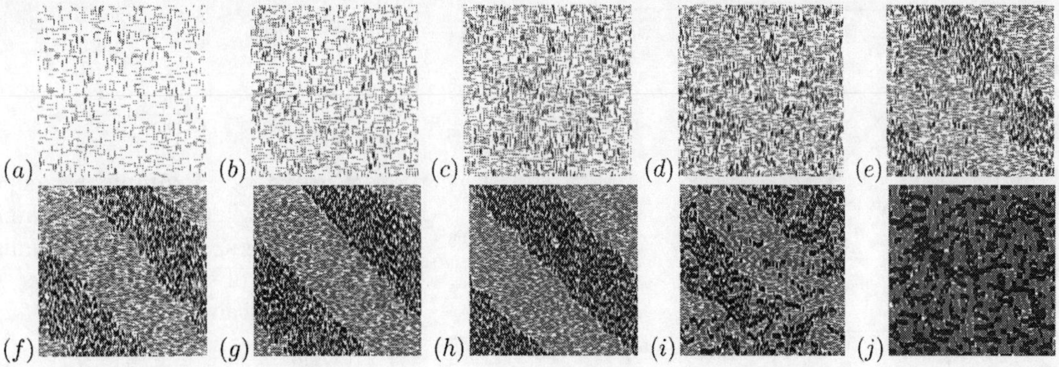

Figure 2. Examples of systems at $t_{MC} = 10^7$ for different values of the packing fractions, p. $k = 9$, $L = 256$. (a) $p = 0.1$, (b) $p = 0.2$, (c) $p = 0.3$, (d) $p = 0.4$, (e) $p = 0.5$, (f) $p = 0.6$, (g) $p = 0.7$, (h) $p = p_j \approx 0.743$, (i) $p = 0.75$, (j) $p = 0.925$.

Figure 3 demonstrates some examples of variations of the relative number of contacts, n^*_{xy}, (a) and of the normalised number of clusters, n, (b) vs MC steps, t_{MC}, for different values of the packing fraction, p, with a fixed value of k, $k = 9$. Packing fractions $p = 0.5, 0.6, 0.7$ show pattern formation, whereas no pattern formation was observed for smaller values of packing fractions Both $n^*_{xy}(t_{MC})$ and $n(t_{MC})$ curves show a stepwise decrease between $t_{MC} = 10^5$ and $t_{MC} = 10^6$ when their packing fractions correspond to pattern formation.

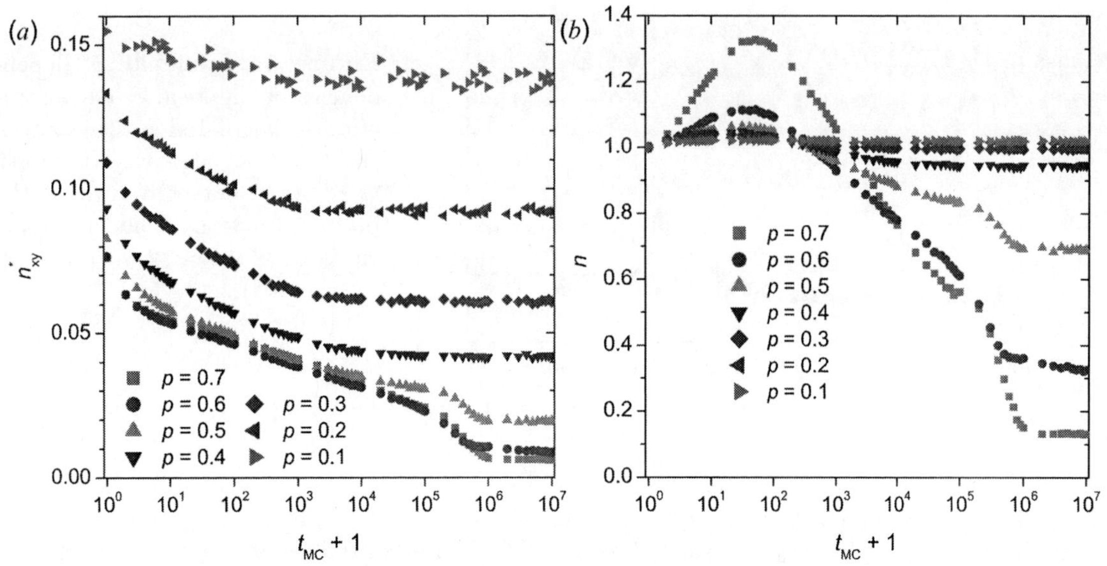

Figure 3. Examples of the dependencies of the relative number of contacts, n^*_{xy}, (a) and the normalised number of clusters, n, (b) vs MC steps, t_{MC}, for different values of the packing fraction, p, with a fixed value of k, $k = 9$.

Figure 4 demonstrates how the relative number of contacts, n_{xy}^*, and the normalised number of clusters, n, varies with the packing fraction, p, for $k = 12$. The occurrence of stripe domains is not accompanied by any characteristic change in the curves. Hence, these quantities are of little value for monitoring self-organisation and its characterisation.

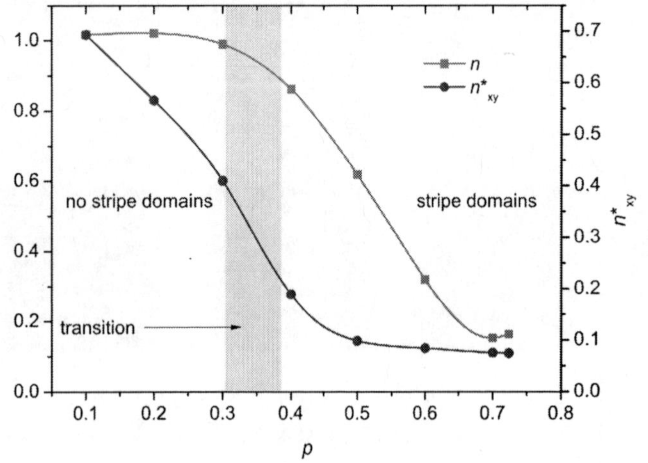

Figure 4. Examples of dependencies of the relative number of contacts, n_{xy}^*, and normalised number of clusters, n, vs the packing fraction, p, for $k = 12$.

By contrast, the local order parameter offers clear evidence of long-range pattern formation (figure 5). Significant growth of the order parameter after 10^5 MC steps is a result of cluster coarsening. This effect is pronounced for packing fractions $p = 0.5, 0.6, 0.7$ but almost imperceptible for smaller packing fractions values.

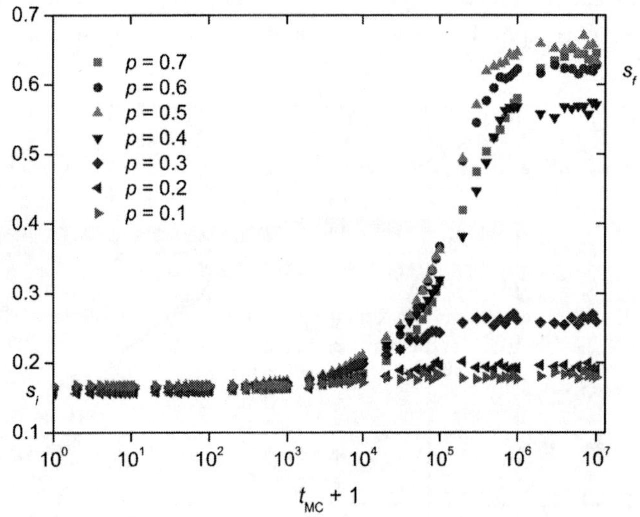

Figure 5. Examples of dependencies of the local order parameter, s, calculated in windows of size 64×64 sites vs MC steps, t_{MC}, for different values of the packing fraction, p, and the fixed value of k, $k = 12$. $s_i = s(0)$, $s_f = s\left(10^7\right)$.

The transition from a homogeneous steady-state to a patterned one appears to be continuous, i.e., the stripe domains become less pronounced for smaller initial packing fractions. For some initial packing fractions, the steady-state looks quite homogeneous (figure 2(d)) whereas imperfect stripe domains start to form at larger initial packing fractions (figure 2(e)). All the quantities of interest change smoothly as the packing fraction changes (figure 3). However, near the critical packing fraction, p_{down}, changes in the local order parameter are more visible (figure 6). This suggests the local order parameter is a useful quantity, suitable for building a phase diagram. For this purpose, we used the relative local order parameter, $s^* = s_f/s_i$, where $s_i = s(0)$, $s_f = s\left(10^7\right)$.

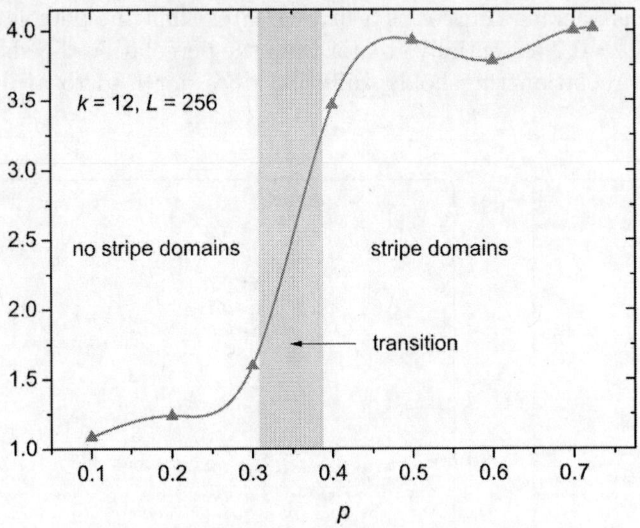

Figure 6. Example of the dependency of relative local order parameter, $s^* = s_f/s_i$, vs the packing fraction, p, for $k = 12$.

Figure 7 presents a phase diagram for the systems under study in the (k, p)-plane. In the phase diagram, the region with diagonal stripe formation in steady state at $t_{MC} \gtrsim 10^7$ is marked as SD (this region is coloured from red to light blue in the online version); the region that corresponds to an absence of stripe domains is marked as NSD (this region is coloured in lilac in the online version); the transient region is marked as T (this region is shown in dark blue in the online version). No diagonal stripes can be observed at small packing fractions below some critical value, p_{down}. This critical value, p_{down}, decreases with increased value of k.

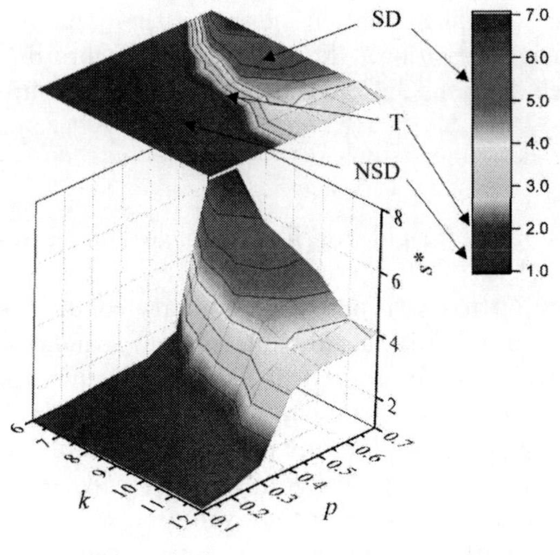

Figure 7. Phase diagram in the (k, p)-plane; SD corresponds to the values of k and p when stripe domain formation occurs; NSD corresponds to the values of k and p when no stripe domain formation is observed; T corresponds to the transient region.

It was argued in [18] that diagonal stripe domains can be produced only by "myopic particles. By contrast, "blind particles demonstrate segregation into two regions of arbitrary shapes. Nevertheless, this problem needs additional study.

Pattern formation affects the connectivity between domains, and this can lead to a change in electrical conductivity. The conventional model is more suitable for describing the initial reorganisation of the system under consideration. When the packing fraction slightly exceeds the percolation threshold, the system in its initial state, i.e., a disordered system, is a conductor. For this particular case, reorganisation of the system leads to a decrease in the electrical conductivity due to decay of the percolation cluster. Figure 8 demonstrates the conductor–insulator phase transition for $k = 10$ and $p = 0.5$. This packing fraction is a little greater than the percolation threshold, $p_c \approx 0.47$. By contrast, for $p = 0.7$ (essentially greater than the percolation threshold) and $p = 0.4$ (somewhat below the percolation threshold), diffusion does not lead to a phase transition from conductor to insulator.

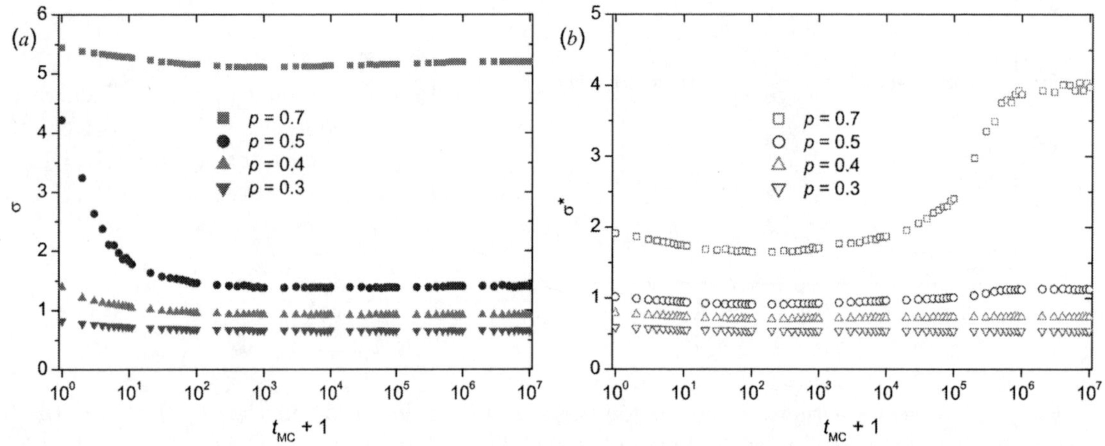

Figure 8. Examples of the temporal dynamics of electrical conductivity for $k = 10$ and different values of the packing fraction, p, $L = 256$. (a) Conventional model, (b) Insulating ends model.

The insulating ends model can catch long-range pattern formation. This model demonstrates quite different behaviour. For small values of the packing fractions, the system is an insulator throughout simulation, whereas for values of the packing fraction close to the jamming concentration, there is an insulator–conductor phase transition. This transition occurs due to the formation of dense stripe domains. Inside such domains, lateral contacts of k-mers ensure electrical conductivity in both directions.

4. Conclusion

Diffusional reorganisation in two-dimensional systems of rigid rod-like particles has been simulated using a lattice approach. Rod-like particles were presented as linear k-mers of two mutually perpendicular orientations on a square lattice with periodic boundary conditions (torus). An initial homogeneous distribution of k-mers was produced using random sequential adsorption. The packing fraction, p, was varied in the range from 0.1 to its jamming value, p_j. By means of the Monte Carlo technique, translational diffusion of the k-mers was simulated as a random walk. In our model, a particle obligatorily moves when it has any possibility of changing its location. For $k \geqslant 6$, the system tends toward a well-organised nonequilibrium steady state in the form of diagonal stripes, when the packing fraction exceeds a critical value.

Additionally, we studied the behaviour of high-density systems ($p > p_j$) produced using random sequential adsorption with diffusion and additional adsorption. In such systems, diffusivity is quite small. This results both in a significant increase of the relaxation time and in a suppression of reorganisation. Even for $p \approx 1.02p_j$, stripe domain formation was not observed within 10^8 MC steps.

It i of note, that pattern formation was absent both in dilute systems and in high-density systems. Stripe domain formation was observed only for systems of intermediate density. There is a curious analogy with social systems, viz., social self-organisation is possible when the average number of social links per person is large enough (high population density) and individual freedom is simultaneously ensured.

Figure 9 presents the phase diagram obtained using both models.

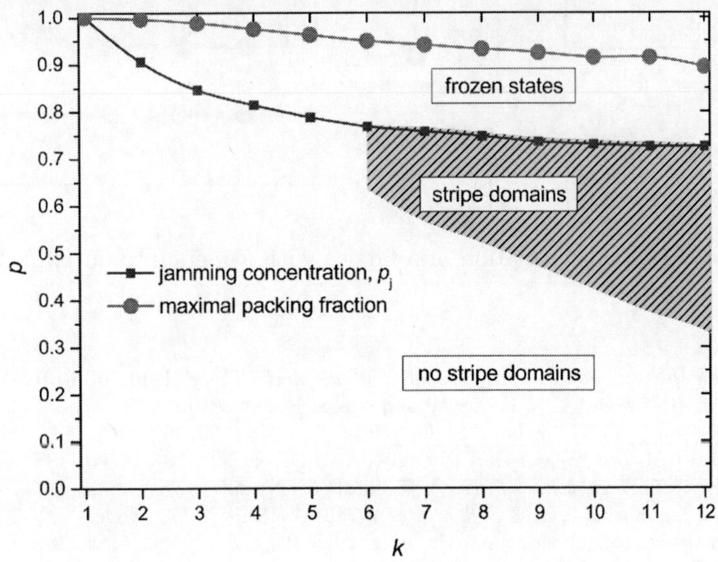

Figure 9. Phase diagram in the (k, p)-plane.

Appendix

To transform the system under consideration into an RRN, we associated each bond between two empty sites of the lattice with conductivity σ_m, each bond between two sites belonging to the same k-mer with conductivity σ_k, each bond perpendicular to the k-mer with conductivity $2\sigma_t$, and each end bond of the k-mer with conductivity $2\sigma_e$ (figure A1(a)). The resulting conductivities between the two different sites i and j were calculated as

$$\sigma_{ij} = \frac{2\sigma_i\sigma_j}{\sigma_i + \sigma_j}.$$

All possible combinations are presented in figure A1(b).

In the so-called conventional model, $\sigma_k = \sigma_t = \sigma_e = 10^6$ a.u. and $\sigma_m = 1$ a.u. In the so-called insulating ends model, $\sigma_k = \sigma_t = 10^6$ a.u. and $\sigma_m = \sigma_e = 1$ a.u.

Acknowledgments

The authors acknowledge that the reported study was funded by the Russian Foundation Basic Research according to research project No. 18-07-00343 (Y.Y.T. and V.V.C.), by the Ministry of Education and Science of the Russian Federation, Project No. 3.959.2017/4.6 (Y.Y.T., V.V.L., and V.V.C.), and by the National Academy of Sciences of Ukraine, Projects No. 43/18-H and 15F (0117U006352) (N.I.L.).

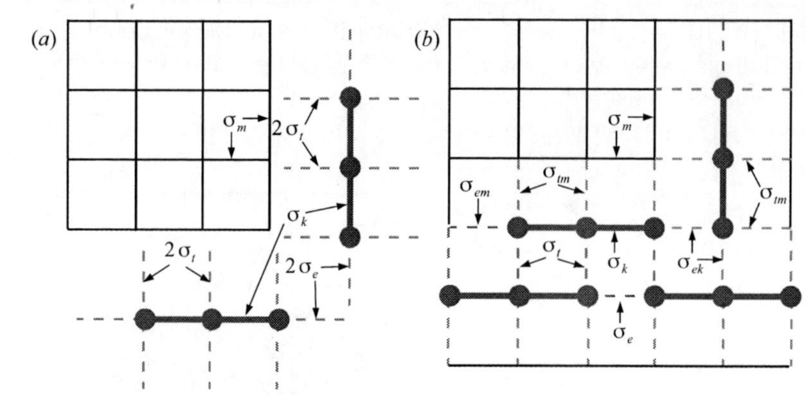

Figure A1. Transformation of a lattice with k-mers into an RRN ($k = 3$).

References

[1] Narayan V, Menon N and Ramaswamy S 2006 *J. Stat. Mech. Theor. Exp.* **2006** P01005 ISSN 1742-5468
[2] Galanis J, Harries D, Sackett D L, Losert W and Nossal R 2006 *Phys. Rev. Lett.* **96** 028002 ISSN 0031-9007
[3] Aranson I S, Volfson D and Tsimring L S 2007 *Phys. Rev. E* **75** 051301 ISSN 1539-3755
[4] Galanis J, Nossal R, Losert W and Harries D 2010 *Phys. Rev. Lett.* **105** 168001 ISSN 0031-9007
[5] Yadav V and Kudrolli A 2012 *Eur. Phys. J. E* **35** 104 ISSN 1292-895X
[6] Börzsönyi T and Stannarius R 2013 *Soft Matter* **9** 7401–7418 ISSN 1744-6848
[7] Müller T, de las Heras D, Rehberg I and Huang K 2015 *Phys. Rev. E* **91**(6) 062207 ISSN 2470-0045
[8] González-Pinto M, Borondo F, Martínez-Ratón Y and Velasco E 2017 *Soft Matter* **13**(14) 2571–2582 ISSN 1744-6848
[9] Aranson I S and Tsimring L S 2006 *Rev. Mod. Phys.* **78**(2) 641–692 ISSN 0034-6861
[10] Evans J W 1993 *Rev. Mod. Phys.* **65**(4) 1281–1329 ISSN 0034-68611
[11] Centres P M and Ramirez-Pastor A J 2015 *J. Stat. Mech. Theor. Exp.* **2015** P10011 ISSN 1742-5468
[12] Kuriata A, Polanowski P and Sikorski A 2016 *Macromol. Theor. Simul.* **25** 360–368 ISSN 1521-3919
[13] Budinski-Petković L, Lončarević I, Dujak D, Karač A, Šćepanović J R, Jakšić Z M and Vrhovac S B 2017 *Phys. Rev. E* **95**(2) 022114 ISSN 2470-0045
[14] Lebovka N I, Tarasevich Y Y, Gigiberiya V A and Vygornitskii N V 2017 *Phys. Rev. E* **95**(5) 052130 ISSN 2470-0045
[15] Tarasevich Y Y, Laptev V V, Burmistrov A S and Lebovka N I 2017 *J. Stat. Mech.: Theory Exper.* **2017** 093203 ISSN 1742-5468
[16] Fusco C, Gallo P and Petri A and Rovere M 2001 *J. Chem. Phys.* **114** 7563–7569 ISSN 0021-9606
[17] Selinger R B and Stanley H E 1990 *Phys. Rev. A* **42**(8) 4845–4852 ISSN 1050-2947
[18] Patra S, Das D, Rajesh R and Mitra M K 2018 *Phys. Rev. E* **97**(2) 022108 ISSN 2470-0045
[19] Landau D P and Binder K 2014 *A Guide to Monte Carlo Simulations in Statistical Physics* 4th ed (Cambridge: Cambridge University Press) ISBN 9781107074026
[20] Stauffer D and Aharony A 1992 *Introduction to Percolation Theory* (London: Taylor & Francis)
[21] Hoshen J and Kopelman R 1976 *Phys. Rev. B* **14**(8) 3438–3445 ISSN 2469-9950
[22] Frank D J and Lobb C J 1988 *Phys. Rev. B* **37**(1) 302–307 ISSN 2469-9950

Clusters of Local Bond-Orientational Order in Liquid Gallium: Studies of *ab initio* and Classical MD Simulations

Lin-Yuan Chen and Ten-Ming Wu*
Institute of Physics, National Chiao-Tung University, Hsinchu 300, Taiwan. R.O.C

Email: tmw@faculty.nctu.edu.tw

Abstract. The structure of liquid Ga at ambient pressure (AP) were generated separately via *ab initio* and classical molecular dynamics simulations, and analysed with a cluster approach developed in terms of the local bond-orientational order (LBOO) parameters. Two types of cluster structures were found in both simulated liquids, one typified by sixfold orientational symmetry and the other showing fourfold orientational symmetry: The static structure factors (SSF) of the former are akin to that of the hard-sphere (HS) fluid; however, the SSFs of the latter display either an asymmetric first peak with a shape different from that of the HS fluid or even a shoulder on the high-q side of the first peak, similar as the anomalous SSF of liquid Ga at AP.

1. Introduction

Gallium has a complex phase diagram, which contains multiple stable and metastable solid phases [1]. Besides the polymorphism of solid, Ga at AP has a liquid phase characterized with a low melting temperature, a high boiling point, and a density higher than its stable solid phase (α-Ga). Supercooled liquid Ga exists in a broad temperature range and crystallizes in several metastable solid states at different temperatures, among which β-Ga is the first one and occurs at 257 K. Just above the melting temperature (T_m = 302 K), liquid Ga is well known for its anomalous SSF, which has a highly asymmetric first peak by displaying a shoulder on its high-q side [2,3]. Because of its non HS-like SSF, Waseda categorized liquid Ga as the third type of liquid structures, which have the Q_2/Q_1 ratios significantly higher than that of the HS fluid [2], where Q_1 and Q_2 are the first- and second-peak positions of a SSF, respectively. The anomalous SSF of liquid Ga has been explained by several scenarios, including the ledge-shape repulsive core in an early study [4], the interplay of two characteristic lengths determined by the effective HS diameter and the Friedel wavelength [3,5], the partial-covalency dimers such as in α-Ga [6,7], and clusters primarily composed of special four-atom units [8]. Hence, the physical cause for the high-q shoulder in the SSF of liquid Ga has been a long-standing debating issue, since its early observation in experiments [9,10].

The SSF of liquid Ga at 323K and AP, including the high-q shoulder, can be reproduced by performing classical molecular dynamics (CMD) simulations with an interatomic pair potential $\phi(r)$ derived from a first-principles pseudopotential theory [11]. The pair potential $\phi(r)$ consists of a ledge-shape repulsive core $\phi_0(r)$, which is much softer than the Lennard-Jones potential, and the long-range Friedel oscillations, decaying in distance with r^{-3}, induced by conduction electrons [12]. However, as performing CMD simulation with the soft-core $\phi_0(r)$ potential only at the same temperature and number density, no high-q shoulder was found in the SSF of this soft-core fluid [13]. Thus, it was concluded that the high-q shoulder in the SSF of the $\phi(r)$ liquid was resulted from a modulation caused by the Friedel oscillations on the structure determined by the soft-core potential $\phi_0(r)$. In real space, a comparison of the radial distribution functions of the liquids with and without the Friedel oscillations manifested that the high-q shoulder was associated with the medium range order beyond the first shell of the radial distribution function. On the other hand, also based on the $\phi(r)$ model, the high-q shoulder in its SSF was attributed to clusters containing special units of four atoms [8]. As quenching the system into supercooled and glassy states, the amounts of the special four-atom units increased considerably, whereas the SSF changed to that of β-Ga [14]. However, the algorithm for systematically searching these particular clusters in a liquid is complicated, because the clusters were estimated to involve more than 20 atoms.

Recently, an algorithm for searching clusters of LBOO in simple liquids was developed [15]. A LBOO cluster in a liquid is defined as the neighboring seeds and their neighbors, where a seed is an atom having at least n_b local-orientational bonds and the neighbors of a seed are atoms within a cutoff distance r_c from the seed. This cluster approach has been applied to study liquid structures generated via *ab initio* molecular dynamics (AIMD) simulations by using the VASP code [16,17] for 1331 Ga atoms at 323 K and AP, where the electronic structures were calculated with a projector augmented wave potential in the local-density approximation. The simulation results indicated that liquid Ga at AP contained two types of cluster structures, one characterized by sixfold orientational symmetry and the other showing fourfold orientational symmetry. In simulated liquids, the clusters with sixfold orientational symmetry were dominated and their SSFs behaved similarly as that of a HS-fluid, whereas the SSFs of clusters with fourfold orientational symmetry exhibited an asymmetric first peak, which is different in shape from that of a HS fluid, or even displayed a shoulder on the high-q side of the first peak as the clusters were in high LBOO.

The CMD simulation with the $\phi(r)$ potential and the AIMD simulation for liquid Ga were performed at the same temperature and pressure; their details have been described in Refs. [13] and [15], respectively. It is interesting to compare the characters of the liquid structures generated by the two MD simulations, especially the LBOO clusters. This paper is organized as follows: In Sec. II, the definition of a LBOO cluster is introduced and the cluster SSF is also defined. In Sec. III, the structures of the two simulated liquids are compared, and so are the SSFs of the LBOO clusters found in the two simulated liquids. Our conclusions are given in Sec. IV.

2. LBOO Cluster

In a simple liquid, the neighbors of an atom are defined as those atoms within a cutoff distance r_c from the central one, where r_c is usually set at the first minimum of the radial distribution function of the liquid. The structure of the neighbors around atom i can be characterized by a set of ($2l$+1) order parameters $q_{lm}(i)$, with l a positive integer and m an integer subject to $-l \leqq m \leqq l$, which is defined as [18,19]

$$q_{lm}(i) = \frac{1}{N_b(i)} \sum_{j=1}^{N_b(i)} Y_{lm}(\hat{r}_{ij}), \tag{1}$$

where the sum runs over all neighbors $N_b(i)$ of the atom, $Y_{lm}(\hat{r}_{ij})$ is the spherical harmonics, and $\hat{r}_{ij}$ is a unit vector in the direction from atom i to its neighbor j. The LBOO parameter $q_l(i)$ of atom i is given as

$$q_l(i) = \left(\frac{4\pi}{2l+1} \sum_{m=-l}^{l} |q_{lm}(i)|^2 \right)^{1/2}, \tag{2}$$

where $q_l(i)$ measures the degree of l-fold orientational symmetry among the atom's neighbors. Thus, associated with the local structure of each atom, one can define a normalized complex vector $\tilde{q}_l(i)$ with

(2l+1) components, which are obtained by dividing $q_{lm}(i)$ with a normalization factor so that the absolute square of $\tilde{q}_l(i)$ is one [18].

The correlation between the local structures of two neighboring atoms i and j can be quantified by the dot product $\tilde{q}_l(i) \cdot \tilde{q}_l(j)$ of their normalized complex vectors, where the dot product is defined in Ref. [19]. As usual, if the dot product $\tilde{q}_l(i) \cdot \tilde{q}_l(j) \geq 0.5$, the two neighboring atoms are considered to connect through a local-orientational bond. By our definition, if the local-orientational bond number of an atom exceeds a threshold value n_b, this atom is identified as a seed. Hence, a LBOO cluster is defined as a composition of the adjacent seeds and all their neighbors; the cluster is specified with two integers: the LBOO n_b and the number of seeds n_s in the cluster. In a liquid, a cluster ensemble is a collection of all LBOO clusters specified with the two integers of n_b and n_s, and is, therefore, indexed with $\{n_b, n_s\}$.

The structure of a LBOO cluster ensemble indexed with $\{n_b, n_s\}$ can be described with the average cluster SSF, which is defined with the equation [8]

$$S_{cl}(q) = 1 + \left\langle \frac{1}{M} \sum_{i \neq j}^{M} \sum_{j=1}^{M} \frac{\sin(qr_{ij})}{qr_{ij}} \right\rangle_c , \tag{3}$$

where M is the atom number of a cluster and the brackets with a subscript c denote an average for all LBOO clusters in the ensemble. As a reference, the SSF of a HS fluid displays a symmetric first peak and has the Q_2/Q_1 ratio close to 1.86 [3]. Two quantities used for examining the deviation of $S_{cl}(q)$ of a LBOO cluster ensemble from that of a HS fluid were the Q_2/Q_1 and the q_R/q_L ratios [15], where q_R and q_L are, respectively, the right-half and left-half widths of the first peak at an altitude of one half between the first maximum and first minimum of $S_{cl}(q)$. Thus, the q_R/q_L ratio measures the asymmetry of the first peak in $S_{cl}(q)$.

3. Results of AIMD and CMD Simulations

The SSFs and the radial distribution functions of the liquids generated separately via AIMD and CMD simulations are presented in Fig. 1, where the simulated SSFs are also compared with the experimental results of liquid Ga [2]. For each simulated liquid, the SSF exhibits a shoulder on the high-q side of its first peak, and the Q_2/Q_1 ratio is approximately 1.94, close to the value of liquid Ga. The q_R/q_L ratios of the SSFs of the classical and the *ab initio* liquids are near 1.61 and 2.60, respectively, while the q_R/q_L ratio estimated for the SSF of liquid Ga is 2.20, which is referred to the experimental value. These indicate that the SSF of the *ab initio* liquid has a stronger high-q shoulder than that of the classical liquid.

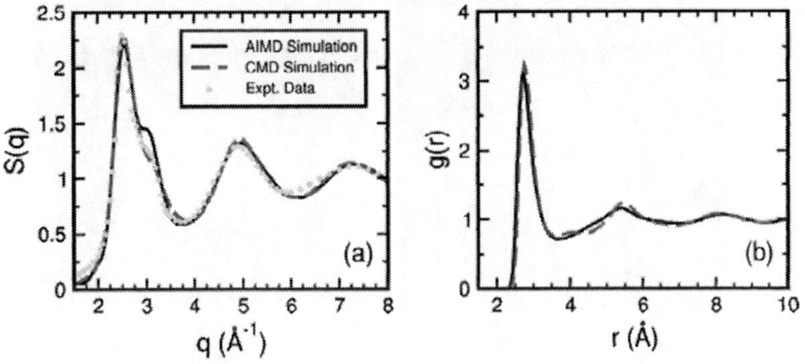

Figure 1 (a) Static structure factors and (b) radial distribution functions of simulated liquids at 323 K and AP. The black-solid and red-dash lines were obtained via AIMD and CMD simulations, respectively. The green circles in (a) are the experimental data of liquid Ga taken from Ref. [2]. The radial distribution functions of the two liquids have almost the same first minimum at 3.61 Å.

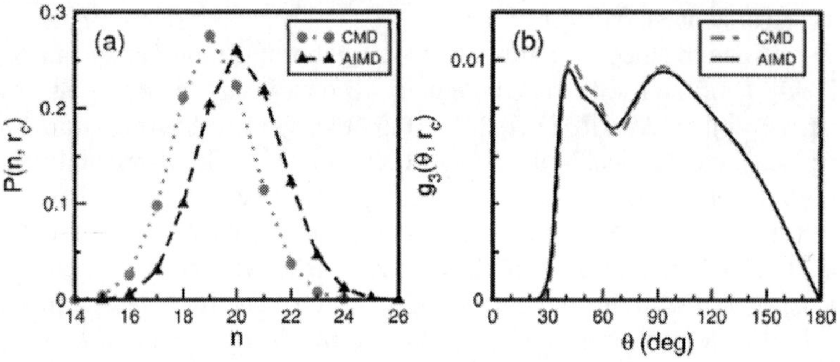

Figure 2 (a) The coordination-number distributions and (b) the bond-angle distributions of simulated liquids at 323 K and AP. The distributions were calculated with r_c = 4.65 Å. The black-solid line and triangles were obtained via AIMD simulation; the red-dash line and circles were via CMD simulation. All distributions are normalized.

According to the definitions given in Ref. [20], the coordination-number distribution $P(n, r_c)$ and the bond-angle distribution $g_3(\theta, r_c)$ of the *ab initio* and the classical liquids were calculated, and the results evaluated with r_c = 3.61 Å, which is set at the first minima of the radial distribution functions of the two liquids, have been shown in Fig. 3 of Ref. [15]. Here, we present in Fig. 2 the $P(n, r_c)$ and $g_3(\theta, r_c)$ distributions of the two simulated liquids calculated with r_c = 4.65 Å, which includes the first attractive well of $\phi(r)$. The coordination numbers of atoms in the classical liquid fall in a range between 14 and 24, with the most probable one occurred at 19, whereas the coordination-number distribution of the *ab initio* liquid shifts toward one more atom, with the most probable coordination number at 20, because the number density of the *ab initio* liquid (ρ = 0.0525 $Å^{-3}$) is slightly higher than that of the classical liquid (ρ = 0.05 $Å^{-3}$).

With r_c = 4.65 Å, the bond-angle distributions of the two simulated liquids are almost identical and display two peaks, which are near 40° and 90°, and a shoulder around 50°, with the heights of the two peaks almost equal and the magnitudes of the shoulder somewhat smaller. Thus, the bond-angle distribution of r_c = 4.65 Å is quite different in feature from that of r_c = 3.61 Å. The 90° peak in the bond-angle distribution shown in Fig. 2(b) provides a strong indication for the present of local structures with fourfold orientational symmetry in each simulated liquid.

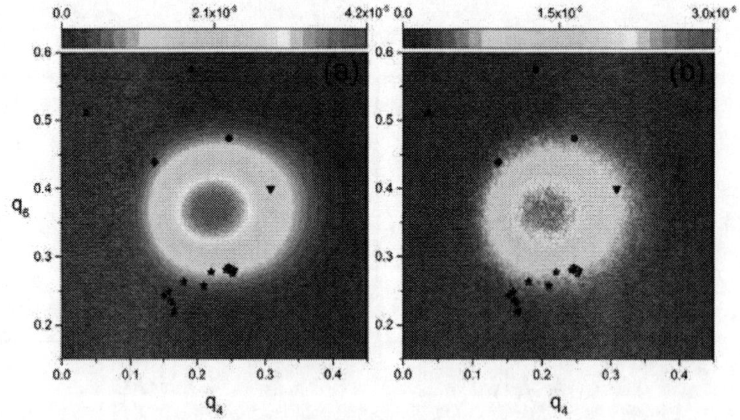

Figure 3 2-D map for the q_4-q_6 distribution of each simulated liquid at 323 K and AP: (a) the classical liquid and (b) the *ab initio* liquid. The distributions were calculated with r_c = 3.61 Å. The symbols indicate the (q_4, q_6) points of several crystals: bcc (▲), fcc (■), α-Ga (●), β-Ga (▼), Ga III (◆), and Ga II (★) [15].

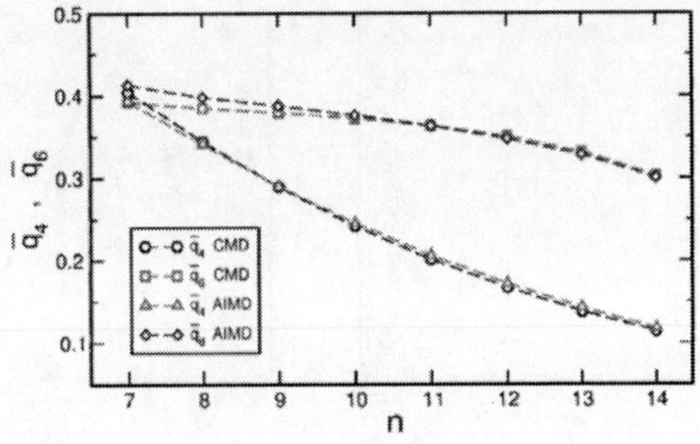

Figure 4 The dependences of $\bar{q}_4$ and $\bar{q}_6$ on coordination number n of an atom in a liquid. The symbols are the results of simulated liquids. The dash lines serve to guide the eye.

With $r_c = 3.61$ Å, the LBOO parameters $q_6(i)$ and $q_4(i)$ of atoms in the *ab initio* and the classical liquids were calculated, and the 2-D map for the q_4-q_6 distribution of each liquid is presented in Fig. 3. The 2-D maps of the two liquids are similar in position and shape, except for higher resolution for the classical liquid due to more atoms involved in CMD simulation. In each 2-D map, we also indicate the (q_4, q_6) points of several crystals, including α-Ga, β-Ga, Ga II and Ga III, with different symbols. Intriguingly, the points of Ga solids lie on the rim of the q_4-q_6 distribution of each simulated liquid but are located at different directions from the centre of the distribution. Particularly, because of its complicated structure, Ga II [21] has 14 points of (q_4, q_6), which are distributed within a region of lower q_6 in the 2-D map. According to the 2-D maps in Fig. 3, it is speculated that the local structure of liquid Ga at AP is constrained by the structures of its crystalline counterparts.

Relative to the Lennard-Jones liquid [22], the q_4-q_6 distribution of each simulated liquid has a centre shifting toward higher q_4 value, causing a broader q_4-distribution. The higher q_4 values come from atoms with lower coordination numbers. This can be verified by examining the averages of $q_6(i)$ and $q_4(i)$ for atoms having the same coordination number, which are denoted as $\bar{q}_6$ and $\bar{q}_4$, respectively [15]. As shown in Fig. 4, the dependences of $\bar{q}_6$ and $\bar{q}_4$ on coordination number n of an atom are almost identical for the two simulated liquids. For each liquid, both $\bar{q}_6$ and $\bar{q}_4$ increase with decreasing coordination number n, and the increase of $\bar{q}_4$ is more significant.

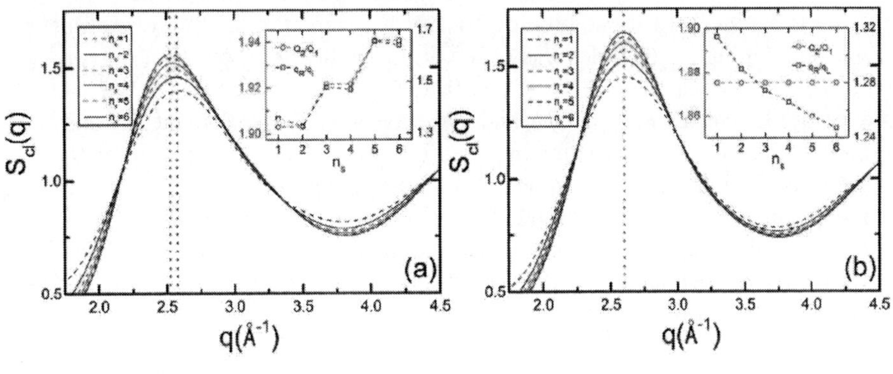

Figure 5 SSFs of cluster ensembles in medium LBOO ($n_b = 5$) in the classical liquid: (a) q_4- clusters and (b) q_6-clusters. In each panel, the main plot presents the variation of the cluster SSF with the number of seeds n_s up to 6. The black and blue dot lines indicate the first-peak position of the cluster SSF of $n_s = 1$ and $n_s = 6$, respectively. The insert shows the variations of the Q_2/Q_1 (circles) and q_R/q_L (squares) ratios of the cluster SSF with n_s, and their values are referred to the left and right vertical axes, respectively.

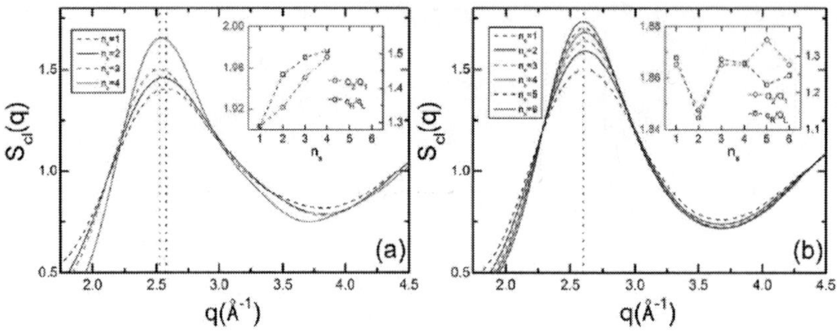

Figure 6 SSFs of cluster ensembles in high LBOO (n_b = 8) in the classical liquid: (a) q_4- clusters and (b) q_6-clusters. The plots are similar as Fig. 5 except that the results in (a) are presented only up to n_s = 4 and the red dot line indicates the first-peak position of the cluster SSF with n_s = 4.

The SSFs of LBOO cluster ensembles found in the *ab initio* liquid have been displayed and discussed in Ref. [15]. Here, we present the results of the classical liquid and compare their characters with those of the *ab initio* liquid. In Fig. 5, the SSFs of the q_4- and q_6-cluster ensembles of n_b = 5 defined with r_c = 3.61 Å are displayed, where n_b = 5 is about one half of the most probable coordination number of atoms in the liquid. The first-peak position of the SSF of q_4-clusters shifts toward a smaller q-value as increasing the number of seeds n_s in a cluster, but that of the q_6-clusters is insensitive to n_s. Further, for the q_4-clusters, both Q_2/Q_1 and q_R/q_L ratios of the cluster SSF, after a proper rescale, increase with n_s in a similar manner, and as n_s = 6 the Q_2/Q_1 ratio reaches the value (1.94) of liquid Ga. In general, the behaviours of the cluster SSF of the q_4-cluster ensembles in the classical liquid are consistent with those in the *ab initio* liquid [15]. For the q_6-clusters, as more seeds are added to the clusters, the Q_2/Q_1 ratio of the cluster SSF keeps at a constant near 1.875 up to n_s = 6, whereas the q_R/q_L ratio decreases monotonically with increasing n_s. Thus, the SSFs of the q_6-clusters in the classical liquid behave somewhat differently from those found in the *ab initio* liquid and are even more HS-like, which indicates that the extent of q_6-cluster structures in the classical liquid mixing with other *l*-fold orientational symmetries is less than that of the q_6-clusters in the *ab initio* liquid.

Fig. 6 presents the SSFs of the q_4- and q_6-cluster ensembles with n_b = 8 in the classical liquid. The SSF of the q_4-cluster ensemble was found to deviate even more from that of a HS fluid: The Q_2/Q_1 ratio at n_s = 4 is near 1.97, and the first peak of the cluster SSF becomes asymmetric as n_s increases. As searching in the classical liquid, no high-q shoulder was found in the SSF of q_4-clusters. However, in the *ab initio* liquid a high-q shoulder was observed in the SSF of a q_4-cluster involving three seeds [15]. In Fig. 6(b), as n_s increases up to 6, the Q_2/Q_1 ratio of the cluster SSF of the q_6-cluster ensemble keeps within a range between 1.88 and 1.84, and may have a value less than the HS one (1.86), whereas the q_R/q_L ratio exhibits a small fluctuation between 1.3 and 1.1. The fluctuations of Q_2/Q_1 and q_R/q_L are possibly resulted from the rare occurrence of q_6-clusters with n_b = 8 in the classical liquid. Thus, the SSFs of the q_4-cluster ensembles in high LBOO behave quite differently from those of the q_6-cluster ensembles; however, their differences found in the classical liquid are not so much as those appearing in the *ab initio* liquid. This explains why the high-q shoulder observed in the SSF of the *ab initio* liquid is stronger than that in the SSF of the classical liquid.

It is interesting to investigate further the structure of highly LBOO q_4-cluster, which SSF displays a high-q shoulder. Found in the *ab initio* liquid [15], the cluster is composed of three seeds, indicated with *A*, *B*, and *C* in Fig. 7(a), where the three seeds form two bonds at nearly 90° and the bond length between seed *B* and *C* is 2.715 Å and that between seed *B* and *A* is 3.105 Å. By referring to the pair potential $\phi(r)$, the two bond lengths are within the soft repulsive core, whereas the length between seed *A* and *C* is 4.125 Å, which is within the region of the first attractive well of $\phi(r)$. The geometry of the right triangle formed by the three seeds provides an insight into the bond-angle distribution shown in Fig.

2(b). On the other hand, obtained via the AIMD simulation, the charge density around the cluster in the *ab initio* liquid is presented in Fig. 7(b), where the 2-D map is plotted in the plane of the three seeds. Clearly, there is no pile up of charge in the center between any two seeds of A, B, and C, which indicates that the bonding among the three seeds is not covalent-like as the dimers in α-Ga [23].

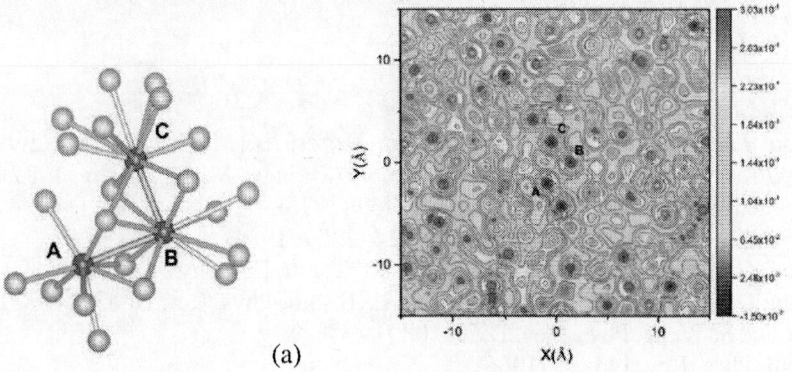

(a) (b)

Figure 7 (a) Structure of q_4-cluster with $n_b = 8$ and containing three seeds. The cluster was found in the *ab initio* liquid. The red and blue circles indicate the seeds and their neighbors of the cluster, respectively. (b) Contours for the charge density (in the unit of $e\text{Å}^{-3}$) around the cluster in the *ab initio* liquid. The 2-D map of charge density is plotted in the plane of the three seeds, indicated with A, B, and C, of the cluster.

4. Conclusions

In this paper, we have investigated cluster structures in liquid Ga via AIMD and CMD simulations at the same temperature and pressure. The liquid generated by each of the two simulations can produce a SSF that behaves similarly as the anomalous SSF of liquid Ga at AP, including the shoulder on the high-q side of its first peak, whereas the shoulder in the SSF of the *ab initio* liquid is somewhat stronger. Both simulated liquids exhibit similar behaviours in the radial distribution function, the bond-angle distribution, and loose pack structure with low coordination number. Analysed with the q_4 and q_6 LBOO parameters, both simulated liquids contained cluster structures with fourfold orientational symmetry and with sixfold orientational symmetry, where the two types of cluster structure display different features in their cluster SSFs. The SSFs of cluster structures with sixfold orientational symmetry are akin to that of a HS fluid by showing a symmetric first peak. However, the SSFs of clusters with fourfold orientational symmetry displays an asymmetric first peak or even a shoulder on the high-q side of the first peak as the clusters are in high LBOO. Thus, the two types of cluster structure play contrasting roles in the appearance of the high-q shoulder in the SSF of the simulated liquid: the clusters with fourfold orientational symmetry favour its appearance but the clusters with sixfold orientational symmetry suppress it.

According to our experiences, the CMD simulation with the $\phi(r)$ potential is efficient in time and can be performed with larger system sizes; however, the calculated results depend on the force field of the $\phi(r)$ model. On the contrary, the AIMD simulation does not need any force field model and the results are more close to realistic liquid Ga, but it is drawback in limited system sizes and long computational time due to its demanding on computer resource. Through both AIMD and CMD simulations capture the anomalous structure of liquid Ga at AP, there are still minor differences between the radial distribution functions generated by the two simulations. Produced by the CMD simulation, a small hump showed up in the region between the first minimum and the second maximum of the radial distribution function, where this region is associated with the first attractive well of the $\phi(r)$ potential. But, there was no hump observed in the corresponding region of the radial distribution function generated by the AIMD simulation, which is agree with the experimental results [15]. The results of the classical liquids indicate that this small hump in the radial distribution function was influenced by the

Friedal oscillations in the $\phi(r)$ potential [13]. Probably, some modification in the first attractive well of the $\phi(r)$ potential will make the structure of the classical liquid more close to realistic liquid Ga. If so, the CMD simulation would provide a useful method to study this peculiar metallic liquid.

Acknowledgement
TMW acknowledge supports from Ministry of Science and Technology, Taiwan R.O.C. under Grant No. MOST 106-2112-M-009-017.

References
[1] L. Bosio, J. Chem. Phys. **68**, 1221 (1978).
[2] Y. Waseda, *The Structure of Non-crystalline Materials* (McGraw-Hill, New York, 1980).
[3] J. Hafner, *From Hamiltonians to Phase Diagrams* (Springer-Verlag, Berlin, 1987).
[4] M. Silbert and W.H. Young, Phys. Lett. A **58**, 469 (1976).
[5] J. Hafner and G. Kahl, J. Phys. F: Met. Phys. **14**, 2259 (1984).
[6] X.G. Gong, G.L. Chiarotti, M. Parrinello, and E. Tosatti, Europhys. Lett. **21**, 469 (1993).
[7] X.G. Gong, G.L. Chiarotti, M. Parrinello, and E. Tosatti, Phys. Rev. B. **43**, 14277 (1991).
[8] S.F. Tsay and S. Wang, Phys. Rev. B **50**, 108 (1994).
[9] P. Ascarelli, Phys. Rev. **143**, 36 (1966).
[10] A.H. Narten, J. Chem. Phys. **56**, 1185 (1972).
[11] S.F. Tsay, Phys. Rev. B **48**, 5945 (1993).
[12] T.M. Wu, AIP Conf. Proc. **1518**, 411 (2013).
[13] K.H. Tsai, T.M. Wu, and S.F. Tsay, J. Chem. Phys. **132**, 034502 (2010).
[14] S.F. Tsay, Phys. Rev. B **50**, 103 (1994).
[15] L.Y. Chen, P.H. Tang, and T.M. Wu, J. Chem. Phys. **145**, 024506 (2016).
[16] G. Kresse and J. Hafner, Phys. Rev. B **47**, 558 (1993).
[17] G. Kresse and J. Furthmüller, Phys. Rev. B **54**, 11169 (1996).
[18] P.R. ten Wolde, M.J. Ruiz-Montero, and D. Frenkel, J. Chem. Phys. **104**, 9932 (1996).
[19] S. Auer and D. Frankel, J. Chem. Phys. **120**, 3015 (2004).
[20] S. Munejiri, T. Masaki, T. Itami, F. Shimojo, and K. Hoshino, Phys. Rev. B **77**, 014206 (2008).
[21] O. Degtyareva, M.I. McMahon, D.R. Allan, and R.J. Nelmes, Phys. Rev. Lett. **93**, 205502 (2004).
[22] W. Lechner and C. Dellago, J. Chem. Phys. **129**, 114707 (2008).
[23] M. Bernasconi, G.L. Chiarotti, and E. Tosatti, Phys. Rev. B. **52**, 9988 (1995).

IOP Conf. Series: Journal of Physics: Conf. Series **1136** (2018) 012017 doi:10.1088/1742-6596/1136/1/012017

Stability of charges in titanium compounds and charge transfer to oxygen in titanium dioxide

D Koch, P Golub, S Manzhos

Department of Mechanical Engineering, National University of Singapore, 9 Engineering Drive 1, Singapore 117576, Singapore

mpemanzh@nus.edu.sg

Abstract. We investigate the charge density distribution in titanium dioxide, molecular titanium complexes and a variety of periodic titanium compounds using delocalization indices and Bader charge analysis. Our results are in agreement with previous experimental and theoretical investigations on the charge stability and deviation from formal oxidation states in transition metal compounds. We present examples for practically relevant redox processes, using molecular titanium dioxide model systems, that illustrate the failure of formal oxidation states to account for some redox phenomena. We observe a pronounced charge stability on titanium for trial systems which are expected to be mainly ionic. No environment tested by us is capable to reduce the local titanium charge remainder below one electron.

1. Introduction

The rationalization of redox processes in terms of formally defined charge transfers between the involved reactants is, by definition, based on the change of the oxidation states of the involved species. Although the way formal oxidation states (FOS) are determined can be justified by a variety of valid arguments [1], the implications that come with them, especially with regard to redox processes, might be counter-intuitive and of little practical use. Quantum theory of atoms in molecules (QTAIM) on the other hand provides a way to quantify the electron density distribution along bonds and assign charges to the atomic centers [2]. It takes into account the continuity of the electron density, as opposed to the necessarily integer charges in FOS. Changes in the electron density distribution upon formation of new bonds can in that way be assigned to electron fluxes between different atoms, a perspective on redox reactions which aligns better with reality than the standard definition of reduction and oxidation states which dates from before quantum mechanics came into the picture but still has substantial influence on the way redox reactions are commonly understood.

Recently, a work from the group of G. Ceder revealed the oxygen redox activity in cation-disordered and Li-excess cathode materials [3]. Electron transfer on oxygen in this cases can happen due to the creation of unoccupied oxygen states into which electrons are transferred upon Li insertion, an unusual process, since in most oxide materials the oxygen-centered ligand group orbitals of the interconnected MO_n polyhedra lie lower in energy than the transition metal-centered orbitals. This view on reduction and oxidation is in accordance with the formal definition of oxidation states. On the other hand, every electron transfer from an intercalant to the host lattice is going to affect the electron density distribution around all crystal atoms and will lead to changes in the QTAIM charge states of the species, which is, together with other electron density-based measures as e.g. spin density differences, commonly used to rationalize electron transfers accompanying insertion reactions [4]. According to these measures, a degree of oxygen reduction can be seen in non-defected oxides [5].

We have addressed the well-known, but nevertheless potentially problematic discrepancy between charge and oxidation states before [5, 6]. In this work, we want to underpin previously reported findings of relative charge stability in transition metal compounds with different FOS of the metal center and demonstrate the charge state changes of oxygen in oxides using simple molecular and periodic model systems. The charge states were on one hand obtained from the commonly used Bader analysis method [7]. On the other hand it is useful to exploit the concept of delocalization indices (DIs) [8], which are, although indirect, also able to provide more insight into charge transfer processes. Delocalization indices, denoted as $\delta(A,B)$, formally provide a measure for the density of the Fermi hole, created by electrons within one space domain A, that is present within another space domain B. Since a Fermi hole can be related to a certain electron and describes the decrease in the probability to find another same-spin electron nearby [9], its distribution pattern mimics the charge density distribution pattern of that electron [10]. Thus one can alternatively say that DIs measure the extent to which electrons in a given space domain A are delocalized into another space domain B, and vice versa.

The expression for the DIs can be obtained from the electron pair density $\Gamma(\vec{r_1},\vec{r_2})$, which, assuming spin-orbitals φ of a Hartree-Fock (HF) wave function, is

$$\Gamma(\vec{r_1},\vec{r_2}) = \frac{1}{2}\sum_{\mathrm{i}}\sum_{\mathrm{j}}(\varphi_{\mathrm{i}}^*(\vec{r_1})\varphi_{\mathrm{i}}(\vec{r_1})\varphi_{\mathrm{j}}^*(\vec{r_2})\varphi_{\mathrm{j}}(\vec{r_2}) - \varphi_{\mathrm{i}}^*(\vec{r_1})\varphi_{\mathrm{i}}(\vec{r_2})\varphi_{\mathrm{j}}^*(\vec{r_2})\varphi_{\mathrm{j}}(\vec{r_1})). \quad (1)$$

The first term is a simple product of individual electron densities $\rho(\vec{r_1})\rho(\vec{r_2})$, the last term is similar to the exchange term of the HF theory, correcting this product to the proper number of electron pairs $N\,(N-1)/2$ in the system. Formally it acts equivalently to a Fermi hole, removing the same amount of electron density [10]. Therefore its integration over two given space domains provides information about how much of the Fermi hole from the first domain delocalizes into the second domain and vice versa, that is the corresponding DI. In Fulton formulation [11]

$$\delta(A,B) = \int_{\mathrm{A}} d\vec{r_1} \int_{\mathrm{B}} d\vec{r_2}\, n_{\mathrm{i}}^{1/2} n_{\mathrm{j}}^{1/2} \varphi_{\mathrm{i}}^*(\vec{r_1})\varphi_{\mathrm{i}}(\vec{r_2})\varphi_{\mathrm{j}}^*(\vec{r_2})\varphi_{\mathrm{j}}(\vec{r_1}), \quad (2)$$

where n denotes the orbital occupancy.

Given that the chosen space domains may be associated with atoms, like quantum atoms in QTAIM [2], the value of the DIs between two such domains is naturally related to the covalent contribution to the interaction between the corresponding atoms. It was shown that with space domains taken as QTAIM atoms, the values of the DIs are related to the formal bond order [12]. Results produced by DFT in Kohn-Sham formulation are generally very similar to the results from HF calculations [13, 14]. Due to the missing electron correlation, they lead to some overestimation of the results compared to fully correlated methods. It was shown that the DIs obtained both at the DFT and HF levels of theory are sufficient to gain chemical insight [14].

Since the population of QTAIM basins – that is the one-electron density integrated over the basin – characterizes the effective atomic charges [2], both concepts, QTAIM basin populations and DIs, provide the ionic and covalent components of the chemical bond, and in combination are therefore giving a measure for its characterization.

In our investigations, we focused on titanium compounds due to their pronounced covalent character and especially on the practically relevant TiO_2. Titanium dioxides are promising electrode materials for Li- and post-Li ion batteries [15] and changes in their electronic structure upon metal insertion and the rationalization of the occurring redox processes are of broad interest. Titania compounds are large-bandgap semiconductors and p-doping (e.g. with Ca) can create low-lying states in the valence band, which would be expected to be occupied by upon metal intercalation [16].

We first compare DIs and Bader charge analysis on rutile TiO_2, followed by a discussion of the relative changes of QTAIM charge in $LiTiO_2$ and Li_2TiO_2 molecules as simplest model systems for lithiated titanium dioxide for which both of these stoichiometries were previously reported in experimental works [17]. Subsequently, we want to demonstrate the relative QTAIM charge stabilities in different complexes and crystals even for widely differing FOS, as well as for interstitial Ti in p-doped crystals. The aim of this work is to illustrate how direct analysis of charge density can be used to rationalize redox processes in a way uncoupled from the formal oxidation states and to illustrate this on practically important phenomena (such as lithiation) which cannot be properly described with FOS. Specifically, the lithiated TiO_2 molecules allow us to clearly demonstrate a degree of oxygen reduction for which FOS fail to account.

2. Computational Details

In order to evaluate DI's and QTAIM basin populations for rutile TiO_2 in section 3.1, projector-augmented wave (PAW) [18] calculations were performed as implemented in ABINIT [19], utilizing the generalized gradient approximation (GGA) with the Perdew-Burke-Ernzerhof (PBE) [20] functional. A plane-wave energy cutoff of 20 Hartree was used and a 4×4×8 k-point grid. A $3s^2 3p^6 4s^2 3d^2$ valence configuration was used for titanium, and a $2s^2 2p^4$ valence configuration was taken for oxygen. DIs were evaluated with the Dgrid-4.7

IOP Conf. Series: Journal of Physics: Conf. Series **1136** (2018) 012017 doi:10.1088/1742-6596/1136/1/012017

program [21]. Molecular complexes were calculated with the quantum chemical program package Gaussian 09 [22] using the Becke three-parameter Lee-Yang-Parr (B3LYP) DFT functional [23] and a *cc*-pVDZ basis set [24], relaxing all molecular structures and using the lowest-energy spin state.

The periodic structures in section 3.3 were computed with the Vienna Ab Initio Simulation Package (VASP 5.3) [25] using the PBEsol functional [26], the PAW method [18, 27] and plane-wave basis sets (500 eV cutoff energy). The used supercells were optimized using a conjugate-gradient algorithm and the valence configurations for the occurring elements were: $3s^23p^63d^24s^2$ (Ti), $4s^24p^5$ (Br), $5s^25p^5$ (I), $2s^22p^4$ (O), $2s^22p^2$ (C), $3s^23p^64s^2$ (Ca) and $2s^2$ (Be). All supercells were chosen to have dimensions of around 12x12x12 Å^3 and a 3x3x3 Monkhorst-Pack [28] k-point mesh was chosen in all cases. Bader charge analyses were performed using a grid-based method as implemented in the Bader v0.95a code [29]. Spin polarization was not found to be of importance for any periodic system.

3. Results

3.1. Titanium dioxide charges using delocalization indices and electron transfer to oxide ions

In rutile TiO_2, the titanium atom is coordinated octahedrally to six oxygen atoms, three at a distance of 1.95 Å, and other three at a distance of 1.98 Å. The corresponding DI's are equal to 0.55 and 0.48 indicating that there are approximately 3*0.55+3*0.48=3.09 electron pairs involved in six Ti-O bonds. The QTAIM population of the titanium basin is equal to 19.77, indicating transfer of 2.23 electrons from the titanium basin to the closest oxygen basins (1.11 electrons to each), which is consistent with the previous calculations: +2.23 in the titanium basin from the PBE calculations and +2.43 from the HSE06 [30] calculations [31]. Both indicators are best of all consistent with the charge state of titanium being less than +4, while the charge state of oxygen can be assumed between -1 and -2. This aligns also very well with our previous findings of Bader charges on the order of 2.5 for rutile TiO_2 and PBE [6]. We will further focus on Bader charge analysis, which is a reliable measure of charge distribution in molecular and bulk titanium dioxide systems, as we showed in that previous investigation [6], and is consistent with the results obtained from the DI formalism.

The attachment of one and two Li atoms to a single TiO_2 molecule leads to the Bader charges (expressed as atomic charges with respect to the neutral atoms) listed in Table 1. The obtained bent geometries for $LiTiO_2$ (tetragon-like arrangement of the four atoms) and Li_2TiO_2 (pentagon-like) correspond to the predicted ground-state structures, while the linear mono- and dilithiated TiO_2 are saddle points of the potential energy surface (the spin states of the mono- and dilithiated compounds are doublet and triplet, respectively, in both geometric arrangements). The latter are included since linear O-Ti-O arrangements are common structural motifs in crystalline titania phases. We use the mono- and dilithiated molecules as simplest models for the local electron transfer in bulk TiO_2 (e.g. rutile or anatase which both include linear and bent O-Ti-O units). Lithiation of bulk TiO_2 creates localized in-gap states [32], which is why we deem a single molecule a meaningful case study for local electron transfers in lithiated, solid TiO_2. However, the more pronounced electron delocalization in periodic systems does not allow for a quantitative transferability of the following results to the solid-state case, but is supposed to provide mechanistic insight and values at the limit of fully localized charges. Representations of all four molecules together with charge density isosurface plots are shown in Fig. 1 (visualization was done with the program VESTA [33]).

Table 1. Bader charges of atoms in TiO_2, $LiTiO_2$ and Li_2TiO_2 in linear and bent configuration calculated with B3LYP/*cc*-pVDZ.

	Bader charge [$\|e\|$]			
Compound	**Ti**	**O1**	**O2**	**Li**
TiO_2 (bent)	+2.11	-1.06	-1.06	-
$LiTiO_2$ (bent)	+1.66	-1.33	-1.32	+0.98
Li_2TiO_2 (bent)	+1.74	-1.40	-1.42	+0.54
TiO_2 (linear)	+2.43	-1.22	-1.22	-
$LiTiO_2$ (linear)	+1.97	-1.26	-1.70	+0.98
Li_2TiO_2 (linear)	+1.53	-1.75	-1.74	+0.97

As can be seen from Table 1, the charges on oxygen upon attachment of Li, which is directly bound to it, change by -0.27 and -0.07 |*e*| from non- to monolithiated and mono- to dilithiated bent TiO_2 respectively, while the charge on Ti decreases by 0.45 |*e*| and then increases by 0.08 |*e*| due to existence of a non-nuclear attractor between the two Li ions reducing the charge donation to the TiO_2 fragment (see inset of Fig. 1). The charge in the additional basin of the non-nuclear attractor was divided between the two Li centers and was used to assign a Bader charge to both ions. The delocalization of charge over the whole molecule leads to a transfer of electron density to Ti and each O which are comparable in magnitude.

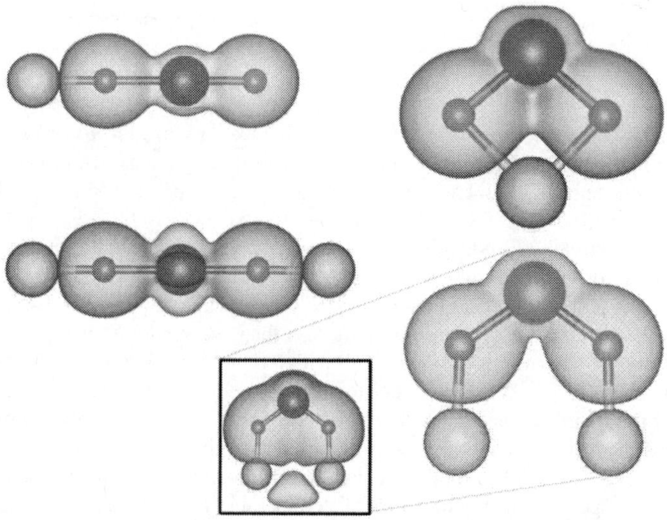

Figure 1. Geometries of four optimized lithiated TiO_2 molecules. From top left to bottom right: $LiTiO_2$ (linear), $LiTiO_2$ (bent), Li_2TiO_2 (linear), Li_2TiO_2 (bent). Isosurface plots of the charge densities in every molecule (isosurface value: 0.03 a.u.) are shown in grey. Ti atoms are represented as blue, oxygen as red and lithium as green coloured spheres. The inset shows an isosurface plot with a lower isovalue (0.006 a.u.) for bent Li_2TiO_2 in order to visualize the non-nuclear attractor between the two Li ions.

In the linear lithium titanium dioxides, both attached lithium atoms are predicted to lose most of their valence density, transferring a charge of -0.90 |*e*| on Ti and -0.53 |*e*| on each oxygen. In other words, a 'whole' electron is transferred on both oxygen atoms and one electron on Ti upon dilithiation, based on the local change of electron density. Analogous trends, which can be traced back to the covalency of the Ti-O bonds, have been reported previously by Albaret *et al.* connecting charge transfers in $Ti_nO_m^x$ (n = 1,2,3; m-n=0,1; x=-1,0,1) to changes in orbital hybridization [34]. From a charge-density point of view, the oxygen ions are being 'reduced'. This on one hand is not well handled by the FOS formalism and on the other hand this shows that oxygen redox activity in battery electrode materials, which has been rationalized in recent literature by invoking O defects, must be present in undefected oxides.

3.2. Charge stability in molecular titanium complexes

In a recent work, Wolczanski presented a way to estimate the relative charge states of transition metal ions in carbonyl complexes, utilizing the weakening of the C-O bond upon reduction of the carbonyl ligands [35]. Using the average of reported C-O stretch frequencies in different Fe complexes and assuming a stable +2 |*e*| charge state on the iron center, he could deduce the charge state of a variety of transition metal carbonyl complexes by comparison of the C-O stretch frequencies, leading to a charge state estimation of +3.2 |*e*| for Ti in $[Ti(CO)_6]^{2-}$. The results of this charge distribution via reporters (CDVR) method were backed up by Mulliken and Natural Bond Order analyses. We decided to verify if Bader charge analyses are in accordance with the results obtained from the formalism in ref. 35.

Fig. 2 shows the average C-O stretch frequencies (without considering degeneracies) versus the assumed CO ligand charges (with assumed fixed +2 charge state of Fe) for $[Fe(CO)_4]^{2-}$ (q_{CO} = -1 |*e*|), $Fe(CO)_5$ (q_{CO} = -0.4 |*e*|) and $[Fe(CO)_6]^{2+}$ (q_{CO} = 0 |*e*|) calculated with DFT. Although the experimentally observed frequencies could not be matched accurately (overestimation on the order of 100 cm^{-1} (7 %)), which might be caused by non-vacuum frequency values used in the data basis of ref. 35, a linear trend can be observed. Using the average C-O stretch in $[Ti(CO)_6]^{2-}$ at 1875 cm^{-1} and comparing with the linear trend line in Fig. 2 (red), the computed charge state on Ti is +3.5 |*e*|. The charge state deduced from the CDVR method was given as +3.2 |*e*|, but the actual interpolated value obtained by Wolczanski was 3.7-3.8 |*e*|. It was modified due to consistency considerations with regard to the other investigated complexes. With this initially obtained value however, a reasonable level of agreement can be observed.

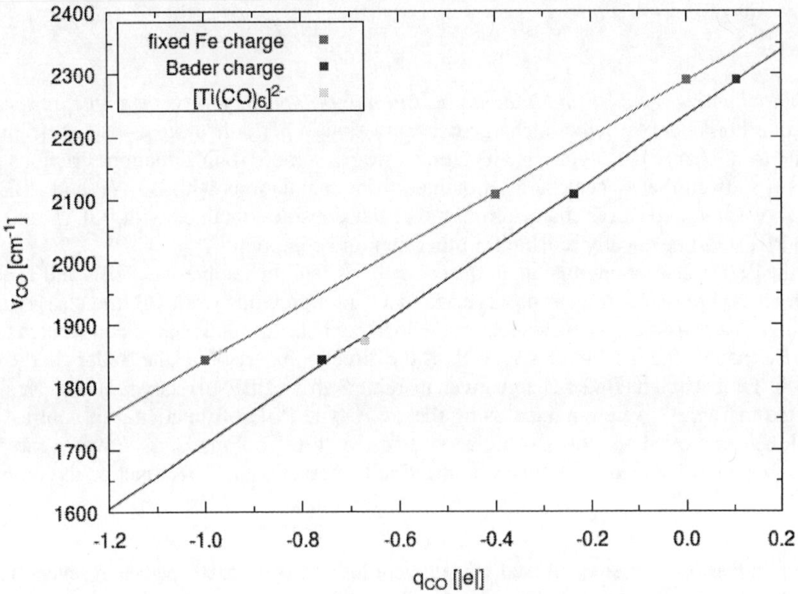

Figure 2. Computed average vibrational frequencies of the C-O stretch in $[Fe(CO)_4]^{2-}$, $Fe(CO)_5$ and $[Fe(CO)_6]^{2+}$ against assumed fixed CO ligand charges (red color, q = -1.0 |*e*|, -0.4 |*e*|, 0.0 |*e*|) or calculated Bader charges from DFT (blue). The (ν, q_{Bader}) pair for $[Ti(CO)_6]^{2-}$ is indicated in green and is in reasonable agreement with the blue linear trend line.

Up to this point, the ligand charges were based on an assumed charge state of Fe which has been fixed to +2. Bader charge analyses of the three iron complexes indicate charges of +1.02 |*e*| in $[Fe(CO)_4]^{2-}$, +1.18 |*e*| in $Fe(CO)_5$ and +1.37 |*e*| in $[Fe(CO)_6]^{2+}$ on Fe, while the titanium charge state is +2.03 |*e*| in $[Ti(CO)_6]^{2-}$. Plotting again symmetric C-O stretch frequencies, this time versus CO Bader charges for all four complexes, the linear relationship in Fig. 2 (blue) was obtained. It can be deduced that CO as a charge reporter ligand is indeed capable to give an estimate of the relative charge state, but the QTAIM Bader charge on Fe in the reference complexes is closer to +1 than +2, which is the reason for the overestimation of the Ti charge state compared to theory. Further, the charges on Fe are not expected to be perfectly constant, which contributes to a small further mismatch with the DFT results.

Charge states of +2.03 |*e*|, +2.11 |*e*| and +2.43 |*e*| for the vastly different molecular complexes $[Ti(CO)_6]^{2-}$ as well as bent and linear TiO_2, respectively (cf. Table 1), indicate that a substantial charge remainder on the Ti center is favored. This is not surprising, since density delocalization and the following decrease in kinetic energy is the main driving force for bond formation [36]. The requirement for a ligand to abstract all (most) of the Ti valence density to observe complete iconicity of Ti in the context of the QTAIM analysis, does not seem to be met by common ligands in molecular Ti complexes as indicated in Table 2. Moreover, a stable charge remainder of at least one electron remains in these complexes.

IOP Conf. Series: Journal of Physics: Conf. Series **1136** (2018) 012017 doi:10.1088/1742-6596/1136/1/012017

Table 2. Bader charges on titanium for several molecular complexes and complex ions calculated with B3LYP/*cc*-pVDZ, unless indicated differently.

Compound	Bader charge (Ti) [|*e*|]	Compound	Bader charge (Ti) [|*e*|]
TiF_4	+2.46	$Ti(CN)_4$	+2.92
$TiCl_4$	+2.72	$[Ti(CO)_6]^{2-}$	+2.03
$TiBr_4$	+2.66 (+2.56[b])	$Ti(TCNE)_2$ (tetragonal)[a]	+1.78
$[TiF_6]^{2-}$	+2.75	$Ti(TCNE)_2$ (tetrahedral)[a]	+2.30

[a]TCNE: tetracyanoethylene; used as bidentate ligand here.
[b]Value obtained with PBE/*cc*-pVDZ.

3.3. *Solid titanium halogenides and titanium insertion in p-doped crystals: incomplete charge transfer*

Naturally the question occurs whether a charge delocalization in periodic systems can contribute to a further lowering of the local charge density around Ti atoms towards a more ionic situation. Previous investigations of our group [6] showed that Bader charges obtained from calculations with B3LYP and PBE do not differ substantially, so we assumed Bader charges from molecular densities obtained with B3LYP and from periodic densities with PBEsol to be equally reliable for other titanium compounds.

In Table 3 the Bader charges on titanium in the crystalline titanium halogenides $TiBr_4$ and TiI_4 (space group: $P2_1/a3$ in both cases) are listed. As was observed for TiO_2 in a previous work [6], the charge on Ti increases from a molecular to a periodic system, since more delocalized background charge contributes to the Ti basin. This seems to be true for the halogenides as well, as the direct comparison of the Bader charge obtained with PBE for $TiBr_4$ in Table 2 and its Bader charge given in Table 3 (from PBEsol) shows. However, since the Bader charges reported in Table 3 were obtained using the GGA-type PBEsol functional in contrast to the hybrid B3LYP functional employed for the molecular systems, a direct comparison between Tables 2 and 3 is eventually not meaningful, due to the more pronounced self-interaction in GGA functionals compared to hybrid ones.

Table 3. Bader charges on titanium in crystalline halogenides and as inserted species in several crystalline host structures (Bader charges for interstitial Ti).

Compound	Bader charge (Ti) [|*e*|]	Compound	Bader charge (Ti) [|*e*|]
$TiBr_4$	+1.70	Ti in Be-doped diamond-C ($TiBe_3C_{213}$)	+1.46
TiI_4	+1.50	Ti in Ca-doped rutile-TiO_2 ($Ca_4Ti_{29}O_{64}$)	+2.09
Ti in I (TiI_{54})	+1.39		

Another approach that was tried to nudge a further reduction of electron density on Ti was the insertion of Ti as interstitial atom in crystalline compounds whose Fermi levels are expected to be substantially lower than the newly created interstitial Ti states and having unoccupied states around the Fermi level. We chose C-diamond (space group: Fd-3m) as insulator p-doped with substitutional Be, rutile-TiO_2 (space group: $P4_2/mnm$) as semiconductor p-doped with substitutional Ca and metallic iodine (space group: Im-3m). Titanium was expected to donate its valence electrons to the low-lying states around the Fermi level on one hand and reduce strain energy of the host lattice by reducing its radius which could be accomplished by further ionization. As can be seen in Table 3, this is not the case. Higher-lying Ti levels are still occupied and there is a substantial electron density remainder on the interstitial Ti in all cases. Relative charge stability and charge remainders on transition metal centers in compounds commonly classified as ionic were reported in the past [37], and the same trend can be observed, besides the molecular systems in Table 2, also for the two types of environment (cation in titanium halogenide vs. interstitial solute in doped semiconductor) of Ti in the periodic systems presented in Table 3.

4. Conclusions

We provided, using QTAIM methods, additional insight into the charge density distributions in common titanium compounds, focusing on titanium dioxide, halogenides and a hexacarbonyl complex. Of course the Bader charge analyses do not technically contradict the definition of formal oxidation states and in this framework they could be simplistically explained by a pronounced covalent character of the investigated compounds. Nevertheless the observed charge stability in a variety of titanium compounds and the Ti atom's apparent resistance to strong electron deficiency leads to counter-intuitive implications when it comes to charge density fluxes in redox processes, such as the lithiation of TiO_2, a reaction occurring in titania electrodes of Li-ion batteries. We could theoretically reproduce experimentally observed trends [35] with regard to the charge state in titanium hexacarbonyl and show that Bader charge analysis results can in principle be backed up with concrete experimental evidence and are not just a computational quirk.

We previously reported that the QTAIM charge in the solid state is increased compared to the molecular setup due to a higher delocalized background charge in TiO_2 [6]. We obtained a similar result for titanium halogenides and tested the capability of p-doped periodic systems to ionize interstitially inserted Ti atoms, which is poor.

We furthermore presented the use of delocalization indices to evaluate the charge state of titanium in rutile TiO_2, supporting the claimed reliability of Bader charges by a more sophisticated systematic theoretical analysis method and providing insight into the peculiarities of the electron density distribution in this compound. From the results of the limited number of trial systems utilized, there seems to be a tendency in titanium compounds to keep at least one electron localized around titanium, although Bader charges of +4 $|e|$ and more are not prohibitive and are possible in general (e.g. $q = +5.8\ |e|$ on S in SF_6).

Acknowledgements
This work was supported by the Ministry of Education of Singapore (grant no. MOE2015-T2-1-011).

References

[1] Karen P, McArdle P and Takats J 2016 Comprehensive definition of oxidation state (IUPAC Recommendations 2016) *Pure Appl. Chem.* **88** 831
Resta R 2008 Charge states in transition *Nature* **453** 735
[2] Bader R F W 1985 Atoms in molecules *Acc. Chem. Res.* **18** 9
[3] Seo D-H, Lee J, Urban A, Malik R, Kang S and Ceder G 2016 The structural and chemical origin of the oxygen redox activity in layered and cation-disordered Li-excess cathode materials *Nat. Chem.* **8** 692
[4] Yaresko A N, Antonov V N, Eschrig H, Thalmeier P and Fulde P 2000 Electronic structure and exchange coupling in α'-NaV_2O_5 *Phys. Rev. B* **62** 15538
Arrouvel C, Parker S C and Islam M S 2009 Lithium insertion and transport in the TiO_2–B anode material: A computational study *Chem. Mater.* **21** 4778
[5] Kulish V and Manzhos S 2017 Comparison of Li, Na, Mg and Al-ion insertion in vanadium pentoxides and vanadium dioxides *RSC Adv.* **7** 18643
[6] Koch D and Manzhos S 2017 On the charge state of titanium in titanium dioxide *J. Phys. Chem. Lett.* **8** 1593
[7] Bader R F W 1998 Atoms in molecules. In *Encyclopedia of Computational Chemistry* (Hoboken: John Wiley & Sons, Inc) p 64
[8] Bader R F W and Stephens M E 1975 Spatial localization of the electronic pair and number distributions in molecules *J. Am. Chem. Soc.* **97** 7391
[9] McWeeny R 1960 Some recent advances in density matrix theory *Rev. Mod. Phys.* **32** 335
[10] Bader R F W, Streitwieser A, Neuhaus A, Laidig K E and Speers P 1996 Electron delocalization and Fermi hole *J. Am. Chem. Soc.* **118** 4959
[11] Fulton R L and Mixon S T 1993 Comparison of covalent bond indices and sharing indices *J. Phys. Chem* **97** 7530
[12] Fradera X, Austen M A and Bader R F W 1999 The Lewis model and beyond *J. Phys. Chem.* **103** 304
[13] Kar T, Angyan J G and Sannigrahi A B 2000 Comparison of ab initio Hartree-Fock and Kohn-Sham orbitals in the calculation of atomic charges, bond index, and valence *J. Phys. Chem. A* **104** 9953

[14] Poater J, Sola M, Duran M and Fradera X 2002 The calculation of electron localization and delocalization indices at the Hartree-Fock, density functional and post-Hartree-Fock levels of theory *Theor. Chem. Acc.* **107** 362

[15] Liu J, Wang J, Ku Z, Wang H, Chen S, Zhang L, Lin J and Shen Z X 2016 Aqueous rechargeable alkaline $Co_xNi_{2}–xS_2/TiO_2$ Battery *ACS Nano* **10** 1007

Oh S-M, Hwang J-Y, Yoon C S, Lu J, Amine K, Belharouak I and Sun Y-K 2014 High electrochemical performances of microsphere C-TiO_2 anode for sodium-ion battery *ACS Appl. Mater. Interfaces* **6** 11295

[16] Lueder J, Cheow M H and Manzhos S 2017 Understanding doping strategies in the design of organic electrode materials for Li and Na ion batteries: an electronic structure perspective *Phys. Chem. Chem.Phys.* **19** 13195

Legrain F and Manzhos S 2015 Aluminum doping improves the energetics of lithium, sodium, and magnesium storage in silicon: a first-principles study *J. Power Sources* **274** 65

Lueder J, Legrain F, Chen Y and Manzhos S 2017 Doping of active electrode materials for electrochemical batteries: an electronic structure perspective *MRS Commun.* **7** 523

[17] Borghols W J H, Lutzenkirchen-Hecht D, Haake U, van Eck E R H, Mulder F M and Wagemaker M 2009 The electronic structure and ionic diffusion of nanoscale $LiTiO_2$ anatase *Phys. Chem. Chem. Phys.* **11** 5742

Huang Y F, Tang H, Li J, Hu J, Zhu X Q, Zhang W J, Shen L and Qing L X 2012 Preparation of Lithium Orthosilicate Ceramic Pebbles by Molten Spray Method Process *Adv. Mater.* **412** 111

[18] Bloechl P E 1994 Projector augmented-wave method *Phys. Rev. B: Condens. Matter. Mater. Phys.* **50** 17953

[19] Gonze X, Beuken J-M, Caracas R, Detraux F, Fuchs M, Rignanese G-M, Sindic L, Verstraete M, Zerah G, Jollet F, Torrent M, Roy A, Mikami M, Ghosez P, Raty J-Y and Allan D C 2002 First-principles computation of material properties: The ABINIT software project *Comput. Mater. Sci* **25** 478

Gonze X, Amadon B, Angade P-M, Beuken J-M, Bottin F, Boulanger P, Bruneval F, Caliste D, Caracas R, Cote M, Deutsch T, Genovese L, Ghosez P, Giantomassi M, Goedecker S, Hamann D R, Hermet P, Jollet F, Jomard G, Leroux S, Mancini M, Mazevet S, Oliveira M J T, Onida G, Pouillon Y, Rangel T, Rignanese G-M, Sangalli D, Shaltaf R, Torrent M, Verstraete M J, Zerah G and Zwanziger J W 2009 ABINIT: First-principles approach to material and nanosystem properties *Comput. Phys. Commun.* **180** 2582

[20] Perdew J P, Burke K and Ernzerhof M 1996 Generalized gradient approximation made simple *Phys. Rev. Lett.* **77** 3865

[21] Kohout M 2012 *DGrid-4.7* (Radebeul)

[22] Frisch M J, Trucks G W, Schlegel H B, Scuseria G E, Robb M A, Cheeseman J R, Scalmani G, Barone V, Petersson G A, Nakatsuji H, Li X, Caricato M, Marenich A, Bloino J, Janesko B G, Gomperts R, Mennucci B, Hratchian H P, Ortiz J V, Izmaylov A F, Sonnenberg J L, Williams-Young D, Ding F, Lipparini F, Egidi F, Goings J, Peng B, Petrone A, Henderson T, Ranasinghe D, Zakrzewski V G, Gao J, Rega N, Zheng G, Liang W, Hada M, Ehara M, Toyota K, Fukuda R, Hasegawa J, Ishida M, Nakajima T, Honda Y, Kitao O, Nakai H, Vreven T, Throssell K, Montgomery Jr V, Peralta J E, Ogliaro F, Bearpark M, Heyd J J, Brothers E, Kudin K N, Staroverov V N, Keith T, Kobayashi R, Normand J, Raghavachari K, Rendell A, Burant J C, Iyengar S S, Tomasi J, Cossi M, Millam J M, Klene M, Adamo C, R. Cammi, Ochterski J W, Martin R L, Morokuma K, Farkas O, Foresman J B and Fox D J 2016 *Gaussian 09* Revision D.01 (Wallingford CT: Gaussian, Inc.)

[23] Lee C, Yang W and Parr R G 1988 Development of the Colle-Salvetti correlation-energy formula into a functional of the electron density *Phys. Rev. B* **37** 785

Becke A D 1993 Density-functional thermochemistry. III. The role of exact exchange *J. Chem. Phys.* **98** 5648

Devlin F J, Finley J W, Stephens P J and Frisch M J 1995 Ab initio calculation of vibrational absorption and circular dichroism spectra using density functional force fields: A comparison of local, nonlocal, and hybrid density functionals *J. Phys. Chem.* **99** 16883

[24] Balabanov N B and Peterson K A 2005 Systematically convergent basis sets for transition metals. I. All-electron correlation consistent basis sets for the 3d elements Sc–Zn *J. Chem. Phys.* **123** 64107

Balabanov N B and Peterson K A 2006 Basis set limit electronic excitation energies, ionization potentials, and electron affinities for the 3d transition metal atoms: coupled cluster and multireference methods *J. Chem. Phys.* **125** 74110.

Dunning Jr T H 1989 Gaussian basis sets for use in correlated molecular calculations. I. The atoms boron through neon and hydrogen *J. Chem. Phys.* **90** 1007

[25] Kresse G and Furthmüller J 1996 Efficient iterative schemes for ab initio total-energy calculations using a plane-wave basis set *Phys. Rev. B* **54** 11169

[26] Perdew J P, Ruzsinszky A, Csonka G I, Vydrov O A, Scuseria G E, Constantin L A, Zhou X and Burke K 2008 Restoring the density-gradient expansion for exchange in solids and surfaces *Phys. Rev. Lett.* **100** 136406

[27] Kresse G and Joubert D 1999 From ultrasoft pseudopotentials to the projector augmented-wave method *Phys. Rev. B* **59** 1758

[28] Monkhorst H J and Pack J D 1976 Special points for Brillouin-zone integrations *Phys. Rev. B* **13** 5188

[29] Tang W, Sanville E and Henkelman G 2009 A Grid-based Bader analysis algorithm without lattice bias *J.Phys. Condens. Matter* **21** 84204

Henkelman G, Arnaldsson A and Jónsson H 2006 A Fast and robust algorithm for Bader decomposition of charge density *Comput. Mater. Sci.* **36** 354

Sanville E, Kenny S D, Smith R and Henkelman G 2007 Improved grid-based algorithm for Bader charge allocation *J. Comput. Chem.* **28** 899

Yu M, Trinkle D R 2011 Accurate and efficient algorithm for Bader charge integration *J. Chem. Phys.* **134** 64111

[30] Krukau A V, Vydrov O A, Izmaylov A F and Scuscria G E 2006 Influcncc of thc cxchangc screening parameter on the performance of screened hybrid functionals *J. Chem. Phys.* **125** 224106

[31] Vasquez G C, Karazhanov S Zh, Maestre D, Cremades A, Piqueras J and Foss S E 2016 Oxygen vacancy related distortions in rutile TiO_2 nanoparticles: A combined experimental and theoretical study *Phys.Rev. B* **94** 235209

[32] Legrain F, Malyi O I and Manzhos S 2015 Insertion energetics of lithium, sodium, and magnesium in crystalline and amorphous titanium dioxide: A comparative first-principles study *J. Power Sources* **278** 197

[33] Momma K and Izumi F 2011 VESTA 3 for three-dimensional visualization of crystal, volumetric and morphology data *J. Appl. Crystallogr.* **44** 1272

[34] Albaret T, Finocchi F and Noguera C 1999 First principles simulations of titanium oxide clusters and surfaces *Faraday Discuss.* **114** 285

[35] Wolczanski P 2017 Flipping the oxidation state formalism: charge distribution in organometallic complexes as reported by carbon monoxide *Organometallics* **36** 622

[36] Schmidt M W, Ivanic J and Ruedenberg K 2014 Covalent bonds are created by the drive of electron waves to lower their kinetic energy through expansion *J. Chem. Phys.* **140** 204104

[37] Raebiger H, Lany S and Zunger A 2008 Charge self-regulation upon changing the oxidation state of transition metals in insulators *Nature* **453** 763

Christensen A and Carter E A 2000 First-principles characterization of a heteroceramic interface: $ZrO_2(001)$ deposited on an α-$Al_2O_3(1102)$ substrate *Phys. Rev. B* **62** 16968

IOP Conf. Series: Journal of Physics: Conf. Series **1136** (2018) 012018 doi:10.1088/1742-6596/1136/1/012018

H-bond symmetrization in high pressure methane hydrate

Sofiane Schaack[1], Philippe Depondt[1], Fabio Finocchi[1]

[1]Sorbonne Université, CNRS UMR 7588, Institut des Nanosciences de Paris ,INSP, 75005 Paris, France

E-mail: schaack@insp.jussieu.fr

Abstract. First-principle molecular dynamics simulations of methane hydrate MH-III, including the quantum properties of the hydrogen nuclei, were carried out at pressures in the 5-65 GPa range, in order to observe the H-bond symmetrization at high pressure and at room temperature. According to our simulations, the symmetrization transition takes place around 40 GPa and is little dependent on isotope substitution. We find that, consistently with the rather complex crystal structure of MH-III, the transition is much more convoluted in the hydrate than in ices VII and X. In methane hydrate, due to the presence of non equivalent O ions with distinct O-O distances, the dissimilar H bonds symmetrize in a pressure domain rather than at a single critical pressure.

1. Introduction

Methane hydrate (MH) is a clathrate compound in which methane molecules are trapped within a solid cage of water[1, 2]. MH could have occurred in the nebula from which ice giants Uranus and Neptune were formed and is still present on Titan [3, 4]. Significant deposits of methane clathrate hydrates have been also found on Earth, in cold upper sediments on the sea and ocean floors, which could be destabilized as the oceans warm up and release large quantities of greenhouse methane gas [5]. The previous facts motivated many researches on the stability of methane hydrates under various conditions.

The interaction between the host molecules and the cage is essentially repulsive; this is why methane hydrates were initially thought to be stable only at relatively modest pressure, below a few GPa. However, quite recently [6, 7], several phases of MH, differing by structure and composition, were found to be stable at very high pressures, beyond 10 GPa and up to $\sim$ 100 GPa. At those pressures, not only could the interaction between the cage and the methane molecules be significantly altered as ice cages shrink, but the ice cage itself could undergo deep modifications. Indeed, the hydrogen bonds between the water molecules that form the cage may become symmetric and the ice cage eventually evolves towards a non-molecular form, which modifies the mechanical properties of the hydrate, as well as the molecule-cage interaction.

The hydrogen bond symmetrization under pressure was first observed in ice phase X. Along the transition from phase VII (H-bonds, with proton disorder) to phase X (symmetric bonds, proton ordered) the quantum properties of the hydrogen nuclei play an important role [8, 9]; the most striking one regards the computed transition pressure at room temperature, which is reduced from approximately 100 GPa within the classical frame down to 65 GPa, in agreement with the experimental findings, when nuclear quantum effects are included. Among the latter, the zero-point energy (ZPE) is crucial in two respects: first, at the onset of proton tunneling between the two potential wells that correspond to the asymmetric H bond; second, in lowering the critical pressure for symmetrization with respect to the classical frame. In methane hydrate, a similar process is expected in the high-pressure MH-III phase, which is the stable phase of the system at ambient temperature and above 2GPa.

The paper is organized as follows: after a brief description of the simulation methods, in section 3 we first describe the order parameter that is commonly used to describe the H-bond symmetrization; then, we analyze the displacements of the proton along the O-O axis or normally to it (section 4); finally, we investigate the role of the interaction between the protons (deuterons) in the cage with the guest methane molecules in Section 5. All along the paper, we compare the behavior of the protons in MH-III to prototypical pure ice.

2. Computational details

Constant volume, constant temperature (NVT) molecular dynamics simulations at room temperature were carried out using the Quantum Thermal Bath method (QTB) [10] in order to include nuclear quantum effects. The QTB method is implemented in the Quantum Espresso [11] package. It is based on a Langevin equation in which the random force $R(t)$ is tailored to fulfill the quantum fluctuation-dissipation theorem [12]: instead of a white noise, one uses a colored noise with the following power spectrum:

$$\left|\tilde{R}(\omega)\right|^2 = 2m\gamma\hbar\omega\left[\frac{1}{2} + \frac{1}{\exp\left(\frac{\hbar\omega}{k_BT}\right) - 1}\right] \tag{1}$$

where m is the nucleus mass, k_B the Boltzmann's constant, T the temperature, $\omega/2\pi$ the frequency, and γ is the friction coefficient entering also the dissipation term in the Langevin equation. QTB is strictly exact only for harmonic systems. While dealing with anharmonicity, semi-classical methods such as QTB are submitted to zero-point energy leakage (ZPEL) which tends to equilibrate the energy distribution following the classical equipartition theorem. Adjusting the friction coefficient γ (eq 1) allows to lower the effect of the ZPEL [13] as the larger γ is the more the classical system is coupled to the quantum bath. Despite the widening of the vibrational peaks, the QTB has proven valuable in several systems [9, 14, 15, 16] and the QTB simulations presenting ZPEL have shown that they provide correct mode frequencies [17]. Rather importantly in the present context, a systematic comparison with the classical case can be carried out by simply replacing the colored QTB noise in equation (1) with the standard white noise $\left|\tilde{R}(\omega)\right|^2 = 2m\gamma\ k_B\ T$, while all the other computational ingredients remain identical [18].

The initial structure was taken from ref. [7]. The simulated samples consist of 16 methane molecules and 32 water molecules in a $2 \times 1 \times 2$ orthorhombic supercell. The methane molecules are hydrogenated (CH_4) while the ice cage is considered with hydrogenated (H_2O) or deuterated (D_2O) water molecules. The electronic structure and atomic forces were obtained within the Generalized Gradient approximation (GGA) [19] to the DFT. Van der Waals interactions were added by following the semi-empirical scheme by Grimme [20]; after extensive tests, a better agreement with experimental lattice constants and compressibility could be obtained with respect to the pure GGA. Ultra-soft pseudo-potentials were used to describe the interaction between the ionic cores and the valence electrons: a plane wave expansion cutoff of $E_{cut} = 40$ Ry ensures total energy and atomic force convergence. The friction coefficient was set to $\gamma = 1$ THz limiting the impact of ZPEL effects in the simulations which tend to underestimate NQE. Their impact was estimated to be at maximum $\sim 7\%$ of the temperature of the different degrees of freedom of MHIII. Simulations were run at constant volume, with lattice parameters chosen to sample pressures from 5 up to 65 GPa by obtaining an isotropic stress tensor within our statistical error. Oxygen and carbon atoms were initially set at their crystallographic positions, while cage hydrogen atoms were let to relax during short simulations with large friction coefficients γ in order to explore different configurations and start with variable orientations for the CH_4 molecules. All atoms are free to move during the simulations. The typical duration time of the simulations is 30 ps.

3. Characterizing the symmetrization transition

Along the VII-X transition of ice under pressure, the effective potential that is felt by the proton in hydrogen bonds is usually described in terms of a one-dimensional double-well potential, along the O-O axis [21, 22]. As pressure increases, the oxygen-oxygen nearest neighbor separation continuously decreases: the distance between the double wells and the barrier is accordingly reduced, making it possible for the proton to tunnel back and forth between the two wells. Upon further contraction of the O-O distance, the proton has its largest probability halfway between the two O ions. The simplest order parameter and the most widely used[8, 9, 14, 16] to estimate the location of a hydrogen nucleus between the two nearest oxygen atoms is:

$$\chi = d(\mathrm{O}^{[1]} - \mathrm{H}) - d(\mathrm{O}^{[2]} - \mathrm{H}) \tag{2}$$

where $d(\mathrm{O}^{[n]} - \mathrm{H})$ is the distance between the selected proton and its nth oxygen nearest neighbor. So defined, this parameter equals zero for proton-symmetric hydrogen bonds.

The order parameter χ was observed to depend on the $\mathrm{O}^{[1]}$-$\mathrm{O}^{[2]}$ distance $d(\mathrm{O}^{[1]} - \mathrm{O}^{[2]})$ in several H-bonded systems: below a threshold ($d(\mathrm{O}^{[1]} - \mathrm{O}^{[2]}) \simeq 2.42$Å), the probability density $P(\chi)$ is maximum at $\chi = 0$ and the H-bond becomes symmetric [22, 21]. As the O-O distances in MH-III are shorter than in ice VII at all pressures, it is reasonable to expect the symmetrization transition to happen at a lower pressure than the ice VII $\rightarrow$ X transition. Although direct measurements of the symmetrization of the hydrogen bonds in MH-III are still lacking, recent X-ray diffraction and Raman spectroscopy experiments [23, 24, 25] indicate a phase transition from MH-III to an High Pressure (HP) phase between 40 and 55 GPa although a precise description of the transition was not provided.

We show in figure 1 the probability distributions of the order parameter χ, averaged over all O-H..O bonds (top panel), and for each different OO distance (bottom panel), in MH-III as computed from our simulations. $P(\chi)$ evolves from a distribution with two maxima, which is characteristic of asymmetric hydrogen bonds, to a single-peak distribution, which denotes the symmetric hydrogen bonds, around 40 GPa. This points towards a symmetrization transition of the ice cage in MH-III around 40 GPa, therefore at a much lower pressure than in pure ice. One can also note that even at a quite low pressure the distribution does not vanish for $\chi = 0$ ($P(\chi = 0) \neq 0$) showing that the ice skeleton is proton disordered even at P $\sim$ 10 GPa.

However, the situation is more complex in MH-III than in pure ice: first, $P(\chi)$ behaves rather similarly in both hydrogenated and deuterated cages. This would imply that isotope effects are practically negligible, and that nuclear quantum effects are not very relevant, in striking contrast with the case of ice VII $\rightarrow$ X transition [22, 9, 6]. Second, there are three non equivalent O-O distances in MH-III (2.811, 2.807 and 2.783 Å detailed in figure 2a) as experimentally measured at 3 GPa [6]. The mean O-O distances along the dynamics can indeed be collected in three distinct groups, two of which ((a), corresponding to O1-O1$'$ and oriented along c; (b), connecting O1 to O6 in the notation of ref. [3]) have the same mean value $\bar{d}(\mathrm{O}^{[1]} - \mathrm{O}^{[2]})$ while the third group (c, corresponding to O1-O2 and O1-O2+, at the vertex of the 6-fold water rings) shows a slightly smaller mean O-O distance at all pressures. The corresponding probability distributions for the order parameter $P^{(\alpha)}(\chi)$, $\alpha = a$, b or c are sketched in the lower panels of figure 1. $P^{(\alpha)}(\chi)$, for $\alpha = a$ and b, shows two maxima at 40.6 GPa and 43.4 GPa for the

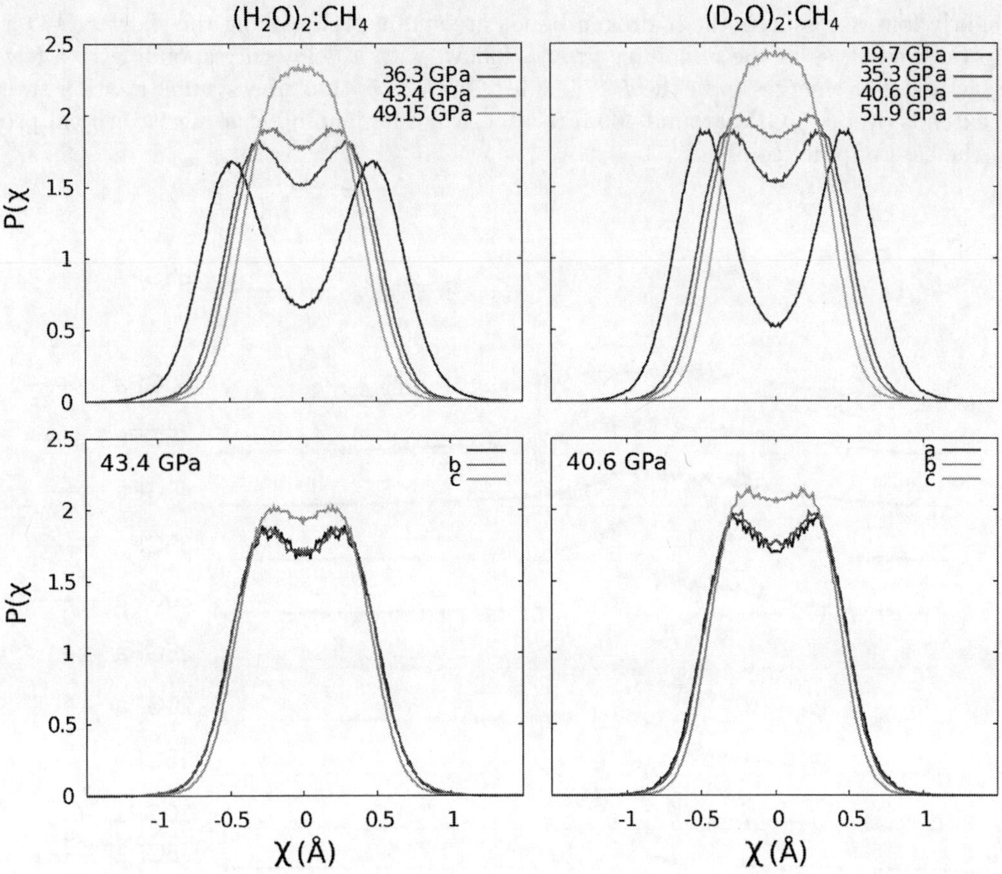

Figure 1. Distributions of the order parameter $P(\chi)$ at several pressures, for the hydrogenated (left) and deuterated (right) ice skeleton. Upper panels: $P(\chi)$ is computed by integrating on all hydrogen bonds. Bottom panels: Distributions of the order parameter $P(\chi)$ for each group of O-O equivalent distances (a, b, c) (see text).

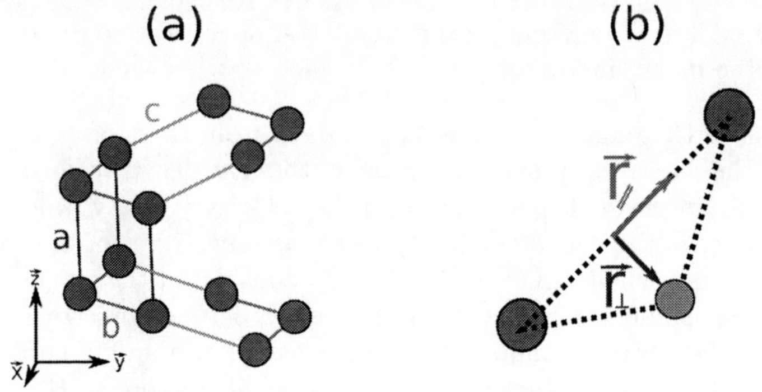

Figure 2. (a) The three different OO distances with $a \geq b > c$. (b) Description of the parallel $r_{\|}$ and normal $r_{\perp}$ proton delocalisation.

deuterated and hydrogenated ice skeletons, respectively, while $P^{(c)}(\chi)$ is flat around $\chi = 0$. The difference between the distributions reveals that the transition takes place in two steps:

at slightly lower pressures, the hydrogen bonds are symmetrized along the shorter O-O axes in group (*c*), and those in the remaining groups follow, with a delay corresponding to a few GPa. The picture that emerges from the $P^{(a)}(\chi)$ distributions is that of a symmetrization transition that extends over $\sim$ 5 GPa around 40 GPa rather than happening at a precise critical pressure, as in the case of pure ice [9].

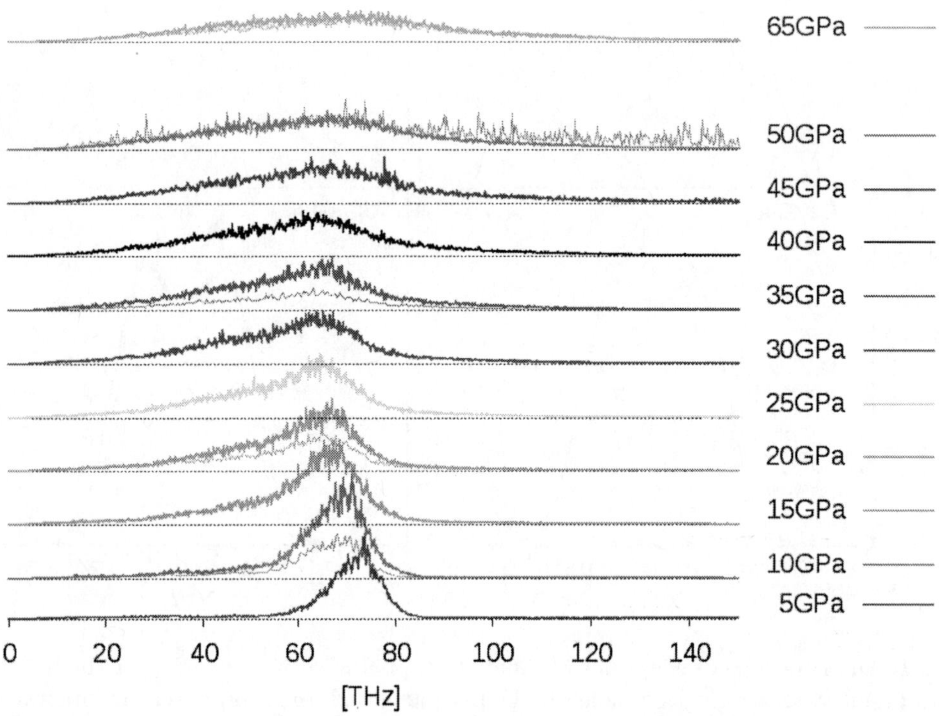

Figure 3. Fourier transforms in the frequency domain of the parallel component of the velocity-velocity time-correlation functions $\langle \dot{r}_{\|}(t)\dot{r}_{\|}(0)\rangle$ along the O-O axis. Thick line are for D_2O-CH_4, thin lines for $H_2O - CH_4$. The frequencies of the latter spectra were divided by $\sqrt{2}$ in order to take into account the mass difference.

In order to confirm the indications that we obtained from the trends of the order parameter χ, we computed the vibrational spectra through the Fourier transforms (FT) of the time correlation functions. In particular, we examined the velocity-velocity time correlation functions $\langle \dot{r}_\alpha(t)\dot{r}_\alpha(0)\rangle$, where r_α is the α component of vector joining the proton (deuterium) in the ice cage and its oxygen first neighbor $O^{[1]} - H$. We then considered the projections of the previous vectors parallel or perpendicular ($r_{\|}$ and $r_{\perp}$, respectively, detailed in Figure 2b) to the $O^{[1]} - O^{[2]}$ axis. The spectra of the parallel components $\dot{r}_{\|}$ show a shallow minimum of the mode's frequency around 35-40 GPa, which is consistent with the softening of the shorter H bonds (figure 3). Such a softening is a signature of the symmetrization transition, as discussed in the case of pure ice [9, 22], which points towards a critical pressure around 35-40 GPa, which is slightly smaller but consistent with the indications from the distribution of the corresponding order parameter χ. As for the $P(\chi)$ distributions, the spectra for hydrogenated and deuterated ice cages little differ once the latter have been renormalized by the mass ratio $\sqrt{2}$, confirming that nuclear quantum effects do not play a crucial role in the symmetrization transition of MH-III. As it downshifts, the OD stretching mode peak intensity flattens, thus the uncertainty on the mode's frequency prevents the difference of $P(\chi)$ distributions in terms of the three OO distances (Figure 2a) to be resolved. The distinction between the proton motion along or normally to the O-O axis translates also in the distributions of the proton (deuterium) localization, which we analyze in the following.

4. Analysis of the proton delocalization

Using simulation data from [9], we compare the hydrogen distribution with that of ice under pressure in which the VII→X transition shows important nuclear quantum effects, in constrast with methane hydrate.

First, we show (figure 4) the hydrogen distribution for both ice and methane hydrate with the same scales and normalization, for comparison. One observes that the OH $r_{\parallel}$ distance is approximately the same in both compounds as are the parallel widths, while the perpendicular extension is larger for the hydrate at all pressures.

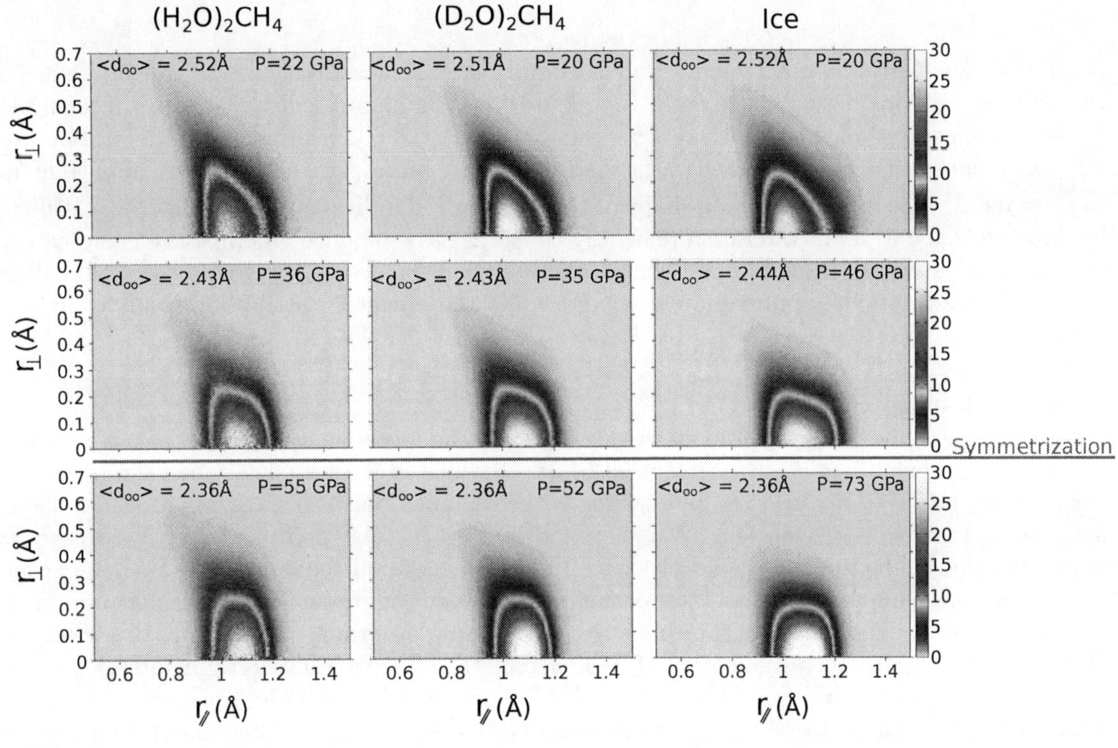

Figure 4. Probability density $P(r_{\parallel}, r_{\perp})$ for $(H_2O)_2 : CH_4$ (left) $(D_2O)_2 : CH_4$ (middle) and ice (right) systems at three pressures : one below the transition, one not very far from it and one above. Pressure values have been selected in order that the mean OO distances $< d_{oo} >$ in the three systems match, as the hydrogen bond symmetrization process is controlled by this parameter.

In order to obtain a more quantitative result one may compute the mean value of $r_{\parallel}$ and parallel $\sigma_{\parallel}$ and perpendicular $\sigma_{\perp}$ standard deviations (figure 5). This clearly confirms a broader distribution in the normal direction than along the O-O axis, and suggests that a purely one-dimensional model hamiltonian, like those usually adopted [22, 21] are unable to catch the complexity of the hydrogen bond in methane hydrate or other highly anisotropic systems. A finer analysis based on the differences of $P(r_{\parallel}, r_{\perp})$ between the hydrogenated and deuterated cages shows that the radial delocalization is larger in the case of $(H_2O)_2 : CH_4$. From the

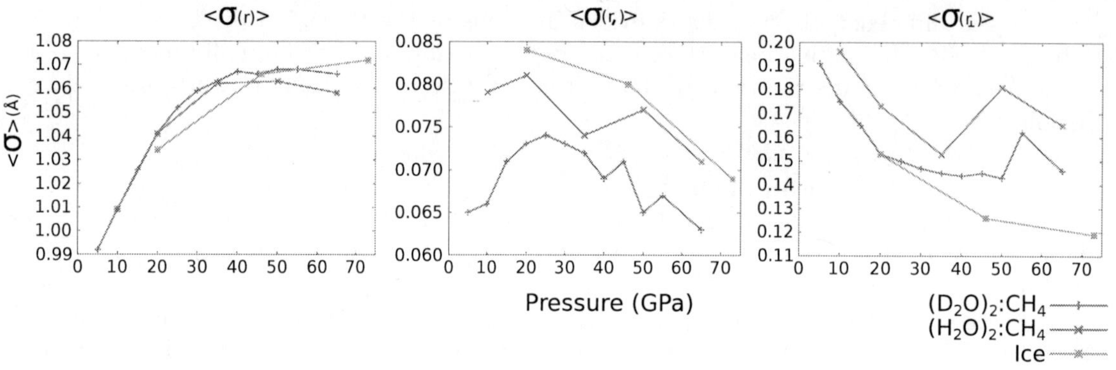

Figure 5. Ice and methane hydrate systems : (left) mean $r_{\|}$, (middle) parallel and (right) perpendicular standard deviations.

theoretical analysis of normal modes and accurate Raman measurements, it can be deduced that the interactions between the water cage and the methane molecules play a significant role in the system's dynamics, starting at a pressure of 20GPa [26]. Figure 4 shows the probability $P(r_{\|}, r_{\perp})$: the normal contributions to the distribution are broad with a significant amplitude up to $r_{\perp} \approx 0.4$ Å. The inverted comma shape of the distribution indicates that the hydrogen atoms, as they move about, tend to retain a relatively constant O-H distance. As pressure is increased, the shape of the distribution changes: the parallel contribution is sensitive to increased tunneling and thus broadens until approximately 40GPa while the normal contribution tends to shrink slightly.

5. Molecule-cage interaction

A step further in the analysis between the ice cage and the methane molecules is taken with the conditional probability $\rho(d_{\rm HH}|r_{\perp})$ (resp. $\rho(d_{\rm DH}|r_{\perp})$) which is computed for the cage hydrogen (deuterium) nuclei: $d_{\rm HH}$ ($d_{\rm DH}$ for deuterium) is the distance between a cage hydrogen nucleus and the closest hydrogen nucleus of a methane molecule. The point is now to assess how the interaction of the methane molecules with the cage acts upon the lateral motion of the cage hydrogen atoms. Figure 6 shows a clear correlation between the presence of a methane hydrogen and the sideways displacement of the cage hydrogen (deuterium) atoms for all examined pressures and without clear indication of a change at the alleged transition pressure of 40GPa. When protons in the CH_4 molecules approach the cage hydrogen (deuterium) below a typical distance ($\sim 2.0\mathring{A}$ at P=10 GPa, $\sim 1.9\mathring{A}$ at P=35 GPa and $\sim 1.7\mathring{A}$ at P=65 GPa), $r_{\perp}$ tends to increase, in order to maximize $d_{\rm HH}$ (resp. $d_{\rm DH}$). This provides an insight into the reason why no clear isotopic effect can be seen at the transition pressure: we know that a small asymmetry in a double well potential can severely reduce tunneling between the two wells [14], now our results show that the effective double well in which the cage hydrogen nuclei reside is perturbed by the molecular hydrogen nuclei. This shows that the normal delocalization is a result of the repulsive interaction between the hydrogen (deuterium) in the cage and those in the methane molecule.

Other simulations were carried out, removing the methane molecules while keeping the oxygen atoms fixed to their original positions and letting the cage hydrogen atoms move freely: the result thereof is that these hydrogen atoms escape from the O-O axes, which shows that the repulsive interactions between the hydrogen of the host CH_4 molecules and the cage hydrogen (resp. deuterium) is also necessary for the stability of the H bonds in the latter frame. It seems that the effect of the environment on the O-H-O groups, due to the presence of the methane

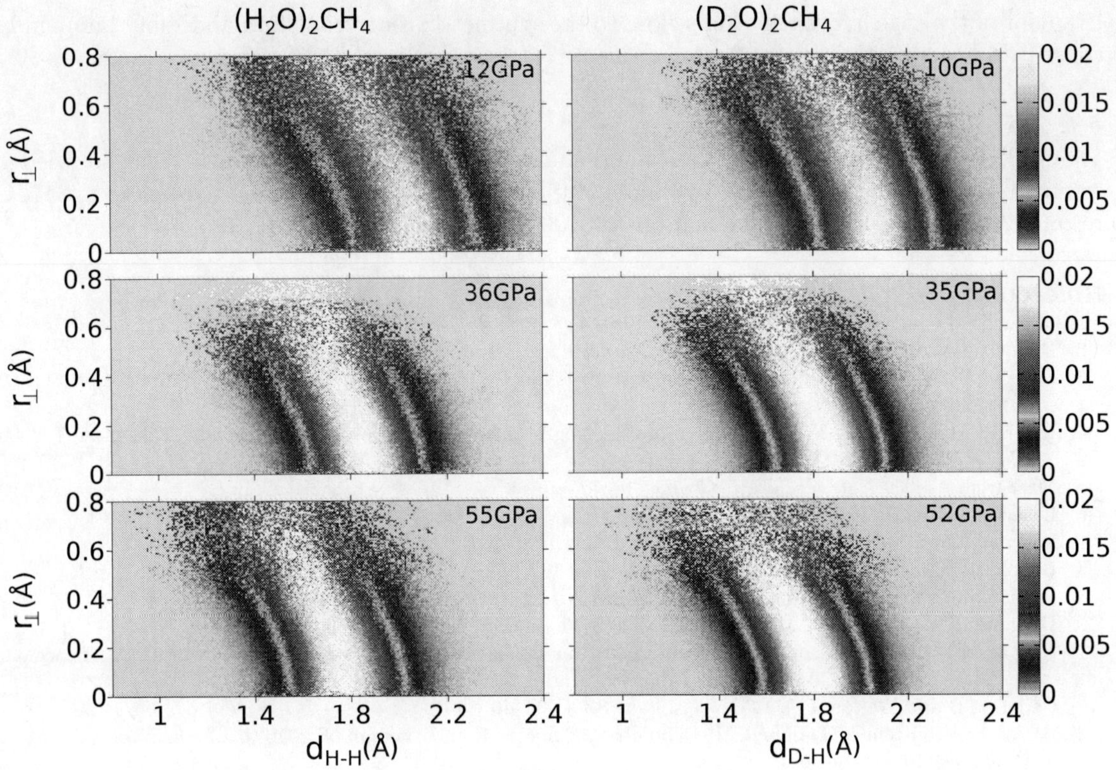

Figure 6. Conditional probability $\rho(d_{\mathrm{HH}}|r_\perp)$ for a hydrogenated cage (left panels) or deuterated cage (right panels).

molecules is to keep the hydrogen atoms close enough to the straight line between the oxygen atoms, but not on it.

6. Concluding remarks

According to previous suggestions, we confirm that, assuming the MH-III structure remains stable at these pressures, a symmetrization transition should occur at approximately 35-40 GPa, much below that in ice VII→X transition due to the smaller O-O distances in methane hydrate with respect to pure ice at all pressures. Moreover, in MH-III, we observe two main groups of hydrogen bonds corresponding to distinct O-O distances, which symmetrize at slightly different pressures. This behaviour contrasts with that of the ice VII where all hydrogen bonds are equivalent and symmetrize at the same pressure. Appropriate spectral experiments could provide a confirmation thereof.

Another difference between ice and MH-III is the isotope effect. While in pure ice the H → D substitution significantly upshifts the symmetrization transition pressure, we found that the symmetrization of the hydrogen bonds occurs at almost same pressures in the hydrogenated and deuterated cages of methane hydrate. Both the lower symmetry of MH-III with respect to pure ice and the interaction between the confined methane molecules and the protons (deuterons) in the H bonds are at the root of a more complex behavior. In particular we observe in methane hydrate an increase of the lateral disorder of the cage hydrogen (deuterium) atoms, as compared with ice VII and X, which points towards a reduction of nuclear quantum effects. As a matter of fact, protons in MH-III cages are more delocalized than deuteriums in both longitudinal and normal directions along the O-O axis. Our results thus exhibit a rather complex and/or subtle behavior of the cage hydrogen atoms close to the symmetrization transition and should hopefully constitute a strong enticement for experimental endeavours in the same pressure range.

6.1. Acknowledgments

We would like to thank L.E. Bove, U. Ranieri and R. Gaal for useful discussions. We acknowledge critical reading of the manuscript by Simon Huppert. This work was granted access to the HPC resources of CINES under the allocation A0010906719 made by GENCI.

References

[1] Sloan J E D and Koh C 2007 *Clathrate hydrates of natural gases* (CRC press)
[2] Demirbas A 2010 *Methane gas hydrate: as a natural gas source* (Springer)
[3] Loveday J, Nelmes R, Guthrie M, Belmonte S, Allan D, Klug D, Tse J and Handa Y 2001 *Nature* **410** 661–663
[4] Hirai H, Hasegawa M, Yagi T, Yamamoto Y, Nagashima K, Sakashita M, Aoki K and Kikegawa T 2000 *Chem. Phys. Lett.* **325** 490–498
[5] Mascarelli A 2009 *Nat. rep. Clim. Change* 46–49
[6] Loveday J, Nelmes R, Guthrie M, Klug D and Tse J 2001 *Phys. Rev. Lett.* **87** 215501
[7] Loveday J and Nelmes R 2008 *Phys. Chem. Chem. Phys.* **10** 937–950
[8] Benoit M, Marx D and Parrinello M 1998 *Nature* **392** 258
[9] Bronstein Y, Depondt P, Finocchi F and Saitta A M 2014 *Phys. Rev. B* **89** 214101
[10] Dammak H Chalopin Y L M H M G J J (2009) *Phys. Rev. Lett.* **103** 190601
[11] Giannozzi P, Baroni S, Bonini N, Calandra M, Car R, Cavazzoni C, Ceresoli D, Chiarotti G L, Cococcioni M, Dabo I and et al 2009 *J. Phys. Condens. Matter* **21** 395502
[12] Callen H B and Welton T A 1951 *Phys. Rev.* **83**(1) 34–40
[13] Brieuc F, Bronstein Y, Dammak H, Depondt P, Finocchi F and Hayoun M 2016 *J. Chem. Theory Comput.* **12** 5688–5697
[14] Bronstein Y, Depondt P, Bove L, Gaal R, Saitta A M and Finocchi F 2016 *Phys. Rev. B* **93** 024104
[15] Dammak H, Antoshchenkova E, Hayoun M and Finocchi F 2012 *J. Phys. Condens. Matter* **24** 435402
[16] Bronstein Y, Depondt P and Finocchi F 2017 *Eur. J. Mineral.* 1–11
[17] Calvo F, Van-Oanh N T, Parneix P and Falvo C 2012 *Phys. Chem. Chem. Phys.* **14** 10503–10506
[18] One can easily verify that the colored noise in equation 1 reduces to the classical white noise with an additional term $\frac{\hbar\omega}{2}$ accounting for the zero point energy when $\frac{\hbar\omega}{k_BT} \to 0$
[19] Perdew J P, Burke K and Ernzerhof M 1996 *Phys. Rev. Lett.* **77** 3865
[20] Grimme S 2006 *J. Comput. Chem.* **27** 1787–1799
[21] Lin L, Morrone J A and Car R 2011 *J. Stat. Phys.* **145** 365–384
[22] Schweizer K S and Stillinger F H 1984 *J. Chem. Phys.* **80** 1230–1240
[23] Machida S I, Hirai H, Kawamura T, Yamamoto Y and Yagi T 2006 *Phys. Earth Planet. Inter.* **155** 170–176
[24] Machida S I, Hirai H, Kawamura T, Yamamoto Y and Yagi T 2007 *Phys. Chem. Miner.* **34** 31–35
[25] Tanaka T, Hirai H, Matsuoka T, Ohishi Y, Yagi T, Ohtake M, Yamamoto Y, Nakano S and Irifune T 2013 *J. Chem. Phys.* **139** 104701
[26] Schaack S, Ranieri U, Depondt P, Gaal R, Kuhs W F, Falenty A, Gillet P, Finocchi F and Bove L E *To be published*

IOP Conf. Series: Journal of Physics: Conf. Series **1136** (2018) 012019 doi:10.1088/1742-6596/1136/1/012019

Prediction of the three-dimensional structures of histone deacetylase 1 complexed with romidepsin and FK-A5

A Oda[1,2,3], K Kato[1], M Morino[2], T Nakayoshi[1,2], S Fukuyoshi[2], K Saijo[4], C Ishioka[4] and E Kurimoto[1]

[1]Faculty of Pharmacy, Meijo University, 150 Yagotoyama, Tempaku-ku, Nagoya, Aichi 468-8503, Japan
[2]Institute of Medical, Pharmaceutical and Health Sciences, Kanazawa University, Kakuma-machi, Kanazawa, Ishikawa 920-1192, Japan
[3]Institute for Protein Research, Osaka University, 3-2 Yamadaoka, Suita, Osaka 565-0871, Japan
[4]Institute of Development, Aging and Cancer, Tohoku University, 4-1 Seiryo-machi, Aoba-ku, Sendai, Miyagi 980-8575, Japan

E-mail: oda@meijo-u.ac.jp

Abstract. Romidepsin is an anticancer drug that inhibits histone deacetylase (HDAC) in human cells. Recently, we found that romidepsin and its analog, FK-A5, inhibit phosphoinositide 3-kinase. Thus, romidepsin and FK-A5 are expected to be promising HDAC/PI3K dual inhibitors for anticancer therapy. In this study, the HDAC1–inhibitor complex structures were predicted by using computational docking and molecular dynamics (MD) simulations. Because romidepsin and FK-A5 are large cyclic molecules, large numbers of conformers can be obtained for these molecules by computational chemical treatment. The docking poses were extracted by comparing romidepsin and FK-A5 because similar compounds are recognized by proteins in similar binding modes. MD simulations were conducted for the selected docking poses, and the protein–ligand interactions were analyzed. The computational results are expected to be useful for the rational drug design of HDAC inhibitors.

1. Introduction

In current cancer treatment, molecular-targeted drugs play important roles [1, 2]. Such drugs are compounds designed to bind to specific molecular drug targets and not to any other biomolecules. Therefore, molecular-targeted drugs generally have fewer side effects and greater efficacy than conventional drugs. Especially, dual (or multiple) inhibitors, which inhibit two (or more) drug target biomolecules simultaneously, have been developed. Recently, some multiple inhibitors have been marketed. For example, the breast cancer drug lapatinib is a dual inhibitor that inhibits epidermal growth factor receptor and HER2 [3]. Many of multiple inhibitors are kinase inhibitors. The active sites of kinases are called “kinase pockets.” The similarity of kinase pockets plays an important role in the multiple inhibitior functions of these drugs. As an alternative to the kinase inhibitors, we developed dual inhibitors that inhibit both a kinase and non-kinase, namely, phosphoinositide 3-kinase (PI3K) and histone deacetylase (HDAC) [4, 5].

PI3K is an enzyme that phosphorylates phosphatidylinositol 4,5-bisphosphate in the cell membrane to generate phosphatidylinositol 3,4,5-trisphosphate [6], and PI3K is one of the most promising drug targets for cancer treatment. On the other hand, HDACs deacetylate the acetylated lysine located in the histone tail [2, 7]. HDACs are important enzymes for transcriptional regulation, and 11 types of HDACs (and 7 types of sirtuins) in humans have been reported [8, 9]. The human HDACs and sirtuins are classified into four classes (classes I to IV). Class I HDACs are distributed throughout the entire body, and HDAC1 in particular is an important enzyme involved in more than half the activity of HDACs. HDACs are known as molecular targets for anticancer drugs. In fact, the inhibitors of class I HDACs are marketed as anticancer drugs: HDACs are though to suppress the transcription of cancer suppressor genes and HDAC inhibitors are considered to activate these genes[9].

Previously, we constructed an assay system to evaluate the inhibitory activities of PI3K inhibitors [4]. Using our assay system, we screened the chemical library of the Screening Committee of Anticancer Drugs supported by a Grant-in-Aid for Scientific Research on Innovative Areas and Scientific Support Programs for Cancer Research from the Ministry of Education, Culture, Sports, Science, and Technology, Japan [4, 5]. Via our screening, the HDAC inhibitor romidepsin and its analogs were identified as candidate compounds with PI3K inhibitory activities. Romidepsin is a natural bicyclic depsipeptide produced by *Chromobacterium violaceum*, and is known as a class I HDAC inhibitor. It was approved by the FDA for the treatment of cutaneous T-cell lymphoma and peripheral T-cell lymphoma. When a PI3K inhibitor was used in combination with an HDAC inhibitor, the cytocidal effect was enhanced [10]. Therefore, HDAC/PI3K dual inhibitors are promising compounds for the treatment of intractable cancer. Figure 1 depicts HDAC/PI3K dual inhibitor candidates found by our team. FK-A5 is the romidepsin analog that has the highest inhibitory activity for PI3K.

The 3D structures of the PI3K–romidepsin and PI3K–FK-A5 complexes have been predicted [11]. Although computational docking [12] is generally used for predicting the structures of protein–ligand complexes, appropriate docking poses cannot be obtained by simple docking studies for romidepsin and FK-A5 because of their macrocyclic structures. Therefore, we introduced the concept of ligand-based drug design (LBDD) [13] to select appropriate docking poses. Reasonable complex structures that were consistent with the experimental inhibitory activities were predicted. In this study, 3D structures of HDAC1 in complex with romidepsin or FK-A5 were predicted. Recently, in silico drug design trials of HDAC inhibitors have been reported [14, 15]. In these studies, the computational results were consistent with the experimental investigations, and the development of new HDAC inhibitors and detection of important substructures of HDAC can be performed. Therefore, in silico drug design techniques, such as docking and MD simulations, are expected to be useful for the investigations of the complex structures between HDAC and inhibitors. The predicted structures obtained by this study will be useful for the further modification of the candidate compounds.

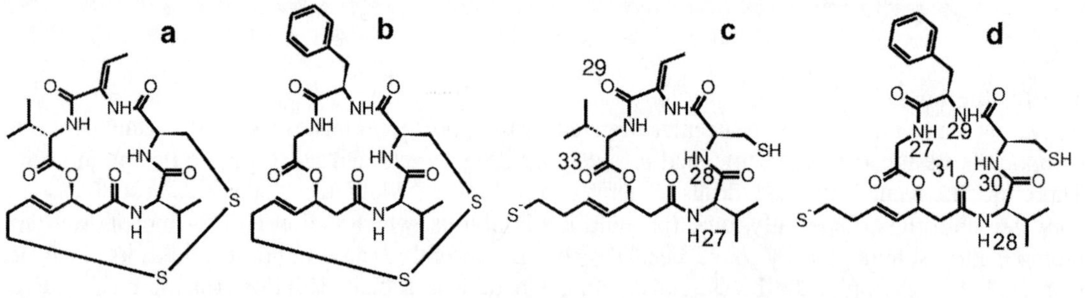

Figure 1. HDAC/PI3K dual inhibitors: (a) romidepsin, (b) FK-A5, (c) active form of romidepsin, and (d) active form of FK-A5.

2. Computational methods

2.1. Protein and ligand structures
The 3D structure of HDAC1 was retrieved from the Protein Data Bank (PDB) (PDB ID: 4BKX) [16]. Because the entry 4BKX includes HDAC1 and metastasis-associated protein MTA1, MTA1 was excluded from the PDB structure and only HDAC1 coordinates were extracted from 4BKX. The hydrogen atoms and missing heavy atoms were complemented by leap module of AmberTools12 and AMBER12 [17]. In this study, romidepsin and FK-A5 were used as ligands. Romidepsin is a prodrug that is activated in the human body by reduction of its disulfide bonds. The longer thiol moiety of the active form is coordinated to zinc in the active site of HDAC by forming a thiolate ion, and bioreductively activated romidepsin inhibits HDAC. In this study, the inhibitory mechanism of FK-A5 was considered to be the same as that of romidepsin. The active forms of romidepsin and FK-A5 are shown in Figure 1c and d, respectively.

2.2. Computational docking
For computational docking, the GOLD program [18] was used. Because the conformational search for ligand molecules in GOLD is conducted by an empirical procedure, conformational searches for romidepsin and FK-A5 including macrocyclic moieties by GOLD were insufficient. In fact, GOLD trials for the docking of romidepsin and FK-A5 into HDAC1 without any preliminary processing cannot generate appropriate docking poses. Therefore, preliminary conformational searches using LMOD [19] based on normal mode analyses were performed in this study, and the conformers generated by LMOD were used for GOLD trials. LMOD searches for possible conformations of molecules according to physicochemical theory. This method is quite different from the conformational search procedure in GOLD. Using LMOD, 12 and 15 conformers were generated for romidepsin and FK-A5 respectively. All of these conformers were used for GOLD. The ligand-binding pocket of HDAC1 was detected by HBOP and HBSITE [20, 21]. In addition, a distance restriction of 1.5–3.5 Å was adopted between zinc of HDAC1 and the sulfur of thiolate.

Many possible docking poses were obtained by computational docking trials. From these probable candidates needed to be selected. The inhibitory activities of several romidepsin analogs for HDAC1 have been experimentally assessed. Similar binding modes were expected to be observed in HDAC1–romidepsin and HDAC1–FK-A5 complexes, because the interactions between HDAC1 and romidepsin were expected to be similar to those between HDAC1 and FK-A5. Although computational docking is one of the structure-based drug design (SBDD) methods, the basic concept of LBDD, in which similar ligands are recognized by a target protein in similar binding modes, was used here. Thus, docking poses with similar binding modes between romidepsin and FK-A5 were extracted.

2.3. Molecular dynamics simulation
The docking poses generally include some structural distortion because a certain amount of tolerance is considered in computational docking. MD simulations of predicted protein–ligand complexes can relax the distorted structures, and refined structures can be obtained. In this study, MD simulations of HDAC1–inhibitor complexes were performed using the AMBER12 program [17]. The ligand topologies were constructed by antechamber [22] in AmberTools. The cyclic boundary box including TIP3P explicit water was spaced with at least 8-Å margins from the protein surface, and the cut-off distance for van der Waals interactions was set at 10 Å. The AMBER ff12SB [23] and general AMBER force fields [24] were used for the protein (including zinc ion) and ligands, respectively. Before MD simulations, 1000-step structural minimizations of water and counter ion (Cl^-) were performed, and 2500-step minimizations of whole systems were carried out. Using the minimized structures, temperature-increasing MD simulations were performed for 20 ps, with the temperature raised from 0 to 300 K. Equilibrating MD simulations were then performed for 20 ns at 300 K with a time step of 2 fs. After the MD simulations, 2500-step minimizations of the whole systems were conducted. The minimized structures were used for the illustration of the complex structures, and the data analyses were mainly performed by using MD trajectories.

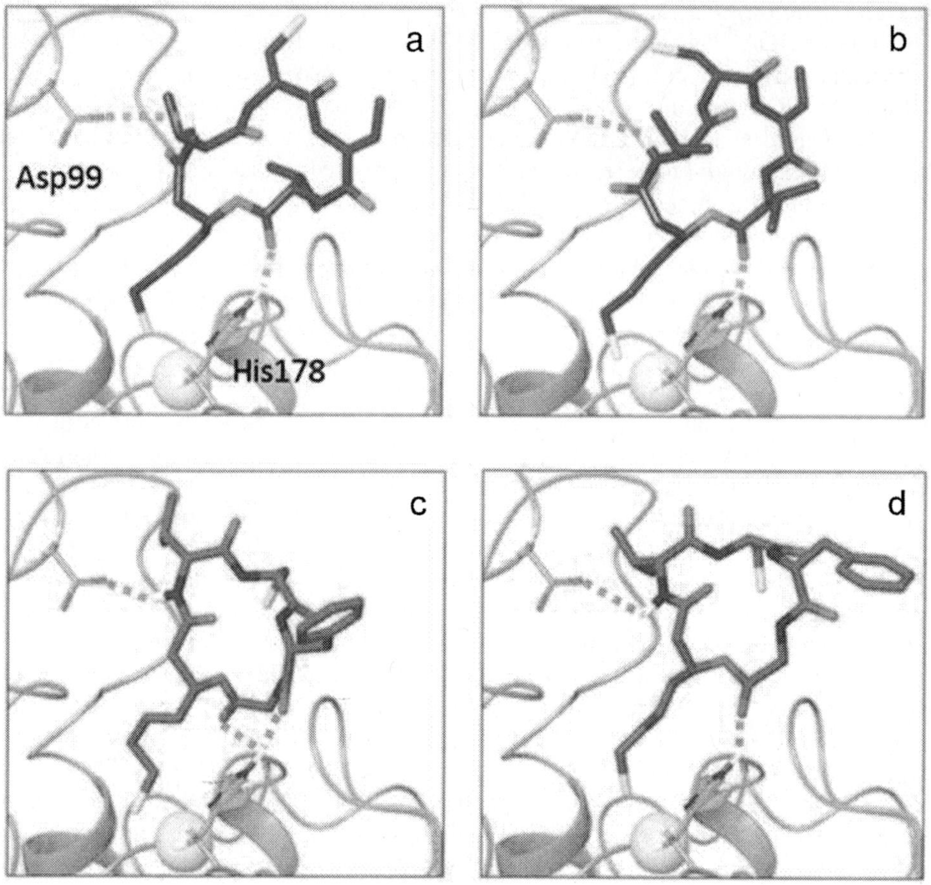

Figure 2. Selected docking poses: (1) pose 1 of romidepsin, (2) pose 2 of romidepsin, (3) pose I of FK-A5, and (4) pose II of FK-A5.

3. Results and discussion

3.1. Docking poses

In total, 74 and 87 docking poses were obtained by the computational docking trials for romidepsin and FK-A5 respectively. As mentioned above, probable poses were selected by the similarity of protein–ligand interaction modes based on the LBDD concept. In the obtained docking poses, similar hydrogen bonding patterns were observed in HDAC1–romidepsin and HDAC1–FK-A5 complexes. The hydrogen-bonding pattern included two hydrogen bonds: one between Asp99 of HDAC1 and hydrogen in ligands, and the other between His178 and oxygen in ligands. The numbers of poses including the common hydrogen-bonding pattern were 21 and 11 for romidepsin and FK-A5, respectively. From these docking poses, two representative poses were selected respectively for romidepsin and FK-A5: poses 1 and 2 for romidepsin and poses I and II for FK-A5. 21 and 11 poses for romidepsin and FK-A5 were clustered into two clusters each by the orientation of macrocycles. For each cluster, the pose with the best GOLD score was selected as the representative pose. These poses are shown in Figure 2. HDAC1 is represented by ribbons, and the residues that form hydrogen bonds with ligands are illustrated by sticks. The sphere represents the zinc atom in the active site of HDAC1. The Zn–S distances were 2.49 and 2.06 Å for romidepsin poses 1 and 2 respectively. For FK-A5, the Zn–S distances were 2.44 and 2.50 Å for poses I and II respectively. Therefore, the sulfurs atoms of ligands were located deep in the active site of HDAC1 in all of these representative poses.

3.2. Molecular dynamics simulations

Because two candidate poses were selected from the GOLD results of each inhibitor, MD simulations were performed to elucidate which candidate is appropriate. The convergences of MD simulations were evaluated by root mean square deviations (RMSDs). For the main chain RMSD calculations, the initial structures of MD simulations were used as references, and the coordinates of the main chain heavy atoms were used. Figure 3 shows the main chain RMSDs for MD trajectories of four complexes.

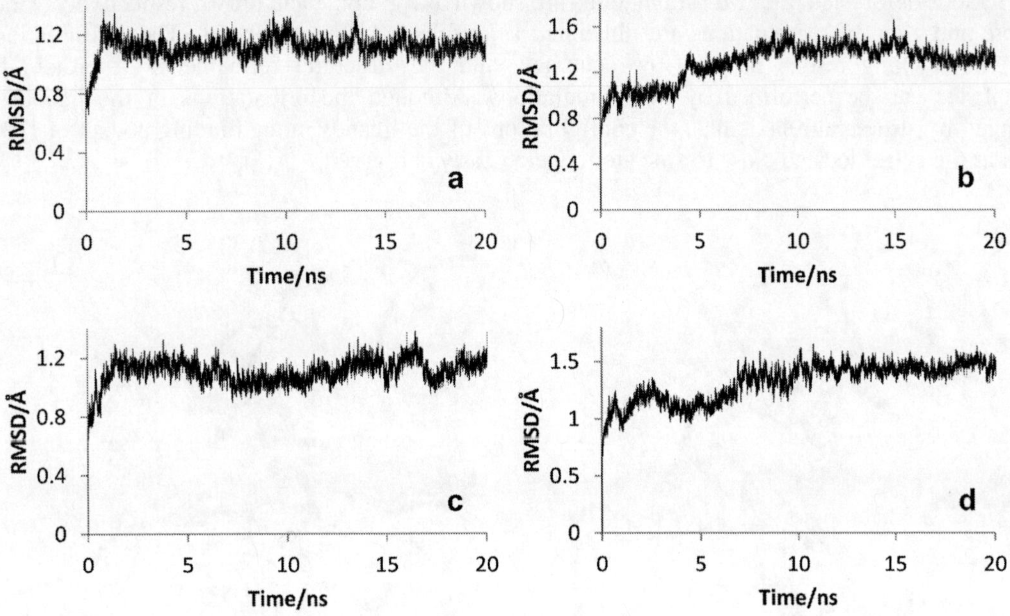

Figure 3. RMSDs of main chain: (a) romidepsin 1, (b) romidepsin 2, (c) FK-A5 I, (d) FK-A5 II.

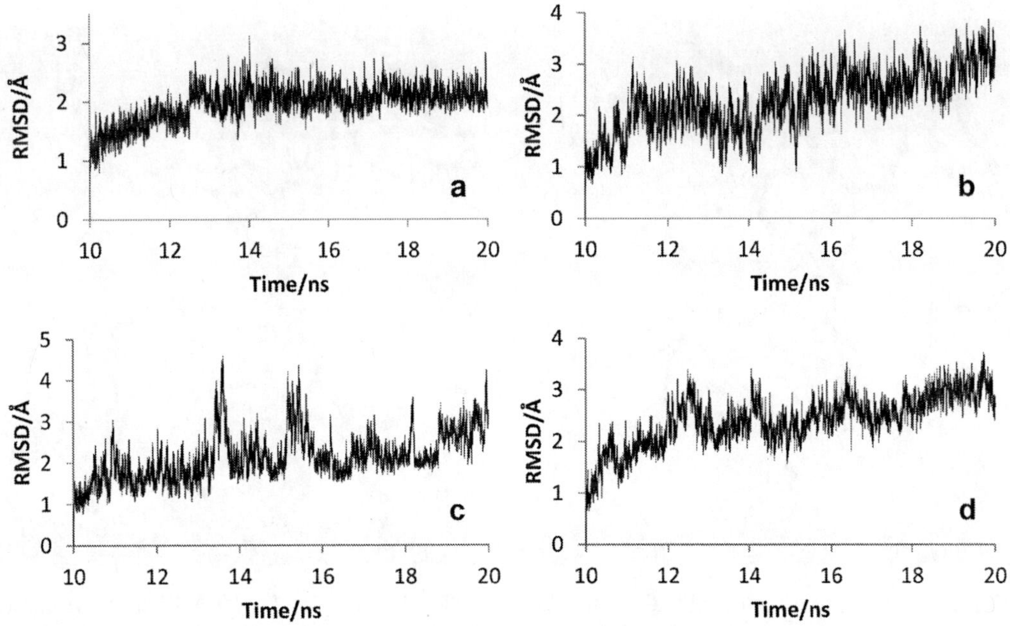

Figure 4. RMSDs of ligands: (a) romidepsin 1, (b) romidepsin 2, (c) FK-A5 I, (d) FK-A5 II.

IOP Conf. Series: Journal of Physics: Conf. Series **1136** (2018) 012019 doi:10.1088/1742-6596/1136/1/012019

On the other hand, for the ligand RMSD, the 3D structures at 10 ns were used as references, and the last 10 ns trajectories were used to reduce the bias of the initial structures. In Figure 4, ligand RMSDs are illustrated. As shown in this figure, RMSDs indicate that the MD simulations of these four complexes converged in the last part of trajectories, and the analyses of MD trajectories can be performed. Comparisons of the 3D structures between the initial and final structures of 20-ns MD simulations are illustrated in Figure 5. The initial and final structures are superposed. HDAC1 chains and ligands before and after MD simulations are shown as light blue and brown respectively. Zinc ions before and after MD simulations are illustrated as purple and orange spheres. These results indicate that the docking results have a low accuracy, and the structural refinements of protein–ligand complexes can be performed by MD simulations. Although the orientations of the ligands were changed by structural relaxations, the configurations of the ligands in the binding pocket of HDAC1, such as the sulfur located close to zinc atom, were nearly conserved.

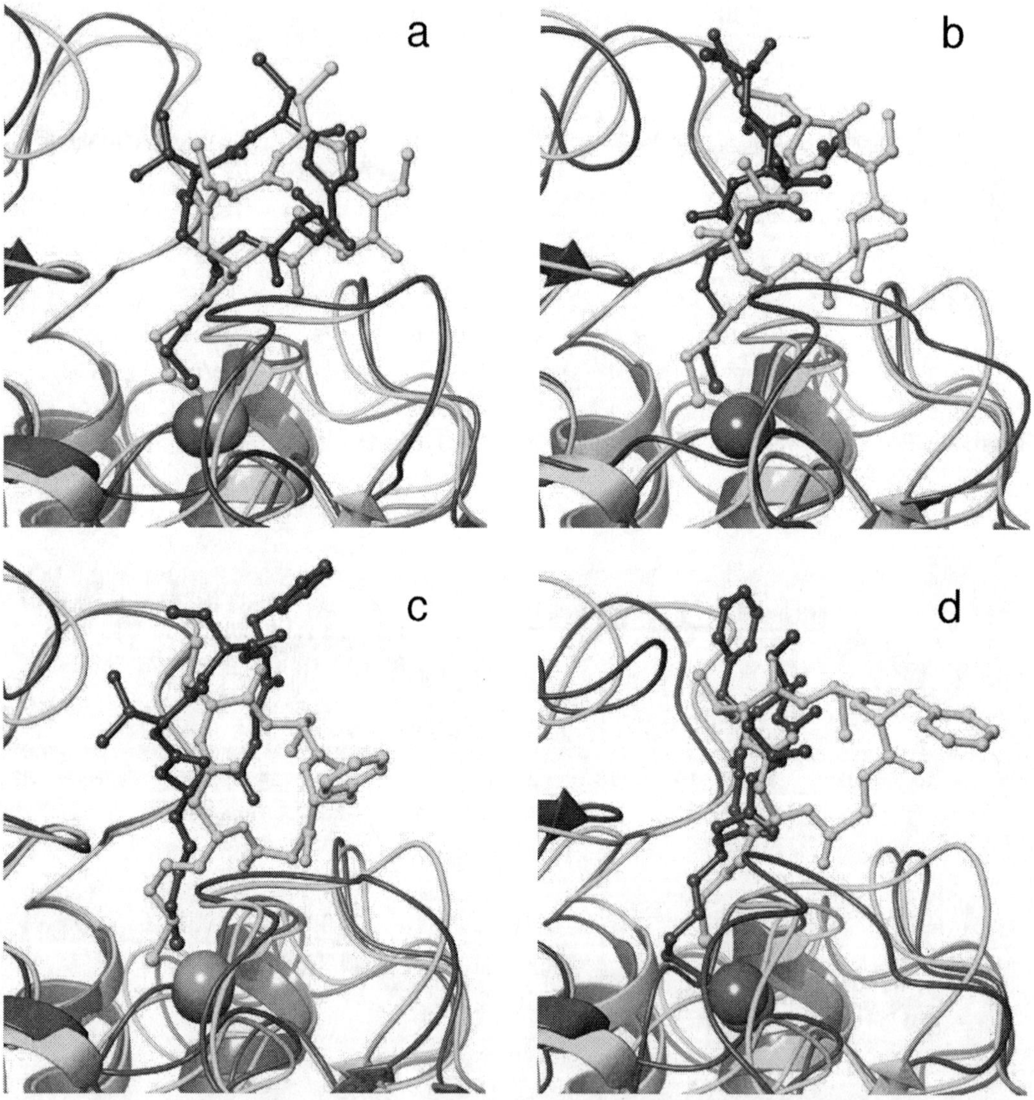

Figure 5. 3D structures of HDAC1–inhibitor complexes before and after MD simulations: (a) romidepsin 1, (b) romidepsin 2, (c) FK-A5 I, and (d) FK-A5 II. Light blue and brown represent HDAC1 before and after MD simulations respectively. HDAC1 and ligands are illustrated by ribbon and ball-and-stick models, respectively.

3.3. Hydrogen bonds between HDAC1 and ligands

For the MD trajectories, the rates of occurrence of hydrogen bonds were analyzed. To avoid any bias from the initial structures, the last half of the MD trajectories were used for the analyses. For the definition of the hydrogen bonds, the default setting of ptraj module of AmberTools was used (distance < 3.5 Å). In Tables 1 and 2, the rates of occurrence of the hydrogen bonds greater than 10% are shown. The starting and ending points of the arrows indicate the hydrogen donors and acceptors respectively. The atom numbers of ligands are illustrated in Figure 2c and 2d. The common hydrogen bonds in HDAC1–romidepsin and HDAC1–FK-A5 complexes mentioned in subsection 3.1 (hydrogen bonds related to Asp99 and His178) are underlined in Tables 1 and 2. There were no MD trajectories in which both of two common hydrogen bonds (hydrogen bonds related to Asp99 and His178) were conserved. The His178 hydrogen bond was conserved in the MD simulation of romidepsin pose 1, and the Asp99 hydrogen bond remained in the MD simulation of romidepsin pose 2. For the MD simulation of FK-A5 pose I, the rates of occurrence of almost all hydrogen bonds were considerably reduced, and the maximum value of the rate of occurrence was only 5.8%. In contrast, many hydrogen bonds were observed in the MD trajectory of FK-A5 pose II and the common hydrogen bond Asp99 $O_{\delta2} \leftarrow (N_{28})H$ was conserved. These results suggest that FK-A5 pose II can form hydrogen bonds more stable than FK-A5 pose I. For the ligand recognition of a protein, hydrogen bonds play important roles in determination of the orientation and configuration of a ligand. Thus, FK-A5 pose I, in which the hydrogen bonds between HDAC1 and ligand were hardly observed, appears to be inappropriate for the predicted complex structure, whereas FK-A5 pose II is plausible. Because the binding modes of romidepsin analogs for HDAC1 are assumed to be similar according to the experimental inhibitory activities [4, 5], the hydrogen bond observed in FK-A5 pose II is also considered to be important for

Table 1. Hydrogen bonds between HDAC1 and romidepsin after MD simulations.

Romidepsin pose 1		Romidepsin pose 2	
Hydrogen bond	Occurrence rate	Hydrogen bond	Occurrence rate
His178 $H_{\varepsilon2} \rightarrow O_{33}$	66.7%	Asp99 $O_{\delta2} \leftarrow (N_{28})H$	82.4%
Phe205 O $\leftarrow (N_{27})H$	34.4%	Asp99 $O_{\delta2} \leftarrow (N_{27})H$	75.6%

Table 2. Hydrogen bonds between HDAC1 and FK-A5 after MD simulations.

FK-A5 pose I		FK-A5 pose II	
Hydrogen bond[a]	Occurrence rate	Hydrogen bond	Occurrence rate
No hydrogen bonds[a)]		Asp99 $O_{\delta2} \leftarrow (N_{30})H$	70.5%
		His178 $H_{\varepsilon2} \rightarrow O_{31}$	64.8%
		Asp99 $O_{\delta2} \leftarrow (N_{28})H$	64.7%
		Asp99 $O_{\delta1} \leftarrow (N_{29})H$	61.2%
		Asp99 $O_{\delta1} \leftarrow (N_{27})H$	49.9%
		Asp99 $O_{\delta2} \leftarrow (N_{29})H$	37.3%
		Asp99 $O_{\delta1} \leftarrow (N_{28})H$	31.6%
		Asp99 $O_{\delta2} \leftarrow (N_{27})H$	23.6%
		Asp99 $O_{\delta1} \leftarrow (N_{30})H$	20.6%
		Tyr204 OH $\rightarrow O_{31}$	15.2%

[a] No hydrogen bonds with a rate of occurrence >10% were observed in pose I.

the romidepsin complex. Because the hydrogen bond related to Asp99, which was observed in FK-A5 pose II, was detected only in romidepsin pose 2, pose 2 is more reasonable than pose 1. Therefore, the MD simulations suggest that romidepsin pose 2 and FK-A5 pose II are the assumed complex structures, which are consistent with the experimental results.

3.4. Other interactions

The average of Zn–S distances for the last 10 ns trajectories of MD simulations were 1.98, 1.95, 1.95, and 1.97 Å for romidepsin poses 1 and 2, and FK-A5 poses I and II respectively. These distances converged by the 20 ns MD simulations. The crystal structure of the complex between HDAC8 and largazole, which is one of the depsipeptide drugs the same as romidepsin, has been reported (PDB ID: 3RQD) [25]. In the HDAC8–largazole crystal structure, the Zn–S distance was 2.3 Å. Because the distances between the coordinating sulfur of ligand and the zinc atom of proteins are reported to be generally 2.0–2.3 Å [26], Zn–S distances observed for HDAC1–romidepsin and HDAC1–FK-A5 complexes in this study seem to be reasonable. In this study, AMBER ff12SB parameters for zinc were used, and special treatment for coordinate bond was not applied. Although the classical MD simulations cannot investigate the electron state related to the coordinate bond between the metal ion and the ligands, the results of MD simulations indicate that the structural predictions can be performed by the classical approximation of protein–ligand complexes.

In addition to the hydrogen bonds and Zn–S, tentative π–π interaction between the imidazole ring of His28 in HDAC1 and the benzene ring of FK-A5 was observed in FK-A5 pose II. Because classical MD simulations using AMBER ff12SB force field cannot evaluate π–π interaction, the tentative π–π interaction was presumed from the positional relation between aromatic rings in 3D structures obtained by MD simulations. Because the benzyl group is included only in FK-A5, the tentative π–π interaction was one of the distinct differences between HDAC1–romidepsin and HDAC1–FK-A5 complexes. Figures 6 and 7 illustrate interactions between HDAC1 and the ligand at the active site for HDAC1–romidepsin and HDAC1–FK-A5 respectively. In these figures, hydrogen bonds with rates of occurrence <50% are omitted. The experimentally determined inhibitory activity of FK-A5 for HDAC1 was higher than that of romidepsin, which may have been caused by the tentative π–π interaction. Around the benzyl group of FK-A5 in the predicted complex structure, additional space was observed and the introduction of substitution groups to this benzyl group for further drug design may be possible. According to the results of our previous study [11], we assumed that this benzyl group also plays an important role in PI3K inhibition because of large hydrophobic effects. Thus, modification of the benzyl group is expected to improve the inhibitory activities for both PI3K and HDAC1, and better HDAC/PI3K dual inhibitors can be developed.

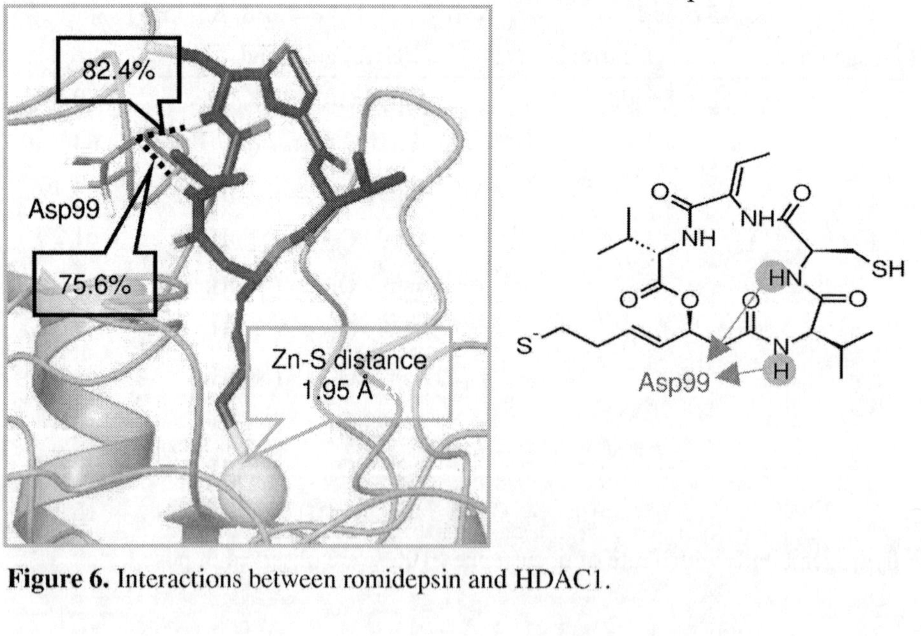

Figure 6. Interactions between romidepsin and HDAC1.

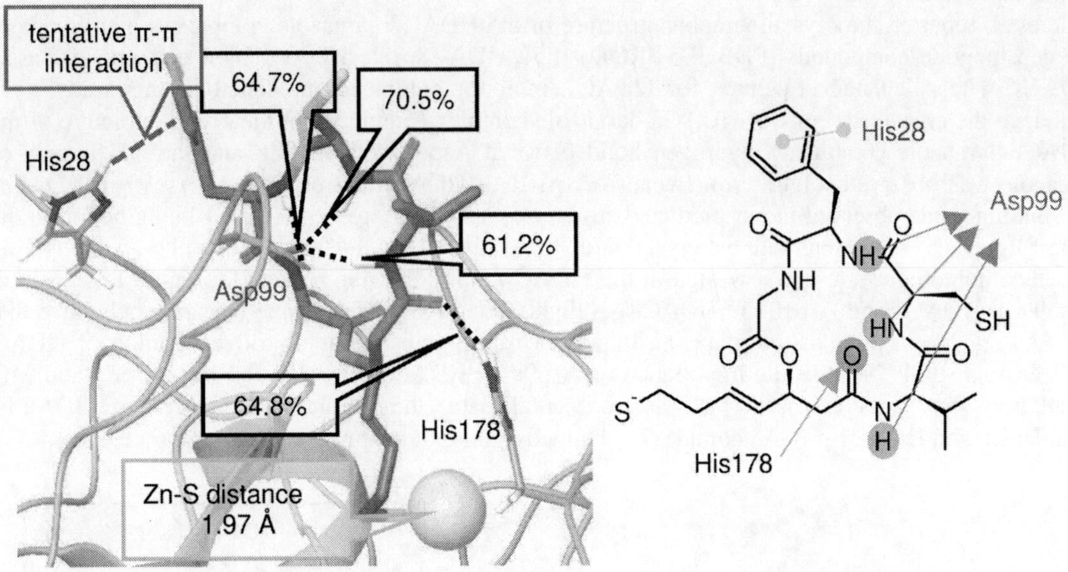

Figure 7. Interactions between FK-A5 and HDAC1.

As mentioned above, romidepsin pose 2 and FK-A5 pose II seem to be appropriate for complex structures because the hydrogen bond related to Asp99 was shared between them. In Figure 8, these structures are superimposed. As shown in this figure, not only the hydrogen bond but also the locations of ligands in the binding pocket of HDAC1 were shared for romidepsin and FK-A5. In addition, the orientations of macrocyclic structures were roughly consistent between romidepsin and FK-A5. Although the similar docking poses were selected for romidepsin and FK-A5, large structural relaxations for ligands were occurred by MD simulations (shown in Figure 5). The results of MD simulations suggest that the binding modes of the two inhibitors were similar even though the large structural relaxations. This supports our predictions of the structures of HDAC1–romidepsin and HDAC1–FK-A5 complexes.

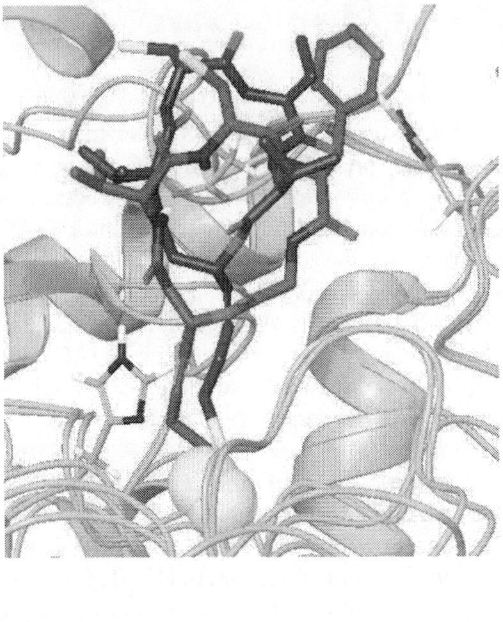

Figure 8. Superposition of 3D structures of HDAC1–romidepsin and HDAC1–FK-A5 complexes obtained by MD simulations. Dark gray and magenta represent romidepsin and FK-A5 respectively.

IOP Conf. Series: Journal of Physics: Conf. Series **1136** (2018) 012019 doi:10.1088/1742-6596/1136/1/012019

3.5. Comparison with HDAC8

Cole et al. reported the crystallographic structure of the HDAC8–largazole complex, which is one of the depsipeptide compounds (PDB ID: 3RQD) [25]. HDAC8, like HDAC1, is a member of class I HDACs. The calculated structures for HDAC1–inhibitor complexes predicted in this study were similar to the crystal structure of HDAC8–largazole complex (Figure 9). In the crystal structure of the HDAC8–largazole complex, a hydrogen bond between Asp101 of HDAC8 and the NH moiety of largazole was observed (Figure 9b). Because Asp101 of HDAC8 corresponds to Asp99 in HDAC1, the binding mode observed in our predicted structures, namely, Asp99 of HDAC1 being bound to the NH of ligands, is consistent with the crystal structure of the HDAC8–largazole complex. Although the sequence homologies are not very high in the HDAC family, the Asp residue located in this position (Asp99 in HDAC1 and Asp101 in HDAC8) is highly conserved. Thus, some researchers believe that the Asp residue in this position plays an important role in substrate/inhibitor recognition of HDAC [27]. In this study, because the importance of Asp99 in HDAC1 was clarified by docking and MD simulations and was consistent with experimental data, the predicted structures of HDAC1–romidepsin and HDAC1–FK-A5 complexes obtained in this study appear to be appropriate.

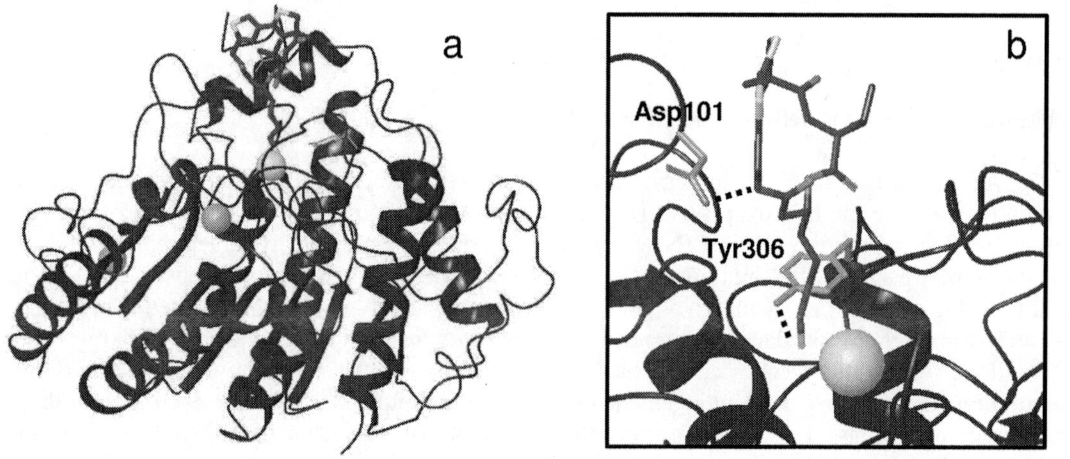

Figure 9. Experimental structure of complex between HDAC8 and largazole (PDB ID: 3RQD): (a) whole complex, (b) ligand binding pocket.

4. Conclusion

In this study, the 3D structures of HDAC1 complexed with romidepsin and FK-A5, which were developed by our team as HDAC/PI3K dual inhibitors, were predicted by using computational docking and MD simulations. Because multiple docking poses are generally constructed by computational docking trials, many possible docking poses were generated for romidepsin and FK-A5, which include macrocyclic structures. For the pose selection, the concept of LBDD (similar compounds are recognized by a target protein in similar binding modes) was used, and the docking poses with common binding modes were selected for romidepsin and FK-A5. The selected docking poses were validated by MD simulations, and the most likely complex structures between HDAC1 and inhibitors were predicted. Therefore, reasonable predictions of protein–ligand complexes can be obtained by the combined use of SBDD and LBDD, even for relatively large ligand molecules such as macrocyclic compounds.

The experimentally determined HDAC1 inhibitory activities IC_{50} of romidepsin and FK-A5 were 3.6 and 2.4 nM, respectively [4]. In the complex structures predicted in this study, the number of hydrogen bonds for HDAC1–FK-A5 was greater than that for HDAC1–romidepsin, and the π–π interaction between the aromatic ring of HDAC1 and the ligand was only observed in HDAC1–FK-A5 (between the imidazole ring of His28 and the benzyl group). These interactions seem to influence the higher binding affinity of FK-A5 for HDAC1. The benzyl group of FK-A5 is also considered to play an important role in PI3K–FK-A5 complex. This indicates that benzyl is a key moiety for further compound design. Therefore, the results of this study will be useful for rational drug design and the development of HDAC/PI3K dual inhibitors.

Acknowledgments
This work was supported by Grants-in-Aid for Scientific Research (15H01064 and 17K08257) from the Japan Society for the Promotion of Science.

References
[1] Widakowich C, de Azambuja E, Gil T, Cardoso F, Dinh P, Awada A and Piccart-Gebhart M 2007 *Int. J. Biochem. Cell Biol.* **39** 1375–87
[2] Hoshino I and Matsubara H 2010 *Surgery Today* **40** 809–15
[3] Xia W, Mullin R J, Keith B R, Liu L, Ma H, Rusnak D W, Owens G, Alligood K J and Spector N L 2002 *Oncogene* **21** 6255–63
[4] Saijo K, Katoh T, Shimodaira H, Oda A, Takahashi O and Ishioka C 2012 *Cancer Sci.* **103** 1994–2001
[5] Saijo K, Imamura J, Narita K, Oda A, Shimodaira H, Katoh T and Ishioka C 2015 *Cancer Sci.* **106** 208–15
[6] Fruman D A, Meyers R E and Cantley L C 1998 *Annu. Rev. Biochem.* **67** 481–507
[7] de Ruijter A J M, Gennip A II, Caron II N, Kemp S and van Kuilenburg A B P 2003 *Biochem J.* **370** 737–49
[8] Dokmanovic M, Clarke C and Marks P A 2007 *Mol. Cancer Res.* **5** 981–989.
[9] Marks P A and Xu W S 2009 *J. Cell. Biochem.* **107** 600–608.
[10] Yoshioka T, Yogosawa S, Yamada T, Kitawaki J, Sakai T 2013 *Gynecol. Oncol.* 129 425–32
[11] Oda A, Saijo K, Ishioka C, Narita K, Katoh T, Watanabe Y, Fukuyoshi S and Takahashi O 2014 *J. Mol. Graph. Model.* **54** 46–53
[12] Kitchen D B, Decornez H, Furr J R and Bajorath J 2004 Nat. Rev. Drug Discovery 3 935–49
[13] Acharya C, Coop A, Polli J E and MacKerell A D Jr. 2011 *Curr. Comput.-Aided Drug Des.* **7** 10–22.
[14] Wang D F, Helquist P, Wiech N L and Wiest O 2005 *J. Med. Chem.* **48** 6936–6947
[15] Bermúdez-Lugo J A, Perez-Gonzalez O, Rosales-Hernández M C, Ilizaliturri-Flores I, Trujillo-Ferrara J and Correa-Basurto J 2012 *J. Mol. Model.* **18** 2301–2310.
[16] Millard C J, Watson P J, Celardo I, Gordiyenko Y, Cowley S M, Robinson C V, Fairall L, Schwabe J W R 2013 *Mol. Cell* **51** 57–67
[17] Case D A, Darden T A, Cheatham T E III, Simmerling C L, Wang J, Duke R E, Luo R, Walker R C, Zhang W, Merz K M, Roberts B, Hayik S, Roitberg A, Seabra G, Swails J, Goetz A W, Kolossváry I, Wong K F, Paesani F, Vanicek J, Wolf R M, Liu J, Wu X, Brozell S R, Steinbrecher T, Gohlke H, Cai Q, Ye X, Wang J, Hsieh M, Cui G, Roe D R, Mathews D H, Seetin M G, Salomon-Ferrer R, Sagui C, Babin V, Luchko T, Gusarov S, Kovalenko A and Kollman P A 2012 *AMBER 12* (San Francisco: University of California, San Francisco)
[18] Jones G, Willett P, Glen R C, Leach A R and Taylor R 1997 *J. Mol. Biol.* **267** 727–48
[19] Kolossváry I and Guida W C 1996 *J. Am. Chem. Soc.* **118** 5011–19
[20] Yamaotsu N, Oda A and Hirono S 2008 *Biol. Pharm. Bull.* **31** 1552–8
[21] Oda A, Yamaotsu N and Hirono S 2009 *J. Comput. Chem.* **30** 2728–37
[22] Wang J, Wang W, Kollman P A and Case D A 2006 *J. Mol. Graph. Model.* **25** 247–60.
[23] Maier J A, Martinez C, Kasavajhala K, Wickstrom L, Hauser K E and Simmerling C 2015 *J. Chem. Theory Comput.* **11** 3696–713
[24] Wang J, Wolf R M, Caldwell J W, Kollman P A and Case D A 2004 *J. Comput. Chem.* **25** 1157–74
[25] Cole K E, Dowling D P, Boone M A, Phillips A J and Christianson D W 2011 *J. Am. Chem.Soc.* 133 12474–7
[26] Chakrabarti P 1989 *Biochemistry* **28** 6081–5.

[27] Vannini A, Volpari C, Gallinari P, Jones P, Mattu M, Carfí A, De Francesco R, Steinkühler C and Di Marco S 2007 *EMBO Rep.* **8** 879–84

Computational studies on cyclic imide formation mechanism of glutamic acid residues catalyzed by two water molecules

T Nakayoshi[1,2], K Kato[2,3], S Fukuyoshi[1], O Takahashi[4], E Kurimoto[2] and A Oda[1,2,5]

[1]Institute of Medical, Pharmaceutical and Health Sciences, Kanazawa University, Kakuma-machi, Kanazawa, Ishikawa 920-1192, Japan
[2]Faculty of Pharmacy, Meijo University, 150 Yagotoyama, Tempaku-ku, Nagoya, Aichi 468-8503, Japan
[3]Department of Pharmacy, Kinjo Gakuin University, 2-1723 Omori, Moriyama-ku, Nagoya, Aichi 463-8521, Japan
[4]Faculty of Pharmaceutical Sciences, Tohoku Medical and Pharmaceutical University, 4-4-1 Komatsushima, Aoba-ku, Sendai, Miyagi 981-8558, Japan
[5]Institute for Protein Research, Osaka University, 3-2 Yamadaoka, Suita, Osaka 565-0871, Japan.

E-mail: oda@meijo-u.ac.jp

Abstract. Aspartic acid (Asp) residues in peptides and proteins are prone to nonenzymatic stereoinversion and/or isomerization to form three types of isomerized Asp residues (L-β-Asp, D-α-Asp, and D-β-Asp) via a five-membered ring succinimide intermediate. These isomerized Asp residues are detected more frequently in aged tissues. However, stereoinversion and/or isomerization of glutamic acid (Glu) residues having a chemical structure similar to Asp residues are hardly detected in proteins. In this study, we investigate computationally the formation mechanism of the cyclic imide, i.e., amino-glutarimidyl (Agl) from Glu residues, with water molecules as catalyst. We study the reaction mechanism by using quantum chemical B3LYP/6-31+G(d,p) density functional theory calculations. All calculations are performed by using model compounds in which a Glu residue is capped with acetyl and methylamino groups on the N- and C-termini, respectively. Agl formation consists of the three steps of iminolization, cyclization, and dehydration, and two water molecules acting as proton-relay mediators catalyze all three steps. The calculated activation energy for Agl formation from Glu residues is 30.3 kcal mol^{-1}, which is somewhat greater than found experimentally for Asp-residue stereoinversion. This calculation suggests that *in vivo* Glu-residue stereoinversion is unlikely to occur because of the high activation barrier.

1. Introduction

Aspartic acid (Asp) residues in peptides and proteins are spontaneously prone to isomerization [1-7]. The three-dimensional structure of proteins is maintained by interactions between amino acid residues. Asp-residue stereoinversion and/or isomerization disturb these interactions, and these reactions may cause undesirable conformational changes in proteins [6,8,9]. Nonenzymatic Asp-residue stereoinversion and/or isomerization proceeds via the five-membered ring aminosuccinimidyl (Asu) intermediate and yield three types of isomerized Asp residues (L-β-Asp, D-α-Asp and D-β-Asp; see Scheme 1) [1-7]. Three types of isomerized Asp residues have been detected from various aging tissues, and the amounts detected increase with aging. Therefore, Asp-residue isomerization is

IOP Conf. Series: Journal of Physics: Conf. Series **1136** (2018) 012020 doi:10.1088/1742-6596/1136/1/012020

suspected as one cause of protein denaturation. Fujii et al. showed experimentally that the activation energy for Asp-residue stereoinversion in distilled water is 25.7–29.0 kcal mol^{-1} [10]. Recently, we reported that water molecules can catalyze Asp-residue stereoinversion [11-13]. The Asu intermediate is considered to form by nucleophilic attack of the side-chain carboxyl carbon by the amide nitrogen of the adjacent residue on the C-terminal side, i.e., the ($N + 1$) residue. However, the nucleophilicity of the amide nitrogen is known not to be very high. Via a quantum chemical calculation, we investigated the cyclization pathway in which Asp residue directly forms Asu intermediates, and the calculated activation barrier is much higher than the experimental value [11]. We have proposed a novel Asu-intermediate formation mechanism via iminol tautomer. Specifically, the ($N + 1$)-residue amide bond changes to iminol tautomer and iminol nitrogen attacks of the side-chain carboxyl carbon of the Asp residue. It is expected to proceed the Asu-intermediate formation easily because the electron density on the iminol nitrogen increases by the resonance effect. The calculated activation barrier of Asu-intermediate formation from Asp residue via iminol tautomer is consistent with the experimental values [11].

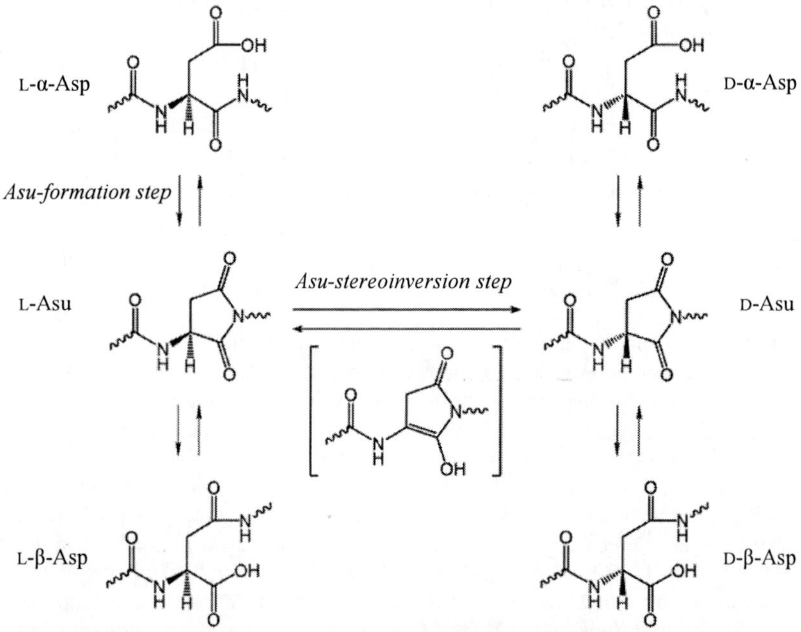

Scheme 1. Nonenzymatical stereoinversion and isomerization pathway of Asp residue via Asu intermediate.

Glutamic acid (Glu) is an acidic amino acid constituting proteins and has a chemical structure similar to that of Asp. In the isomerization mechanism of Glu presumed from the analogy with the Asp isomerization mechanism, the Glu residue is considered to form a cyclic imide intermediate, i.e., an aminoglutarimidyl (Agl) intermediate similar to the Asp residue and to yield three types of Glu isomers (L-γ-Glu, D-α-Glu, D-γ-Glu; see Scheme 2). However, few reports claim to detect D-Glu residues in proteins [14]. It remains unclear why Glu-residue stereoinversion is less likely to occur compared with Asp-residue stereoinversion. Based on computations, we proposed the Agl-residue stereoinversion mechanism model catalyzed by a water molecule and estimated its activation energy [15]. We investigated the possibility that water molecules catalyze Agl-stereoinversion and reported that the activation barriers for the stereoinversions of Asu and Agl are nearly equal [15]. Therefore, we consider that the difficulty of Glu-residue stereoinversion may be due to the Agl-intermediate formation step and not to the Agl-stereoinversion step. In the present study, we investigate the reaction mechanisms of Agl-intermediate formation catalyzed by two water molecules by using quantum chemical calculations, and the resulting activation barrier is compared with the experimentally determined activation energy of Asp stereoinversion.

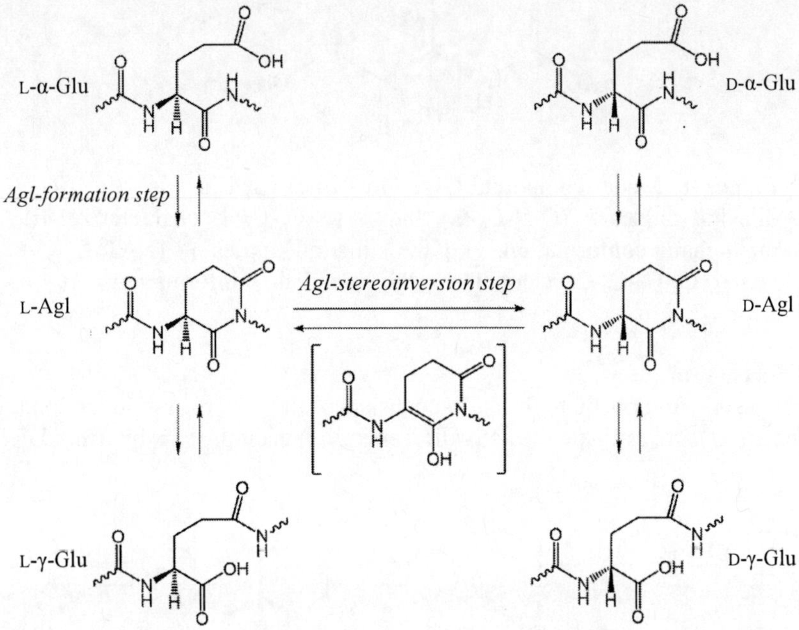

Scheme 2. Presumed stereoinversion and isomerization pathway of Glu residue via Agl intermediate.

2. Computational methods

All calculations were performed using Gaussian 09 [16]. In this study, all calculations were performed using model compound shown in Figure 1, in which a Glu residue was capped with an acetyl (Ace) and a methylamino (Nme) groups on the N- and C-termini, respectively. The dihedral angles φ (C-N-C_α-C) and ψ (N-C_α-C-N) characterize the main-chain conformation, and the χ_1 (N-C_α-C_β-C_γ) and χ_2 (C_α-C_β-C_γ-C_δ) dihedral angles characterize the side-chain conformation. In the model compound, the Glu-residue side chain is protonated because the protonated form is more likely to undergo nucleophilic attack by the amide nitrogen of the ($N + 1$) residue to form the Agl intermediate. Catak et al. reported that the calculated activation energies for Asu-intermediate formation are 53.3 and 46.1 kcal mol^{-1} in the absence of catalyst and presence of one water molecule [17], however the deviation from the experimental value is large. We have reported that Asp-residue stereoinversion can proceeds catalyzed by two water molecules [11] and its activation barrier falls within experimental range [10]. Thus, we assumed that two water molecules also catalyze the Glu-residue stereoinversion and use as initial structure a reactant complex in which two water molecules are arranged at appropriate positions around the Glu residue.

Energy-minimum and transition-state (TS) geometries are optimized with no constraints by density functional theory (DFT) calculations using the B3LYP exchange-correlation functional [18] and the 6-31+G(d,p) basis set using Gaussian 09. The vibrational frequency calculations were performed for all optimized geometries to confirm them as energy-minimum geometries (with no imaginary frequency) or TS (with a single imaginary frequency) and to correct the relative energies for the zero-point energy. Intrinsic reaction coordinate (IRC) calculations were performed from the TSs, followed by full geometry optimizations, to confirm that each TS connects two energy-minimum geometries.

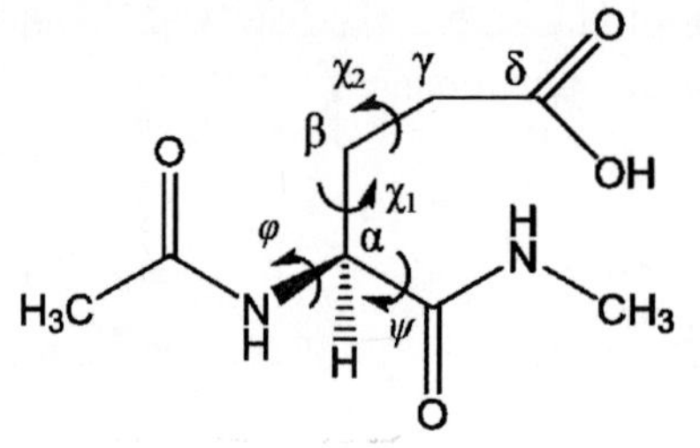

Figure 1. Model compound (Ace-Glu-Nme) used in this study. The dihedral angles φ (C-N-C_α-C) and ψ (N-C_α-C-N) characterize the main-chain conformation, and the dihedral angles χ_1 (N-C_α-C_β-C_γ) and χ_2 (C_α-C_β-C_γ-C_δ) characterize the side-chain conformation.

3. Results and Discussion

We presume that the Agl-intermediate formation consists of three steps: iminolization, cyclization, and dehydration (Scheme 3) and estimate the activation energy in each process by using DFT calculations.

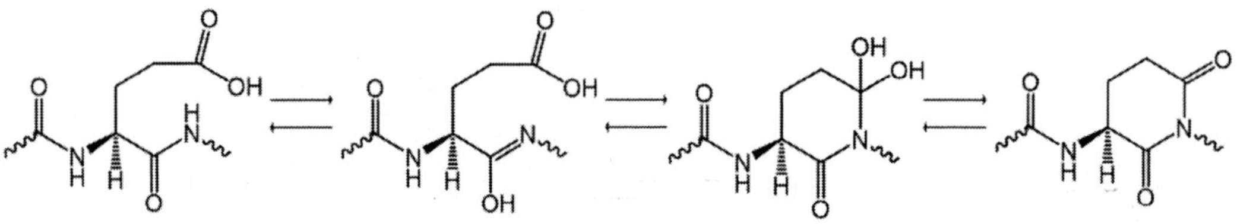

Scheme 3. Agl-intermediate-formation mechanism from Glu residue, which constitutes the three steps of iminolization, cyclization, and dehydration.

3.1 Iminolization on C-terminal side amide group of Asp

Figure 2 shows the geometries calculated for the reaction system for the iminolization step. The reactant complex (1-AM) is composed of the Glu residue (amide form) and two water molecules (Figure 2a). In 1-AM, the hydrogen bridge forms among the C-terminal amide group, two water molecules, and the side-chain carboxyl group. One of the water molecules forms an additional hydrogen bond with the amide oxygen of the ($N-1$) residue. In this process, the ($N+1$)-residue amide NH proton is pulled out and transferred to the Glu-residue amide C=O oxygen. Iminolization proceeds by quadruple proton transfer mediated by the Glu-residue side-chain carboxyl group and two water molecules, and transition state 1-TS forms (Figure 2b). When 1-TS forms from 1-AM, the change in all dihedral angles defined for Glu residues is small, and no significant conformational change occurs (Table 1). The calculated activation barrier for iminolization is 21.0 kcal mol^{-1}, and it is considered that 1-AM can undergo sufficient iminolization under physiological conditions. As a result of quadruple proton transfer mediated by two water molecules and the Glu-residue side-chain carboxyl group, iminolization completes and the iminol tautomer 1-IM forms (Figure 2c). The 1-IM complex is composed of Glu residue (iminol tautomer) and two water molecules. When 1-IM forms from 1-TS, a small change occurs in all dihedral angles defined for the Glu residue (Table 1). The hydrogen-bond bridge observed in 1-AM remains throughout the iminolization. It is considered that the hydrogen-bond bridge serves to immobilize the two water molecules and the Glu residue. The immobilization of a reactant complex seems to lead to the catalytic actions of the water molecules and hinders a large conformational change during the iminolization. The relative energy of 1-IM with respect to 1-AM is calculated to be 14.0 kcal mol^{-1}.

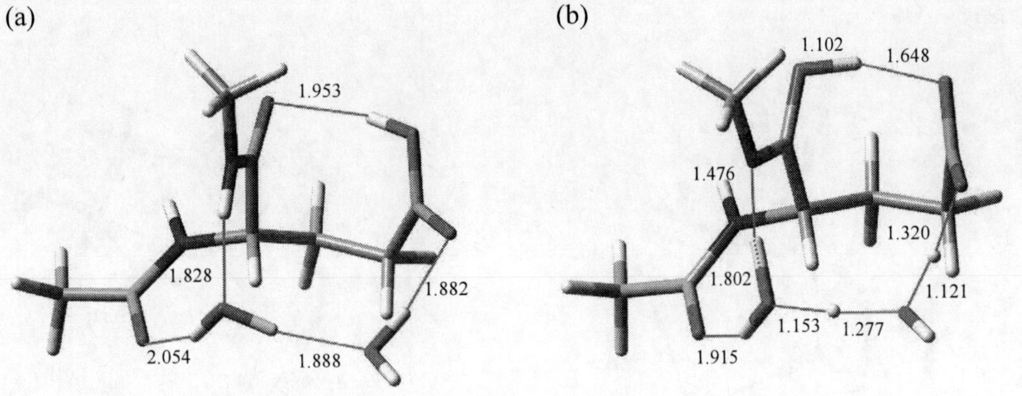

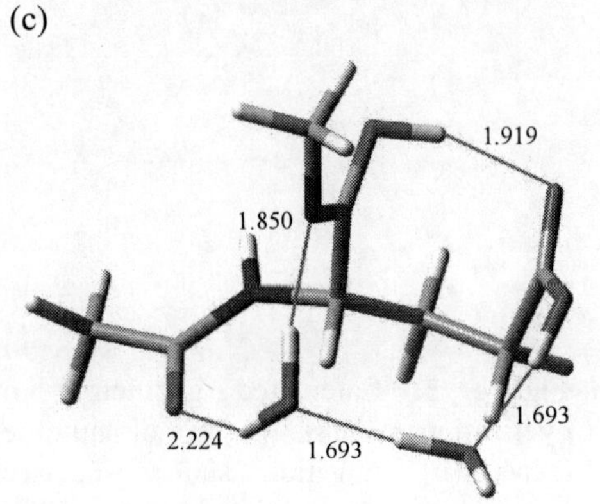

Figure 2. Calculated geometries of iminolization system: (a) Glu residue (Amide form) 1-AM, (b) transition state 1-TS, (c) iminol tautomer 1-IM. Selected interatomic distances are shown in Å.

3.2. Cyclization of iminol tautomer

Figure 3 shows the calculated geometries of the reaction system. The iminol tautomer 2-IM, which differs from 1-IM, forms by C_α–C bond rotation of the Glu-residue main chain and water-molecule migration from 1-IM (Figure 3a). The 2-IM complex is composed of a Glu residue (iminol tautomer) and two water molecules. In 2-IM, two water molecules connect iminol OH and the Glu side-chain carboxyl C=O by a hydrogen-bond bridge. One of the water molecules forms an additional hydrogen bond with the amide oxygen of the $(N-1)$ residue. In this process, nucleophilic attack by the $(N+1)$ residue iminol nitrogen and proton transfer from the Glu-residue iminol C–OH proton to the side-chain carboxyl group occurs by using two water molecules as catalyst, and the transition state 2-TS forms (Figure 3b). When 2-TS forms from 2-IM, the change in dihedral angles φ and χ_1 is small; however, ψ and χ_2 change by 82° and 63°, respectively (Table 1). The large conformational change in this process is attributed to the approach of the side-chain carboxyl group to the main chain, and the conformation of the main and side chains changes remarkably. In 2-TS, N–C bonds are being newly formed by nucleophilic attack of the $(N+1)$-residue iminol nitrogen and the Glu-residue side chain (the interatomic N–C distance is 1.690 Å). The calculated activation energy for the cyclization process is 30.3 kcal mol^{-1} when using 1-AM as energy reference. As the nucleophilic attack occurs by the iminol nitrogen and triple proton transfer mediated by two water molecules, cyclization completes and the tetrahedral *gem*-diol intermediate 2-TH forms (Figure 3c). When 2-TH forms from 2-TS, the change in all dihedral angles defined for the Glu residue is less than 10° (Table 1). In 2-TS, C–N bond forms by cyclization, and the interatomic distance shrinks to 1.480 Å. The calculated relative energy of 2-TH with respect to 1-AM is 9.65 kcal mol^{-1}.

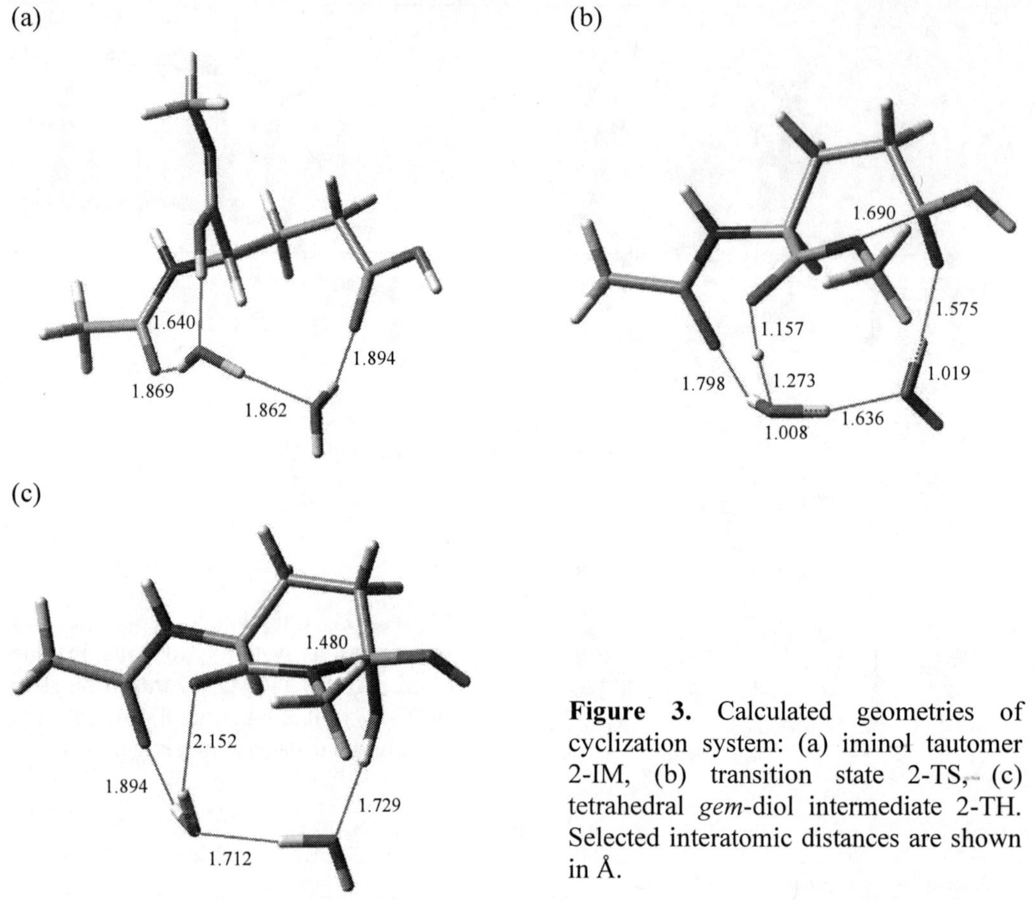

Figure 3. Calculated geometries of cyclization system: (a) iminol tautomer 2-IM, (b) transition state 2-TS, (c) tetrahedral *gem*-diol intermediate 2-TH. Selected interatomic distances are shown in Å.

3.3. Dehydration of tetrahedral intermediate

Figure 4 shows the calculated geometries of the reaction system. The tetrahedral *gem*-diol 3-TH forms from 2-TH by water migration (Figure 4a). The 3-TH complex is composed of a *gem*-diol intermediate and two water molecules. In 3-TH, two OH groups in *gem*-diol and two water molecules are connected by a hydrogen-bond bridge. In this process, triple proton transfer mediated by two water molecules occurs and one of the *gem*-diol C-O bonds is cleaved, and the transition state 3-TS forms (Figure 4b). When 3-TS forms from 3-TH, the change in all dihedral angles defined for the Glu residue is small (Table 1); however, the length of the C–O bond that was cleaved increases from 1.457 to 1.873 Å. The calculated activation energy for the dehydration process is 28.4 kcal mol^{-1} with respect to 1-AM. As triple-proton transfer and cleaving of one of the *gem*-diol C–O bonds proceeds, the Agl intermediate 3-GI yields (Figure 4c). When 3-GI forms from 3-TS, the *gem*-diol OH oxygen receives a proton from a water molecule and one water molecule is newly eliminated, accompanying the *gem*-diol C–O bond cleavage. Changes are small for the dihedral angles defined for the main and side chains of the Glu residues for 3-TH formation from 3-TS (Table 1). However, since the C_δ carbon transforms from sp^3 to sp^2 hybrization, the dihedral angle χ_2 changes by 32° over the dehydration process. The Agl-intermediate piperidinedione ring forms a boat-type conformation. The calculated relative energy of 3-GI with respect to 1-AM is −3.98 kcal mol^{-1}.

IOP Conf. Series: Journal of Physics: Conf. Series **1136** (2018) 012020 doi:10.1088/1742-6596/1136/1/012020

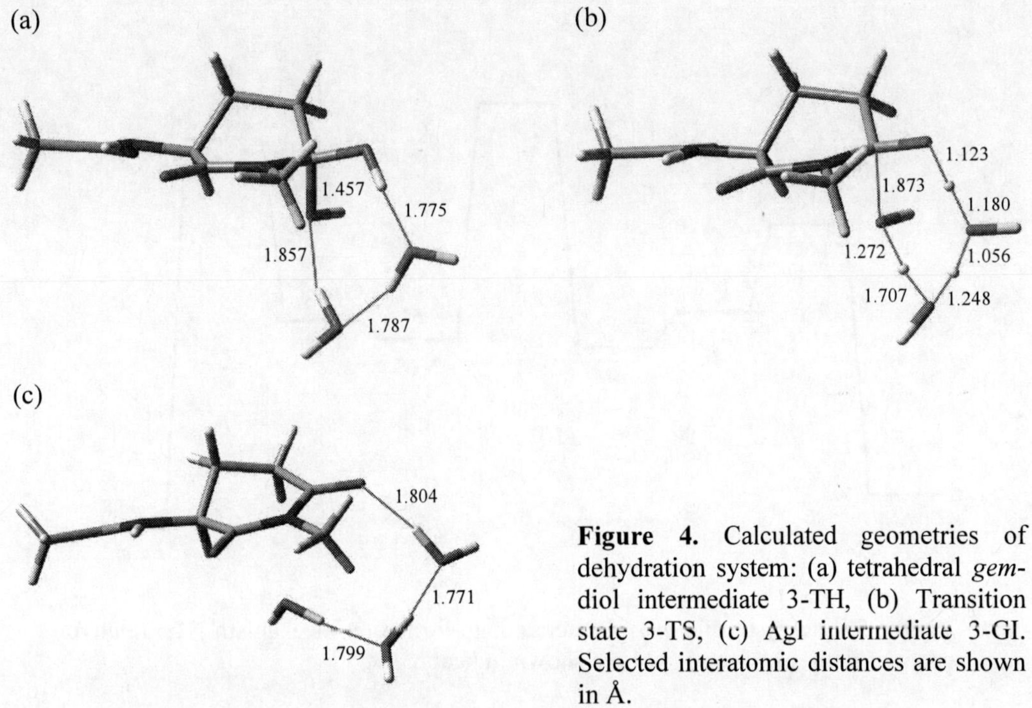

Figure 4. Calculated geometries of dehydration system: (a) tetrahedral *gem*-diol intermediate 3-TH, (b) Transition state 3-TS, (c) Agl intermediate 3-GI. Selected interatomic distances are shown in Å.

Table1. Dihedral angles φ, ψ, χ_1, and χ_2 (degrees) of optimized geometries for cyclization mechanism via iminol tautomer intermediate.

	φ	ψ	χ_1	χ_2
1-AM	−111	95	−160	−58
1-TS	−101	85	−161	−58
1-IM	−104	96	−165	−66
2-IM	−109	−80	178	65
2-TS	−91	−162	169	6
2-TH	−86	−171	172	−2
3-TH	−160	−170	177	−17
3-TS	−163	−167	180	−30
3-GI	−161	−165	−178	−49

4. Conclusion

In this study, we investigated the Agl-intermediate formation pathway from Glu residues as a part of Glu-residue stereoinversion. The reaction pathway analyses indicate that our presumed pathway, which includes the three steps of iminolization, cyclization, and dehydration, is appropriate for Agl formation. We show that Agl-intermediate formation can be potentially catalyzed by two water molecules, which act as proton-transfer mediators. Although Agl-intermediate formation consists of three steps, significant conformational changes of the main and side chains occur only in the cyclization process. Therefore, it is considered that Glu-residue stereoinversion only occurs in an environment where such a conformational change is possible. Figure 5 shows the entire energy profile throughout the iminolization-cyclization-dehydration process. The calculated activation energy required for Glu-residue cyclization is 30.3 kcal mol^{-1}; however, this value slightly exceeds the experimentally observed activation energy of Asp-residue stereoinversion, which is reported to be 25.7–29.0 kcal mol^{-1} [10]. The result of this calculation supports the fact that few reports have detected D-Glu residues.

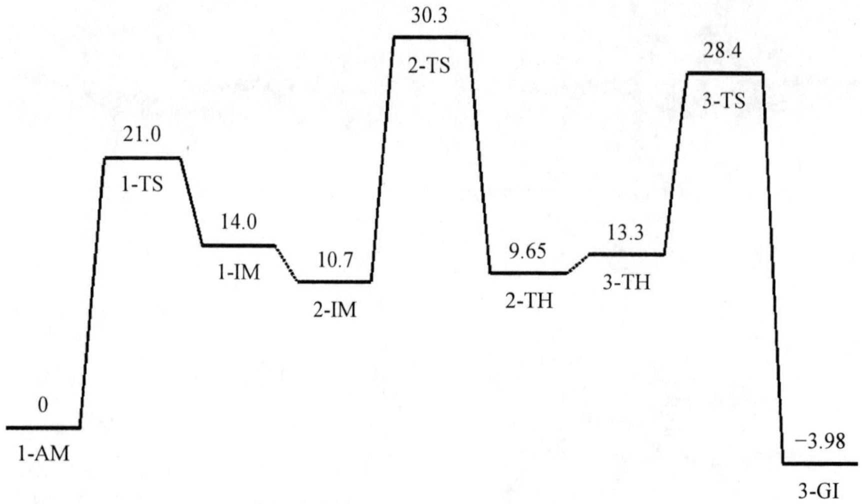

Figure 5. Energy profile of Agl-intermediate formation mechanism. The relative energies with respect to 1-AM are shown in kcal mol^{-1}.

Acknowledgments
This work was supported by grants-in-aid for scientific research [15H01064] and [17K08257] from the Japan Society for the Promotion of Science. The authors would like to thank Enago (www.enago.jp) for the English language review.

References

[1] Fujii N and Saito T 2004 Homochirality and life. *Chem. Rec.* **4** 267-278
[2] Fujii N 2002 D-amino acids in living higher organisms. *Orig. Life Evol. Biosph.* **32** 103-127
[3] Fujii N, Kaji Y and Fujii N 2011 D-amino acids in aged proteins: analysis and biological relevance. *J. Chromatogr. B.* **879** 3141-3147
[4] Fujii N, Kawaguchi T, Sasaki H and Fujii N 2011 Simultaneous stereoinversion and isomerization at the Asp-4 residue in βB2-crystallin from the aged human eye lenses. *Biochemistry* **50** 8628-8635
[5] Ritz-Timme S and Collins MJ 2002 Racemization of aspartic acid in human proteins. *Ageing. Res. Rev.* **1** 43-59
[6] Fujii N, Ishibashi Y, Sato K, Fujino M and Harada K 1994 Simultaneous racemization and isomerization at specific aspartic acid residues in αB-crystallin from the aged human lens. *Biochimi. Biophys. Acta.* **1204** 157-163
[7] Hooi MYS, Raftery MJ and Truscott RJW 2013 Interconversion of the peptide isoforms of aspartate: Stability of isoaspartates. *Mech. Aging. Dev.* **134** 103-109
[8] Oda A, Kobayashi K and Takahashi O 2010 Molecular-dynamics simulations for amyloid β_{1-42} monomer with D-aspartic acid residues using continuous solvent. *Chem & Biodivers.* **7** 1357-1363
[9] Sakaue H, Kinouchi T, Fujii N, Takata T and Fujii N 2017 Isomeric replacement of a single aspartic acid induces a marked change in protein function: the example of ribonuclease A. *ACS Omega* **2** 260-267
[10] Kuge K, Fujii N, Miura Y, Tajima S, Saito T 2004 Kinetic study of racemization of aspartyl residues in synthetic elastin peptides. *Amino Acids* **27** 193-197

[11] Takahashi O and Oda A 2012 Amide-iminol tautomerization of the C-terminal peptide groups of aspartic acid residues: two-water-assisted mechanism, cyclization from the iminol tautomer leading to the tetrahedral intermediate of succinimide formation, and implication to peptide group hydrogen exchange. In: Jones JE and Morano DM editor. *Tyrosine and Aspartic Acid: Properties, Source and Health Benefits.* pp 131-147

[12] Takahashi O, Kobayashi K and Oda A 2010 Modeling the enolization of succinimide derivatives, a key step of racemization of aspartic acid residues: importance of a Two-H_2O mechanism. *Chem & Biodivers.* **7** 1349-1356

[13] Takahashi O 2013 Two-water-assisted racemization of the succinimide intermediate formed in proteins. A computational model study. *Health* **5** 2018-2021

[14] Hooi MYS and Truscott RJW 2011 Racemisation and human cataract. D-Ser, D-Asp/Asn and D-Thr are higher in the lifelong proteins of cataract lenses than in age-matched normal lenses. *AGE* **33** 131-141

[15] Fukuyoshi S, Nakayoshi T, Takahashi O and Oda A 2017 Theoretical study on keto-enol tautomerization of glutamimide for exploration of the isomerization reaction pathway of glutamic acid in proteins using density functional theory. *Mol. Phys.* **115** 560-565

[16] Frisch MJ, Trucks GW, Schlegel HB, Scuseria GE, Robb MA, Cheeseman JR, Scalmani G, Barone V, Mennucci B, Petersson GA, Nakatsuji H, Caricato M, Li X, Hratchian HP, Izmaylov AF, Bloino J, Zheng G, Sonnenberg JL, Hada M, Ehara M, Toyota K, Fukuda R, Hasegawa J, Ishida M, Nakajima T, Honda Y, Kitao O, Nakai H, Vreven T, Montgomery Jr JA, Peralta JE, Ogliaro F, Bearpark M, Heyd JJ, Brothers E, Kudin KN, Staroverov VN, Keith T, Kobayashi R, Normand J, Raghavachari K, Rendell A, Burant JC, Iyengar SS, Tomasi J, Cossi M, Rega N, Millam JM, Klene M, Knox JE, Cross JB, Bakken V, Adamo C, Jaramillo J, Gomperts R, Stratmann RE, Yazyev O, Austin AJ, Cammi R, Pomelli C, Ochterski JW, Martin RL, Morokuma K, Zakrzewski VG, Voth GA, Salvador P, Dannenberg JJ, Dapprich S, Daniels AD, Farkas O, Foresman JB, Ortiz JV, Cioslowski J, Fox DJ. Gaussian 09, Revision C.01 (Gaussian, Inc., Wallingford, CT, 2010).

[17] Catak S, Monard G, Aviyente V, Ruiz-López MF 2009 Deamidation of Asparagine Residues: Direct Hydrolysis versus Succinimide-Mediated Deamidation Mechanisms. *J. Phys. Chem.* 113 1111-1120.

[18] Becke AD 1993 Density-functional thermochemistry. III. The role of exact exchange. *J. Chem. Phys.* **98**, 5648-5652.

IOP Conf. Series: Journal of Physics: Conf. Series **1136** (2018) 012021 doi:10.1088/1742-6596/1136/1/012021

Structural differences between the ligand-binding pockets of estrogen receptors alpha and beta

K Kato[1,2], K Fujii[3], T Nakayoshi[3], Y Watanabe[4], S Fukuyoshi[3], K Ohta[5], Y Endo[5], N Yamaotsu[6], S Hirono[6], E Kurimoto[1] and A Oda[1,3,7]

[1]Graduate School of Pharmacy, Meijo University, 150 Yagotoyama, Tempaku-ku, Nagoya, Aichi 468-8503, Japan
[2]Department of Pharmacy, Kinjo Gakuin University, 2-1723 Omori, Moriyama-ku, Nagoya, Aichi 463-8521, Japan
[3]Institute of Medical, Pharmaceutical and Health Sciences, Kanazawa University, Kakuma-machi, Kanazawa, Ishikawa 920-1192, Japan
[4]Faculty of Pharmacy, Showa University, 1-5-8 Hatanodai, Shinagawa-ku, Tokyo 142-8555, Japan
[5]Faculty of Pharmaceutical Sciences, Tohoku Medical and Pharmaceutical University, 4-4-1 Komatsushima, Aobaku, Sendai, Miyagi 981-8558, Japan
[6]School of Pharmacy, Kitasato University, 5-9-1 Shirokane, Minato-ku, Tokyo 108-8641, Japan
[7]Institute for Protein Research, Osaka University, 3-2 Yamadaoka, Suita, Osaka 565-0871, Japan

E-mail: oda@meijo-u.ac.jp

Abstract. The estrogen receptor (ER) is a member of the nuclear receptor superfamily and has two subtypes: ERα and ERβ. Inhibition of ERα is an effective therapeutic strategy in breast cancer. In contrast, ERβ is the presumed drug target of various autoimmune diseases. Many experimental structures of ERα/β have been reported; however, their structures vary due to ligand variability. Here, we performed structural bioinformatics studies for three-dimensional structures of ERs retrieved from the protein data bank (PDB) and clarify the detailed structural differences between ERα and ERβ. In total, 48 structures registered in the PDB were analyzed by HBOP and HBSITE, which we developed to identify ligand-binding cavities in proteins. PDB entries were clustered by number, shape, size, and location of the ligand-binding pockets. In addition, C-terminal domain shapes, which are divided into "agonist form" and "antagonist form," were also used for clustering. We classified 27 entries of ERα into five clusters and 21 entries of ERβ into seven clusters (a total of 12 clusters). Differences in the pockets and hydrogen bonds with ligands were observed between ERα and ERβ and occurred even within the same ER species. Therefore, we conclude that the structure of ERs is diverse and is affected by ligands.

1. Introduction

Estrogen receptors (ERs) are nuclear hormone receptors that mediate the physiological effects of estrogens. Two types of ER exist, ERα and ERβ, whose function and distribution differ from each

IOP Conf. Series: Journal of Physics: Conf. Series **1136** (2018) 012021 doi:10.1088/1742-6596/1136/1/012021

other [1–3]. ERα is mainly expressed in the uterus, prostatic stroma, ovarian theca cells, Leydig cells in the testis, epididymis, bone, breast, various regions of the brain, liver, and white adipose tissue. In contrast, ERβ is expressed in the colon, prostatic epithelium, testis, ovarian granulosa cell, bone marrow, salivary gland, vascular endothelium, and certain regions of the brain [3,4]. The transcriptional function of ERα includes involvement in osteoporosis, cardiovascular disease, and breast cancer. On the other hand, ERβ is the presumed drug target for various autoimmune diseases [5]. Therefore, ERβ-selective ligands are predicted to be an effective therapy for autoimmune diseases. Recently, we developed ERβ-selective compounds and clarified the structure–activity relationships between these ligands [6–8]. In these studies, several compounds, including carborane or bis (4-hydroxyphenyl) methane moieties as a hydrophobic core, were found to be selective ERβ ligands. To develop such ligands, it is essential to understand the detailed structural differences between ERα and ERβ. Despite the functional differences, ERα and ERβ share homology, and the degree of amino acid sequence similarity is approximately 97% in the DNA-binding domain and approximately 56% in the ligand-binding domain [9]. Actually, aligned amino acid sequences of ligand-binding domains indicates that those sequences are well conserved (Fig. 1). To date, many experimental structures have been determined; however, ligand-binding pocket structures are diverse because of the wide variety of ligands [10–12]. In addition, the shape of C-terminus is thought to be important for ER activation [13]. Ligand-based selectivity studies have been conducted; however, there is insufficient information to develop ER-selective ligands [14], and structure-based clustering for ER has not been performed. In this study, we classified ERα and ERβ according to number, shape, size, location of the ligand-binding pockets, and shape of the C-terminal domains. Our results indicate that ERs can form diverse structures and the structures are affected by ligands. Furthermore, this structural information could be beneficial for the development of ER subtype-selective ligands.

```
      292                 302                                                  333
ERα   W P S P L M I K R S K K N S L A L S L T A D Q M V S A L L D A E P P   I L Y S E Y D P
ERβ                       H A P R V R E L L L D A L S P E Q L V L T L L E A E P P H V L I S R       P
                          251                                                  285

      334                                351 353                               375
ERα   T R P F S E A S M M G L L T N L A D R E L V H M I N W A K R V P G F V D L T L H D Q V
ERβ   S A P F T E A S M M M S L T K L A D K E L V H M I S W A K K I P G F V E L S L F D Q V
      286                                303 305                               328

      377             387       394                                            419
ERα   H L L E C A W L E I L M I G L V W R S M E H P G K L L F A P N L L L D R N Q G K C V E
ERβ   R L L E S C W M E V L M M G L M W R S I D H P G K L I F A P D L V L D R D E G K C V E
      329             339       346                                            371

      420                                                                      462
ERα   G M V E I F D M L L A T S S R F R M M N L Q G E E F V C L K S I I L L N S G V Y T F L
ERβ   G I L E I F D M L L A T T S R F R E L K L Q H K E Y L C V K A M I L L N S S M Y P L V
      372                                                                      414

      463                                                                      505
ERα   S S T L K S L E E K D H I H R V L D K I T D T L I H L M A K A G L T L Q Q Q H Q R L A
ERβ   T A T Q D A D S S R K L A H   L L N A V T D A L V W V I A K S G I S S Q Q Q S M R L A
      415                                                                      456

      506                       521   524                                      548
ERα   Q L L L I L S H I R H M S N K G M E H L Y S M K C K N V V P L Y D L L L E M L D A H R
ERβ   N L L M L L S H V R H A S N K G M E H L L N M K C K N V V P V Y D L L L E M L N A H V
      457                       472   475                                      499

      549       554
ERα   L H A P T S
ERβ   L
      500
```

Figure 1. Alignment of amino acid sequences of ERα and ERβ. The sequences of ligand-binding domain of ERs aligned using BLAST are shown. Black square indicates identity region.

IOP Conf. Series: Journal of Physics: Conf. Series **1136** (2018) 012021 doi:10.1088/1742-6596/1136/1/012021

2. Methods

To compare the three-dimensional structures of ERs, we retrieved 48 crystal structures, comprising 27 entries for ERα and 21 entries for ERβ, from the protein data bank (PDB; Tables 1 and 2). The HETATM, such as water molecules and ligands, in these entries were deleted for analysis. Ligand-binding cavities of ERs were identified using HBOP and HBSITE programs [15,16], from which hydrophobic potentials [17] were calculated around the proteins and the hydrophobic areas were detected as tentative binding pockets. The number, shape, size, and location of the identified cavities were compared among the PDB structures, and the structural features of the ligand-binding pockets were used to elucidate the differences between ERα and ERβ. To investigate ligand interactions for ERα and ERβ, we analyzed the amino acid residues forming hydrogen bonds with ligands in the PDB structures. In addition, ER structures were clustered according to the C-terminal structures important for transcriptional activity.

Table 1. Retrieved ERα structures from the PDB

1A52	1GWR	1L21	1XP9
1QKT	1SJ0	1XQC	2AYR
2OUZ	2POG	2Q70	2R6W
2R6Y	3DT3	3ERT	1G50
1ERE	1GWQ	3ERD	1PCG
1QKU	1XP1	1XP6	1XPC
1ERR	1R5K	1UOM	

Table 2. Retrieved ERβ structures from the PDB

1NDE	1U3Q	1U9E	1U3S
1X7B	2JJ3	2J7X	1U3R
1QKM	1QKN	1X7J	1X76
1X78	1YY4	1YYE	2GIU
2NV7	1L2J	2J7Y	2QTU
2Z4B			

3. Results and Discussion

3.1. Pocket number and size of ERs

The ER pockets identified using HBOP and HBSITE are shown in Tables 3 and 4. The hydrophobic pockets detected by HBOP and HBSITE are "HBSx," in which "x" is the rank of pockets ordered by the degree of hydrophobicity. In the table, the volume of pocket is written in Å^3. Some pockets detected by HBOP and HBSITE were connected and formed one pocket; therefore, they were defined as a single pocket and indicated by same mark († or ‡). The under bar indicates the pockets in which ligands were actually bound in the PDB structures. In this study, "ligand-binding pockets" refers to the tentative binding pockets detected by HBOP and HBSITE, and the actual pocket in which the ligand was experimentally observed is defined as the "actual pocket." In Tables 3 and 4, HBSx is the ligand-binding pocket, and the underlined pocket is the actual pocket. For ERα, 19 entries had three ligand-binding pockets, seven entries had four, and another one had five. The minimum volume of the actual

IOP Conf. Series: Journal of Physics: Conf. Series **1136** (2018) 012021 doi:10.1088/1742-6596/1136/1/012021

pocket was 152 Å^3, and the maximum volume was 238 Å^3 (average = 205 Å^3). Similarly, for ERβ, three to five ligand-binding pockets were detected. Twelve entries had three ligand-binding pockets, seven entries had four, and two entries had five. The minimum volume of the actual pocket was 115 Å^3 and the maximum volume was 245 Å^3 (average = 165.8 Å^3). The standard deviation of the actual pocket volume of ERα and ERβ was 27.75 and 31.11, respectively. These results indicate that the cavities of ERs can form diverse shape by ligand, and ERs have some cavities besides the actual pocket. Even within the same ER subtype, the number and size of the cavities were different in the PDB structures. Such differences within the same subtypes may be affected by the ligands.

Table 3. Number and size of pockets in ERα[a, b, c]

PDB ID	HBS1	HBS2	HBS3	HBS4	HBS5	No. of Pockets
1A52	168	210†	61	20†	—	3
1ERE	192	179	10†	13†	—	3
1ERR	128	195	25	—	—	3
1G50	178	169	50	11	10	5
1GWQ	152	205	25	—	—	3
1GWR	185	149	13	27	—	4
1L2I	205	153	14	—	—	3
1PCG	182	156	32	—	—	3
1QKT	138	237	48	—	—	3
1QKU	188	24†	78†	29‡	13‡	3
1R5K	167	217	48	24	—	4
1SJ0	147	222	39	—	—	3
1UOM	137	201	46	13	—	4
1XP1	141	233	42	16	—	4
1XP6	145	238	36	—	—	3
1XP9	133	217	23	10	—	4
1XPC	137†	221	35	10†	10	4
1XQC	120	186	56	—	—	3
2AYR	139	226	44	—	—	3
2OUZ	130	200	44	10	—	4
2POG	149	199	46	—	—	3
2Q70	140	216	43	—	—	3
2R6W	126	210	14†	30†	10	4
2R6Y	143	199	47	—	—	3
3DT3	140	237	27	—	—	3
3ERD	163	134†	10‡	10†	26‡	3
3ERT	151	220	50	—	—	3

[a] The volume of the detected pockets is shown in Å^3.
[b] Pockets with the same marks († or ‡) were defined as a single pocket because they were connected and formed one pocket.
[c] Underline indicates the actual pocket in the PDB structure.

Table 4. Number and size of pockets in ERβ[a,b,c]

PDB ID	HBS1	HBS2	HBS3	HBS4	HBS5	HBS6	No. of Pockets
1L2J	157	237	10	49†	70†	22†	4
1NDE	156	245	27	16†	13	14†	5
1QKM	134	115	15	12	—	—	4
1QKN	154	188†	52†	23	—	—	3
1U3Q	135	92	11	10	—	—	4
1U3R	160	111	15	—	—	—	3
1U3S	160	121	11	—	—	—	3
1U9E	148	124	23	—	—	—	3
1X7B	138	139	45	—	—	—	3
1X7J	148	133	40	—	—	—	3
1X76	145	130	41	—	—	—	3
1X78	144	124	34	—	—	—	3
1YY4	166	107	25	13	—	—	4
1YYE	142	127	55	—	—	—	3
2GIU	125†	192	17	19	12†	—	4
2J7X	169	153	23	—	—	—	3
2J7Y	156	155	18	—	—	—	3
2JJ3	168	151	18†	28	14†	—	4
2NV7	148	130	29	—	—	—	3
2QTU	163	167	20†	28	21†	—	4
2Z4B	139	158	16	25	13	—	5

[a] The volumes of the detected pockets are shown in Å^3.
[b] Pockets with same marks († or ‡) were defined as a single pocket because they were connected and formed one pocket.
[c] Underline indicates the actual pocket in the PDB structure.

We classified the PDB entries by number, shape, size, and location of the ligand-binding pockets. As shown in Figs. 2 and 3, ERα and ERβ were clustered into five and seven groups, respectively. Therefore, all ER structures were classified into 12 groups. For ERα, the entries included in groups A and B had three ligand-binding pockets, the entries in groups C and D had four, and one entry in group E had five. Furthermore, the volumes of the actual pockets were different; that of group B was larger than that of group A, and that of group C was larger than that of group D. For ERβ, the entries included in groups F and G had three ligand-binding pockets, the entries in groups H, I, and J had four, and the entries in groups K and L had five. The shapes of the ligand-binding pockets were different between groups F and G, and were also different between groups H, I, and J. The volumes of the actual pockets differed between groups K and L (K > L). There were large differences in shape and number of cavities not only between different subtypes but also between entries within the same subtypes. Although entries with similar ligands tend to have similar structures of actual pockets, some exceptions were found. For example, 1ERE and 1QKT which have same ligand (estrogen) were classified into groups A and B, respectively. These exceptions might be caused by the flexibility of actual pocket, but the structural flexibility cannot be evaluated by X-ray structures. We will investigate the structural flexibility of ERs using molecular dynamics simulations. Although few atypical forms of ligand-binding pockets were observed, the shape and size of cavities for ERs mainly depend on ligands.

group	(a)	(b)	PDB ID	No. of pockets
A		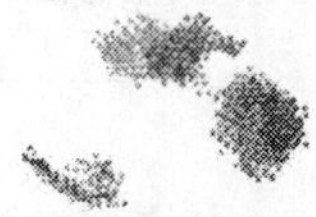	1ERE, 1GWQ, 1L2I, 1PCG, 1QKU, 3ERD	3
B	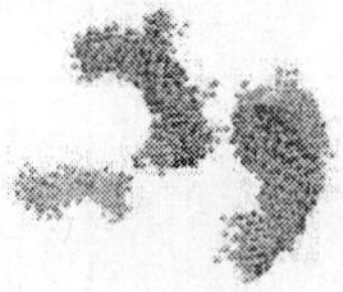		1A52, 1ERR, 1QKT, 1SJ0, 1XP6, 1XQC, 2AYR, 2POG, 2Q70, 2R6Y, 3DT3, 3ERT	3
C		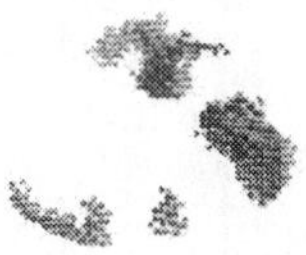	1R5K, 1UOM, 1XP1, 1XP9, 1XPC, 2OUZ, 2R6W	4
D			1GWR	4
E			1G50	5

Figure 2. Clusters of ERα classified by their pockets. Both (a) and (b) show grid points with high hydrophobic potentials conducted by HBOP and HBSITE analysis and were identified as ligand-binding pockets. (b) shows the view from underneath (a).

3.2. Hydrogen bonding between ERs and ligands

The specific recognition between protein and ligand mainly depends on hydrogen bonds. We investigated hydrogen bonds between ERs and ligands to elucidate ligand-recognition mechanisms for development of subtype-selective ligands. In ERα, Asp351, Glu353, Leu387, Arg394, Gly521, and His524 form hydrogen bonds with ligands (Fig. 4 and Table 5). In particular, Glu353, Arg394, and His524 are considered to be important residues for ligand binding because hydrogen bonds between

group	(a)	(b)	PDB ID	No. of pockets
F			1X78, 1U3R, 1U3S, 1U9E, 1X7B, 1X7J, 1X76, 1YYE, 2J7X, 2J7Y, 2NV7	3
G			1QKN	3
H	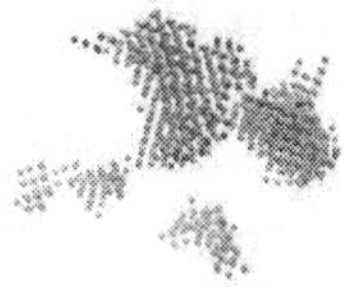		1QKM, 1YY4, 2GIU, 2JJ3, 2QTU	4
I	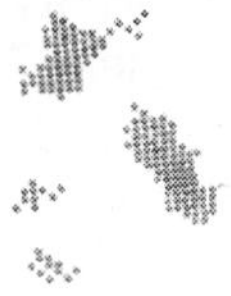		1U3Q	4
J			1L2J	4
K			1NDE	5
L			2Z4B	5

Figure 3. Clusters of ERβ classified by their pockets. Both (a) and (b) are grid points with high hydrophobic potentials conducted by HBOP and HBSITE and were identified as ligand-binding pockets. (b) shows a view from underneath (a).

these residues and ligands are observed in the majority of ERα structures. These residues were previously suggested to be important for ligand binding [18,19]. Hydrogen bond formation between residues and ligands in ERβ are described in Fig. 5 and Table 6. Asp351, Glu353, Leu387, Arg394,

IOP Conf. Series: Journal of Physics: Conf. Series **1136** (2018) 012021 doi:10.1088/1742-6596/1136/1/012021

Gly521 and His524 in ERα correspond to Asp303, Glu305, Leu339, Arg346, Gly472 and His475, respectively. Although no interaction was observed at Leu339, other residues corresponding to ERα residues formed hydrogen bonds with ligands. Hydrogen bond formation at Glu305, Arg346, and His475 was frequently observed as same as corresponding residues of ERα. Since those hydrogen-bonded residues are conserved among ERs, these are considered to be particularly important biologically. However, only half the expected bonds at Arg346 were formed. Furthermore, in ERα, Asp351 formed hydrogen bonds in 15 of the 27 entries, whereas in ERβ, Asp303 formed hydrogen bonds in only two entries. Therefore, the major differences in residues forming hydrogen-bond with ligands are aspartate (351 in ERα and 303 in ERβ) and Arginine (394 in ERα and 346 in ERβ). ERα had more entries forming hydrogen bond in both residues. These differences could be useful for the development of subtype-selective ligands.

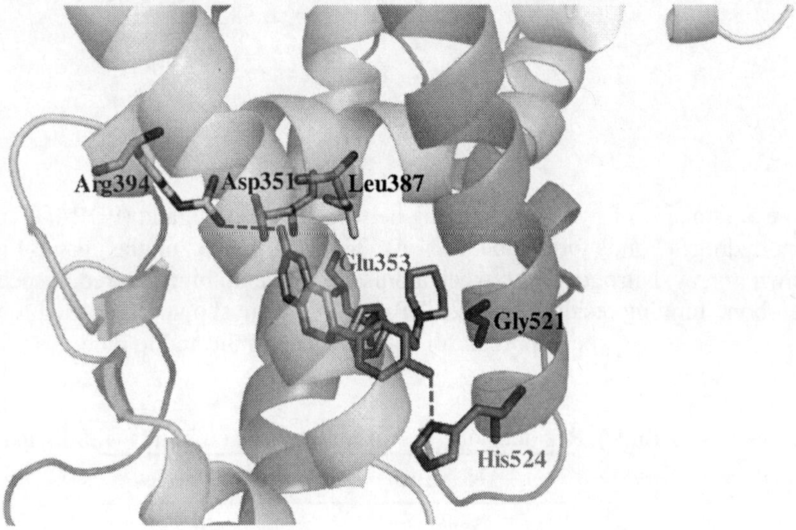

Figure 4. Example of hydrogen bonding between ERα and ligand (PDB ID: 1SJ0). Red lines represent hydrogen bonds formed between Arg394 and His524 and the ligand. Ligand carbon atoms are shown in gray. Nitrogen and oxygen atoms are shown in blue and red, respectively. Conserved hydrogen-bond forming residues between ER α and ERβ are shown in red, and residues considered to be important for selectivity are indicated in blue.

Table 5. Residues of ERα forming hydrogen bonds with ligands

Residue	No. of entries [a]
Asp351[b]	15
Glu353[b]	26
Leu387	1
Arg394[b]	25
Gly521	6
His524[b]	20

[a] 27 entries in total

[b] Conserved hydrogen-bond forming residues between ERα and ERβ are underlined by solid line, and residues considered to be important for selectivity are underlined by dotted line.

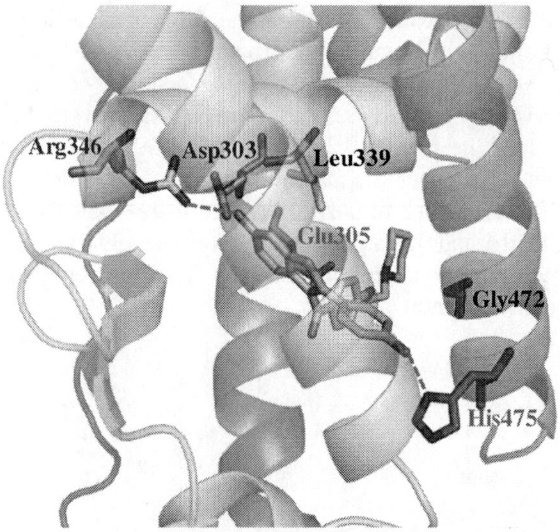

Figure 5. Example of hydrogen bonding between ERβ and ligand (PDB ID: 1QKN). Red lines represent hydrogen bonds formed between Arg346 and His475 and the ligand. Ligand carbon atoms are shown in gray. Nitrogen and oxygen atoms are shown in blue and red, respectively. Conserved hydrogen-bond forming residues between ERα and ERβ are shown in red, and residues considered to be important for selectivity are indicated in blue.

Table 6. Residues of ERβ forming hydrogen bonds with ligands

Residue	No. of entries [a]
Asp303[b]	2
Glu305[b]	21
Leu339	0
Arg346[b]	10
Gly472	1
His475[b]	18

[a] 21 entries in total.
[b] Conserved hydrogen-bond forming residues between ERα and ERβ are underlined by solid line, and residues considered to be important for selectivity are underlined by dotted line.

3.3. C-terminal structure of ERs

We focused on the C-terminal structure of ERs because the agonist and antagonist forms were observed in the C-termini of the experimental structures [13]. Fig. 6 shows examples of ER C-terminal structures shown as top and side views. The ER C-terminal region formed a helix structure that is exposed to solvent in the inactive form (Fig. 6A). Agonist binding causes C-terminal helix modification and decreases exposure to the solvent (Fig. 6B). One ERα, PDB ID: 1A52, showed a fully-extended form (Fig. 6C), which may have been an artifact. The agonist-binding crystal structures were in their active forms, and antagonist binding was in its inactive form (Table 7). Partial agonist-binding structures almost establish inactive forms. Whether ERs form active or inactive forms depends on ligands; therefore, structural information is useful for agonist/antagonist design. In addition, because the partial agonists could be candidates for tissue-selective ER modulators, detailed research for inactive forms, such as hydrogen bond analyses in this study, may assist molecular targeted drug design for ERs to reduce side effects.

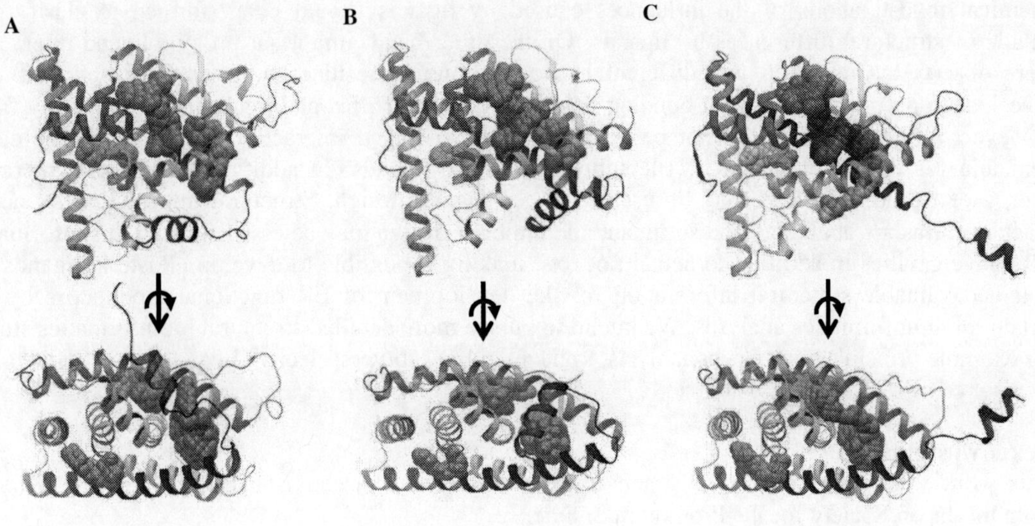

Figure 6. Examples of inactive, active, and fully-extended forms are shown as top and side views. The C-terminal helix is show in purple. (A) Inactive form (PDB ID: 1ERR), (B) active form (PDB ID: 1ERE), and (C) fully-extended form (PDB ID: 1A52).

Table 7. Number of active, inactive, and fully-extended forms

Subtype	Ligand	Active form	Inactive form	Extended form
	Agonist	7	0	1
ERα	Antagonist	0	6	0
	Partial agonist	1	12	0
	Agonist	13	2	0
ERβ	Antagonist	0	2	0
	Partial agonist	1	3	0

IOP Conf. Series: Journal of Physics: Conf. Series **1136** (2018) 012021 doi:10.1088/1742-6596/1136/1/012021

4. Conclusion

In this study, we clarified the structural differences between ERα and ERβ crystal structures registered in the PDB. The ligand-binding pockets of ERs are diverse and can be changed by ligands; therefore, more comprehensive structural analyses such as molecular dynamics simulations are required for structure-based drug design (SBDD). In addition, those ligand-free structures have not been determined. It is essential for investigation of ligand-independent ERs structure and conformation change by ligand binding. We will analyze the conformation change using molecular dynamics simulations in the near future. This analysis may reveal important residue for selectivity. Furthermore, chemical investigations of the influences caused by ligands should be performed to clarify more details of structural differences by ligands. On the other hand, important protein–ligand interactions were observed. Conserved and different hydrogen bonding residues between ERα and ERβ were revealed. In particular, hydrogen bonding between Asp351 of ERα and ligands was frequently found; however, Asp303 of ERβ did not participate greatly in ligand interactions. This difference can be available for the development of ER subtype-selective ligands. In addition to this, the interaction energy of ligands and ERs may be useful for new ligand design. Structural information on active–inactive forms are abundant, and stimulant and inhibitor design may be assisted using this information. ERs have cavities in addition to actual pockets, making it possible to develop allosteric ligands. We obtained valuable structural information for the development of ER functional modulators by only structural bioinformatics analyses. We intend to pursue more detailed structural bioinformatics studies, for example principle component analysis, to develop ER subtype-selective ligands using SBDD.

Acknowledgments

This work was supported by the grants-in-aid for scientific research [15H01064] and [17K08257] from the Japan Society for the Promotion of Science.

References

[1] Kuiper G G, Enmark E, Pelto-Huikko M, Nilsson S and Gustafsson J A 1996 Cloning of a novel receptor expressed in rat prostate and ovary *Proc. Natl. Acad. Sci. U. S. A.* **93** 5925–5930

[2] Minutolo F, Macchia M, Katzenellenbogen B S and Katzenellenbogen J A 2011 Estrogen receptor β ligands: Recent advances and biomedical applications *Med. Res. Rev.* **31** 364–442

[3] Couse J F, Lindzey J, Grandien K, Gustafsson J A and Korach K S 1997 Tissue distribution and quantitative analysis of estrogen receptor-α (ERα) and estrogen receptor-β (ERβ) messenger ribonucleic acid in the wild-type and ERα-knockout mouse *Endocrinology* **138** 4613–4621

[4] Dahlman-Wright K, Cavailles V, Fuqua S A, Jordan V C, Katzenellenbogen J A, Korach K S, Maggi A, Muramatsu M, Parker M G and Gustafsson J A 2006 International union of pharmacology. LXIV. Estrogen receptors, *Phamacol. Rev.*, **58** 773–781

[5] Shim G J, Wang L, Andersson S, Nagy N, Kis L L, Zhang Q, Mäkelä S, Warner M and Gustafsson J A 2003 Disruption of the estrogen receptor β gene in mice causes myeloproliferative 71 disease resembling chronic myeloid leukemia with lymphoid blast crisis *Proc. Natl. Acad. Sci. U. S. A.* **100** 6694–6699

[6] Ohta K, Ogawa T, Kaise A, Oda A and Endo Y 2014 Aliphatic Substitution of o-Carboranyl Phenols Enhances Estrogen Receptor Beta Selectivity *Chem. Pharm. Bull.* **62** 386–391

[7] Ohta K, Ogawa T, Oda A, Kaise A and Endo Y 2015 Design and synthesis of carborane-containing estrogen receptor-beta (ERβ)-selective ligands *Bioorg. Med. Chem. Lett.* **25** 4174–4178

[8] Kojima T, Ogawa T, Kitao S, Sato M, Oda A, Ohta K and Endo Y 2015 Estrogenic activity of bis(4-hydroxyphenyl)methanes with cyclic hydrophobic structure *Bioorg. Med. Chem.* **23** 6900–6911

[9] Ogawa S, Inoue S, Watanabe T, Hiroi H, Orino A, Hosoi T, Ouchi Y and Muramatsu M 1998 The complete primary structure of human estrogen receptor β (hERβ) and its heterodimerization with ERα in vivo and in vitro *Biochem. Biophys. Res. Commun.* **243** 122–126

[10] Tanenbaum D M, Wang Y, Williams S P and Sigler P B 1998 Crystallographic comparison of the estrogen and progesterone receptor's ligand binding domains *Proc. Natl. Acad. Sci. U. S. A.* **95** 5998–6003

[11] Anstead G M, Carlson K E and Katzenellenbogen J A 1997 The estradiol pharmacophore: Ligand structureestrogen receptor binding affinity relationships and a mode for the receptor binding site *Steroids* **62** 268–303

[12] Pike A C, Brzozowski A M, Hubbard R E, Bonn T, Thorsell A G, Engström O, Ljunggren J, Gustafsson J A and arlquist M 1999 Structure of the ligand-binding domain of oestrogen receptor beta in the presence of a partial agonist and a full antagonist *Embo. J.* **18** 4608–4618

[13] Brzozowski A M, Pike A C, Dauter Z, Hubbard R E, Bonn T, Engström O, Ohman L, Greene G L, Gustafsson J A and Carlquist M 1997 Molecular basis of agonism and antagonism in the oestrogen receptor *Nature* **389** 753–758

[14] Huang W, Wei W, Yang Y, Zhang T and Shen Z 2015 Discovery of novel selective ERα/ERβ ligands by multi-pharmacophore modeling and virtual screening *Chem. Pharm. Bull.* **63** 780–791

[15] Yamaotsu N, Oda A and Hirono S 2008 Determination of ligand-binding sites on proteins using long-range hydrophobic potential *Biol. Pharm. Bull.* **31** 1552–1558.

[16] Oda A, Yamaotsu N and Hirono S 2009 Evaluation of the searching abilities of HBOP and HBSITE for binding pocket detection *J. Comput. Chem.* **30** 2728–2737

[17] Israelachvili J N and Pashley R M 1982 The hydrophobic interaction is long range, decaying exponentially with distance *Nature* **300** 341–342

[18] Ekena K, Katzenellenbogen J A and Katzenellenbogen B S 1998 Determinants of ligand specificity of estrogen receptor-α: Estrogen versus androgen discrimination *J. Biol. Chem.* **273** 693–699

[19] Watanabe C, Fukuzawa K, Tanaka S and Aida-Hyugaji S 2014 Charge clamps of lysines and hydrogen bonds play key roles in the mechanism to fix helix 12 in the agonist and antagonist positions of estrogen receptor α: intramolecular interactions studied by the ab initio fragment molecular orbital method *J. Phys. Chem.* **118** 4993–5008

IOP Conf. Series: Journal of Physics: Conf. Series **1136** (2018) 012022 doi:10.1088/1742-6596/1136/1/012022

Molecular dynamics study of binary POPC bilayers: molecular condensing effects on membrane structure and dynamics

Hiroaki Saito[1,*], Tetsuya Morishita[2], Taku Mizukami[3], Ken-ichi Nishiyama[4], Kazutomo Kawaguchi[5], and Hidemi Nagao[5]

[1]RIKEN, Quantitative Biology Center, Suita, Japan,
[2]CD-FMat, and MathAM-OIL, National Institute of Advanced Industrial Science and Technology (AIST), Tsukuba, Japan,
[3]Japan Advanced Institute of Science and Technology (JAIST), Nomi, Japan,
[4]Cryobiofrontier Research Center, Faculty of Agriculture, Iwate University, Morioka, Japan,
[5]Institute of Science and Engineering Kanazawa University, Kanazawa, Japan.

[*]hiroaki.saito@riken.jp

Abstract. Molecular dynamics (MD) simulations of binary 1-palmitoyl-2-oleoyl-*sn*-glycero-3-phosphocholine (POPC) bilayers containing cholesterol (CHOL), ceramide (CER), diacylglycerol (DAG), or sphingomyelin (SM) were carried out to investigate effects of these molecules on the structure and dynamics of membranes. Clear condensing effects were observed in all POPC binary bilayers, and those molecular condensing effects were ranked in the order of POPC/CHOL, POPC/CER, POPC/DAG, and POPC/SM bilayer. The differences of condensing effects are attributable to the factors such as molecular shapes, type of polar head groups and hydrogen bond networks in the binary lipid bilayers. The rigid sterol ring of CHOL is due to the ordering of acyl chains of POPC in the membrane. The lack of bulky PC head group of lipids results in the lipids condition in which DAG, CER and CHOL are buried in the hydrophobic region of the bilayer, leading to the condensed membranes. CHOL and CER are closely arranged in each binary bilayer due to the better intermolecular affinities of these lipids. The hydrogen bond network can also be a factor for the condensing of membranes and the decrease of lateral diffusion of lipids in the binary lipid bilayers.

1. Introduction

Addition of different molecules to membranes largely changes the structural and dynamical properties of the membranes, and these changes alter the functions of membranes such as signaling and molecular translocation across the membranes. Especially cholesterol (CHOL) makes membrane dense and changes the membrane state to the liquid-ordered phase above 20-30 mol% of cholesterol [1,

2]. It has been known that such molecular condensing effects also occur by adding other lipids such as ceramide (CER) [3-5], diacylglycerol (DAG) [6, 7] and sphingomyelin (SM)[8, 9]. However, the differences of molecular condensing effects by these lipids on the lipid structure and dynamics, and the molecular mechanisms in membranes have not been well known. In this study, we thus carried out molecular dynamics (MD) simulations of binary 1-palmitoyl-2-oleoyl-*sn*-glycero-3-phosphocholine (POPC) bilayers containing the CHOL, CER, DAG, or SM, and systematically investigated the effects of these lipids on structure and dynamics of the membranes. We then discussed the factors to yield the differences of membrane condensing in the binary lipid bilayers.

2. Material and method

2.1. Molecular structures of lipids

Five kinds of lipids, POPC, DAG, SM, CER, and CHOL were used in the MD study. The binary lipid bilayers consist of the POPC and another lipid (i.e. DAG, SM, CER, or CHOL). Figure 1 shows the molecular structures of the lipids used in the study. POPC has a 16:0 palmitic acid (PA) and an 18:1 oleic acid (OA) linking to the glycerol backbone and has a phosphorylcholine (PC) head group. DAG has a hydroxyl group instead of the PC head group of the POPC. SM has an 18:1 sphingosine (SPH) and a PA with an amide and a hydroxyl group at the linkage of the PA and SPH. CER has a hydroxyl group instead of the PC head group of the SM. CHOL has a rigid sterol ring linking to a hydroxyl group and has a short hydrocarbon chain. DAG, SM, CER and CHOL have some hydroxyl or amide groups around the head or linkage to the acyl chains of the lipids, and these can be some donor/acceptor sites to form hydrogen bonds with surrounding POPC or themselves in the membrane.

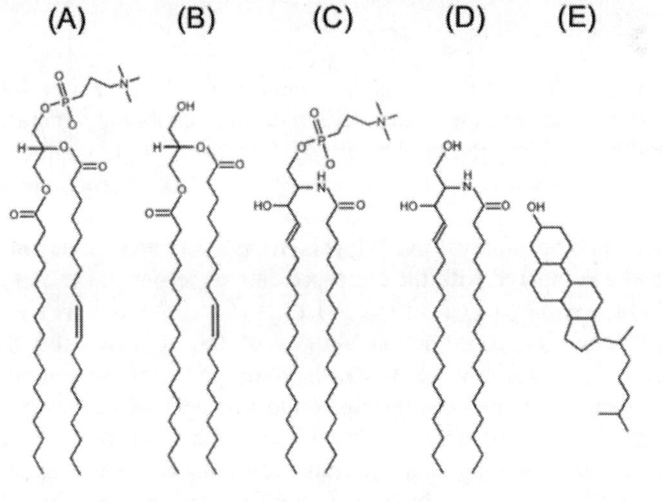

Figure 1. Molecular structures of (A) 1-palmitoyl-2-oleoyl-*sn*-glycero-3-phosphocholine (POPC), (B) diacylglycerol (DAG), (C) sphingomyelin (SM), (D) ceramide (CER), and (E) cholesterol (CHOL).

2.2. Molecular dynamics simulations

Initial bilayer structures of a pure POPC and four binary POPC mixtures were prepared for MD simulations. The pure POPC bilayer was used as a reference system. Each system consists of 128 lipids and 8,192 water molecules. 64 lipids and 4,096 water molecules were packed into each half region of the MD box. In the binary POPC bilayer, thirteen POPC were replaced to the other lipids (CHOL, DAG, CER, or SM) in each leaflet of the bilayer, corresponding to about 20 mol % of molecular addition. The initial coordinates of each system were prepared by CAHRMM-GUI [10].

MD simulations of the pure POPC bilayer and four binary POPC bilayers (POPC/CHOL, POPC/DAG, POPC/DAG and POPC/SM) were run under the constant temperature ($T = 303.15$ K) and pressure ($P = 1$ atm) condition for 200 ns. Nose-Hoover thermostat [11, 12] and Parinnelo-Rahman type borostat [13] were adopted to control the system temperature and the pressure. The periodic lengths of thermostat and barostat were both 1 ps. CHARMM36 force field [14, 15] and TIP3P water

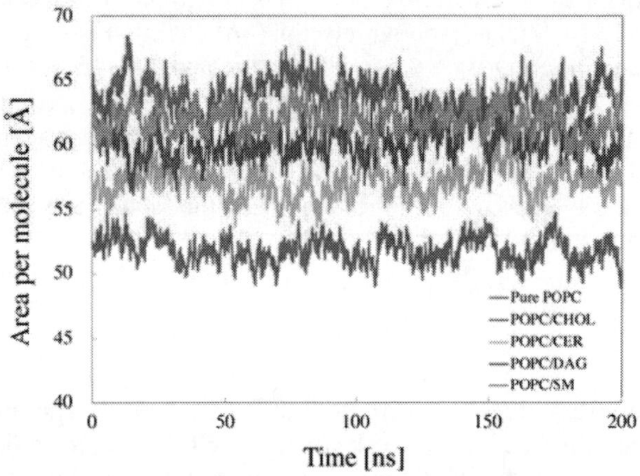

Figure 2. Area per molecule as a function of MD time for the pure POPC bilayer and the binary POPC bilayers (POPC/CHOL, POPC/CER, POPC/DAG, and POPC/SM).

model [16] were used for lipids and water molecule, respectively. Cutoff for the van der Waals (vdW) interaction was applied using a switching scheme, within a range of radius from 10 to 12 Å. Particle mesh Ewald (PME) method [17] was adopted for the calculation of the coulomb electrostatic interactions. The time step for numerical integration of equations of motion was 2 fs, and the coordinates of the systems were stored every 1 ps. Figure 2 shows the area per molecule as a function of MD time for each system. The plots show that all molecular areas are sufficiently equilibrated after 100 ns. Thus, the last 100 ns of the MD runs were used for the analysis. All MD simulations were done by GROMACS 5.1.2 [18].

2.3. *Analysis*

Time averaged area per molecule A_{mol}, membrane thickness $d_{P\text{-}P}$, and order parameter $-S_{CD}$ for each system were calculated to investigate the effects of molecular addition on the membrane structure. A_{mol} at each MD step was simply estimated by dividing the *xy* area of MD box by the lipid number in the leaflet of bilayer. The values of $d_{P\text{-}P}$ were calculated as the time averaged distances between positions of the phosphorus atoms in each leaflet of the bilayer. The order parameters $-S_{CD}$ of two acyl chains (*sn*-1 and *sn*-2) of POPC in the pure and the binary lipid bilayers were estimated. Some of these parameters on membrane structure can be compared with the corresponding experimental values.

The average number of hydrogen atoms within 3Å cutoff length from the acceptor site for hydrogen bonds of the lipid was defined as a hydrogen coordination number of the site, and this can be calculated by the radial distribution function (RDF). In this study, the hydrogen coordination numbers for all possible acceptors of the lipids were calculated to investigate the hydrogen bond networks and the stabilities of the hydrogen bonds in the membrane. The total number of these hydrogen coordination numbers, which is calculated by summing up all coordination numbers, was evaluated for each binary bilayer and compared each other. The total hydrogen coordination numbers for between additional molecule and POPC and for between additional molecules were also evaluated to investigate the hydrogen bond networks between those lipids in the membrane.

Some experimental observations have shown that the addition of CHOL/CER to lipid bilayer systems causes formation of CHOL/CER enriched micro-domains in the membrane [1-5]. This is because these lipids have better affinity with them than the lipids of host membrane, and this difference of molecular interactions could result in the ununiform distribution of CHOL/CER in the binary POPC bilayer plane. Two-dimensional RDF for the centers of mass of additional lipids was calculated to assess the arrangements of the lipids in the membrane plane.

Mean square displacement (MSD) of the centers of mass of lipids in the membrane plane directions (*x*-*y* plane) was calculated to examine the effects of molecular addition on the lateral diffusions of lipids in the binary lipid bilayers. In the estimation of MSD, the lateral diffusions of lipids were sampled in each leaflet of the bilayers, and the center of mass motion of the leaflet was removed. The

Table 1. Time averaged area per molecule A_{mol}, membrane thickness $d_{P\text{-}P}$, and order parameters $-S_{CD}$ for pure POPC, and binary POPC bilayers (POPC/CHOL, POPC/CER, POPC/DAG, and POPC/SM). The parenthesis shows the standard deviations of the values. The listed $-S_{CD}$ in the table were average values of $-S_{CD}$ at each carbon of *sn*-1 and *sn*-2 chains.

	A_{mol} [Å^2]	$d_{P\text{-}P}$ [Å]	$-S_{CD}$
Pure POPC	63.5 (1.3)	39.1 (0.7)	0.15
POPC/CHOL	51.8 (1.1)	43.3 (0.7)	0.21
POPC/CER	57.4 (1.2)	41.9 (0.7)	0.18
POPC/DAG	60.4 (1.3)	41.2 (0.7)	0.17
POPC/SM	61.8 (1.1)	39.1 (0.6)	0.16

time length observed for each lipid lateral diffusion was 20 ns, and the MSD curves were averaged using the MD trajectories for the following 80 ns to obtain the converged MSD curves. The MSD of POPC and additional molecules (DAG, SM, CER, and CHOL) in the binary of lipid bilayer were estimated, and the MSD of POPC evaluated in the pure POPC bilayer was used as a reference MSD data in this study. The lateral diffusion coefficients, D_l, of the lipids were estimated by using the gradients of the liner functions which were fitted to the liner portion of the averaged MSD curves for between 10 and 20ns.

3. Results

Table 1 lists the estimated average of A_{mol}, $d_{P\text{-}P}$, and $-S_{CD}$ for each system. The estimated A_{mol} of pure POPC bilayer was 63.5 Å^2, which is comparable with experimental value, 68.3Å^2, at 303K [19]. We observed that the all A_{mol} of the binary lipid bilayers decreased from that of pure POPC system, whereas the all $d_{P\text{-}P}$ of the binary lipid bilayers were increased. We found that the largest changes of those values were shown in the POPC/CHOL bilayer, and then the changes were shown in the order of POPC/CER, POPC/DAG, and POPS/SM. Figures 3 plots the $-S_{CD}$ of each acyl chain (*sn*-1 and *sn*-2) of the POPC in each binary lipid bilayer. We observed that all $-S_{CD}$ of acyl chains of POPC in the binary lipid bilayers increased from those in the pure POPC bilayer. The POPC/CHOL system showed the largest $-S_{CD}$ values in the systems, and these values are comparable with the previous MD studies [20]. Generally, the ordered lipid chains by CHOL are tightly packed in the membrane, leading to condensed membrane [21]. In this study, such membrane condensing effects were clearly shown in the other binary lipid bilayer systems containing CER, DAG, or SM, too. The strengths of the condensing effect were ranked in the order of POPC/CHOL, POPC/CER, POPC/DAG, and POPC/PSM in this MD study.

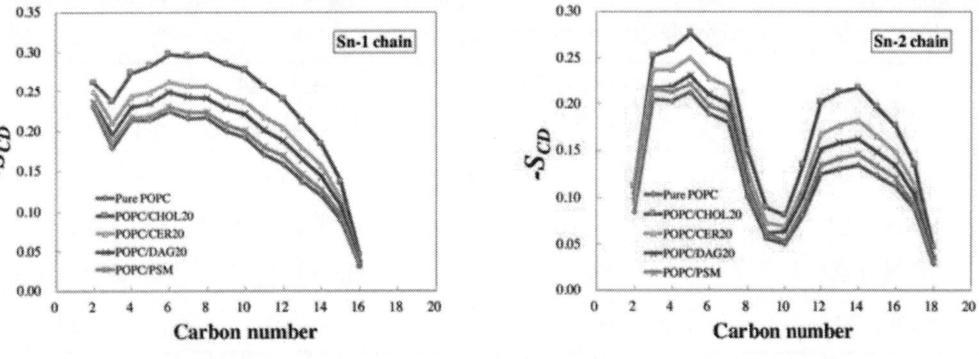

Figure 3. Order parameters, $-S_{CD}$, of each acyl chain (left:*sn*-1 and right:*sn*-2) of POPC in each binary POPC bilayer.

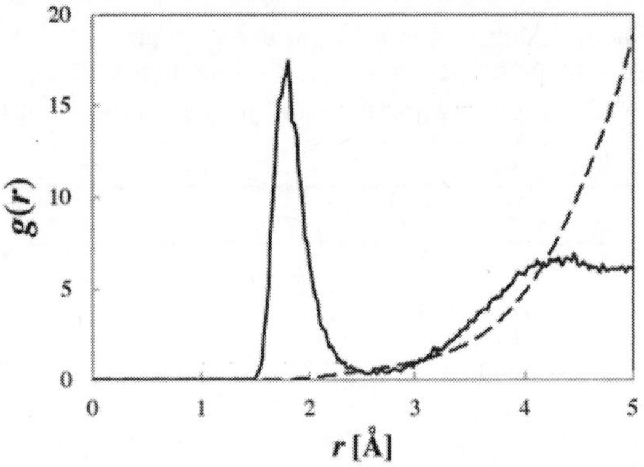

Figure 4. Radial distribution function (RDF) for the hydrogen of hydroxyl of CHOL and the oxygen (O22) of the carbonyl group of POPC in the POPC/CHOL bilayer. The dashed line shows the running coordination number of hydrogen as a function of distance *r*.

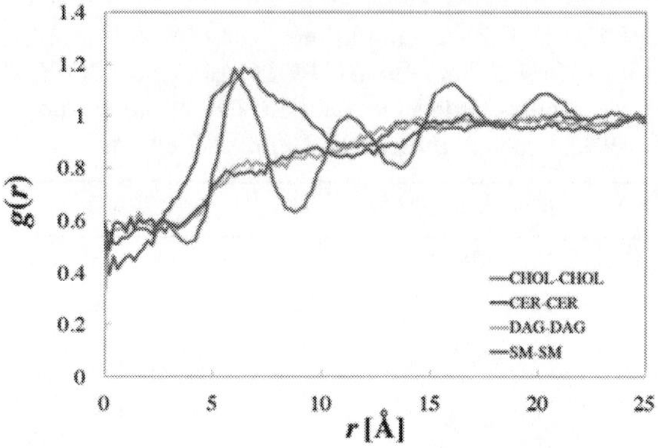

Figure 5. Two-dimensional RDFs between the centers of mass of additional lipids (CHOL, CER, DAG and SM) in each membrane plane.

The hydrogen coordination numbers around the acceptor sites of the POPC were calculated to investigate the hydrogen bond networks in the binary lipid bilayers. Figure 4 shows the RDF for the hydrogen of hydroxyl group of CHOL and the oxygen (O22) of the carbonyl group of POPC in the POPC/CHOL bilayer as an example of stable hydrogen bond formation. The peak of RDF at 1.8 Å shows the forming of hydrogen bond around the oxygen (O22) of the carbonyl group. The time averaged coordinated number for this hydrogen bond was evaluated by counting the hydrogens by the distance *r* of 3 Å and the estimated running coordination number was 3.95 in this graph.

Table 2 summarizes the total coordination numbers of hydrogen in each binary lipid bilayer. In this table, the total coordination numbers for between additional lipid and POPC and for between additional lipids were also listed to assess the contributions of hydrogen bonds to the intermolecular interactions between those lipids in the binary lipid bilayers. The difference of the total coordination numbers depends on the number of possible hydrogen donor/acceptor sites in the systems. While DAG and CHOL have only one hydroxyl group for the formation of hydrogen bond, SM and CER have three and four possible sites for hydrogen bonds, respectively. In the case of CER, the number of hydrogen bonds were much higher in both POPC-CER and CER-CER in the membrane. These properties of hydrogen bond networks should contribute not only to the changes of lipid structure and dynamics but also lipid distribution in the membrane plane.

Table 2. Total hydrogen coordination numbers for between the additional lipid and POPC and for between the additional lipids in the POPC/CHOL, POPC/CER, POPC/DAG and POPC/SM bilayers. The parenthesis shows the standard deviations of the values.

System	Additional lipid – POPC	Additional lipid – Additional lipid	Total
POPC/CHOL	18.0 (2.0)	0.1 (0.2)	18.1 (1.8)
POPC/CER	30.4 (0.6)	10.0 (1.2)	40.4 (1.2)
POPC/DAG	10.5 (0.6)	3.6 (0.2)	14.1 (0.6)
POPC/SM	21.8 (2.4)	1.4 (1.0)	23.1 (1.6)

Figure 5 shows the two-dimensional RDFs between the centers of additional lipids in each binary lipid bilayer. We observed a clear peak of RDF around 6 Å in the POPC/CHOL and POPC/CER bilayer, whereas there are no RDF peaks in the POPC/DAG and POPC/SM bilayer. These results mean that the CHOL and CER rearranged from the randomly arranged initial positions in the membrane plane, and those are closely arranged in each membrane. These should be due to the better intermolecular interaction of those lipids in the membrane

Figures 6 (a)-(b) show that MSD curves of the POPC and additional lipids in each lipid bilayer. The evaluated diffusion coefficients, D_l, of the POPC and additional lipids in each system are listed in Table 3. The evaluated D_l of POPC in pure POPC was 7.49×10^{-8} cm²/s, which is very closed to the experimental value, 8–10 × 10^{-8} cm²/s [22-24]. The calculated diffusion coefficients D_l for the POPC and CHOL in the POPC/CHOL bilayer largely decreased, and those values for each lipid are also in good agreement with the experimental values, 4.46×10^{-8} cm²/s and 3.21×10^{-8} cm²/s, respectively [23]. We found that the D_l of lipids in the POPC/CER and POPC/SM were also decreased. Such reduction of lipid diffusion has also been known to related to the lipid condensing due to the ordering of lipids in the membrane [25]. Our MD study also shows such ordering of lipid chains in the POPC/CER and POPC/SM, being consistent with the experimental observations. In the case of POPC/DAG bilayer, the calculated D_l of lipids was comparable with that of POPC in the pure POPC bilayer.

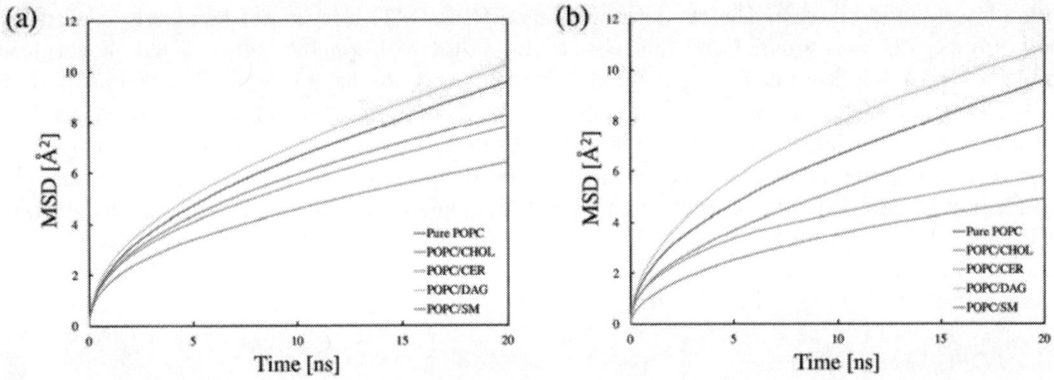

Figure 6. (a) Means square displacement (MSD) of the center of mass of POPC in the pure POPC and each binary lipid bilayer system, (b) MSD of the center mass of POPC and additional lipids (CHOL, CER, DAG, and SM) in the pure POPC and each binary lipid bilayer system.

Table 3. Lateral diffusion coefficients, D_l, of the POPC and the additional molecules in the pure POPC and the binary POPC binary bilayers. The parenthesis shows the standard deviations of the values. Units are in 10^{-8} cm²/s.

	Pure POPC	POPC/CHOL	POPC/CER	POPC/DAG	POPC/SM
POPC	7.49 (1.28)	4.51 (0.60)	5.49 (1.51)	7.89 (1.17)	5.78 (1.48)
Additional molecule	—	3.35 (1.34)	3.59 (1.38)	7.27 (1.69)	6.26 (1.4)

4. Discussions

In this section, we summarize the structure and dynamics of each binary POPC bilayer system and discuss the factors which yield the differences of molecular condensing effects.

4.1. POPC/CHOL bilayer

The POPC/CHOL bilayer shows the largest condensing effects in the binary POPC bilayer systems. Such large condensing effect are attributable to the structural characters of CHOL. CHOL has a small polar (hydroxyl) head group and a rigid sterol ring in the body, and the hydroxyl group of CHOL faces the membrane interface region. Thus, the most of CHOL structure is buried in the hydrophobic region of the bilayer [20], also those are oriented to the membrane normal. The acyl chains of POPC next to the sterol ring of CHOL are ordered to avoid the steric hindrance with the CHOL in the membrane. These highly ordered lipid chains results in the decrease of A_{mol} and increase of $d_{P\text{-}P}$, leading to the tightly packed membrane. The 2D-RDF analysis of centers of CHOLs in the POPC/CHOL binary bilayer shows that CHOLs are closely arranged in the membrane plane. This should be due to that the system gets energetically stable by facing the ordered rigid sterol rings of CHOLs in the membrane. Such arrangement of CHOLs has been observed in the previous DOPC/CHOL and DPPC/CHOL MD simulations [20]. We also observed that the lateral diffusions of both POPC and CHOL were largely decreased due to the tight packing of POPC/CHOL bilayer. These results correspond to the previous MD studies [26] and experimental observations [23].

4.2. POPC/DAG bilayer

Since the structural difference between DAG and POPC is only seen in the head group of the lipids, a simple effect of replacements of a PC head group with a hydroxyl group can be observed in the POPC/DAG bilayer. We observed that the POPC/DAG bilayer showed a moderate condensing effect in the binary lipid bilayers. The condensing of the POPC/DAG bilayer can be mainly caused by the lack of bulky PC head group, DAG lipids are buried in the hydrophobic region of the bilayer, leading to the condensed packing of the membrane. This result was comparable with the previous MD study [26]. The RDF analysis showed that DAG can form some hydrogen bonds between the hydroxyl group of and surrounding POPC in the membrane. However, since the total number of the hydrogen coordination was the smallest in the binary lipid bilayer system, the contribution to the membrane condensing and the diffusion of lipids in the membrane should be minor. In fact, the diffusion constant of POPC and DAG in the POPC/DAG bilayer were comparable with that of POPC in the pure POPC bilayer. The decrease of bulky PC head groups in the system could also be a factor to determine the lateral diffusion of lipids in the POPC/DAG bilayer.

4.3. POPC/SM bilayer

The POPC/SM bilayer shows the smallest condensing effect in the binary lipid bilayers. This is because that the SM has a PC head group, whereas the CHOL, DAG and CER have a small polar (hydroxyl) head group. Since the PC head group of SM has a comparable area with that of POPC, the change of membrane area of the POPC/SM bilayer results in minor. On the other hand, SM has an amide, hydroxyl and carbonyl groups at the linkage of the PA and SPH, and these can be donor/acceptor sites for forming of hydrogen bonds. The total number of hydrogen coordinate around the sites was larger than those of POPC/CHOL and POPC/DAG systems. These hydrogen bonds could contribute to the condensing of the membrane and the decrease of the lateral diffusion of lipids in the membrane.

4.4. POPC/CER bilayer

A simple effect of replacements of a PC head group of SM with a hydroxyl group can be observed in the POPC/CER bilayer. The POPC/CER bilayer showed the second largest condensing effect in the binary lipid bilayers. Since the lack of bulky PC head group, the most of CER structure is also buried in the hydrophobic region of the bilayer. This "Umbrella effect" in the POPC/CER should be a factor of the condensing of the membrane [20, 26]. Also, CER has four donor/acceptor sites for forming hydrogen bonds. Thus, the POPC/CER bilayer has the largest hydrogen coordination number both between POPC-CER and CER-CER. These hydrogen bonds should contribute to the condensing of the membrane and the large decrease of lateral diffusions of lipids in the membrane. The 2D-RDF analysis of centers of CERs showed that the CERs were arranged closely in the membrane. This should be due to the buried condition of CER in the membrane and the hydrogen bond networks between CERs. Such arrangement of CERs could be a seed for the CER enriched micro-domain in biomembrane [3-5].

5. Conclusion

Molecular dynamics (MD) simulations of the binary POPC bilayers containing cholesterol (CHOL), ceramide (CER), diacylglycerol (DAG), or sphingomyelin (SM) were carried out to investigate the effects of these lipids on structure and dynamics of POPC bilayer. The membrane structure parameters such as area per lipid (A_{lipd}), membrane thickness ($d_{P\text{-}P}$) and order parameter ($-S_{CD}$) for each system were calculated and compared with each other. The clear condensing effects (the decreases of A_{lipd}, and the increases of $d_{P\text{-}P}$ and $-S_{CD}$) were observed in all binary POPC bilayers, and the strengths of the condensing effect were ranked in the order of POPC/CHOL, POPC/CER, POPC/DAG, and POPC/PSM in this MD study.

We found that the differences of condensing effects are attributable to the factors such as molecular shape, type of polar head group of the lipids the hydrogen bond networks in the system. The lipid chains surrounding the rigid sterol ring of CHOL in the membrane are highly ordered to avoid the steric barrier, resulting in the large decrease of A_{lipid} and increase of $d_{P\text{-}P}$ of the POPC/CHOL bilayer. The DAG, CER and CHOL have a small polar (hydroxyl) head group instead of the PC head group. Due to the lack of bulky PC head group, these molecules are buried in the hydrophobic region of the bilayer, leading to the condensed packing of membranes. The POPC/CER bilayer has the largest hydrogen coordination number both between POPC-CER and CER-CER. These hydrogen bond networks would contribute to the membrane condensing and the arrangement of CER in the membrane. We observed that the condensing of membranes leads to the decrease of lateral diffusion of lipids in the binary lipid bilayer. Such decreased lateral diffusions of the lipids were good agreement with the experimental observations. The lateral diffusion of lipids in the POPC/DAG bilayer were comparable with that in the pure POPC bilayer. This could be determined with the small number of hydrogen bonding, the moderate condensing effect, and the decrease of bulky PC head groups in the POPC/DAG bilayer.

6. Acknowledgment

This work was supported by JSPS KAKENHI (Grant Numbers 15KT0073, 16H01374, 16K15083, 17H02209 to K.N., 26400429 to H.N, 17K05620 to T.M. and 26102722, 16K05648 to H.S.). The computations in this study were performed using the advanced center for computing and communication of RIKEN, the research center for computational science of Institute for Molecular Science (IMS), the research center for advanced computing infrastructure of Japan Advanced Institute of Science and Technology (JAIST), and the center for computational science of Tsukuba university.

References

[1] Ohvo-Rekila H, Ramstedt B, Leppimaki P and Slotte J. P. 2002 *Prog Lipid Res.* **41** 66-97.
[2] Vist, M.R. and Davis J.H. 1990 *Biochemistry* **29** 451-64.
[3] Zou S, Johnston L. J. 2010 *Current Opinion in Colloid & Interface Science* **15** 489-498.
[4] Zhang Y, Li X, Becker K.A. and Gulbins E 2009 *Biochim Biophys Acta* **1788** 178-83.
[5] Bollinger C.R., Teichgraber V, Gulbins E 2005 *Biochim Biophys Acta* **1746** 284-94.
[6] Kawashima Y, Miyazaki E, Muller M, Tokuda H and Nishiyama K 2008 *J Biol Chem* **283** 24489-96.
[7] Goni F.M. and Alonso A.1999 *Prog Lipid Res* **38** 1-48.
[8] Ramstedt B. and Slotte J.P. 2002 *FEBS Lett* **531** 33-7.
[9] Barenholz Y. and Thompson T.E. 1999 *Chem Phys Lipids* **102** 29-34.
[10] Lee J., Cheng X. Swails J. M. Yeom M. S. Eastman P. K. Lemkul J. A. Wei S. Buckner J. Jeong J. C. Qi Y. Jo S. Pande V. S. Case D. A. Brooks C. L., MacKerell A. D., Klauda J. B. and Im W. 2016 *J. Chem. Theory Comput.* **12** 405-13.

IOP Conf. Series: Journal of Physics: Conf. Series **1136** (2018) 012022 doi:10.1088/1742-6596/1136/1/012022

[11] Nosé S. 1984 *Molecular Physics* **52** 255-268.
[12] Hoover W.G. 1985 *Physical Review A* **31** 1695-1697.
[13] Parrinello M. and Rahman A.1981 *Journal of Applied Physics* **52** 7182-7190.
[14] Lim J.B., Rogaski B. and Klauda J.B., 2012 *J Phys Chem B* **116** 203-10.
[15] Klauda J.B., Venable R.M. ,Freites J.A., O'Connor J.W., Tobias D.J., Mondragon-Ramirez C., Vorobyov I., MacKerell A.D. and Pastor R.W. 2010, *J. Phys. Chem. B* **114** 7830-43.
[16] Jorgensen, W.L., Chandrasekhar J., Madura J.D., Impey Roger W. and Klein, M.L.1983 *J. Chem. Phys.* **79** 926-935.
[17] Essmann U., Perera L., Berkowitz M.L., Darden T., Lee H. and Pedersen L.G. 1995 *J. Chem. Phys.* **103** 8577-8593.
[18] Abraham M.J., Murtola T, Schulz R, Páll S., Smith J.C., Hess B. and Lindahl E. 205 *SoftwareX*, **1-2** 19-25.
[19] Kucerka N., Tristram-Nagle S., and Nagle J.F. 2005 *J Membr Biol* **208** 193-202.
[20] Dai J., Alwarawrah M. and Huang J.Y. 2010 *J. Phys. Chem. B*, 2010. **114** 840-848.
[21] Rog T., Pasenkiewicz-Gierula M., Vattulainen I. and Karttunen M. 2009 *Biochim Biophys Acta* **1788** 97-121.
[22] Gaede H.C. and Gawrisch K. 2003 *Biophys J* **85** 1734-40.
[23] Shin Y.K. and Freed J.H. 1989 *Biophys J.* **55** 537-50.
[24] Kuo A.L. and Wade C.G.1979 *Biochemistry* **18** 2300-2308.
[25] Lindblom G., Oradd G. and Filippov A. 2006 *Chem Phys Lipids* **141**179-84.
[26] Alwarawrah M., Dai J. and Huang J., 2012 *J. Chem. Theory Comput.* **8** 749-758.

IOP Conf. Series: Journal of Physics: Conf. Series **1136** (2018) 012023 doi:10.1088/1742-6596/1136/1/012023

Effect of the Arg456His mutation on the three-dimensional structure of cytochrome P450 1A2 predicted by molecular dynamics simulations

Y Watanabe[1,2], K Kato[3], S Fukuyoshi[2], M Hiratsuka[4], N Yamaotsu[5], S Hirono[5], H Gouda[1] and A Oda[2, 3, 6*]

[1] School of Pharmacy, Showa University, 1-5-8 Hatanodai, Shinagawa-ku, Tokyo 142-8555, Japan
[2] Faculty of Pharmacy, Institute of Medical, Pharmaceutical and Health Science, Kanazawa University, Kakuma, Kanazawa, Ishikawa 920-1192, Japan
[3] Graduate School of Pharmacy, Meijo University, 150 Yagotoyama, Tempaku-ku, Nagoya, Aichi 468-8503, Japan
[4] Graduate School of Pharmaceutical Sciences, Tohoku University, 2-1-1 Katahira, Aoba-ku, Sendai, Miyagi 980-8577, Japan
[5] School of Pharmacy, Kitasato University, 5-9-1 Sirokane, Minato-ku, Tokyo 108-8641, Japan
[6] Institute for Protein Research, Osaka University, 3-2 Yamadaoka, Suita, Osaka 565-0871, Japan

*E-mail: oda@meijo-u.ac.jp

Abstract. Cytochrome P450 1A2 (CYP1A2), one of the major drug-metabolizing enzymes among CYPs, has many variant alleles. The genetic polymorphism of CYP1A2 is thought to cause individual differences in the pharmacokinetics of medicines. CYP1A2.8 (the Arg456His mutant), a CYP1A2 variant, has decreased enzymatic activity. In our previous work, to understand why the Arg456His mutant lost its enzymatic activity, we constructed an Arg456His mutant with the hydrogen on the epsilon nitrogen of the histidine (HIE) and performed a 300-ns MD simulation. After the MD simulation, the Arg456His mutant with HIE showed large differences in static structure and flexibility compared with the wild type, which would cause the decreased activity. In the work described here, we constructed an additional Arg456His mutant with positively charged histidine and performed a 300-ns MD simulation to consider the effect of the protonation state of the histidine. Comparing these two Arg456His mutants revealed differences in their static structures, flexibilities, and interactions, suggesting that the protonation state of the His456 residue has a considerable influence on the physiological properties of CYPs. Our results indicate that the Arg456His mutation causes the CYP1A2 structure to unfold regardless of the protonation state of His456.

1. Introduction

Cytochrome P450 (CYP), a superfamily of heme-containing multifunctional enzymes, is prevalent in the natural world, existing in a wide range of organisms, such as bacteria, plants, and animals [1]. Although CYPs play many important roles in the synthesis of biologically active substances, metabolic activation of some mutagenic compounds, and detoxification of xenobiotics, CYPs have

IOP Conf. Series: Journal of Physics: Conf. Series **1136** (2018) 012023 doi:10.1088/1742-6596/1136/1/012023

attracted clinical attention mainly for their drug-metabolizing function [2–4]. Therefore, genetic polymorphism in CYPs is viewed as a pharmacotherapeutic challenge, as this variation can affect their efficacy and result in adverse side effects of medicine [5]. Major drug-metabolizing CYPs such as CYP2C9, CYP2D6, and CYP3A4/5 have many variant alleles, which are important from the viewpoint of pharmacokinetics [6]. In drug metabolism reactions, CYPs catalyze the monooxygenation of drugs using oxygen molecule. The proposed reaction mechanism by CYPs is shown in scheme. 1 [7]. First, a water molecule, the sixth ligand of the heme iron is replaced by the substrate. The binding of the substrate changes the six-coordinate high-spin Fe^{III} to the five-coordinate, low spin Fe^{III}. Next, first electron from an electron carrier, such as nicotinamide adenine dinucleotide phosphate reduces the five-coordinate Fe^{III} to the Fe^{II}. Then, the binding of the oxygen molecule converts the five-coordinate Fe^{II} to the six-coordinate Fe^{III}. After the reduction by second electron, two proton cleaves the O—O double bond and generate a water molecule and an active species so-called compound I. Although this oxidizing intermediate has not been elucidated experimentally, it is proposed that the oxygen atom is inserted in the substrate and generate the final oxidized product.

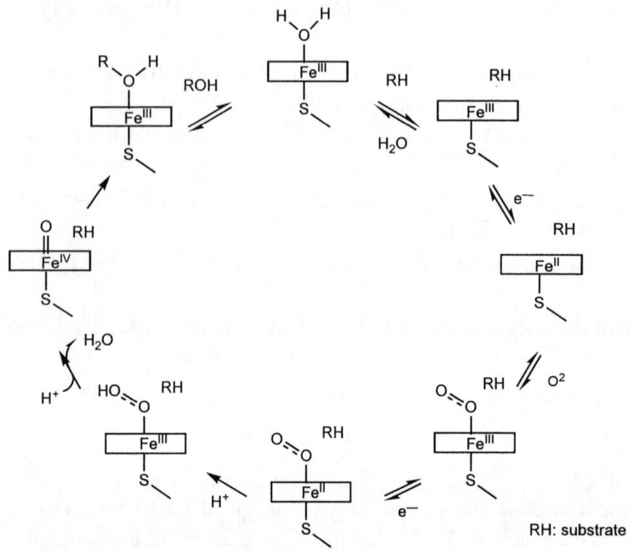

Scheme 1. Catalytic cycle of heme molecule for monooxygenation reaction.

Thus, the catalytic reaction requires two electrons from its electron carrier and electron transfers are mediated by the membrane-bound flavoprotein cytochrome P450 redox partner (CPR) [8]. Thus, binding and interaction with CPR are necessary for the catalytic reaction of CYPs, and putative CPR-binding sites in CYPs have been studied [9–12]. Although the potential binding site is not yet verified, it is likely located proximal to the C, K, and L helices of the CYP protein. Moreover, in the control of the heme catalytic cycle, water molecules play important roles [8] as the sixth ligand for the heme and are required for heme inactivation. When the substrate binds to the active site and approaches the heme iron, CYPs undergo a conformational change to form a substrate-bound complex structure. Formation of this complex changes the spin state from a low spin (inactive) to a high spin (active). In the P450-BM3 X-ray crystal structure, water molecule is expected to participate in proton transfer for the hydrolysis reaction [13]. Schlichting et al. conducted cryo-crystallography of P450cam bound to the substrate camphor [7] and showed that water molecules formed hydrogen bond networks with some residues. Moreover, in the hydrolysis reaction of camphor by P450cam, Parasad et al. confirmed via NMR that the hydrogen atom added to camphor was derived from water molecules [14]. Crystal structure analyses revealed that some mammalian CYPs, including CYP2C19 (PDB ID: 4GQS) and CYP3A4 (PDB ID: 4I3Q), have water molecules in their active sites. However, due to the high mobility of water molecules, their functions are difficult to define using current experimental techniques.

CYP1A2, a major drug-metabolizing subtype among CYPs, has many variant alleles including some low-active mutants [15]. The genetic polymorphism of CYP1A2 may cause individual differences in the pharmacokinetics of clinically important medicines such as asthma remedies (theophylline), antipsychotics (amitriptyline, imipramine, and olanzapine), and antiarrhythmic drugs (mexiletine and propranolol). One of the CYP1A2 variants, CYP1A2.8 (the Arg456His mutant), has been reported to decrease enzymatic activity [15,16]. The Arg456His mutation is located at the KL loop proximal to Cys458, which is the fifth ligand for heme. According to previously reported sequence analyses of CYPs, the heme-binding domain is considered to be the most conserved region [17–19], and the amino acid sequence around the cysteine (FXXGXRXCXG) is highly conserved in most CYPs [18]. In the crystal structure analyses of Hasemann et al., arginine was sometimes substituted by another basic residue, such as histidine in P450terp and P450cam, and lysine in P450BM3, suggesting that the basic residues (arginine/histidine/lysine) fold the Cys-pocket into position by interacting their positively charged side chains with the propionate group of heme [19]. In the natural human CYP1A2.8 mutant, the Arg456His substitution dramatically reduced the heme peak at 450 nm [16], indicating a disturbance of the normal function of heme. A quantum mechanics/molecular mechanics (QM/MM) study of the P450cam–camphor complex by Guallar et al. suggested that the His355 of P450cam, corresponding to the Arg456 of CYP1A2, forms a hydrogen bond with heme and contributes to the stabilization of the oxygenated form [20]. In our previous work, to understand why the Arg456His mutant lost its enzymatic activity, we constructed an Arg456His mutant with the hydrogen on the epsilon nitrogen of the histidine (HIE) and performed a 300-ns MD simulation [21]. HIE should be suitable for interaction with the surrounding residues, including the heme molecule. After the MD simulation, the Arg456His mutant with HIE showed large differences in static structure and flexibility compared to the wild type, which would explain its decreased activity. In the present study, in order to consider the effect of the protonation state of histidine, we constructed an additional Arg456His mutant structure with positively charged histidine (HIP), which would have more electrostatic interaction with the surrounding residues than HIE. We performed 300-ns MD simulations at 300 K for the Arg456His mutant with HIP and compared the results with those from the Arg456His mutant with HIE.

2. Methods

Initial 3D structures of the Arg456His mutant with HIE were constructed from the optimized structure of wild-type CYP1A2 obtained by 100-ns MD simulation [21], which was constructed from the crystal structure (PDB ID: 2HI4) [22]. Likewise, the initial structure of Arg456His with HIP was also constructed from the optimized wild-type structure. Hereafter we refer to these, the Arg456His mutant with HIE and the Arg456His mutant with HIP, as the Arg456Hie mutant and the Arg456Hip mutant, respectively. First, all crystal water molecules and the ligand were removed from the 2HI4 crystal structure of wild-type CYP1A2. To obtain the equilibrium state for the ligand-free form, a 100-ns MD simulation was performed at 300 K under constant pressure [21]. Using the energy-minimized structure of wild-type CYP1A2, initial structures of the Arg456His mutants were constructed by manually substituting Arg to His. Calculation settings were the same as used in our previous studies of the ligand-free CYP1A2 [21], CYP2B6 [23], CYP2C9, CYP2C19 [24], and CYP2D6 [25]. The system was solvated using the TIP3P model [26] and calculated under cyclic boundary conditions. The cyclic boundary box was spaced with at least 8Å margins from the protein surface, the cutoff distance for van der Waals interactions was set at 10 Å, and the particle mesh Ewald method was used for calculating electrostatic interactions [27]. Chloride ions were arranged to neutralize the system. For the force field parameter around the heme iron, our determined parameters were used, which represent the sextet state with five-coordinate iron(III) [28]. When the substrate is recognized by CYP, forming the sextet and five-coordinate iron is essential for the catalytic reaction. For the rest of the heme molecule, parameters determined by Giammona were used [29]. AMBER ff12SB force field equipped within AMBER12 package [30] and general AMBER force field (GAFF) [31] were used for amino acids and the ligand, respectively. The tleap module of AmberTools was used to supplement hydrogen atoms not observed in the crystal structure. The time step of all MD simulations was 2 fs. The SHAKE algorithm was used to constrain bond stretching of hydrogen atoms [32].

First, the solvent was relaxed and the whole system was minimized, then temperature-increasing MD simulations were performed for 20 ps, with the temperature raised from 0 to 300 K. Equilibrating MD simulations were then performed for 300 ns under constant pressure at 300 K. For the NPT ensemble, the Langevin thermostat was used to maintain the temperature of the system. To validate the 3D structures of the Arg456Hie and Arg456Hip mutants after MD simulations, root mean square deviations (RMSDs) for the main-chain atoms were evaluated along the calculated MD trajectories. As reference structures for RMSDs, the 3D structures after temperature-increasing MD simulations were used. To investigate the structural flexibilities of residues, root mean square fluctuations (RMSFs) were calculated for all Cα atoms using the last 20-ns MD trajectories. As reference structures for RMSFs, the average structures of 280-300 ns trajectories were used. Calculations of RMSDs and RMSFs were performed using the cpptraj module of AmberTools12. The hydrogen bond networks were also analyzed using 280–300 ns trajectories. Regarding hydrogen bond criteria, the distance between two heavy atoms and the angle between the acceptor, hydrogen, and donor atoms were determined with cutoffs of 3 Å and 120°, respectively.

3. Results and discussion

We conducted 300-ns MD simulations for both Arg456His mutants and calculated the RMSDs to validate the 3D structures (figure 1). In our previous work [21], 100-ns MD simulation was conducted for the wild-type CYP1A2 structure and the optimized structure was used as the initial structures of two Arg456His mutants. As shown in fig. 2a, the RMSD of the wild-type CYP1A2 converged around 50-100 ns, suggesting the 3D structure of the wild-type CYP1A2 reached its equilibrium state by 100-ns MD simulation.

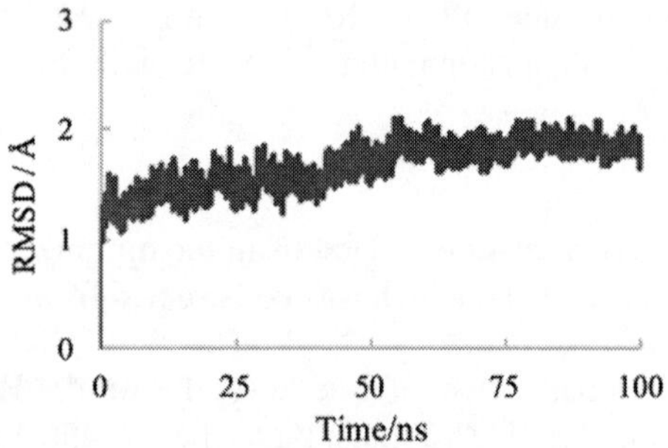

(a) Wild-type CYP1A2

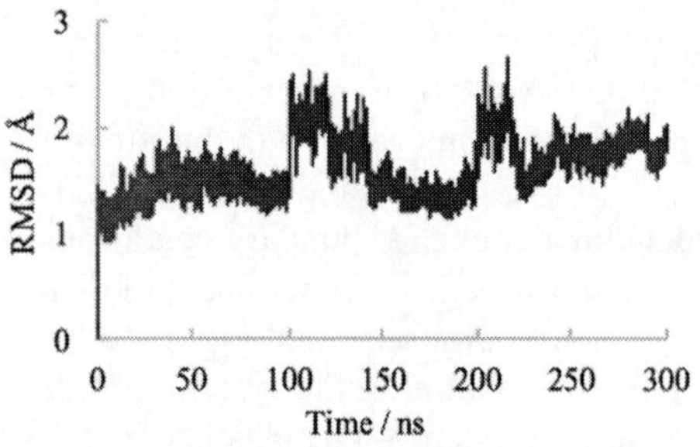

(b) Arg456Hie

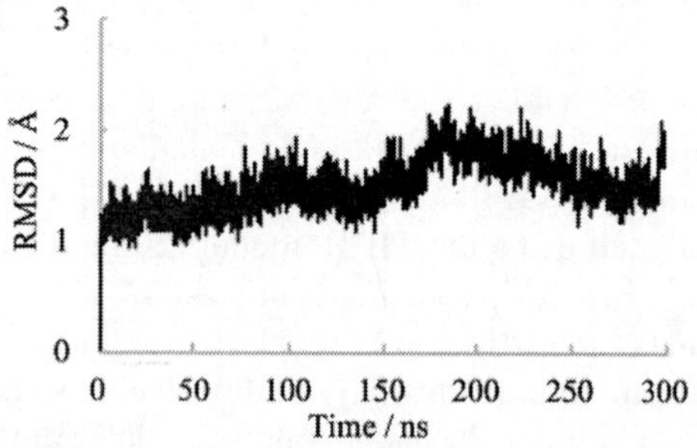

(c) Arg456Hip

Figure 1. Root mean square deviations (RMSDs) of the main-chain atoms for (a) the wild-type CYP1A2, (b) the Arg456Hie mutant and (b) the Arg456Hip mutant.

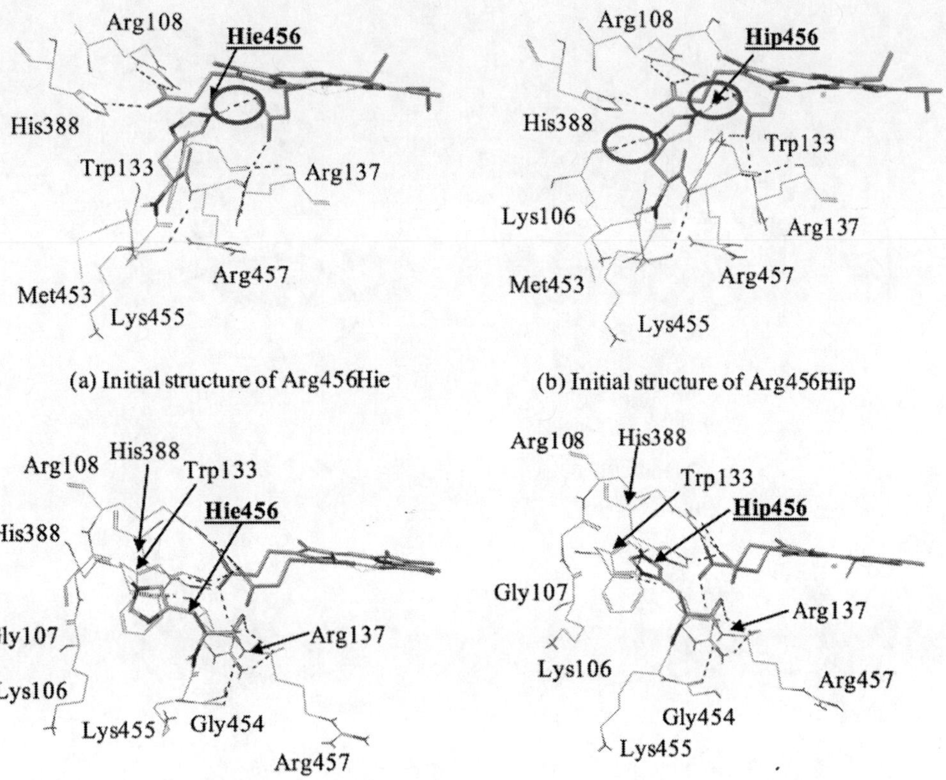

(a) Initial structure of Arg456Hie (b) Initial structure of Arg456Hip

(c) Optimized structure of Arg456Hie (d) Optimized structure of Arg456Hip

Figure 2. Interaction around His456 in the initial and optimized structures. Initial structures of Arg456Hie (a) and Arg456Hip (b) mutants. Optimized structures of Arg456Hie (c) and Arg456Hip (d) mutants. Residues within 3 Å are labeled. Hydrogen bonds shown with black dashed line.

Table 1. Average distance between the propionate oxygen atom of the heme and the nearest nitrogen atom of the 456th residue, calculated using last 20 ns of trajectories.

Protein name	Heavy atom	Distance/ Å
Wild type (the ligand-free form) [21]	HEME@O1D–ARG456@NH1	5.07
Wild type (the substrate-bound form) [33]	HEME@O2D–ARG456@NH1	2.84
Wild type (the inhibitor bound form) [33]	HEME@O2D–ARG456@NH1	5.17
Arg456Hie mutant	HEME@O1A–HIP@NE2	5.21
Arg456Hip mutant	HEME@O1A–HIP@ND1	5.28

On the contrary, the RMSD of the Arg456Hie mutant repeated rise and fall around 1.1 to 2.6 Å. Although the RMSD of the Arg456Hip mutants was less than 2.3 Å and not largely fluctuated, it is also rising. From these results, it seems that 3D structures of the Arg456Hie and Arg456Hip mutants did not reach their equilibrium states by 300-ns MD simulations. Although the 3D structures of two Arg456His mutants were not yet relaxed by 300 ns MD simulations, these results suggest that the Arg456His mutation have large influence on the CYP1A2 structure independently of the His456 protonation state . Moreover, although the RMSD of the Arg456Hie mutant showed the existence of a metastable state during 100–145 and 200–250 ns [21], the RMSD of Arg456Hip did not show this

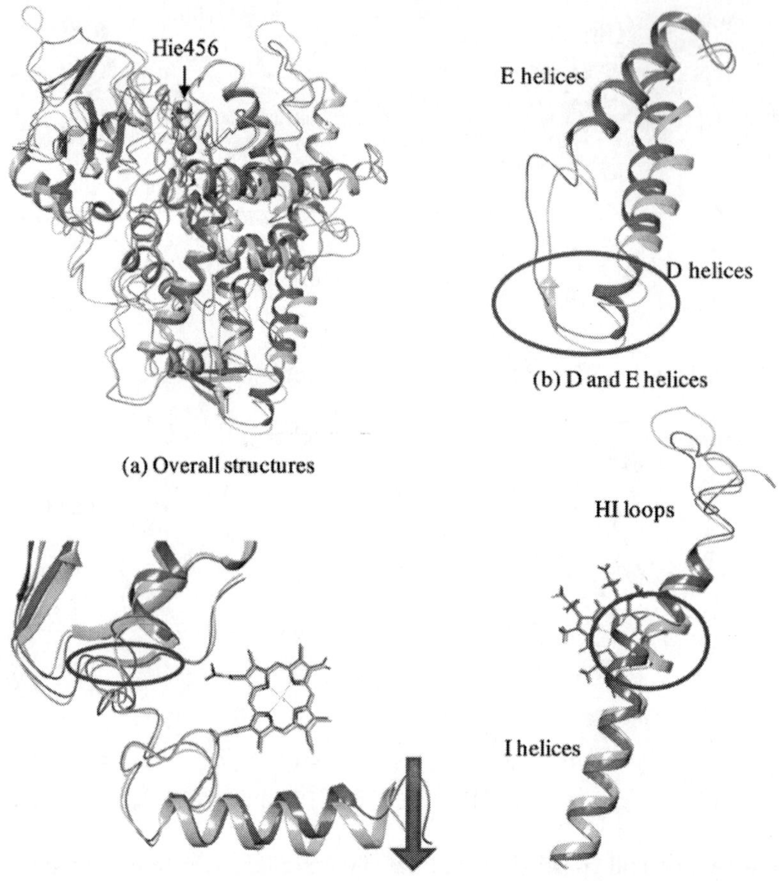

Figure 3. Superimposition of static structures for the Arg456Hie (blue) and Arg456Hip (pink) mutants. (a) Overall structures, (b) D and E helices, (c) KL loops and C helices, and (d) HI loops and I helices.

metastable feature. This suggests that the protonation state of His456 has an influence on both the stability and energy state of the CYP1A2 structure. Figure 2 shows the initial and optimized 3D structures of the Arg456Hie and Arg456Hip mutants. In the initial structures, the side chains of both Hie456 and Hip456 interacted with the propionate oxygen of heme (figure 2a and 2b). In the Arg456Hip mutant only, the imidazole side chain of Hip456 formed a hydrogen bond with the charged amino group of Lys106. On the other hand, in the optimized structures, neither Hie456 or Hip456 interacted with the heme molecules. Moreover, although the imidazole side chain of Hie456 formed a hydrogen bond with Ser129 (figure 2c), that of Hip456 did not interact with any residues (figure 2d). Thus, Hie456 and Hip456 clearly had different interactions, which could spread across the 3D structures. In our previous work on wild-type CYP1A2, the interaction between Arg456 and heme was also investigated [21]. In the ligand-free form of CYP1A2, the distance between Arg456 and the heme was 5.07 Å, and the interaction was weak (table 1). However, in the substrate-bound CYP1A2, the distance between Arg456 and the heme was shortened to 2.84 Å, and the interaction became strong [33]. On the other hand, in the complex of CYP1A2 with the inhibitor, the distance between Arg456 and the heme was over 5 Å [33]. These results suggest that the interaction between Arg456 and heme might be important to start the catalytic reaction in the substrate-bound form. At the least, the interaction between Arg456 and heme was one feature of the substrate-bound form. Therefore, it is not surprising that His456 did not interact with the heme in the ligand-free Arg456His mutants.

Figure 3 shows structural differences between the Arg456Hie and Arg456Hip mutants. After 300-ns MD simulations, RMSDs between the optimized structures for the Arg456Hie and Arg456Hip mutants were 2.39 Å, and the overall structures were not so different (figure 3a). However, large structural differences were observed in D helices (figure 3b), KL loops and C helices (figure 3c), and I helices (figure 3d). In the Arg456Hip mutants, the length of the D helix was shortened and a small β strand was formed in the DE loop (figure 3b). In the 3D structure of wild-type CYP1A2, the D helix is located at the protein surface and away from the Arg456His mutated position. This suggests that the protonation state of His456 affected the distant substructure. In addition, in the Arg456Hip mutant, the C helix was apart from the heme and partial β sheets were formed in the KL loop (figure 3c). The C helix reportedly constitutes a protein–protein interface with CPR [9–12]. Therefore, misalignment of the C helix would affect the electronic state of the heme and change its catalytic activity. Furthermore, structural differences were also observed in the I helices, considered to be among potential substrate recognition sites [17]. While the I helix of the Arg456Hip mutant was a straight α helix, that of the Arg456Hie mutant formed a loop and bent at the center. From this result, the protonation state of Arg456His seems to influence substrate recognition.

It is well known that CYP proteins have structural flexibility or conformational flexibility because CYP proteins undergo conformational change by ligand binding, so-called induced fit [34]. Our computational studies about CYP proteins suggested that the structural flexibility have large influence on the substrate recognition and the enzymatic activity [21, 23, 35]. Regarding the wild-type CYP1A2 obtained after 100 ns MD simulation [21], we calculated the RMSF using last 20 ns trajectory and observed highly flexible regions at the CD, FG, and HI loops (figure 4a and 5). To compare the structural flexibilities between the wild type and two Arg456His mutants, RMSFs were calculated using last 20 ns trajectories (figure 5 and table 2). Moreover, to examine whether the choice of last 20 ns trajectory is adequate for RMSF calculation of two Arg456His mutants, RMSFs were also calculated using last 20, 40, and 60 ns (280–300, 260–300, and 240–300 ns) trajectories (figure 6). Values of high RMSF peaks at the BC, CD, FG, and HI loops were also tabulated in table 3. As for the Arg456Hie mutant (figure 6a), although the RMSF value of the BC loop was different by 0.73 Å when 280–300 and 240–300 ns trajectories were compared, overall RMSFs were well similar. On the other hand, regarding the Arg456Hip mutant (figure 6b), RMSF value of the CD loop was significantly higher by 1.29 Å using 240–300 ns trajectory than that of 280–300 ns trajectory. Furthermore, in the Arg456Hip mutant, the RMSF of 240–300 ns trajectory was relatively higher than that of 260–300 ns and 280–300 ns trajectories. Especially, the high RMSF peak at Pro183 was 2.97 Å in 240–300 ns trajectory, whereas the RMSF values were below 0.9 Å in 260–300 ns and 280–300 ns trajectories. Pro183 of the DE loop was exposed to the solvent in the wild type and the RMSF value was only 0.81 Å. This result suggests that 240–260 ns trajectory of Arg456Hip mutant includes some peculiar conformations. As shown in figure 6 and table 3, overall flexibilities of two Arg456His mutants did

Table 2. Values of high RMSF peaks at BC, CD, FG, and HI loops in wild-type Arg456His mutants, calculated using last 20 ns of trajectories. Significantly high values shown in bold.

	Gly102 of BC loop	Ser156 of CD loop	Phe246 of FG loop	Gly299 of HI loop
Wild type	0.71	2,19	1.32	1.58
Arg456Hie mutant	0.75[1]	2.39	1.84	**3.43**[2]
Arg456Hip mutant	**1.80**	**3.15**[3]	1.26	1.51

[1] RMSF value at Asp103.
[2] RMSF value at Ser298.
[3] RMSF value at Ser155.

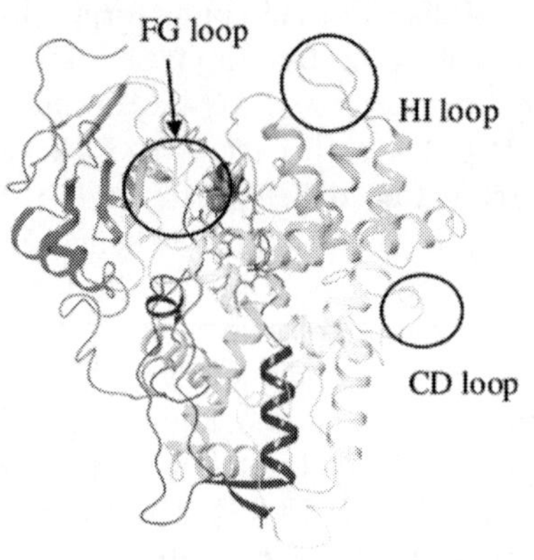

(a) Flexible regions of wild-type CYP1A2

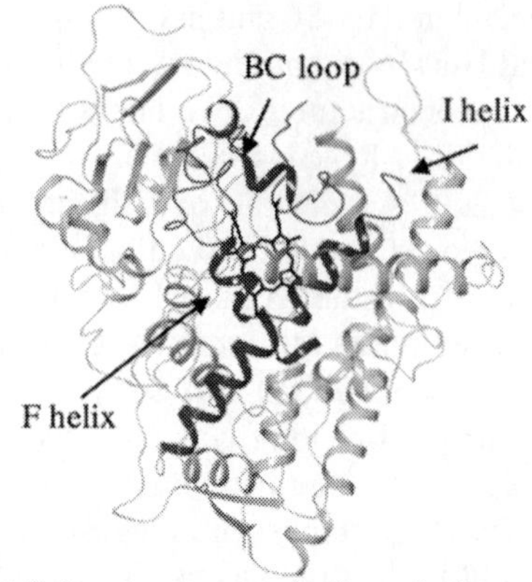

(b) Substructures constituting of substrate recognition site for wild-type CYP1A2

Figure 4. Overall structure of wild-type CYP1A2, from [21]. Highly flexible regions indicated with circles in (a); main substrate recognition sites shown with blue color in (b).

not so change during 240–300 ns trajectory though minor differences were observed at high RMSF peaks. These results would reflect that RMSDs of the Arg456Hie and Arg456Hip mutants have not yet converged by 300 ns simulation. Based on these results, in order to unify the calculation condition with the wild type, we selected the last 20 ns (280–300 ns) MD trajectory for the RMSF calculation and focused on them in following discussion.

As for the structural flexibility of wild-type CYP1A2, high RMSF peaks were observed at the CD, FG, and HI loops (figure 4a), with RMSF values of 2.19 Å at Ser156, 1.32 Å at Phe246, and 1.58 Å at Gly299 [21] (table 2 and figure 5). Therefore, it is not unexpected that these RMSFs were relatively high in the two Arg456His mutants (figure 5). Importantly, the CD loop leads to the C helix, which is thought to be a constituent of the interaction site with CPR. Moreover, the FG and HI loops link to the F and I helices, respectively, which are considered to be substrate recognition sites [17] (figure 4b). Thus, the flexibility of these loops is expected to be important for the catalytic function of CYP1A2. In the Arg456Hie mutants, the RMSF of the HI loop was extremely high (3.43 Å) (figure 5), while in the Arg456Hip mutant, the CD loops were so flexible that the RMSF value was 3.15 Å. Thus, neither the Arg456Hie nor the Arg456Hie mutants seem to have appropriate dynamics for the catalytic function. Moreover, an RMSF peak of 1.80 Å was observed at Gly102 of the BC loop in the Arg456Hip mutant, which was not observed in the wild type, with a value of 0.71 Å at Gly102. As the BC loop also constitutes part of the substrate recognition site, flexibility would impair the ability to recognize some substrates.

In order to identify the differences in interaction of Hie456 and Hip456, hydrogen bond frequency was calculated using the last 20 ns of trajectories (figure 7a and 7b). In the Arg456Hie mutants (figure 7a), the imidazole hydrogen of Hie456 formed hydrogen bonds with the hydroxyl group of Ser129 at a frequency of 19.3% (average distance: 2.89 Å). On the other hand, Hip456 did not interact with Ser129 in the Arg456Hip mutant (average distance: 6.2 Å). Although the Arg456Hip mutants could easily form hydrogen bonds by the charged imidazole, hydrogen bonds were not observed between Hip456 and any residue with a frequency of >1%. This result indicates that interactions of Hie456 and Hip456 were different because of the protonation state. Importantly, in wild-type CYP1A2, the guanidine nitrogen (NH1) interacted with the hydroxyl oxygen of Ser126 (not Ser129) at a frequency of 43.2% (average distance: 2.82 Å), suggesting that interactions by Hie456 and Hip456 were not the same as those of Arg456. In mammalian CYP structures, though arginine of the Cys-pocket is highly conserved [18], in some bacterial CYPs such as P450cam and P450terp, the corresponding arginine is substituted by histidine [19]. However, our results imply that His could not be used for the conserved arginine in the case of human CYP1A2.

IOP Conf. Series: Journal of Physics: Conf. Series **1136** (2018) 012023 doi:10.1088/1742-6596/1136/1/012023

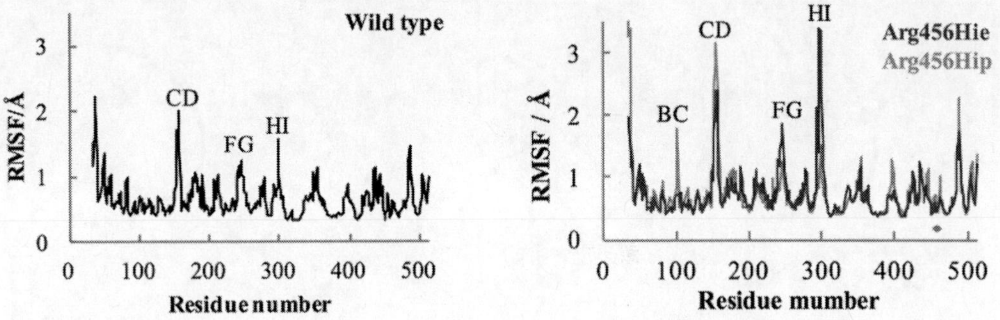

Figure 5. Structural flexibilities for the wild type (black), Arg456Hie (blue) and Arg456Hip (magenta). Highly flexible loops are labeled. The mutated position was marked with a red dot.

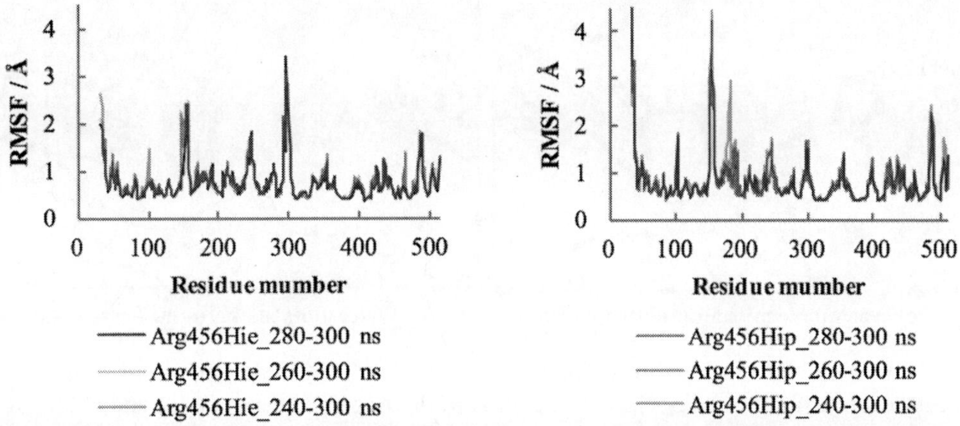

Figure 6. Superimposition of structural flexibilities for Arg456Hie (top) and Arg456Hip (bottom) using last 20, 40, 60 ns MD trajectories Highly flexible loops are labeled.

Table 3. Values of high RMSF peaks at BC, CD, FG, and HI loops in Arg456His mutants, calculated using last 20, 40, 60 ns of trajectories.

	MD trajectory	Gly102 of BC loop	Ser156 of CD loop	Phe246 of FG loop	Gly299 of HI loop
Arg456Hie mutant	280-300 ns	0.75[1]	2.39	1.84	3.43[2]
	260-300 ns	0.83[1]	2.44	1.72	3.06
	240-300 ns	1.48	2.15	1.61	2.81
Arg456Hip mutant	280-300 ns	1.80	3.15[3]	1.26	1.51
	260-300 ns	1.84	4.23[4]	1.73	1.67
	240-300 ns	1.66	4.44[4]	1.61	1.69

[1] RMSF value at Asp103.
[2] RMSF value at Ser298.
[3] RMSF value at Ser155.
[4] RMSF value at Ser154.

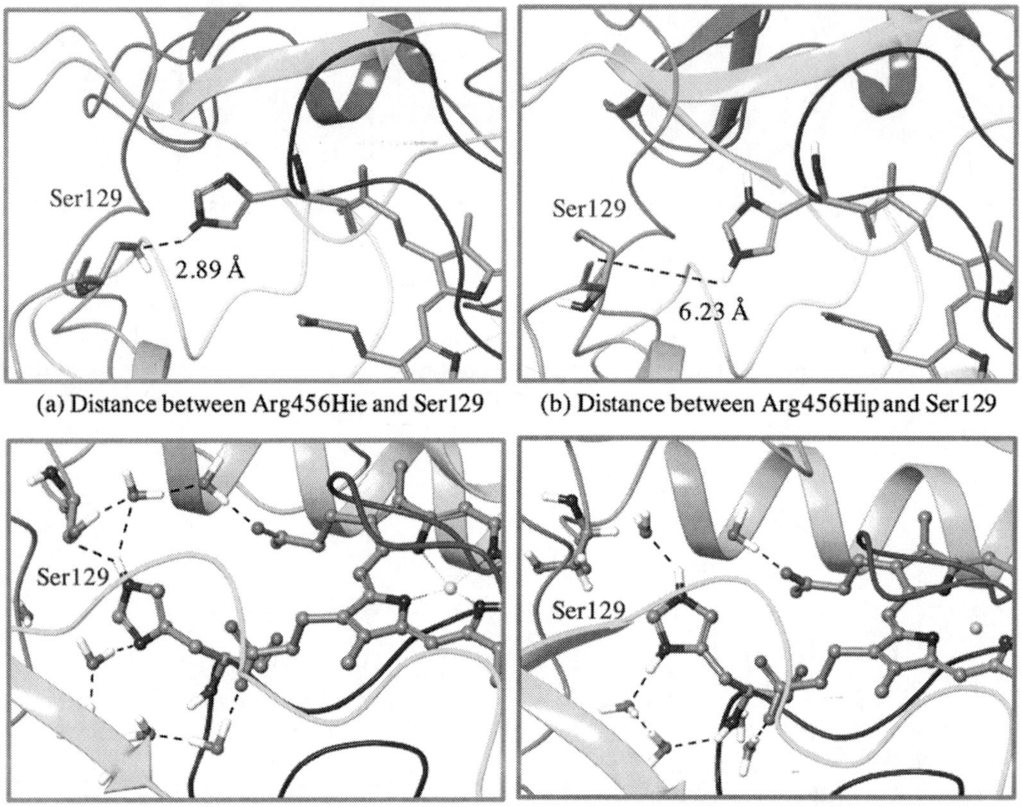

(a) Distance between Arg456Hie and Ser129 (b) Distance between Arg456Hip and Ser129

(c) Water molecules around Arg456Hie (d) Water molecules around Arg456Hip

Figure 7. Interacting residues and water molecules around Arg456 in the optimized structures. Distances between His456 and Ser126 were 2.89 Å in the Arg456Hie mutant (a) and 6.23 Å in the Arg456Hip mutant (b). Water molecules surrounding His456 in the Arg456Hie (c) and Arg456Hip (d) mutants. Residues and water displayed with ball-and-stick diagram, hydrogen bonds shown with black dashed line.

To study the cause of the differing interactions of Hie456 and Hip456, we compared the optimized structures including solvent water (figure 7c and 7d). In the Arg456Hie mutant, Hie456 interacted with one water molecule as well as Ser126 (figure 7c). On the other hand, Hip456 of the Arg456Hip mutant formed hydrogen bonds with two water molecules. This result implies that interaction with water molecules could change the hydrogen bond network of the Arg456Hie and Arg456Hip mutants.

4. Conclusion

Although we conducted 300-ns MD simulations for both Arg456Hie and Arg456Hip mutants, equilibrium structures could not be obtained, indicating that the Arg456His mutation has a large influence on the 3D structure of CYPA2 independently of the protonation state. Moreover, the 3D structures of the Arg456His mutants were quite different from that of the wild-type CYP1A2, suggesting that the mutants could not maintain these structures and have the same enzymatic activities as the wild type. Finally, the Ag456Hie and Arg456Hip mutants were significantly different in their 3D structures, flexibilities, and interactions around His456. Therefore, the Arg456His mutation differently affect on the CYP1A2 structure because of the His456 protonation state. Although the RMSDs have not yet been converged by 300 ns MD simulations, the Arg456His mutation could largely affect the 3D structure of CYP1A2 regardless of the His456 protonation state.

Acknowledgments
This work was supported by a grant-in-aid for scientific research [17K08257] from the Japan Society for the Promotion of Science and Showa University Research Grant for Young Researchers.

References

[1] Nelson DR 1999 Cytochrome P450 and the individuality of species *Arch. Biochem. Biophys.* **369** 1–10
[2] Nebert DW and Russell DW 2002 Clinical importance of the cytochromes P450 *Lancet.* **360** 1155–62
[3] Anzenbacher P and Anzenbacherová E 2001 Cytochromes P450 and metabolism of xenobiotics *Cell. Mol. Life. Sci.* **58** 737–47
[4] Guengerich FP 2008 Cytochrome P450 and chemical toxicology *Chem. Res. Toxicol.* **21** 70–83
[5] Evans WE and Relling MV 1999 Pharmacogenomics: translating functional genomics into rational therapeutics *Science.* **286** 487–91
[6] Bozina N, Bradamante V and Lovrić M 2009 Genetic polymorphism of metabolic enzymes P450 (CYP) as a susceptibility factor for drug response, toxicity, and cancer risk *Arh. Hig. Rada. Toksikol.* **60** 217–42
[7] Groves JT 2005 Models and mechanisms of cytochrome P450 action. *In: Cytochrome P450: Structure, Mechanism, and Biochemistry,* ed P.R. Ortiz de Montellano (New York: Kluwer Achademic/Plenum Publishers) pp.1–43
[8] Chang YT, Stiffelman OB, Vakser IA, Loew GH, Bridges A and Waskell L 1997 Construction of a 3D model of cytochrome P450 2B4 *Protein. Eng.* **10** 119–29
[9] Bridges A, Gruenke L, Chang YT, Vakser IA, Loew G and Waskell L 1998 Identification of the binding site on cytochrome P450 2B4 for cytochrome b5 and cytochrome P450 reductase *J. Biol. Chem.* **273** 17036–49
[10] Hamdane D, Xia C, Im SC, Zhang H, Kim JJ and Waskell L 2009 Structure and function of an NADPH-Cytochrome P450 oxidoreductase in an open conformation capable of reducing cytochrome P450 *J. Biol. Chem.* **284** 11374–84
[11] Shimizu T, Tateishi T, Hatano M and Fujii-Kuriyama Y 1991 Probing the role of lysines and arginines in the catalytic function of cytochrome P450d by site-directed mutagenesis. interaction with NADPH-cytochrome P450 reductase *J. Biol. Chem.* **266** 3372–5
[12] Haines DC, Tomchick DR, Machius M and Peterson JA 2001 Pivotal role of water in the mechanism of P450BM-3 *Biochemistry.* **40** 13456–65
[13] Schlichting I, Berendzen J, Chu K, Stock AM, Maves SA, Benson DE, Sweet RM, Ringe D, Petsko GA and Sligar SG 2000 The catalytic pathway of cytochrome p450cam at atomic resolution *Science.* **287** 1615–22
[14] Prasad B, Mah DJ, Lewis AR and Plettner E 2013 Water oxidation by a cytochrome P450: mechanism and function of the reaction *PLoS ONE.* **25** e61897
[15] Ito M, Katono Y, Oda A, Hirasawa N and Hiratsuka M 2015 Functional characterization of 20 allelic variants of CYP1A2 *Drug. Metab. Pharmacokinet.* **30** 247–52
[16] Saito Y, Hanioka N, Maekawa K, Isobe T, Tsuneto Y, Nakamura R, Soyama A, Ozawa S, Tanaka-Kagawa T, Jinno H, Narimatsu S and Sawada J 2005 Functional analysis of three CYP1A2 variants found in a Japanese population *Drug. Metab. Dispos.* **33** 1905–10
[17] Gotoh O 1992 Substrate recognition sites in cytochrome P450 family 2 (CYP2) proteins inferred from comparative analyses of amino acid and coding nucleotide sequences *J. Biol. Chem.* **267** 83–90
[18] Syed K and Mashele SS 2014 Comparative analysis of P450 signature motifs EXXR and CXG in the large and diverse kingdom of fungi: identification of evolutionarily conserved amino acid patterns characteristic of P450 family *PLoS ONE.* **9** e95616
[19] Hasemann CA, Kurumbail RG, Boddupalli SS, Peterson JA and Deisenhofer J 1995 Structure and function of cytochromes P450: a comparative analysis of three crystal structures

Structure. **3** 41–62

[20] Guallar V, Baik MH, Lippard SJ and Friesner RA 2003 Peripheral heme substituents control the hydrogen-atom abstraction chemistry in cytochromes P450 *Proc. Natl. Acad. Sci. U. S. A.* **100** 6998–7002

[21] Watanabe Y, Fukuyoshi S, Hiratsuka M, Yamaotsu N, Hirono S, Takahashi O and Oda A 2016 Prediction of three-dimensional structures and structural flexibilities of wild-type and mutant cytochrome P450 1A2 using molecular dynamics simulations *J. Mol. Graph. Model.* **68** 48–56

[22] Sansen S, Yano JK, Reynald RL, Schoch GA, Griffin KJ, Stout CD and Johnson EF 2007 Adaptations for the oxidation of polycyclic aromatic hydrocarbons exhibited by the structure of human P450 1A2 *J. Biol. Chem.* **282** 14348–55

[23] Kobayashi K, Takahashi O, Hiratsuka M, Yamaotsu N, Hirono S, Watanabe Y and Oda A 2014 Evaluation of influence of single nucleotide polymorphisms in cytochrome P450 2B6 on substrate recognition using computational docking and molecular dynamics simulation *PLoS ONE*. **9** e96789

[24] Oda A, Yamaotsu N and Hirono S 2004 Studies of binding modes of (S)-mephenytoin to wild types and mutants of cytochrome P450 2C19 and 2C9 using homology modeling and computational docking *Pharm. Res.* **21** 2270–8

[25] Fukuyoshi S, Kometani M, Watanabe Y, Hiratsuka M, Yamaotsu N, Hirono S, Manabe N, Takahashi O and Oda A 2016 Molecular dynamics simulations to investigate the influences of amino acid mutations on protein three-dimensional structures of cytochrome P450 2D6.1, 2, 10, 14A, 51, and 62, *PLoS ONE* **11** e0152946

[26] Joung IS and Cheatham TE III 2008 Determination of alkali and halide monovalent ion parameters for use in explicitly solvated biomolecular simulations *J. Phys. Chem. B* **112** 9020–41

[27] Darden TA, York DM and Pedersen LG 1993 Particle mesh Ewald: An N·log(N) method for Ewald sums in large systems *J. Chem. Phys.* **98** 10089–92

[28] Oda A, Yamaotsu N and Hirono S 2005 New AMBER force field parameters of heme iron for cytochrome P450s determined by quantum chemical calculations of simplified models *J. Comput. Chem.* **26** 818–26

[29] Giammona DA 1984 Ph.D. thesis, University of California, Davis.

[30] Case DA, Darden TA, CheathamTE III, Simmerling CL, Wang J, Duke RE, Luo R, Walker RC, Zhang W, Merz KM, Roberts B, Hayik S, Roitberg A, Seabra G, Swails J, Götz AW, Kolossváry I, Wong KF, Paesani F, Vanicek J, Wolf RM, Liu J, Wu X, Brozell SR, Steinbrecher T, Gohlke H, Cai Q, Ye X, Wang J, Hsieh M-J, Cui G, Roe DR, Mathews DH, Seetin MG, Salomon-Ferrer R, Sagui C, Babin V, Luchko T, Gusarov S, Kovalenko A and Kollman PA AMBER 12 2012 University of California, San Francisco.

[31] Wang J, Wolf RM, Caldwell JW, Kollman PA and Case DA 2004 Development and testing of a general amber force field *J. Comput. Chem.* **25** 1157–74

[32] Ryckaert JP, Ciccotti G and Berendsen HJC 1977 Numerical integration of the cartesian equations of motion of a system with constraints: molecular dynamics of n-alkanes *J. Comput. Phys.* **23** 327–41

[33] Watanabe Y, Fukuyoshi S, Kato K, Hiratsuka M, Yamaotsu N, Hirono S, Gouda H and Oda A 2017 Investigation of substrate recognition for cytochrome P450 1A2 mediated by water molecules using docking and molecular dynamics simulations *J. Mol. Graph. Model.* **74** 326–36

[34] Muralidhara BK, Negi S, Chin CC, Braun W, Halpert JR 2006 Conformational flexibility of mammalian cytochrome P450 2B4 in binding imidazole inhibitors with different ring chemistry and side chains. Solution thermodynamics and molecular modeling *J. Biol. Chem.* **281** 8051–61

Molecular dynamics study of the interaction between PACAP (6'–38') and the N-terminal extracellular domain of the human splice variant hPAC1-R-short, aiming at the development of neuropathic pain medicine

Y Watanabe[1,2] A Oda[2,3,4] and H Gouda[1*]

[1]School of Pharmacy, Showa University, 1-5-8 Hatanodai, Shinagawa-ku, Tokyo 142-8555, Japan
[2]Faculty of Pharmacy, Institute of Medical, Pharmaceutial and Health Science, Kanazawa University, Kakuma, Kanazawa, Ishikawa 920-1192, Japan
[3]Graduate School of Pharmacy, Meijo University, 150 Yagotoyama, Tempaku-ku, Nagoya, Aichi 468-8503, Japan
[4]Institute for Protein Research, Osaka University, 3-2 Yamadaoka, Suita, Osaka 565-0871, Japan

[*]E-mail: godah@pharm.showa-u.ac.jp

Abstract. Chronic pathetic pain is caused by an external injury to the peripheral and central nervous systems. The pathogenic mechanism has not been elucidated and existing analgesics are insufficiently effective for pain relief. In the nervous conduction of pain, pituitary adenylate cyclase-activating polypeptide (PACAP), which is constituted by 38 or 27 amino acid residues, acts as a neurotransmitter. Therefore, the selective receptor, namely pituitary adenylate cyclase-activating polypeptide type I receptor (PAC1-R), is a potential target for alleviating neuropathic pain. Detailed investigation on the interaction between PACAP (6'–38') and PAC1-R would be informative for structure-based drug design of the PAC1-R inhibitor because the short analog PACAP (6'–38') acts as a potent antagonist. In this study, we conducted molecular dynamics simulations to investigate the solution structure of the complex of PACAP (6'–38') and the N-terminal extracellular domain of human splice variant hPAC1-R-short (hPAC1-RS). As a result, several strong hydrogen bonds were observed between acidic residues of hPAC1-RS and basic residues of PACAP (6'–38'), which suggests that these residues are important for peptide binding and ligand selectivity.

1. Introduction

Chronic pathetic pain is defined as a prolonged or recurrent pain that persists long after the usual response time to an acute disease, which is represented by an intractable neurogenic disorder such as postherpetic neuralgia or sciatica [1]. Although the onset of chronic pathetic pain is hypothesized to be caused by an external injury to the central and peripheral nerves [1], the pathogenic mechanism of chronic neuropathy has not been elucidated and existing analgesics such as non-steroidal anti-inflammatory agents are insufficiently effective for pain relief. In the nervous conduction of pain, pituitary adenylate cyclase-activating polypeptide (PACAP), which is constituted by 38 or 27 amino acid residues, and vasoactive intimal peptide (VIP), which consists of 28 residues, act as a neurotransmitter, or nerve modulation factor. PACAP and VIP are known to be important for neuroprotection, neurotrophy and nerve growth [2,3]. PACAP and VIP are bound to the following two subtypes of the Class II G-protein-coupled receptors (GPCR) [4]: subtype I and subtype II. Although the affinity of the subtype II receptor for PACAP is similar to that of VIP, the subtype I receptor selectively binds to PACAP over VIP, with a difference in affinity of more than 1,000-fold [5]. The subtype I PACAP selective receptor, namely the pituitary adenylate cyclase-activating polypeptide type I receptor (PAC1-R), is expressed in the central and peripheral nervous systems and has been suggested as a potential target for the treatment of neuropathic pain [2]. Inhibition of the interaction between PACAP and PAC1-R is expected to reduce neuropathic pain and lead to the development of a new strong painkiller.

Recently, the NMR structure of PACAP (6'–38') in complex with the N-terminal extracellular (EC) domain of the human splice variant hPAC1-R-short (hPAC1-RS) was reported [6]. PACAP (6'–38') is a shorter analog of PACAP-38 and is reported to act as a potent antagonist with a similar affinity to that of a PACAP-38. Therefore, detailed investigation of the interaction between PACAP (6'–38') and the N-terminal EC domain of hPAC1-RS would be informative for structure-based drug design of the PAC1-R inhibitor. In this study, we conducted a molecular dynamics (MD) simulation to investigate the solution structure of the complex of PACAP (6'–38') and the N-terminal EC domain of hPAC1-RS. Using the MD trajectory, the interaction between PACAP and PAC1-R was studied to identify the key residues that are responsible for the strong binding, which would be useful in designing the PAC1-R inhibitor [7]. Moreover, we conducted additional MD simulations of the peptide-unbound N-terminal EC domain of hPAC1-RS to investigate the structural differences that are induced by peptide binding. These results are expected to enhance the understanding of the peptide recognition mechanism and the selectivity of PAC1-R. Furthermore, since PAC1-R was discovered from rat cDNA in 1993 [8], its function and potency have attracted attention in the field of pathophysiology. For example, PAC1-R is highly expressed in the hypothalamus, the brain stem, and the limbic system [9], and is hypothesized to be associated with several diseases, such as psychiatric disorders [10], schizophrenia [11], and Alzheimer's disease [12]. Therefore, the insight about the PAC1-R structure might be lead to the discovery of several effective medicines.

2. Methods

The initial three-dimensional (3D) structure of PACAP (6'–38') and the N-terminal EC domain of the hPAC1-Rs complex (26th to 122th residues) were obtained from the Protein data bank (PDB ID: 2JOD), as determined by NMR spectroscopy [6]. The ligand-free EC domain of hPAC1-RS (26th to 119th residues) was registered as the crystal structure (PDB ID: 3N94) and included some crystal water molecules [13]. These water molecules were removed from the initial structure before minimization. The complex and ligand-free structure, in this article, are denoted as PACAP–PAC1R and ligand-free PAC1R, respectively.

MD simulations were conducted using AMBER14 program [14]. AMBER ff14SB force field [15] was used for amino acid residues. The leap module of AmberTools was used to supplement the missing heavy atoms and hydrogen atoms that were not observed in the experimental structures. The systems were solvated using the TIP3P model [16]. Sodium ions were arranged to neutralize PAC1-R. The cyclic boundary condition was applied with 8-Å margins from the protein surfaces. The cutoff distance for van der Waals interaction was set to 10 Å, and the particle mesh Ewald method was used for calculating electrostatic interactions [17]. The SHAKE algorithm was used to constrain the bond stretching of hydrogen atoms [18]. Before MD simulations, the solvent model was relaxed and the entire systems were minimized. Next, the temperature-increasing MD simulations (from 0 to 300 K) were performed for 20 ps. Finally, equilibrating MD simulations were performed using the Langevin thermostat for 80 ns under constant pressure at 300 K. The time step of all MD simulations was 2 fs.

To validate the 3D structures that were obtained by MD simulations, root mean square deviations (RMSDs) for main-chain heavy atoms were calculated along MD trajectories. The 3D structures that were obtained after temperature-increasing MD simulations were used as reference structures for the

IOP Conf. Series: Journal of Physics: Conf. Series **1136** (2018) 012024 doi:10.1088/1742-6596/1136/1/012024

RMSDs. To investigate the structural flexibilities of residues, root mean square fluctuations (RMSFs) for Cα atoms were calculated using 60–80 ns MD trajectories, in which average structures were used as reference structures. To identify hydrogen bond interactions between PACAP and PAC1R, 60–80 ns MD trajectories (10,000 frames) were also analyzed. For the hydrogen bond criteria, the cutoff distance between two heavy atoms was set to 3 Å and the cutoff angles between the acceptor, hydrogen, and donor atoms were set to 120°. The hydrophobic interactions were determined by measuring distance of two hydrophobic residues with a cut-off distance of 4 Å. RMSDs and RMSFs were calculated by the cpptraj module of AmberTools14. The binding free energy (ΔG_{bind}) calculation was conducted using the molecular mechanics Poisson-Boltzman surface area (MM-PBSA) python script included within the AMBER14 package. The value of the relative dielectric constant of protein was set to 2, and that of the bulk solvent was set to 80. The MM-PBSA calculation was performed using 1,000 snapshots, which were extracted with an interval of 20 ps from the last 20 ns trajectory. The *in silico* alanine scanning protocol was further performed to investigate the contribution of a specific residue to the ΔG_{bind}.

3. Results and Discussion

3.1. Structural differences between the experimental structure and optimized structure that was obtained after MD.

80-ns MD simulations were conducted for both ligand-free PAC1R and PACAP–PAC1R (Fig. 1 and 2). The MD simulations seemed to converge to equilibrium after 50–60 ns. Therefore, the 60–80 ns MD trajectories were used for the analyses in the next sections.

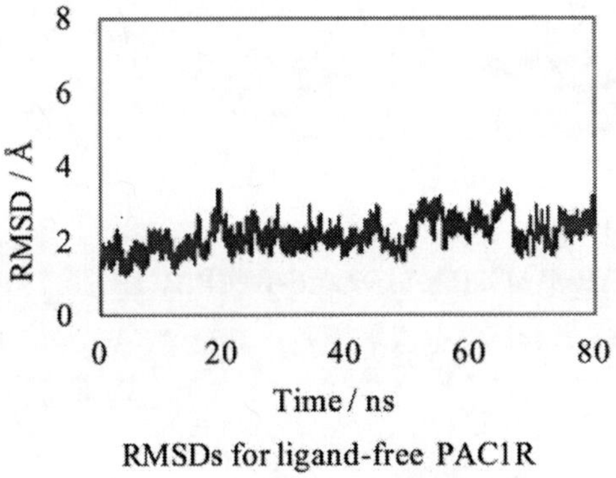

Figure 1. RMSDs for main-chain heavy atoms of ligand-free PAC1R.

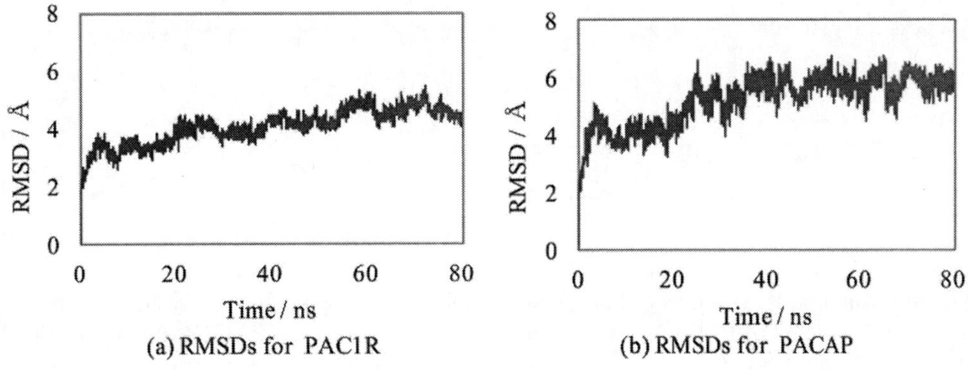

Figure 2. RMSDs for main-chain heavy atoms of PACAP–PAC1R. (a) PAC1R and (B) PACAP.

IOP Conf. Series: Journal of Physics: Conf. Series **1136** (2018) 012024 doi:10.1088/1742-6596/1136/1/012024

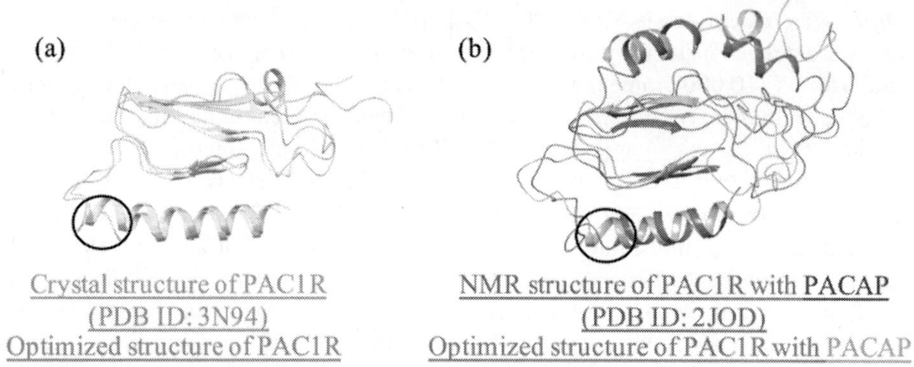

Figure 3. Superimposition of the experimental structures and optimized structures that were obtained after MD. (a) The 3N94 crystal structure of PAC1R (green) and its optimized structure (pale blue). (b) The 2JOD NMR structure of PAC1R (gray) and PACAP (violet) and the optimized structure of PAC1R (blue) and PACAP (pink). The partial loops of Arg40–Leu44 in the experimental structures are circled.

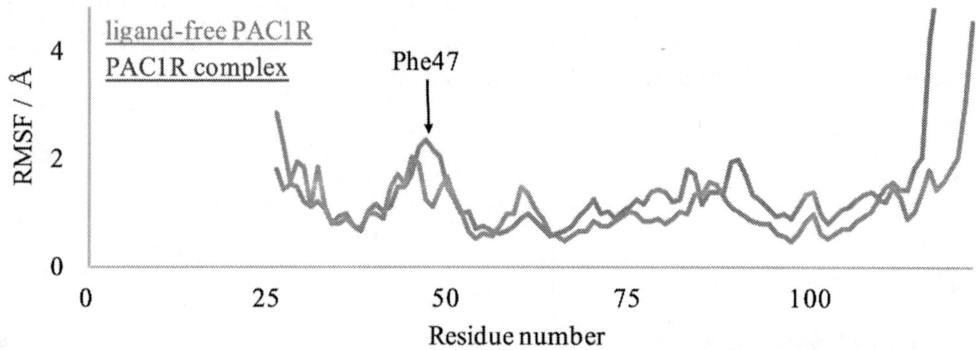

(a) Superimposition of RMSFs for PACAP–PAC1R and ligand-free PAC1R

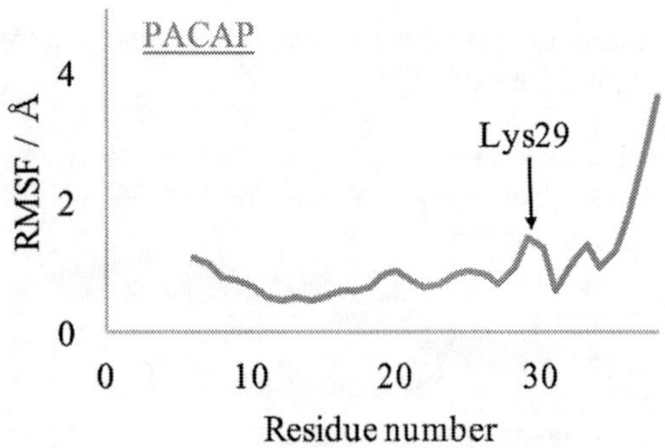

(b) RMSFs for PACAP

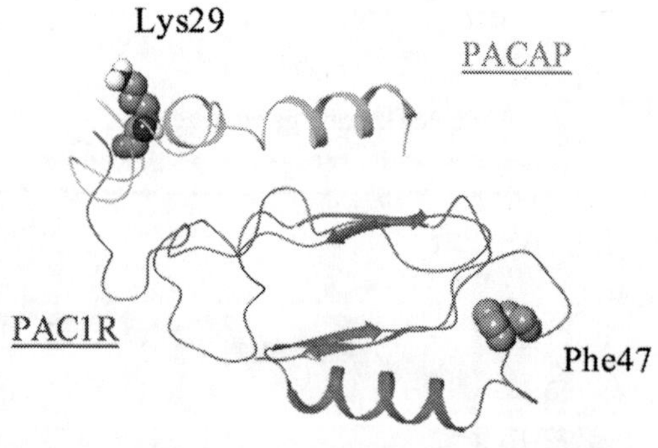

(c) Position of high RMSF peaks in PACAP–PAC1R

Figure 4. Structural flexibilities of PACAP–PAC1R and the ligand-free PAC1R. (a) Superimposed RMSFs for Cα atoms of PAC1R–PACAP (blue) and the ligand-free PAC1R (green). High RMSF peaks are indicated by arrows and residues are labeled. The RMSF values of C-terminal residues are not considered because they were exposed and not involved in the interaction with PACAP. (b) RMSF of PACAP. (c) Residues of high RMSF peaks in the PACAP–PAC1R complex.

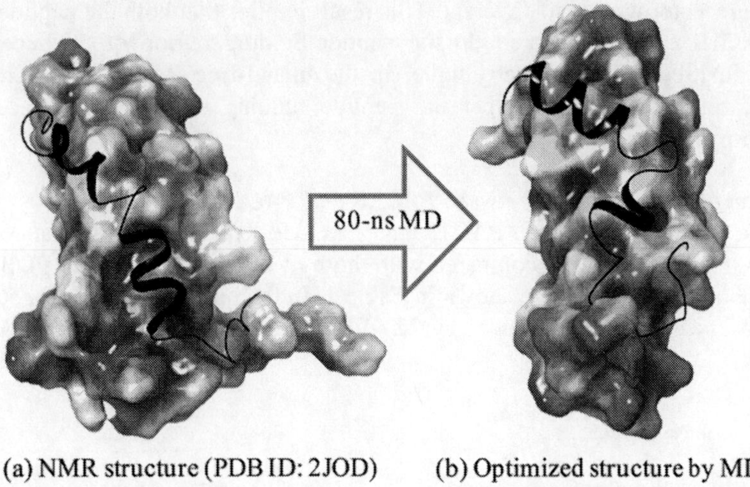

(a) NMR structure (PDB ID: 2JOD) (b) Optimized structure by MD

Figure 5. Molecular surfaces of PACAP–PAC1R, colored according to the electrostatic potential. (a) the NMR structure and (b) the optimized structure by MD. PACAP is shown as a black ribbon.

The RMSD values of ligand-free PAC1R converged to relatively small values of approximately 1.5 ~ 3.3 Å after 50 ns (the final RMSD: 2.60 Å), as shown in Fig. 1, which suggests that the crystal structure of ligand-free PAC1R was well retained during MD simulation. In contrast, the RMSD values of PAC1R in PACAP–PAC1 were approximately 4 Å during 60–80 ns (the final RMSD: 4.37 Å; Fig.2(a)), which is somewhat higher than that of the ligand-free PAC1R. Moreover, the RMSD values of PACAP reached almost 6 Å, though it also reached convergence after 40 ns (the final RMSD: 5.69 Å). Since PACAP is a linear peptide that is constituted by 33 residues (6'–38'), the high RMSD value of PACAP would be acceptable. It was expected that the high RMSD values of PAC1R in PACAP–PAC1R were due to the binding of the flexible PACAP.

Figures 3(a) and 3(b) show superimpositions of the initial and optimized structures of the ligand-free PAC1R and PACAP–PAC1R, respectively. Additionally, the RMSFs of their Cα atoms are shown in Figure 4. According to Fig. 3(a), although significant differences were not observed between the 3N94 crystal structure and the optimized ligand-free PAC1R, the length of the α1 helix was shortened and the residues between Arg40 to Leu44 formed a loop in the optimized structure. Likewise, a loop substructure from Arg40 to Leu44 was also observed in the optimized PACAP–PAC1R (Fig. 3(b)). In both optimized structures, the main-chain oxygen atom of Arg40 did not interact with Leu44 to form the α-helix conformation. Moreover, although the side chains of Arg40 were exposed to the solvent in both initial structures, those of Arg40 interacted with Glu36 in both optimized structures, which is why the α1 helices were partially unfolded. Moreover, in the loop that leads to the α1 helix, the highest RMSF peaks were observed in both the ligand-free PAC1R and PACAP–PAC1R (Fig, 4(a)), which suggests that these regions were originally flexible despite the PACAP binding. The RMSF peak at Phe47 was higher in the ligand-free PAC1R (2.33 Å) than in PACAP–PAC1R (1.23 Å). The RMSF differences around Phe47 were distant from the peptide binding site; thus, it might not be involved in the interaction with PACAP.

The receptor structures were similar as compared with the initial and optimized PACAP–PAC1R structures (Fig. 3(b)), except for the peptide-interacting region. Regarding the optimized PACAP structures (Fig. 3(b)), the α-helix conformation was unfolded and the C-terminal side of the peptide had moved from its initial position. Figure 5 shows the molecular surfaces of the initial and the optimized PACAP–PAC1R structures, colored with the electrostatic potential. As shown in Fig. 3(b) and Fig. 5, the peptide conformation was significantly changed by 80-ns MD and the molecular surfaces of the receptor were also different. This result implies that both the peptide and the interaction surface of PAC1R easily fluctuated. In the peptide-binding region of the receptor (72nd to 122nd residues), the RMSFs were relatively higher in the ligand-free PAC1R structure than in PACAP–PAC1R structure, which suggests that the peptide binding stabilized the receptor structure and restricted its movement.

3.2 Comparison of the interaction between PACAP and PAC1R.

To identify the interaction between PAC1R and PACAP, hydrogen bonds that were observed in the optimized PAC1R–PACAP were compared with those of the initial structure (PDB ID: 2JOD; Fig. 6). The hydrogen bond frequencies are shown in Table 1. In the initial structure (Fig. 6(a)), five hydrogen bonds were observed, in Ser11'–Glu71, Tyr22'–Val92, Lys32'–Thr122, Lys36'–Asp24, and Arg34'–Asp24.

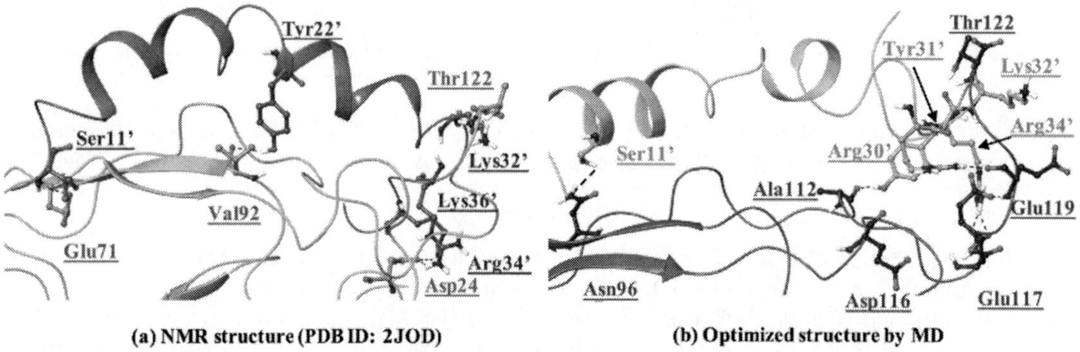

Figure 6. Hydrogen bonds between PACAP and PAC1R of (a) the initial NMR structure (PDB ID: 2JOD) and (b) the optimized structure. Hydrogen bonds > 10 % are shown as black dashed lines.

Table 1. Hydrogen bond frequencies using 60–80 ns MD trajectories of PAC1R–PACAP*

		PAC1R (26–122)						
		Glu79	Asn96	Ala112	Asp116	Glu117	Glu119	Thr122
PACAP (6'–38')	Ser11'		25.8 %					
	Gly28'							9.3 %
	Lys29'							10.0 %
	Arg30'	37.1 %					40.9 %	
	Tyr31'		4.8 %	62.8 %				
	Lys32'					42.2 %		
	Arg34'				63.2 %			

*1,000 frames were calculated and hydrogen bond frequencies of over 3 % are listed.

In contrast, in the optimized structure (Fig. 6(b)), strong hydrogen bonds (of frequency > 20 %) were formed in Ser11'–Asn96 (25.8 %), Arg30'–Glu79 (37.1 %), Arg30'–Glu119 (40.9 %), Tyr31'–Ala112 (62.8 %), Lys32'–Glu117 (42.2 %), and Arg34'–Asp116 (63.2 %). Comparing the hydrogen bonds of the optimized structure with those of the initial structure, although the hydrogen bond patterns were not same, the basic residues of PACAP (Lys32' and Arg34') and the hydrophilic residues (Ser11' of PACAP and Thr122 of PAC1R) were equally involved in the hydrogen-bonding interaction. As shown in Table 1, the C-terminal side of PACAP has 4 basic residues (Lys29', Arg30', Lys32', and Arg34') that form strong hydrogen bonds. This result suggests that the basic residues of PACAP participate in the peptide binding and the selectivity by electrostatic interaction. According to the experimental study, the binding affinity of PAC1R for PACAP-38 is much higher than for PACAP-27 [3]. Moreover, the mutational study by Sun et al. showed that the deletion of the 29'th to 32'th residues of PACAP significantly decreased its binding affinity (inhibition constant $K_i > 15{,}000$ nM) [6]. Additionally, in their single mutational experiments, the dissociation constant K_d for PACAP (6'–38') was 350 nM, whereas those of the Lys29'Glu and Arg30'Ala mutants were 1,000 nM and 1,500 nM, respectively. Thus, our calculation result would be in agreement with those of the experimental results. Regarding the PAC1 residues, the dissociation constant K_d of wild-type PAC1R was 200 nM and that of mutant Glu117Ala was 1,050 nM, which indicate that Glu117 played an important role in the peptide-binding. It is expected that the acidic residues of PAC1R (Asp116, Glu117, and Glu119) also contribute to the peptide binding and the selectivity. In order to investigate the contribution of these polar residues on the binding affinity, we conducted binding free energy (ΔG_{bind}) calculation by MM-PBSA method and *in silico* alanine scanning protocol. In the alanine scanning, each of acidic residues of PAC1R (Asp116, Glu117, Glu119) and basic residues of PACAP (Arg30', Lys32', Arg34') was substituted by alanine because they form strong hydrogen bonds between the receptor–peptide. Moreover, in order to compare with these polar residues, each of Ser120 of PAC1R and Gln33' of PACAP, which do not form hydrogen bonds between the receptor–peptide, was also substituted by alanine. The contribution of a specific residue was estimated by a difference ($\Delta\Delta G_{bind}$) between a ΔG_{bind} value of the mutated complex, in which the specific residue was replaced by alanine, and that of the wild-type complex. The results are shown in Table 2. As expected, we could see that the contribution of Ser120 of PAC1R and Gln33' of PACAP were quite small, because the $\Delta\Delta G_{bind}$ values were estimated to be 0.25 and 0.50 kcal/mol, respectively. By contrast, regarding polar residues forming hydrogen bonds, $\Delta\Delta G_{bind}$ values were estimated to be much greater. For example, the $\Delta\Delta G_{bind}$ value of 12.49 kcal/mol was obtained for the R30'A mutant of PACAP. These results indicate that the polar residues forming strong hydrogen bonds have significant contribution on the binding affinity of the PAC1R–PACAP.

Table 2. Differences of binding free energies between mutated complex and wild-type complex*[1,2]

Alanine mutant	PAC1R				PACAP			
	D116A	E117A	E119A	S120A	R30'A	K32'A	Q33'A	R34'A
$\Delta\Delta G_{bind}$ (kcal/mol)	6.75 (1.85)	10.73 (4.12)	3.79 (1.15)	0.25 (0.41)	12.49 (4.11)	8.34 (4.89)	0.50 (1.48)	7.81 (2.22)

*[1] mean values using 1,000 frames extracted with an interval of 20 ps from 60-80 ns. Values in parenthesis were standard deviations.
*[2] $\Delta\Delta G_{bind}$ no more than 0.5 kcal/mol were shadowed.

Figure 7 shows the hydrophobic interaction between PACAP and PAC1R. In the optimized structure, Tyr10' was located near Met72 and Ala112. Moreover, Tyr22', Val26', and Leu27' interacted with Pro107, Leu74, and Leu80. These hydrophobic residues, except for Leu80 and Ala112, were substituted by alanine scanning and the binding constants of the alanine mutants were investigated [6]. For alanine mutation on PACAP, the values of the inhibition constant Ki for Y10'A, Y22'A, V26'A, and L27'A were 670 nM, > 15,000 nM, > 15,000 nM, and 6,800 nM, respectively (Ki of the wild-type: 350 nM). In contrast, for the alanine mutants of PAC1R, the values of the dissociation constant Kd for M72A, L74A, and P107A were 440 nM, 580 nM, and 840 nM, respectively (Kd of the wild-type: 200 nM). Thus, these hydrophobic interactions between PACAP and PAC1R also contribute greatly to the peptide binding.

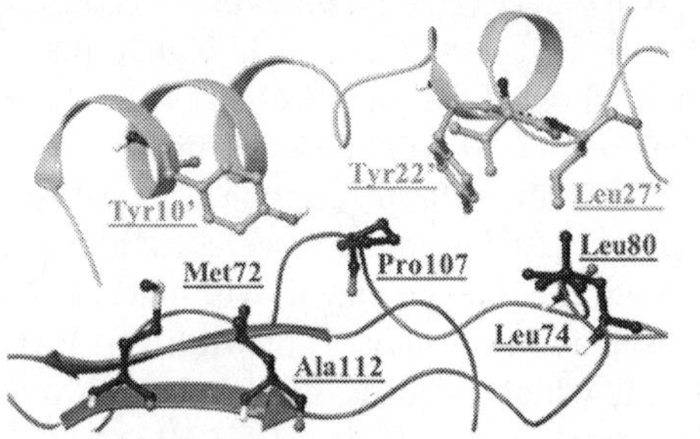

Figure 7. Hydrophobic interaction between PACAP and PAC1R of the optimized structure.

4. Conclusion

In this study, we investigated the interaction between the N-terminal EC domain of PAC1R and the potent antagonist PACAP (6'–38'). Strong hydrogen bonds were observed between the basic residues of PACAP (Arg30', Lys32', Arg34') and the acidic residues of PAC1R (Asp116, Glu117, Glu119). These residues would be important for peptide binding and ligand selectivity. Moreover, Met72, Leu74, Leu80, Pro107, and Ala112 of PAC1R and Tyr10', Tyr22', Val26', and Leu27' of PAC1R also contribute to peptide binding. Our results would be informative for developing a selective PAC1-R inhibitor.

References

[1] Merskey H and Bogduk N 1994 *Classification of Chronic pain* 2nd ed, Merskey H and Bogduk N (Seattle: IASP Press) pp 209–14
[2] Dickinson T and Fleetwood-Walker SM 1999 VIP and PACAP: very important in pain? *Trends Pharmacol. Sci.* **20** 324–9
[3] Laburthe M and Couvineau A 2002 Molecular pharmacology and structure of VPAC Receptors for VIP and PACAP *Regul. Pept.* **108** 165–73
[4] Rosenbaum DM, Rasmussen SG and Kobilka BK 2009 The structure and function of G-protein-coupled receptors *Nature* **459** 356–63
[5] Dickinson T, Fleetwood-Walker SM, Mitchell R and Lutz EM 1997 Evidence for roles of vasoactive intestinal polypeptide (VIP) and pituitary adenylate cyclase activating polypeptide (PACAP) receptors in modulating the responses of rat dorsal horn neurons to sensory inputs *Neuropeptides* **31** 175–85
[6] Sun C, Song D, Davis-Taber RA, Barrett LW, Scott VE, Richardson PL, Pereda-Lopez A, Uchic ME, Solomon LR, Lake MR, Walter KA, Hajduk PJ and Olejniczak ET 2007 Solution structure and mutational analysis of pituitary adenylate cyclase-activating polypeptide binding to the extracellular domain of PAC1-RS *Proc. Natl. Acad. Sci. USA* **104** 7875–80
[7] De Vivo M, Masetti M, Bottegoni G, Cavalli A 2016 Role of molecular dynamics and related methods in drug discovery *J. Med. Chem.* **59** 4035-61
[8] Hashimoto H, Ishihara T, Shigemoto R, Mori K and Nagata S 1993 Molecular cloning and tissue distribution of a receptor for pituitary adenylate cyclase-activating polypeptide *Neuron* **11** 333–42
[9] Otto C, Zuschrattera W, Gass P and Schützb G 1999 Presynaptic localization of the PACAP-typeI-receptor in hippocampal and cerebellar mossy fibres *Mol. Brain Res.* **66** 163–74
[10] Hashimoto H, Shintani N, Tanida M, Hayata A, Hashimoto R and Baba A 2011 PACAP is implicated in the stress axes *Curr. Pharm. Des.* **17** 985–89

[11] Hashimoto R, Hashimoto H, Shintani N, Chiba S, Hattori S, Okada T, Nakajima M, Tanaka K, Kawagishi N, Nemoto K, Mori T, Ohnishi T, Noguchi H, Hori H, Suzuki T, Iwata N, Ozaki N, Nakabayashi T, Saitoh O, Kosuga A, Tatsumi M, Kamijima K, Weinberger DR, Kunugi H and Baba A 2007 Pituitary adenylate cyclase-activating polypeptide is associated with schizophrenia *Mol. Psychiatry* **12** 1026–32

[12] Rat D, Schmitt U, Tippmann F, Dewachter I, Theunis C, Wieczerzak E, Postina R, van Leuven F, Fahrenholz F and Kojro E 2011 Neuropeptide pituitary adenylate cyclase-activating polypeptide (PACAP) slows down Alzheimer's disease-like pathology in amyloid precursor protein-transgenic mice *FASEB J.* **25** 3208–18

[13] Kumar S, Pioszak A, Zhang C, Swaminathan K and Xu HE 2011 Crystal structure of the PAC1R extracellular domain unifies a consensus fold for hormone recognition by class B G-protein coupled receptors *PLoS One* **6** e19682

[14] Case DA, Babin V, Berryman JT, Betz RM, Cai Q, Cerutti DS, Cheatham TE III, Darden TA, Duke RE, Gohlke H, Goetz AW, Gusarov S, Homeyer N, Janowski P, Kaus J, Kolossváry I, Kovalenko A, Lee TS, LeGrand S, Luchko T, Luo R, Madej B, Merz KM, Paesani F, Roe DR, Roitberg A, Sagui C, Salomon-Ferrer R, Seabra G, Simmerling CL, Smith W, Swails J, Walker RC, Wang J, Wolf RM, Wu X and Kollman PA 2014 *AMBER 14* University of California, San Francisco.

[15] Maier JA, Martinez C, Kasavajhala K, Wickstrom L, Hauser KE and Simmerling C 2015 ff14SB: Improving the accuracy of protein side chain and backbone parameters from ff99SB *J. Chem. Theor. Comput.* **11** 3696–713

[16] Joung IS and Cheatham TE III 2008 Determination of alkali and halide monovalent ion parameters for use in explicitly solvated biomolecular simulations *J. Phys. Chem. B* **112** 9020–41

[17] Darden TA, York DM and Pedersen LG 1993 Particle mesh Ewald: An N·log(N) method for Ewald sums in large systems *J. Chem. Phys.* **98** 10089–92

[18] Ryckaert JP, Ciccotti G and Berendsen HJC 1977 Numerical integration of the cartesian equations of motion of a system with constraints: molecular dynamics of n-alkanes *J. Comput. Phys.* **23** 327–41

IOP Conf. Series: Journal of Physics: Conf. Series **1136** (2018) 012025 doi:10.1088/1742-6596/1136/1/012025

Application of order-N first-principles DFT calculations with temperature controlled molecular dynamics to biomolecular system

Takao Otsuka[1] and Makoto Taiji[1]

[1] Laboratory for Computational Molecular Design, Quantitative Biology Center (QBiC), RIKEN, QBiC Building B, 6-2-4 Furuedai, Suita, Osaka 565-0874, Japan

e-mail: totsuka@riken.jp

Abstract. We have performed large-scale first-principle molecular dynamics (FPMD) calculations of the large-scale hydrated DNA system as an example of large-scale biomolecular system, using order-N density functional theory (DFT) code CONQUEST, which is based on the combined method with an order-N DFT method using the density matrix minimization scheme and the extended Lagrangian Born-Oppenheimer molecular dynamics (XL-BOMD) scheme. Our large-scale FPMD calculations by CONQUEST show stable FPMD simulation and good energy conservation in NVE ensemble. We also show temperature controlled MD simulations such as velocity scaling, stochastic velocity scaling, and Nosé-Hoover chain methods. Our protocols using velocity scaling, stochastic velocity rescaling, and Nosé-Hoover chain methods are very effective for temperature control in early stage of order-N FPMD simulations. Our order-N FPMD simulation would be able to realize the large-scale biomolecular simulations.

1. Introduction

Molecular simulation technology is now commonly used to explore biological functions of biomolecular systems. Especially, molecular simulation methodologies are expected to help us understand the mechanism of biological phenomena with dynamical molecular motion, for example, such as enzyme reaction, photoexcitation, or molecular interaction and so on. In these biological phenomena, electron plays an important role so that it needs molecular dynamics (MD) simulation based on quantum mechanics. Recently, quantum mechanics and molecular mechanics (QM/MM) hybrid methods or its combined molecular dynamics (QM/MM-MD) method are also used well in this field in order to elucidate the biological functions and mechanism [1]. However, there are still some problems in those QMMM hybrid method; for example, the artificial boundary problem, which is usually constrained with the artificial boundary between QM and MM regions, and the limitation of the size of QM region, which is also restricted due to the computational costs of QM calculation. Therefore, there are increasing demands for all-atom QM simulations on complex biological systems.

In recent progress for the QM calculations, an order-N (or linear-scaling) methodology for first-principle density functional theory (DFT) have been developed [2]. We have also developed order-N DFT code, CONQUEST, which has high parallel efficiency and enable us to perform DFT studies on very large-scale systems [2,3]. Recently, CONQUEST have been introduced very stable and efficient order-N first-principle MD (FPMD) method combined with the extended Lagrangian Borm-

Oppenheimer MD scheme proposed by Niklasson [4,5] and have shown the reliable and efficient FPMD on large-scale systems [6,7].

In practical application of large-scale FPMD to large-scale biomolecular systems, in many case, the initial structure is still prepared by using classical MD simulations. The equilibrium structure obtained by classical MD simulation is set as ones of large-scale FPMD simulation. In such case, especially huge-scale biomolecular systems, it is very important to achieve the setting temperature rapidly, without taking too much real computational time in large-scale FPMD simulations. In this study, we report further progress and preliminary results for the temperature controlled FPMD simulations on the order-N FPMD code in biomolecular system. As the temperature controlled method, we used velocity scaling method, stochastic velocity rescaling method [25] and Nosé-Hoover chain method [26-28] in NVT ensemble. Recently, two method of temperature control such as the velocity scaling and the Nosé-Hoover chain method were implemented in CONQUEST code [7,8]. In this study, we implemented the stochastic velocity rescaling method in CONQUEST code. Although we have already shown the effective results of the velocity scaling method [7] in biomolecular system, here we applied the stochastic velocity rescaling method and the Nosé-Hoover chain method in CONQUEST code to biomolecular system. Especially, in this study, we focused on early stage of large-scale FPMD steps to investigate the initial behavior of temperature in order-N FPMD simulations. We also investigated the combined protocol of velocity scaling and Nosé-Hoover chain method. We used a test DNA model, which contains a decamer DNA base pairs, Mg counter ion, and a large number of water molecules, consisting of about 3400 atoms. We demonstrate that the FPMD simulations on hydrated DNA system are robust and accurate even in NVT ensemble.

2. Theoretical background

Here we briefly describe some key points of the order-N DFT method and the combined XL-BOMD with temperature controlled method. The details of the order-N methodology used in the CONQUEST code are explained in previous works [8-10].

In CONQUEST, we use the Kohn-Sham density matrix defined as

$$\boldsymbol{\rho}(\boldsymbol{r},\boldsymbol{r}') = \sum_{\boldsymbol{n}} \boldsymbol{f}_{\boldsymbol{n}} \boldsymbol{\psi}_{\boldsymbol{n}}(\boldsymbol{r}) \boldsymbol{\psi}^{*}_{\boldsymbol{n}}(\boldsymbol{r}'),$$

where $\boldsymbol{\psi}_{\boldsymbol{n}}(\boldsymbol{r})$ is the Kohn-Sham orbital for band index $\boldsymbol{n}$, and $\boldsymbol{f}_{\boldsymbol{n}}$ is its occupation number between 0 and 1. The DFT total energy of system can be calculated from density matrix, with the use of the pseudopotential technique and the standard exchange-correlation functionals such as the local density approximation (LDA) or the generalized gradient approximation (GGA). The density matrix is represented by localized orbitals called as “support functions”,

$$\boldsymbol{\rho}(\boldsymbol{r},\boldsymbol{r}') = \sum_{\boldsymbol{i\alpha},\boldsymbol{j\beta}} \boldsymbol{\phi}_{\boldsymbol{i\alpha}}(\boldsymbol{r}) \boldsymbol{K}_{\boldsymbol{i\alpha},\boldsymbol{j\beta}} \boldsymbol{\phi}_{\boldsymbol{j\beta}}(\boldsymbol{r}'),$$

where $\boldsymbol{\phi}_{\boldsymbol{i\alpha}}(\boldsymbol{r})$ are the support function that are non-zero only inside “support region” centered on the atoms $\boldsymbol{i}$, and $\boldsymbol{\alpha}$ runs over the support functions on a given atom. The matrix $\boldsymbol{K}_{\boldsymbol{i\alpha},\boldsymbol{j\beta}}$ is the density matrix expressed by non-orthogonal basis of support functions, that is,

$$\boldsymbol{K} = 3\boldsymbol{LSL} - 2\boldsymbol{LSLSL},$$

where the matrix $\boldsymbol{L}$ and $\boldsymbol{S}$ are an auxiliary density matrix and the support function overlap matrix, respectively. In CONQUEST, we can use two types of basis function; one is B-splines on regular grids [11] and another is numerical pseudo atomic orbitals (PAOs) [12-14]. For the PAO treatment, recently,

IOP Conf. Series: Journal of Physics: Conf. Series **1136** (2018) 012025 doi:10.1088/1742-6596/1136/1/012025

an efficient and accurate scheme for large size basis set have become available in CONQUEST, called the multisite support function (MSSF) method [15,16]. This scheme can perform accurate calculations without increasing the CPU time significantly. The density matrix $\boldsymbol{K}_{i\alpha,j\beta}$ using the support functions is obtained by the conventional diagonalization of Hamiltonian matrix or the order-N method based on the combination of McWeeny's purification transformation [17] and the density matrix minimization (DMM) scheme by Li, Nunes, and Vanderbilt [18]. In the practical linear-scaling computations, we use DMM method introducing the region cut-off for the matrix elements [19,20].

In the MD simulation using CONQUEST, the XL-BOMD scheme proposed by Niklasson [4,5] is introduced into CONQUEST code [6]. The XL-BOMD scheme can maintain the time-reversal symmetry while keeping the efficiency of density matrix reuse. The Born-Oppenheimer Lagrangian in the XL-BOMD scheme is expressed by

$$L^{XBO}(\boldsymbol{X},\dot{\boldsymbol{X}},\boldsymbol{R},\dot{\boldsymbol{R}}) = L^{BO}(\boldsymbol{R},\dot{\boldsymbol{R}}) + \frac{\mu}{2}\mathrm{Tr}[\dot{\boldsymbol{X}}^2] - \frac{\mu\omega^2}{2}\mathrm{Tr}[(\boldsymbol{LS}-\boldsymbol{X})^2],$$

$$L^{BO}(\boldsymbol{R},\dot{\boldsymbol{R}}) = \frac{1}{2}\sum_{I=1}^{N} M_I\dot{\boldsymbol{R}}_I^2 + E_{tot}(\boldsymbol{R}_I),$$

where the potential energy E_{tot} is defined by the self-consistent ground state density matrix at nuclear positions $\boldsymbol{R}_I$ with masses M_I. The matrix $\boldsymbol{X}$ is a sparse matrix with the range of the matrix $\boldsymbol{LS}$, μ is the fictitious electronic mass, and ω is the curvature of the electronic harmonic potential. As shown in previous study [6], we use the $\boldsymbol{LS}$ matrix as an auxiliary degree of freedom in the Born-Oppenheimer Lagrangian and the time-reversible Verlet scheme to obtain the matrix $\boldsymbol{X}$. The matrix $\boldsymbol{X}$ is time-reversible and evolves in a harmonic potential centered on the ground-state $\boldsymbol{LS}$. In the practical numerical propagation of the matrix $\boldsymbol{X}$, the equation of motion with a dissipative force is used to maintain the numerical stability of the matrix $\boldsymbol{X}$ (see the details in ref. [6]).

3. Computational details

As a demonstration model in this study, we used the large-scale hydrated DNA system prepared by our previous study [7,22], which consists of DNA 10 base pairs ($d(CCATTAATGG)_2$ by PDB ID: 1WQZ) of 634 atoms, Mg^{2+} ion of 9 atoms, and H_2O of 932 molecules, the total number of atoms in DNA system is 3,439 atoms, as shown in Figure 1. The initial structure was prepared by using the AMBER9 [23] with the force fields of PARM99 for the DNA atoms TIP3P for water molecules. The system was equilibrated with the constant-pressure MD and then the structure at last step was adopted. For the details of preparation of our DNA system, see our previous related work [22]. The linear-scaling FPMD simulations were performed by using our linear-scaling DFT code CONQUEST with PAO of single-zeta with polarization (SZP) basis set, Perdew-Burke-Ernzerhof (PBE) exchange-correlation functional [24], the DMM with self-consistent field (SCF) calculations, and the numerical integration grid cutoff of 75 Ha. For the setting of XL-BOMD conditions, we used the time-reversible Verlet scheme with a dissipative force correction of the order of 5 [4,5]. The MD time step and the initial temperature for the each atomic velocity were set as 0.5 fs and 300 K for NVE ensemble. For NVT ensemble, we performed three kinds of the temperature controlled MD simulations: MD simulations using the velocity scaling method with the temperature of 300 K and 600K, the stochastic velocity rescaling method with the temperature of 300K, and the Nosé-Hoover chain method with the temperatures of 300 K. As following the previous study [8], we used 5 chained thermostats, the 15th-order Yoshida-Suzuki integrator and the thermostat frequency of 500 cm^{-1} in the Nosé-Hoover chain method. The fictitious thermostat mass of first thermostat is different from those of the rest (see the ref. [8]). In this study, although we limited the computational conditions for DFT calculations (PBE/SZP without dispersion correction) in order to know the energy behaviour of the thermostats, we can

IOP Conf. Series: Journal of Physics: Conf. Series **1136** (2018) 012025 doi:10.1088/1742-6596/1136/1/012025

improve the computational accuracy using such as the MSSF method [15,16] with higher quality of PAO basis set and the DFT-D dispersion correction [29]. In Appendix, we also showed the performance of three thermostats in CONQUEST code using silicon carbide system (see below).

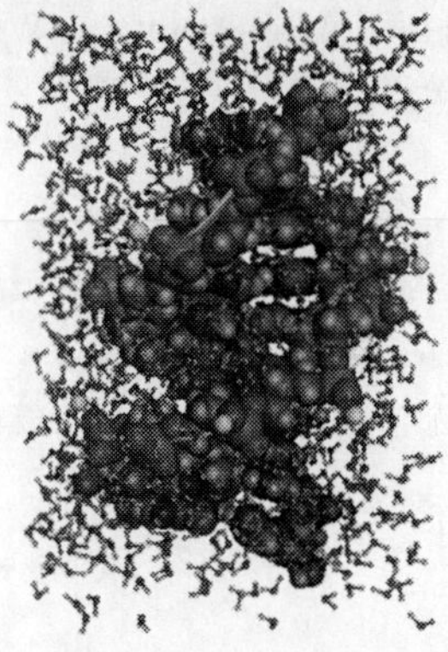

Figure 1. Structure of hydrated DNA (PDB ID: 1WQZ).

4. Results and Discussion

First, we demonstrate our order-N FPMD simulations of hydrated DNA system in NVE ensemble. Figure 2 shows the energy profile obtained by the order-N FPMD simulations. In the early stage of the order-N FPMD simulation until around 50 MD steps, the potential and the kinetic energies of the hydrated DNA system fluctuate considerably. This fluctuations show the difference of the force fields between classical MD simulation and the order-N FPMD simulation. Around 100 MD steps or later, the hydrated DNA structure is relaxed sufficiently by the order-N FPMD simulation, and then towards the equilibration. The structure relaxation and equilibration process of the large-scale biomolecular system does not take many MD steps in the linear-scaling FPMD simulation. It is also seen that the order-N FPMD simulation shows good energy conservation and stable MD simulation in NVE ensemble.

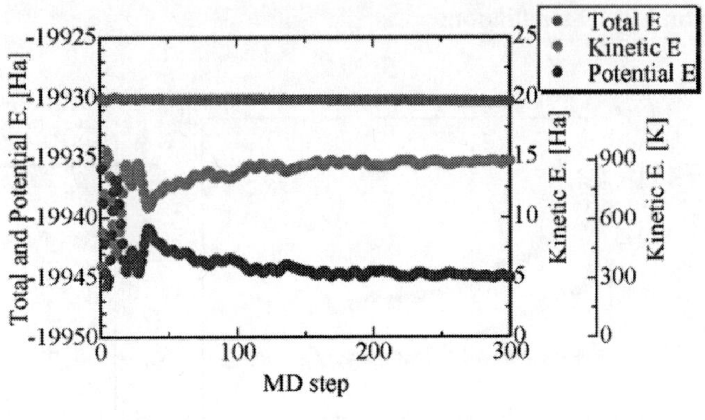

Figure 2. Energy profile calculated by order-N FPMD simulation; total energy (green), potential energy (blue), and kinetic energy (red).

Here we investigate the temperature controlled FPMD simulations by using the velocity scaling, stochastic velocity rescaling, and Nosé-Hoover chain method. In this study, we focused on the temperature behavior in early stage of order-N FPMD simulation. Figure 3 shows the energy profile obtained by the order-N FPMD simulation with the velocity scaling method. We set the controlled temperature as 300 K (Fig. 3(a)) and 600 K (Fig. 3(b)), respectively. In the velocity scaling calculations, we performed the velocity correction for each step. As seen in Figure 3, although the fluctuation of the kinetic energy is large in the early MD steps (that is, 0-50 MD steps), then the kinetic energy converges to around the setting temperature. We can expect that this rapid convergence in the early steps of FPMD simulation would be very useful.

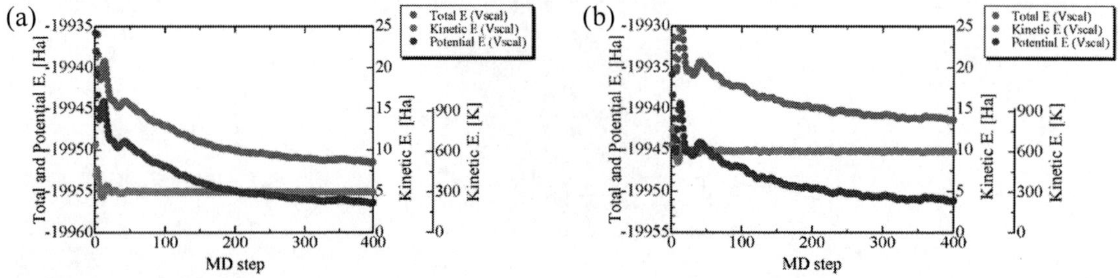

Figure 3. Energy profile calculated by order-N FPMD simulation with velocity scaling method as (a) 300 K and (b) 600 K; total energy (green), potential energy (blue), and kinetic energy of 300 K (pink) and of 600K (red).

Next Figure 4 shows the energy profile obtained by the stochastic velocity rescaling method. In the stochastic velocity rescaling method, we set the temperature as 300K and also performed the velocity correction for each step. From Fig. 4, we can see that the behavior of the energy profile is as almost same as the velocity scaling method as 300K (Fig. 3(a)). Table 1 shows the average and the variance of temperatures by the velocity scaling and the stochastic velocity rescaling methods within 100-300 MD steps. As see in Table 1, the two average temperatures show almost 300K. Although the variance of temperature by the stochastic velocity rescaling method is larger than that by the velocity scaling method, it is very interesting that the energy profile of DNA system in early stage of FPMD simulation is similar behavior, this behavior is different from the case of small SiC system (see Appendix). We do not clearly understand why we have observed such a behavior in DNA system (large-scale biomolecular system), but we can expect that the use of the stochastic velocity rescaling method, which can sample the canonical distribution, would help to analyze the dynamics of a large-scale biomolecular system on FPMD simulations.

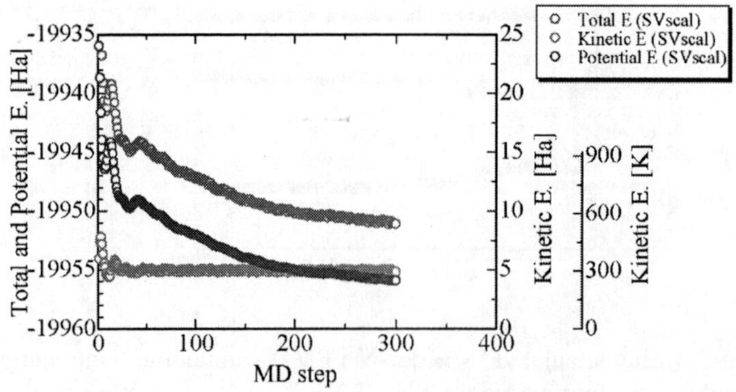

Figure 4. Energy profile calculated by order-N FPMD simulation with stochastic velocity rescaling method as 300 K for 0-300 MD steps (total energy (green open circle), potential energy (blue open circle), and kinetic energy of 300 K (pink open circle)).

Table 1. Average values and the variances of temperatures by velocity scaling and stochastic velocity rescaling methods within 100-300 MD steps.

	Average	Variance
Velocity scaling	301.17 K	1.15
Stochastic velocity rescaling	300.76 K	28.02

Figure 5 shows the energy profile obtained by the Nosé-Hoover chain method. From Fig. 5 we can see that the temperature control by the Nosé-Hoover chain method in the early MD steps of FPMD simulations is relatively slow convergence in present conditions, unlike that of the velocity scaling method. This slow convergence of velocity by the Nosé-Hoover chain method takes many MD steps in FPMD simulations. In order to find the clear reason for this slow convergence of the temperature in the Nosé-Hoover chain method, we think that the evaluation of the calculation conditions (the number of thermostats, the fictitious thermostat masses and the system size) for biomolecular system in Nosé-Hoover chain method is necessary. In this study, we just applied the general parameter reported in previous study [8], hence we cannot describe the further discussion.

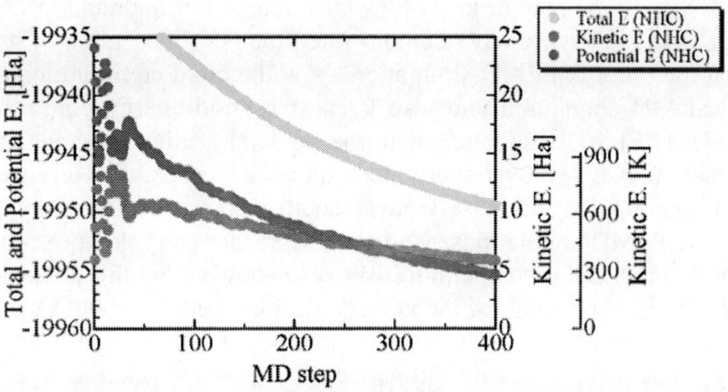

Figure 5. Energy profile calculated by order-N FPMD simulation with Nosé-Hoover chain method as 300 K for 0-400 MD steps (total energy (yellow green), potential energy (aqua blue), and kinetic energy of 300 K (bright pink)).

On the other hand, as one of the possible improvements of this slow convergence of the temperature control in the Nosé-Hoover chain method, we tried combining the velocity scaling method with the Nosé-Hoover chain methods. In early MD steps in order-N FPMD simulations, we used the velocity scaling method, and then we switched to the Nosé-Hoover chain method. Figure 5 show the energy profile combined with these two method. In this energy profile, we set the velocity scaling method from 0 to 100 MD steps, and then the Nosé-Hoover chain method. From Figure 6, the kinetic energy in the Nosé-Hoover chain method at around 200 MD steps and later converged to the setting temperature of 300 K. We can expect that this combined temperature controlled protocol in NVT ensemble should be very effective in the early steps of the temperature controlled FPMD simulations.

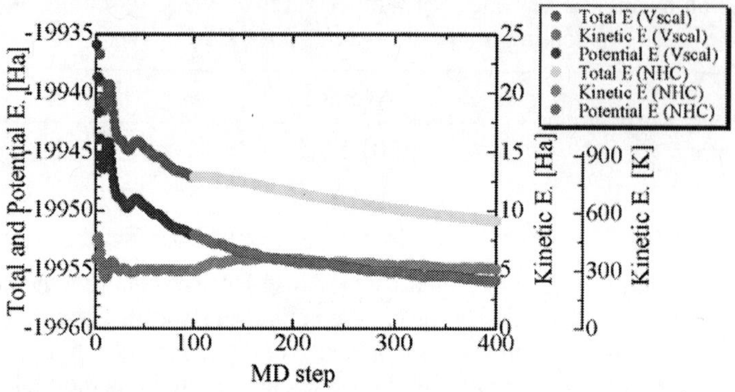

Figure 6. Energy profile calculated by order-N FPMD simulation with the velocity scaling and Nosé-Hoover chain combined method as 300 K for 0-400 MD steps.

5. Conclusion

In this study, we have performed the order-N FPMD simulation of hydrated DNA system using order-N DFT code CONQUEST. First we have demonstrated the NVE ensemble of hydrated DNA system and have found that the order-N FPMD simulation show the rapid equilibration using initial structure obtained by classical MD simulation and also keeps the good energy conservation. We have also investigated the order-N FPMD simulation with the velocity scaling method, the stochastic velocity rescaling method, and the Nosé-Hoover chain method in NVT ensemble. Here we have just tested one case of calculation condition in the Nosé-Hoover chain method. In the initial stage of temperature control in our order-N FPMD simulations, we have found that the velocity scaling method was very effective to converge to the setting temperature. We also showed that the stochastic velocity rescaling method was also the similar behaviour of the velocity scaling method in DNA system. Apart from this, we can expect that the stochastic velocity rescaling method would be one of the possible choices on total FPMD simulations of large-scale biomolecular system, which can sample the canonical distribution. On the other hand, although we need to explore the further calculation conditions of the Nosé-Hoover chain method, the Nosé-Hoover chain method showed the slow convergence compared with the velocity scaling method. We also showed the combined protocol of these temperature control method. These combined temperature controlled protocol in NVT ensemble was very effective for the temperature control in early stage on the order-N FPMD simulations and for reducing the real computational time. We expect that the temperature controlled order-N FPMD simulations by CONQUEST code will be useful for understanding the large-scale biomolecular system. Further numerical tests for above points will be presented in a future publication.

Acknowledgements

This work was partly supported by the JSPS KAKENHI projects: Grant Numbers, 22790122, 16KT0168. The calculations were performed in part on the NIMS material numerical simulator and the COMA (PACS-IX) system of University of Tsukuba. The authors also acknowledge Dr. T. Miyazaki (National Institute for Materials Science) and Prof. D. R. Bowler (University College London) for critical comments and fruitful discussion, and Prof. M. J. Gillan (University College London), Dr. M. Arita (Tokyo University of Science), Dr. N. Okimoto (RIKEN), and Dr. H. Saito (RIKEN) for useful discussions.

Appendix: The behaviour of SiC system by three thermostats in CONQUEST code

Here, we show the order-N FPMD simulations on a silicon carbide (SiC) system using three methods such as the velocity scaling, the stochastic velocity rescaling, and the Nosé-Hoover chain method in NVT ensemble. The SiC system is consisted of 64 atoms. The simulation cell is cubic cell of length 8.716 Å with periodic boundary conditions. The PAO of single-zeta was used with an energy grid cutoff of 80 Ha. These conditions are almost same as the literature of [8]. A time step of 1.0 fs was used and the total MD steps was just 500 steps. The initial temperature was set as 300 K.

Figure A1 show the energy profiles of kinetic energy and the time-averaged temperatures obtained by the order-N FPMD simulations with three methods. From Fig. A1(a), we can see that the fluctuation of kinetic energy by the velocity scaling method is small because the kinetic energy is forced to be exactly equal to target energy (target temperature). While the fluctuation of kinetic energy by the stochastic velocity rescaling method is rather large, where the target energy (target temperature) is selected to be satisfy the canonical equilibrium distribution. As seen in Fig. A1(b), the time-averaged temperatures show the convergence to the target temperature of 300 K. On the other hand, the kinetic energy by the Nosé-Hoover chain method seems to fluctuate slowly around the target temperature.

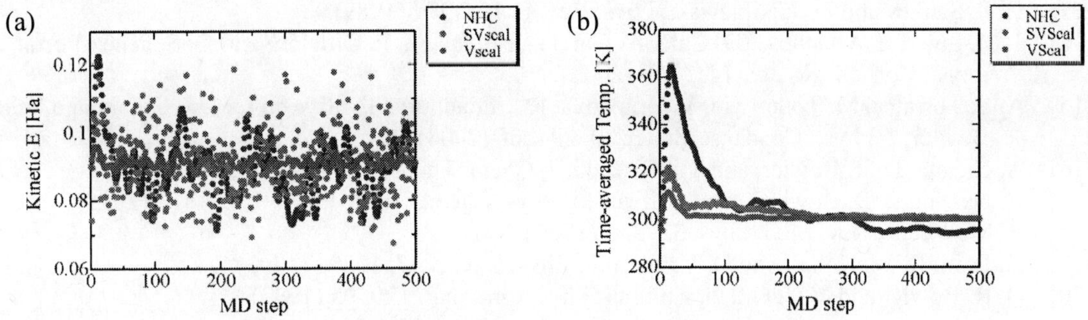

Figure A1. (a) Energy profiles of kinetic energy and (b) Time-averaged temperatures of SiC system by order-N FPMD simulation with three thermostats; Nosé-Hoover chain (blue), stochastic velocity rescaling (red), and velocity scaling (green). The initial temperature is set as 300 K.

Figure A2 show the energy profile of total energy (that is, kinetic energy and potential energy) and the time-averaged total energy obtained by the order-N FPMD simulations with three methods. As expected from the kinetic energy profile, although the fluctuation of total energy by the stochastic velocity scaling method is large, the time-averaged total energy is as almost same behavior as that of the velocity scaling method.

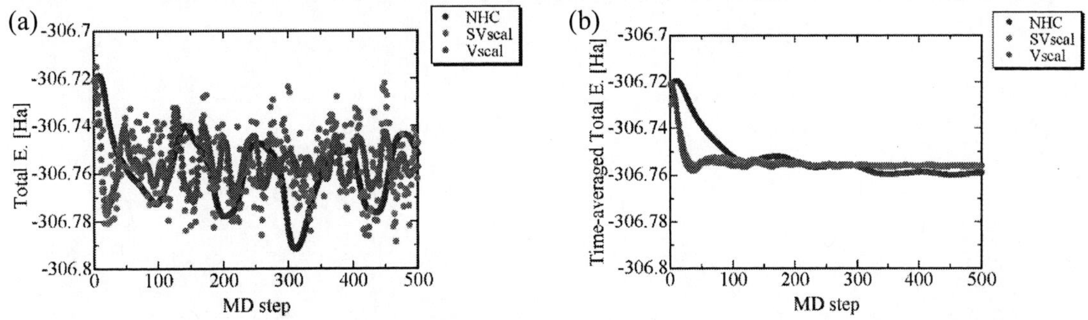

Figure A2. (a) Energy profiles of total energy and (b) Time-averaged total energy of SiC system by order-N FPMD simulation with three thermostats; Nosé-Hoover chain (blue), stochastic velocity rescaling (red), and velocity scaling (green). The initial temperature is set as 300 K.

IOP Conf. Series: Journal of Physics: Conf. Series **1136** (2018) 012025 doi:10.1088/1742-6596/1136/1/012025

References

[1] For example, Novel prize of chemistry 2013, Martin Karplus, Michael Levitt, Arieh Warshel, "for the development of multiscale models for complex chemical systems".
[2] D. R. Bowler and T. Miyazaki, Rep. Prog. Phys. 75, 036503 (2012).
[3] See the web page; http://www.order-n.org/
[4] A. M. N. Niklasson, C. J. Tymczak, M. Challacombe, Phys. Rev. Lett. 97, No. 123001 (2006).
[5] A. M. N. Niklasson, Phys. Rev. Lett., 100, No. 123004 (2008).
[6] M. Arita, D. R. Bowler, T. Miyazaki, J. Chem. Theory Comput., 10, 5419−5425 (2014).
[7] T. Otsuka, M. Taiji, D. R. Bowler, T. Miyazaki, Jpn. J. Appl. Phys. 55, 1102B1 (2016).
[8] T. Hirakawa, T. Suzuki, D. R. Bowler, T. Miyazaki, J. Phys.: Condens. Matter 29, 40591 (2017).
[9] E. Hernández, M. J. Gillan, C. M. Goringe, Phys. Rev. B, 53, 7147−7157 (1996).
[10] D. R. Bowler, T. Miyazaki, M. J. Gillan, J. Phys.: Condens. Matter 14, 2781−2798 (2002).
[11] D. R. Bowler, R. Choudhury, M. J. Gillan, T. Miyazaki, Phys. Status Solidi B, 243, 989−1000 (2006).
[12] E. Hernández, M. J. Gillan, and C. M. Goringe, Phys. Rev. B 55, 13485 (1997).
[13] O. F. Sankey and D. J. Niklewski, Phys. Rev. B 40, 3979 (1989).
[14] J. M. Soler, E. Artacho, J. D. Gale, A. García, J. Junquera, P. Ordejón, and D. Sánchez-Portal, J. Phys.: Condens. Matter 14, 2745 (2002).
[15] A. S. Torralba, M. Todorović, V. Brázdová, R. Choudhury, T. Miyazaki, M. J. Gillan, and D. R. Bowler, J. Phys.: Condens. Matter 20, 294206 (2008).
[16] A. Nakata, D. R. Bowler, and T. Miyazaki, J. Chem. Theory Comput., 10, 4813 (2014).
[17] A. Nakata, D. R. Bowler, and T. Miyazaki, Phys. Chem. Chem. Phys., 17, 31427 (2015).
[18] R. McWeeny, Rev. Mod. Phys. 32, 335 (1960).
[19] X. P. Li, R. W. Nunes, and D. Vanderbilt, Phys. Rev. B 47, 10891 (1993).
[20] D. R. Bowler and M. J. Gillan, Comput. Phys. Commun. 120, 95 (1999).
[21] D. R. Bowler and M. J. Gillan, Chem. Phys. Lett. 325, 473 (2000).
[22] T. Otsuka, T. Miyazaki, T. Ohno, D. R. Bowler, and M. J. Gillan, J. Phys.: Condens. Matter 20, 294201 (2008).
[23] Case D A et al. 2006 AMBER 9 University of California, San Francisco.
[24] J. P. Perdew, K. Burke, M. Ernzerhof, Phys. Rev. Lett., 77 3865 (1996), Errata: Phys. Rev. Lett., 78, 1396 (1997).
[25] G. Bussi, D. Donadio, and M. Parrinello, J. Chem. Phys. 126, 014101 (2007).
[26] S. Nosé, J. Chem. Phys. 81, 511 (1985).
[27] W. G. Hoover, Phys. Rev. A 31, 1695 (1985).
[28] G. J. Martyna, M. E. Tuckerman, D. J. Tobias, and M. L. Klein, Mol. Phys. 87, 1117 (1996).
[29] S. Grimme, J. Comp. Chem. 27, 1787 (2006).

IOP Conf. Series: Journal of Physics: Conf. Series **1136** (2018) 012026 doi:10.1088/1742-6596/1136/1/012026

Theoretical studies on association/dissociation process of plastocyanin and cytochrome *f* in photosynthesis

Isman Kurniawan, Takahiro Matsui, Satoshi Nakagawa, Kazutomo Kawaguchi, Hidemi Nagao

Faculty of Mathematics and Physics, Institute of Science and Engineering, Kanazawa University, Japan

E-mail: nagao@wriron1.s.kanazawa-u.ac.jp, isman@wriron1.s.kanazawa-u.ac.jp

Abstract. We investigate the association/dissociation process of plastocyanin with cytochrome *f* before/after the reaction of the electron transfer from cytochrome *f* to plastocyanin by using all-atom molecular dynamics simulation and our coarse-grained model simulation. The force field parameters of the oxidized and reduced plastocyanin are estimated by quantum chemical calculations and are summarized in this paper. The free energy profiles corresponding to the association/dissociation process are calculated by the thermodynamic integration method with the all-atom molecular dynamics simulations. A flat free energy landscape in the middle range of the association/dissociation process for both cases before and after the reaction is observed. The radial distribution function of the case before the reaction is calculated by the coarse-grained model with the Langevin equation of motion. We estimate the free energy landscape from the radial distribution function. The results by the coarse-grained model may reproduce similar results by all-atom molecular dynamics simulation. We discuss the association/dissociation process of the complex of plastocyanin and cytochrome *f* in relation to the free energy landscape with some physical properties such as the binding free energy, the activation free energy, the energy of the reaction, and so on.

1. Introduction

Plastocyanin is one of the blue copper proteins which have the function of the electron transfer between proteins. The soluble protein of plastocyanin catches one electron from cytochrome *f* in cytochrome b_6f complex. The reduced plastocyanin moves in thylakoid lumen by diffusion and releases the electron to P700 in Photosystem I. The active site of plastocyanin consists of one copper ion, two histidines, one cysteine, and one methionine. The copper ion is coordinately bonded to two nitrogen atoms in the histidines, one sulfur atom in the cysteine, and the sulfur atom in the methionine. The structure of plastocyanin has been analyzed by X-ray analysis[1, 2, 3, 4]. On the other hand, the structure of cytochrome *f* by X-ray analysis has been reported by C. J. Carrell and coworkers [5]. Cytochrome *f* is one of the c-type cytochrome with a heme compound and has two soluble structural domains in the lumen-side segment. The structure of P700 has also been analyzed by X-ray analysis[6, 7].

The reduction and oxidization processes by the electron transfer are rapid with the complex between the soluble plastocyanin and the soluble domain in Cytochrome *f* or the P700, and

IOP Conf. Series: Journal of Physics: Conf. Series **1136** (2018) 012026 doi:10.1088/1742-6596/1136/1/012026

the complex is short-lived and weak. The structure of the weak complex of plastocyanin has been investigated by many groups[8, 9, 10, 11]. The association and/or dissociation processes of plastocyanin has been discussed from the viewpoint of the hydrophobic and the electrostatic interactions between proteins in relation to the possible structures of the weak and short-lived complex[10, 11]. In our previous studies[12, 13], we have presented a simple coarse-grained (CG) model describing protein-protein interaction to investigate the structure and the dynamical properties such as the reaction rate of the electron transfer with protein complexes[13] and so on[12]. The reaction rate of the electron transfer of plastocyanin with cytochrome f has been investigated by our coarse-grained model[13]. We have found the results in the previous work is a good agreement with experimental one[14].

In this study, we focus the association/dissociation processes of plastocyanin with cytochrome f before and after the electron transfer reaction written as PC_{ox}-$Cyt_{red} \rightarrow PC_{red}$-$Cyt_{ox}$, where PC_{ox} and PC_{red} mean plastocyanin of the oxidized and reduced states, respectively, and Cyt_{red} and Cyt_{ox} correspond to reduced and oxidized cytochrome f, respectively. The association process of oxidized plastocyanin and reduced cytochrome f is investigated by all-atom molecular dynamics simulation to estimate the free energy profile corresponding to the association process. We also investigate the dissociation process of reduced plastocyanin and oxidized cytochrome f after the electron transfer by the same procedure of simulations. The force field parameters around the active site in reduced and oxidized states of two proteins for all-atom molecular dynamics simulations are estimated by quantum chemical calculations. We calculate the free energy profile by all-atom molecular dynamics simulation according to the procedure of the estimation of the free energy profile presented in the previous works[15, 16]. We discuss several snapshot structures of the complex between plastocyanin and cytochrome f in the association process by comparing with the results estimated by our coarse-grained model in relation to the stability of the complex with some physical properties such as the binding free energy, the activation energy of the reaction, the free energy of the reaction and so on.

2. Molecular dynamics simulations and coarse-grained modeling

The free energy profiles corresponding to two states before and after the electron transfer are calculated by all-atom molecular dynamics simulation with the thermodynamic integration method[15, 16] and by the coarse-grained model presented in the previous work[13].

2.1. Molecular dynamics simulation

The X-ray crystal structures of plastocyanin solved by Guss, et al. (PDB ID: 1PLC) [17] and the lumen-side domain of cytochrome f solved by Martinez, et al. (PDB ID: 1HCZ) [18] are used as the initial structures for the simulation of association process. These structures correspond to the oxidized plastocyanin and the reduced cytochrome f, respectively. As for the dissociation process, the simulation was performed by using the NMR crystal structure of the complex of poplar plastocyanin with turnip cytochrome f solved by Lange, et al. (PDB ID: 1TKW) [19]. We adapted different crystal structure for association and dissociation processes because the oxidation state of plastocyanin and cytochrome f existed in those processes is different. In the association process, plastocyanin and cytochrome f were found in oxidized and reduced state, respectively, while in the dissociation process, as the reaction has occurred, plastocyanin and cytochrome f were found in reduced and oxidized state, respectively. Thus, we assumed that the initial structure of both proteins used in the simulation is similar to the natural system. Plastocyanin is a water-soluble protein, consists of 99 amino acid residues, and includes Cu^{2+} ion in its active site. The lumen-side domain of cytochrome f consists of 252 amino acid residues and includes a heme and Fe^{2+} in its active site. Plastocyanin and the lumen-side domain of cytochrome f are immersed in a water solvent.

The partial charge distribution around the active sites before and after the electron transfer process was calculated by using quantum calculations. The force field parameters related to the active sites of plastocyanin before and after the electron transfer process were also calculated by using quantum calculations. The quantum calculation was performed by using B3LYP density functional with 6-31G(d,p) basis set in Gaussian package [20]. More detailed procedures to calculate the force field parameters related to the active sites have been presented in Ref. [21]. As for the cytochrome f, the force field parameters of the structure before and after the reaction are set to be similar to those of reduced cytochrome f presented by CHARMM27[22, 23]. However, note that in this simulation, we do not focus on the structural relaxation after the reaction of the electron transfer because we assume that the electron transfer occurred faster than the structural dynamics.

Temperature and pressure are controlled at 300 K by using Nose-Hoover chain and 0.101 MPa by using Andersen algorithm, respectively [24, 25, 26]. The long-range coulombic interactions under the periodic boundary condition are calculated by Particle Mesh Ewald method [27], in which the Ewald dispersion parameter is 0.375×10^{10} m^{-1}, $64\times64\times64$ grids are adopted, and the cutoff distance is 1.2 nm. The Newton's equations of motion are solved with the time step of 2.0 fs. The SHAKE/RATTLE/ROLL algorithm is used for constraint of the bond lengths including the hydrogen atoms [26]. The force field parameters are derived from CHARMM27 [22, 23] and TIP3P [28] for cytochrome f protein and water molecules, respectively. Molecular dynamics (MD) simulation was performed with a constraint on the distance between the mass center of plastocyanin and cytochrome f by using SHAKE and RATTLE methods. The constraint distance was gradually decreased and increased from $r = 31.25\AA$ to $25\AA$ and $45\AA$, respectively. The initial structures with 17 different distance were prepared, which the distance was varied from $25\AA$ to $45\AA$ with the interval of $1.25\AA$.

Thermodynamic integration method [29] is applied to calculate the free energy profile as a function of the distance between the centers of mass of plastocyanin and cytochrome f. We apply same procedure as our previous study [15] to calculate the free energy profile from the trajectories of molecular dynamics simulations. All molecular dynamics simulations are performed by using the MODYLAS program package [30].

2.2. Coarse-grained model

In our coarse-grained model, each amino acid residue is represented as one coarse-grained particle located on α carbon. We solve the Langevin equation of motion of the i-th coarse-grained particle described by

$$m_i \frac{d^2 \boldsymbol{r}_i}{dt^2} = -\frac{\partial V}{\partial \boldsymbol{r}_i} - \gamma \frac{d\boldsymbol{r}_i}{dt} + \boldsymbol{P}_i(t), \tag{1}$$

where m_i, $\boldsymbol{r}_i$, V are represented the mass and coordinate of the i-th coarse-grained particle, and the total potential energy of target system, respectively. We set the mass of coarse-grained particle 137 amu. The friction coefficient represented as γ does not depend on the amino acid residue and is set to 3.1×10^{-13} kg/s. The random force, $\boldsymbol{P}_i(t)$, is the Gaussian white noise described by

$$\langle P_{i\alpha}\rangle = 0, \quad (\alpha = x, y, z), \tag{2}$$

$$\langle P_{i\alpha}(t)P_{j\beta}(t')\rangle = 2\gamma k_{\mathrm{B}} T \delta_{ij}\delta_{\alpha\beta}\delta_{tt'}, \tag{3}$$

where $P_{i\alpha}(t)$ indicates the α component of random force at time t. The total potential energy V is expressed as

$$V = V_{\mathrm{intra}} + V_{\mathrm{inter}}, \tag{4}$$

where V_{intra} and V_{inter} indicate the intramolecular potential energy and the intermolecular potential energy, respectively. The intramolecular potential energy is used Gō-like potential [31]

IOP Conf. Series: Journal of Physics: Conf. Series **1136** (2018) 012026 doi:10.1088/1742-6596/1136/1/012026

described by

$$\begin{aligned} V_{\rm intra} = & \sum_{\rm bonds} K_b(r_i - r_{i0})^2 + \sum_{\rm angles} K_\theta(\theta_i - \theta_{i0})^2 \\ & + \sum_{\rm dihedrals} K_\phi\Big[\{1-\cos(\phi_i - \phi_{i0})\} + \frac{1}{2}\{1-\cos 3(\phi_i - \phi_{i0})\}\Big] \\ & + \sum_{i>j+3}^{\rm nc} \varepsilon^{\rm nc}\Big[5\Big(\frac{r_{ij0}}{r_{ij}}\Big)^{12} - 6\Big(\frac{r_{ij0}}{r_{ij}}\Big)^{10}\Big] + \sum_{i>j+3}^{\rm nnc} \varepsilon^{\rm nnc}\Big(\frac{d}{r_{ij}}\Big)^{12}. \end{aligned} \quad (5)$$

In the equation (5), r_i, θ_i, ϕ_i, r_{ij} represent the i-th virtual bond length between i-th and $(i+1)$-th coarse-grained particles, the virtual bond angle between i-th and $(i+1)$-th virtual bonds, the virtual dihedral angle around the $(i+1)$-th virtual bond, and the distance between i-th and j-th coarse-grained particles, respectively. The subscript "0" means the value at the reference structure. In the fourth and fifth term, nc and nnc indicate native contact and non-native contact, respectively. The other parameters are constant value and are set to $K_b = 100{\rm kcal/mol}\cdot{\rm Å}^2$, $K_\theta = 20{\rm kcal/mol}\cdot{\rm rad}^2$, $K_\phi = 1{\rm kcal/mol}$, $\varepsilon^{\rm nc} = \varepsilon^{\rm nnc} = 3.6{\rm kcal/mol}$, $d = 4.0$Å.

The intermolecular potential energy proposed in our previous study[13] consists of four terms described by

$$V_{\rm inter} = V_{\rm LJ} + V_{\rm DP} + V_{\rm G} + V_{\rm rep}. \quad (6)$$

In the equation (6), $V_{\rm LJ}$, $V_{\rm DP}$, $V_{\rm G}$, and $V_{\rm rep}$ mean the van der Waals interaction between two hydrophobic amino acid residues in different protein molecules, the depletion interaction, two Gaussian potentials, and the repulsive potential between two coarse-grained particles which do not correspond to hydrophobic amino acid residues, respectively. These potential energies are written as

$$V_{\rm LJ} = \sum_{i\alpha} \varepsilon_{i\alpha}^{\rm LJ}\Big[\Big(\frac{R_{i\alpha}}{r_{i\alpha}}\Big)^{12} - 2\mu(\rho_i)\mu(\rho_\alpha)\Big(\frac{R_{i\alpha}}{r_{i\alpha}}\Big)^{6}\Big], \quad (7)$$

$$V_{\rm DP} = -\frac{\pi p_0}{12}\sum_{i\alpha}\mu(\rho_i)\mu(\rho_\alpha)(r_{i\alpha} - l_{i\alpha})^2(r_{i\alpha} + 2l_{i\alpha}), \quad (8)$$

$$V_{\rm G} = \sum_{k=1}^{2}\sum_{i\alpha}\mu(\rho_i)\mu(\rho_\alpha)\varepsilon_{i\alpha}^{G(k)}\exp\Big[-\Big(\frac{r_{i\alpha} - \lambda_{i\alpha}^{(k)}}{\sigma_{i\alpha}^{(k)}}\Big)^2\Big], \quad (9)$$

$$V_{\rm rep} = \sum_{i,\alpha\neq{\rm hydrophobic}} \varepsilon_{\rm rep}\Big(\frac{C}{r_{i\alpha}}\Big)^{12}, \quad (10)$$

where $r_{i\alpha}$ represents the distance between the i-th and α-th coarse-grained particles in different protein molecules. The osmotic pressure, p_0 is estimated by all-atom molecular dynamics simulation and set to 1.37974×10^8 Pa in this work. $\varepsilon_{i\alpha}^{\rm LJ}$, $R_{i\alpha}$, $\varepsilon_{i\alpha}^{G(k)}$, $\lambda_{i\alpha}^{(k)}$, and $\sigma_{i\alpha}^{(k)}$ are parameters obtained by all-atom molecular dynamisc simulations and satisfy with the Lorents-Berthelot mixing roles[32]. In the equation (9), $l_{i\alpha}$ is described by $l_{i\alpha} = (R_{i\alpha} + R_{\rm W})/2^{\frac{1}{6}}$. $R_{\rm W}$ represents van der Waals parameter and is set to 1.4Å. In the equation (10), $\varepsilon_{\rm rep}$ and C are constant parameters and are given as 3.6 kcal/mol and 4.0Å, respectively.

We introduce a function related to the molecular crowding effect as

$$\mu(\rho_i) = \frac{1}{\exp\Big[20(\frac{\rho^*}{\rho_i} - 1)\Big] + 1}, \quad (11)$$

where ρ^*, ρ_i is the critical density of hydrophobic amino acid residue and is the i-th density of hydrophobic amino acid residue in different protein molecules, respectively. ρ_i is expressed as

$$\rho_i = \sum_\alpha \frac{1}{\exp\left[20(\frac{r_{i\alpha}}{\sigma} - 1)\right] + 1}, \tag{12}$$

where σ corresponds to the constant parameter with a distance dimension. We set ρ^* and σ to 6.2 and 13.9Å from our previous study[13]. If the density of particle is higher than the critical density, the attractive term of V_{LJ}, V_{DP}, and V_{G} rapidly become zero.

3. Results and discussion

Table 1 summarizes force field parameters for bonds around the copper ion in the active site of oxidized and reduced states of plastocyanin derived by quantum mechanical calculations. r_c and K_r mean the equilibrium distance and the force constant for the bond stretching, respectively. The bond stretching potential $U_b(r)$ can be expressed as $U_b(r) = K_r(r - r_c)^2$, where r is the distance between two atoms. Other parameters for the angle of plastocyanin are summarized in Table 2. θ_c and K_θ correspond to the equilibrium angle and the force constant for the angle bending, respectively. The potential energy for the angle bending can be expressed as $U_a(\theta) = K_\theta(\theta - \theta_c)^2$, where θ is the angle bending defined by three atoms. We use the parameters summarized in those Tables for all-atom molecular dynamics simulations. Figure 1 and 2 show schematic diagrams of the charge distributions of the oxidized and the reduced plastocyanin, respectively. The schematic diagrams of the charge distribution of the active site of the oxidized and reduced cytochrome f proteins are also shown in Figs. 3 and 4, respectively.

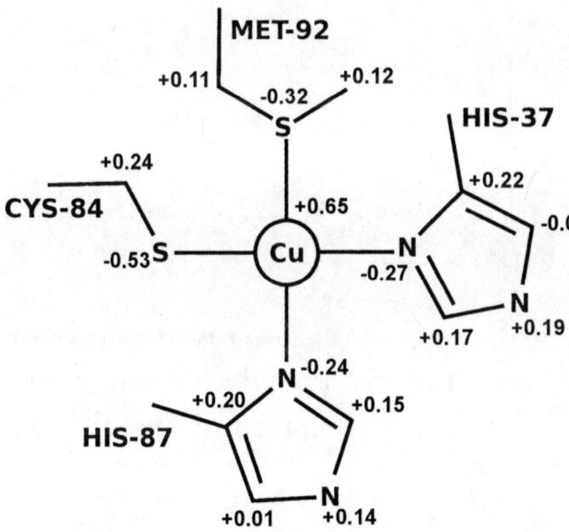

Figure 1. Charge distribution around the active site of oxidized plastocyanin.

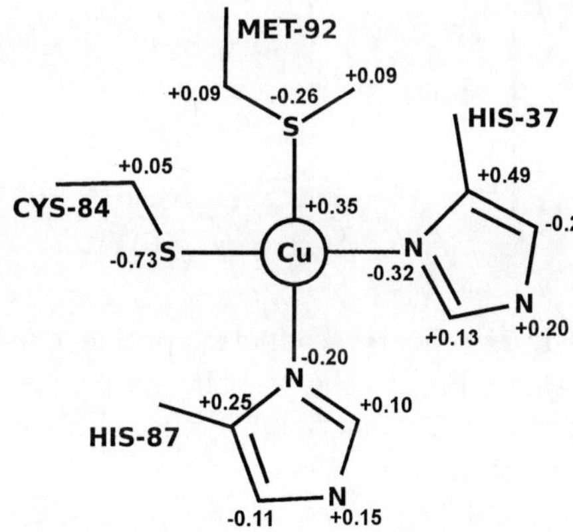

Figure 2. Charge distribution around the active site of the reduced plastocyanin.

Figure 5 shows the free energy profile of the oxidized plastocyanin with the reduced cytochrome f. The horizontal axis in Fig. 5 means the distance between the centers of mass of the protein. From Fig. 5, we can approximately estimate the binding free energy of the state before the electron transfer reaction as about 19.5 kcal/mol, where we assume the free energy value at a distance between the centers of mass of 45Å is the origin of the free energy. The flat free energy landscape is observed around the region from 35Å to 40Å. In the flat free energy landscape, we can find the extremely small activation energy as 1.8 kcal/mol around 35Å in Fig. 5. Figure 6 shows the free energy profile of the state after the electron transfer reaction. The

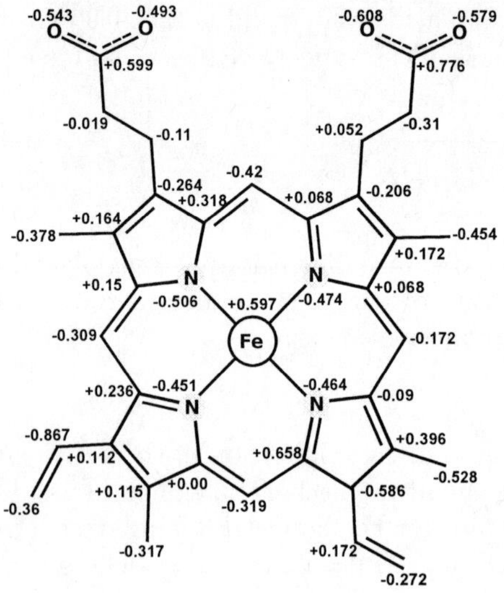

Figure 3. Charge distribution around the active site of oxidized cytochrome *f*.

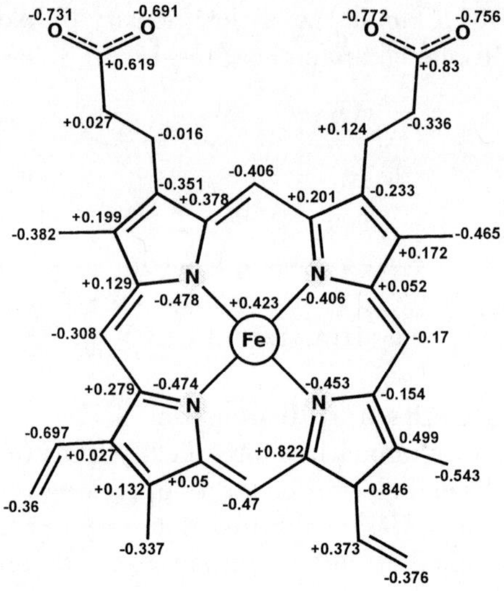

Figure 4. Charge distribution around the active site of the reduced cytochrome *f*.

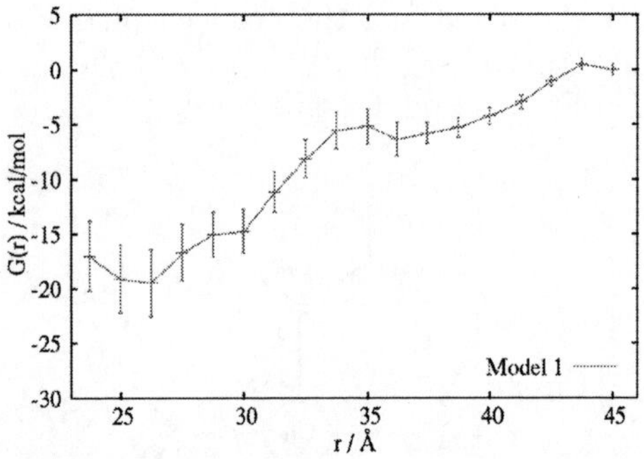

Figure 5. Free energy profile corresponding to the association process before the reaction of the electron transfer.

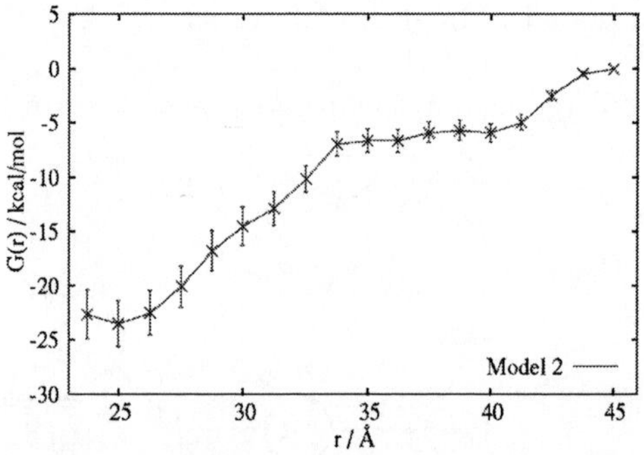

Figure 6. Free energy profile corresponding to the dissociation process after the reaction of the electron transfer.

Table 1. Parameters for bonds around the active site.

	Oxidized State		Reduced State	
	r_c [Å]	K_r [kcal/mol/Å^2]	r_c [Å]	K_r [kcal/mol/Å^2]
Cu-N(His37)	1.975	110.817	1.94	117.21
Cu-S(Cys84)	2.145	138.055	2.20	100.95
Cu-N(His87)	2.021	87.92	2.09	71.91
Cu-S(Met92)	2.988	26.146	2.82	17.41

Table 2. Parameters for angles around the active site.

	Oxidized State		Reduced State	
	θ_c [degree]	K_θ [kcal/mol/rad^2]	θ_c [degree]	K_θ [kcal/mol/rad^2]
N(His37)-Cu-S(Cys84)	130.542	285.948	136.29	237.6
N(His37)-Cu-N(His87)	96.491	349.708	99.91	214.46
N(His37)-Cu-S(Met92)	87.075	307.703	87.26	312.8
S(Cys84)-Cu-N(His87)	134.434	69.157	108.95	120.85
S(Cys84)-Cu-S(Met92)	111.733	294.284	114.22	249.35
N(His87)-Cu-S(Met92)	107.425	55.564	106. 02	199.13
Cu-N(His37)-C_ϵ	122.682	245.728	122.99	81.53
Cu-N(His37)-C_γ	127.68	224.148	120.71	92.79
Cu-S(Cys84)-C_β	113.062	74.816	106.18	36.27
Cu-N(His87)-C_ϵ	120.787	88.285	128.57	208.46
Cu-N(His87)-C_γ	95.602	88.196	117.95	225.52
Cu-S(Met92)-C_ϵ	97.66	62.314	98. 37	106.77
Cu-S(Met92)-C_γ	127.663	253.084	133.52	195.27

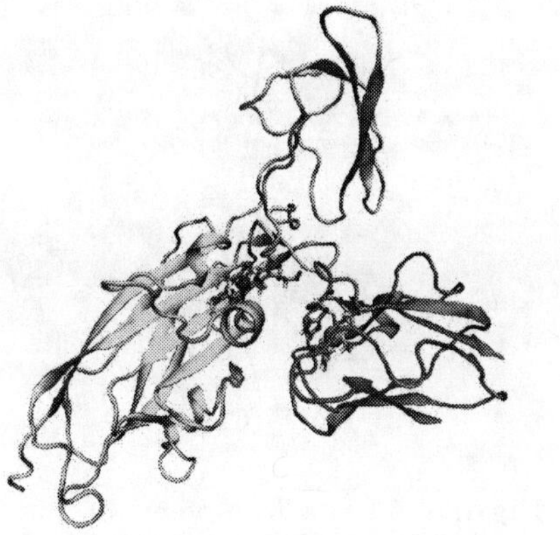

Figure 7. The snapshot of a structure around the region of the lowest free energy. The secondary structure plastocyanin and cytochrome *f* represent by the structure with the color of violet and green, respectively.

Figure 8. The snapshot of a structure around the distance 35Å of the flat free energy landscape in the association process. The secondary structure plastocyanin and cytochrome *f* represent by the structure with the color of violet and green, respectively.

binding free energy is estimated as 23.9 kcal/mol, and we can find the flat free energy landscape around the region from 35Å to 40Å similar to that shown in Fig. 5. However, we can not

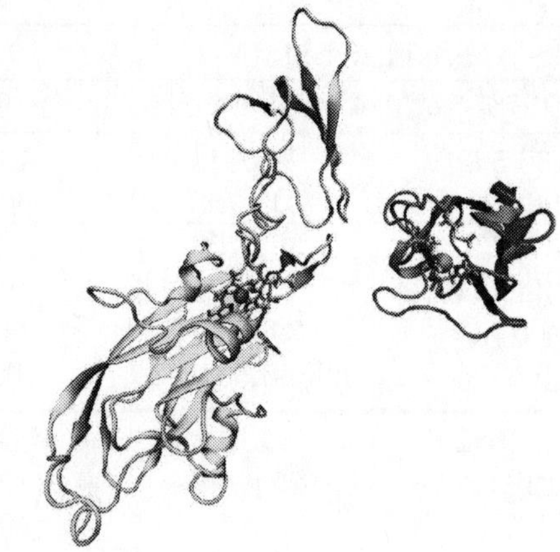

Figure 9. The snapshot of a structure around the distance 40Å of the flat free energy landscape. The secondary structure plastocyanin and cytochrome f represent by the structure with the color of violet and green, respectively.

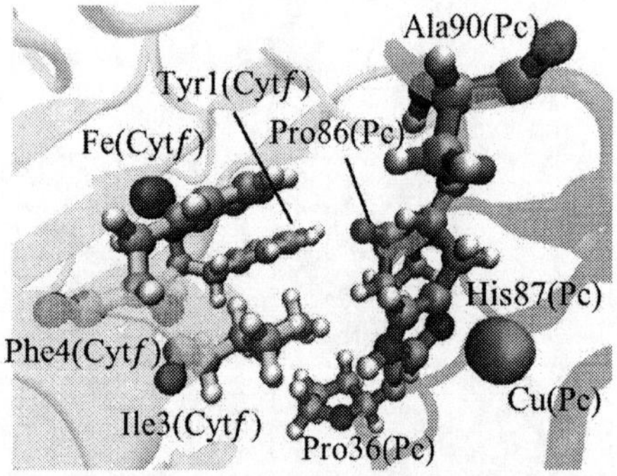

Figure 10. Close view of the binding region of the complex between the oxidized plastocyanin and the reduced cytochrome f of the snapshot shown in Fig. 7.

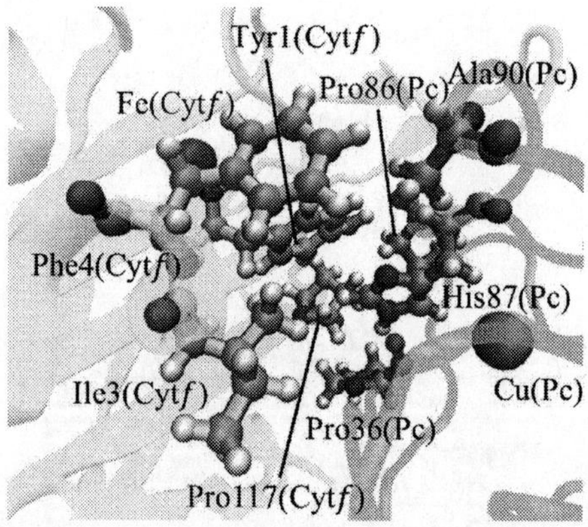

Figure 11. Close view of binding region of the complex between the reduced plastocyanin and the oxidized cytochrome f around the lowest free energy at the distance about 25Å shown in Fig. 6.

find the certain activation energy in the case of the dissociation process between the reduced plastocyanin and the oxidized cytochrome f.

Free energy profile shown in Figs. 5 and 6 is calculated as a function of the distance between the centers of mass of two protein molecules by using the thermodynamic integration (TI) method. Even though their orientation also has an impact on the free energy profile, it is calculated as a function of only the distance by averaging the motion of the orientation. The detail procedures have been reported in our previous studies [15]. Calculated binding free energy is around 20 kcal/mol for both cases. Although it seems to be larger than expected binding free energy, the binding free energy along the distance could be around 20 kcal/mol or more[15, 33, 37]. The free energy profile does not become constant at long range of the distance between those proteins because plastocyanin may not leave the surface of cytochrome f at 45Å.

From the two results shown in Figs. 5 and 6, the free energy of the reduction reaction of plastocyanin becomes about 4.4 kcal/mol. Because no crossing between two free energy landscapes, we can approximately expect the electron transfer reaction from the reduced cytochrome f to the oxidized plastocyanin proceeds soon once the complex is formed. The implicit result of the rapid reaction presented in this work is good agreement with the comments by other works[14, 34, 35, 36].

Figures 7, 8, and 9 show some snapshots corresponding to the configurations in the association process of the oxidized plastocyanin and the reduced cytochrome f. In those figures, the structure of plastocyanin and cytochrome f was represented by the structure with violet and green colors, respectively. A snapshot structure of the complex around the lowest free energy at a distance about 26Å is shown in Fig. 7. The configuration is similar to the crystal structure analyzed by X-ray analysis. The snapshot structures shown in Fig. 8 and 9 are sometimes found in the flat free energy landscape with the distances about 35Å and 40Å, respectively. The configuration shown in Fig. 8 roughly changes to that in Fig. 9 with keeping the distance between the two surfaces.

Figures 10 and 11 show close views for the binding site of the complex structures before and after the reaction around the lowest free energy, respectively. Similar to the previous figures, the structure with violet and green colors represent the structure of plastocyanin and cytochrome f, respectively. From these figures, we can find the distance between two active sites is close in the case of the lowest free energy and can find two regions consisting of hydrophobic and acid patches. In the hydrophobic patch, there are Ala90, Pro86, and Pro36 in plastocyanin binding with Phe4, Ile3 in cytochrome f shown in Fig. 10. On the other hand, in the case after the electron transfer, the distance between the active sites becomes shorter than that before the reaction. These results imply the hydrophobic interaction between the active sites contributes to the stability of the configuration of the complex of plastocyanin and cytochrome f.

Let us discuss the results obtained from the coarse-grained model presented in the previous work[13]. Figure 12 shows the radial distribution function of the distance between the centers of mass. The peak of the radial distribution function is around 27Å. The distance between the centers in the complex structure by NMR analysis is about 26.1$\AA$ (PDB ID: 2PCF)[8]. The result suggests that cytochrome f and plastocyanin have the structure similar to that by experiments and that the structure obtained by the present simulation is a good agreement with the experimental result[8].

The solid line shown in Fig. 13 shows the free energy landscape obtained from the radial distribution function shown in Fig. 12. The vertical axis in Fig. 13 is plotted with the renormalization by the binding free energy. The dashed line means the free energy profile from Fig. 5. The lowest free energy value is observed around the distance 27Å, and we can find the flat free energy landscape around the region from 35Å to 40Å. The flat free energy landscape obtained from the coarse-grained model may become similar to that obtained by all-atom molecular dynamics simulations shown in Fig. 5. On the other hand, from Fig. 13, it is not easy to find the activation energy of the reaction similar to that shown in Fig. 5.

We can find the large difference of the free energy value around shorter distance than the region around the lowest free energy by the coarse-grained model on that by the molecular dynamics simulation from Fig. 13. To compare the landscape by the coarse-grained model with that by all-atom molecular dynamics simulation with better accuracy, it may be necessary to take a larger number of samplings. It may also be better to multiply the binding free energy, obtained by a coarse-grained model, by a constant as the ratio of the total number of the freedom of the

IOP Conf. Series: Journal of Physics: Conf. Series **1136** (2018) 012026 doi:10.1088/1742-6596/1136/1/012026

coarse-grained particle to that of the solute of protein. This is necessary to directly compare the result by the coarse-grained model with that of all-atom molecular dynamics simulation. The at least shape of the free energy landscape obtained from the coarse-grained model corresponding to the dissociation process may become similar to that obtained by all-atom molecular dynamics simulations.

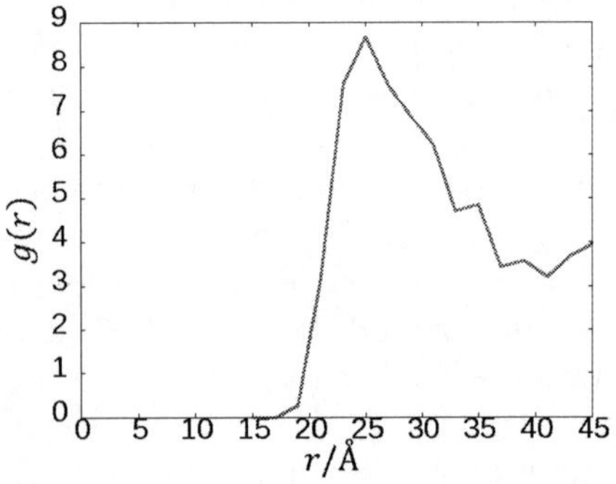

Figure 12. Radial distribution function of the distance between the centers of the oxidized plastocyanin and the reduced cytochrome f.

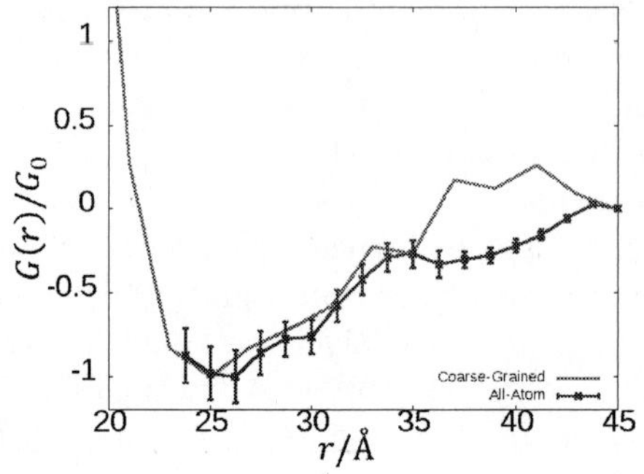

Figure 13. Free energy landscape obtained from Fig. 12, which the error bar represent the standard deviation of the free energy obtained from all-atom simulation.

4. Summary

In this study, we have investigated the association/dissociation processes of plastocyanin with cytochrome f before/after the reaction of the electron transfer from cytochrome f to plastocyanin by using all-atom molecular dynamics simulations and our coarse-grained model simulations.

The force field parameters around the active sites of the oxidized and reduced plastocyanin have been estimated by quantum chemical calculations and have been summarized in this paper. The free energy profiles corresponding to the association/dissociation processes have been calculated by the thermodynamic integration method with the all-atom molecular dynamics simulations. From the simulations, we have found a flat free energy landscape in the middle range of the association/dissociation processes for both cases before and after the reaction. The results suggest that in this flat free energy landscape, plastocyanin dissociates from cytochrome f with gradually changing the conformation of the complex while rotating and keeping the distance between the surfaces of the proteins. The conformation changing in the dissociation process has also been theoretically pointed out by Gumbart etc. [37].

We have also found the small activation free energy about 1.8 kcal/mol in the case before the reaction at a distance between the centers of mass of the proteins around 35Å. However, there is no activation free energy after the reaction. The results from the crossing the free energy profiles for both cases before and after the reaction suggest the reaction of the electron transfer proceeds soon once the complex of the oxidized plastocyanin and the reduced cytochrome f is formed. The binding free energy, the activation energy, and the free energy of the reaction have been estimated by all-atom molecular dynamics simulations. The binding free energies before and after the reactions are about 19.5 kcal/mol around 26Å and about 23.9 kcal/mol around 25Å. Therefore, the free energy of the reaction becomes 4.4 kcal/mol.

The radial distribution function of the case before the reaction has been calculated by the coarse-grained model with the Langevin equation of motion. We have estimated the free energy landscape corresponding to the association/dissociation processes from the radial distribution function and have found the flat free energy landscape in the region from 35Å to 45Å similar to that obtained by the all-atom molecular dynamics simulation. The results by the coarse-grained model may be a good agreement with that by all-atom molecular dynamics simulation. We will present more accurate data by a larger number of sampling for simulations elsewhere[38].

Finally, we have calculated the free energy profiles for both cases before and after the reaction of the electron transfer from reduced cytochrome f to oxidized plastocyanin by all-atom molecular dynamics simulations and the coarse-grained model simulations. We have also discussed the association and/or dissociation processes of the complex of plastocyanin and cytochrome f in relation to the free energy landscape with some physical properties such as the binding free energy, the activation free energy, the energy of the reaction, and so on. The flat free energy landscape has been found in the middle range of the association/dissociation processes.

5. Acknowledgments

This work was partly supported by JSPS KAKENHI Grant-in-Aid for Scientific Research C [grant number 26400429] and Kanazawa University SAKIGAKE Project.

References

[1] Colman P M, Freeman H C, Guss J M, Murata M, Norris V A , Ramshaw J M and Venkatappa M P 1978 *Nature* **272** 319-24
[2] Guss J M and Freeman H C 1983 *J. Mol. Biol.* **169** 521-63
[3] Guss J M, Harrowell P R, Murata M, Norris V A and Freeman H C 1986 *J. Mol. Biol.* **192** 361-87
[4] Redinbo M R, Yeates T O and Merchant S 1994 *J. Bioenerg. Biomembr.* **26** 49-66
[5] Carrell C J, Schlarb B G, Bendall D S, Howe C J, Cramer W A and Smith J L 1999 *Biochem.* **38** 9590-99
[6] Jordan P, Fromme P, Witt H T, Klukas O, Saenger W and Krauss N 2001 *Nature* **411** 909-17
[7] Plato M, Krauss N, Fromme P and Lubitz W 2003 *Chem. Phys.* **294** 483-99
[8] Ubbink M, Ejdebck M, Karlsson B G and Bendall D S 1998 *Structure* **6** 323-35
[9] Hope A B 2000 *Biochim. Biophys. Acta* **1456** 5-26
[10] Crowley P B, Otting G, Schlarb-Ridley B G, Canters G W and Ubbink M 2001 *J. Am. Chem. Soc.* **123** 10444-53
[11] Daz-Moreno I, Daz-Quintana A, De la Rosa M A and Ubbink M 2005 *J. Biol. Chem.* **280** 18908-15
[12] Kawaguchi K, Nakagawa S, Kinoshita S, Wada M, Saito H and Nagao H 2017 *Mol. Phys.* **115** 587-97
[13] Nakagawa S, Kurniawan I, Kodama K, Arwansyah A M S, Kawaguchi K and Nagao H 2018 *Mol. Phys.* **116** 666-77
[14] Illerhaus J, Altschmied L, Reichert J, Zak E, Herrmann R G and Haehnel W 2000 *J. Biol. Chem.* **275** 17590-95
[15] Kawaguchi K, Saito H, Okazaki S and Nagao H 2013 *Chem. Phys. Lett.* **588** 226-30
[16] Kawaguchi K, Saito H and Nagao H 2016 *Mol. Simul.* **42** 896-901
[17] Guss J M, Bartuniket H D and Freeman H C 1992 *Acta. Crystallogr. Sect.* B **48** 790-811
[18] Martinez S E, Huang D, Ponomarev M, Cramer W A and Smith J L 1996 *Protein Sci.* **5** 1081-92
[19] Lange C, Cornvik T, Diaz-Moreno I and Ubbink M. 2005 *Biochim. Biophys. Acta* **1707** 179-188
[20] Frisch M J, Trucks G W, Schlegel H B, Scuseria G E, Robb M A, Cheeseman J R, Scalmani G, Barone V, Mennucci B, Petersson G A, Nakatsuji H, Caricato M, Li X, Hratchian H P, Izmaylov A F, Bloino J, Zheng G, Sonnenberg J L, Hada M, Ehara M, Toyota K, Fukuda R, Hasegawa J, Ishida M, Nakajima T, Honda Y, Kitao O, Nakai H, Vreven T, Montgomery J A Jr, Peralta J E, Ogliaro F, Bearpark M, Heyd J J, Brothers E, Kudin K N, Staroverov V N, Kobayashi R, Normand J, Raghavachari K, Rendell A, Burant J C, Iyengar S S, Tomasi J, Cossi M, Rega N, Millam J M, Klene M, Knox J E, Cross J B, Bakken V, Adamo C, Jaramillo J, Gomperts R, Stratmann R E, Yazyev O, Austin A J, Cammi R, Pomelli C, Ochterski J W, Martin R L, Morokuma K, Zakrzewski V G, Voth G A, Salvador P, Dannenberg J J, Dapprich S, Daniels A D, Farkas , Foresman J B, Ortiz J V, Cioslowski J and Fox D J Gaussian, Inc., Wallingford CT, Gaussian 09, Revision E.01, 2009
[21] Shuku T, Sugimori K, Sugiyama A, Nagao H, Sakurai T and Nishikawa K 2005 *Polyhedron* **24** 2665-70
[22] MacKerell A D, Bashford D, Bellott M, Dunbrack R L, Evanseck J D, Field M J, Fischer S, Gao J, Guo H, Ha S, Joseph-McCarthy D, Kuchnir L, Kuczera K, Lau F T K, Mattos C, Michnick S, Ngo T, Nguyen

IOP Conf. Series: Journal of Physics: Conf. Series **1136** (2018) 012026 doi:10.1088/1742-6596/1136/1/012026

D T, Prodhom B, Reiher W E, Roux B, Schlenkrich M, Smith J C, Stote R, Straub J, Watanabe M, Wiorkiewicz-Kuczera J, Yin D and Karplus M 1998 *J. Phys. Chem.* B **102** 3586-616

[23] Mackerell A D 2004 *J. Comput. Chem.* **25** 1584-1604

[24] Martyna G J, Klein M L and Tuckerman M 1992 *J. Chem. Phys.* **97** 2635-43

[25] Martyna G J 1994 *Phys. Rev.* E *Stat. Phys. Plasmas Fluids Relat. Interdiscip. Topics* **50** 3234-36

[26] Martyna G J, Tuckerman M E, Tobias D J and Klein M L 1996 *Mol. Phys.* **87** 1117-57

[27] Essmann U, Perera L, Berkowitz M L, Darden T, Lee H and Pedersen L G 1995 *J. Chem. Phys.* **103** 8577-93

[28] Jorgensen W L, Chandrasekhar J, Madura J D, Impey R W and Klein M L 1983 *J. Chem. Phys.* **79** 926-35

[29] Zacharias M, Straatsma T P, Mccammon J A and Quiocho F A 1993 *Biochem (Mosc)* **32** 7428-34

[30] Andoh Y, Yoshii N, Fujimoto K, Mizutani K, Kojima H, Yamada A, Okazaki S, Kawaguchi K, Nagao H, Iwahashi K, Mizutani F, Minami K, Ichikawa S, Komatsu H, Ishizuki S, Takeda Y and Fukushima M 2013 *J. Chem. Theory Comput.* **9** 3201-09

[31] Nymeyer H, Gracia A E and Onuchic J N 1998 *Proc. Natl. Acad. Sci. U.S.A. 95*, 5921

[32] Kawaguchi K, Nakagawa S, Kinoshita S, Wada M, Saito H and Nagao H 2017 *Mol. Phys.* **115** 587

[33] Doudou S, Burton N A and Henchman H 2009 *J. Chem. Theory Comput.* **5** 909-18

[34] Gross E L and Pearson, Jr. D C 2003 *Biophys. J.* **85** 2055-68

[35] Pearson, Jr. D C and Gross E L 1998 *Biophys. J.* **75** 2698-711

[36] Gross E L, Curtiss A, Stewart R, Durell S R and White D 1990 *Biochim. Biophys. Acta* **1016** 107-14

[37] Gumbart J C, Roux B and Chipot C 2013 *J. Chem. Theory Comput.* **9** 3789-98

[38] Arwansyah M S, Kurniawan I, Nakagawa S, Kodama K, Kawaguchi K, Nagao H *Mol. Phys.* in preparation.

IOP Conf. Series: Journal of Physics: Conf. Series **1136** (2018) 012027 doi:10.1088/1742-6596/1136/1/012027

Solidification of a simple liquid near wall in high-speed shear flows

Shugo YASUDA and Ryo OOKAWA
Graduate School of Simulation Studies, University of Hyogo, Kobe 6500047, Japan
E-mail: `yasuda@sim.u-hyogo.ac.jp`

Abstract. The viscous-heating-induced solidification of a Lennard-Jones fluid near wall in high-speed shear flows is discovered by the molecular-dynamics simulation. The solidification is characterized by a non-dimensional parameter defined by the ratio of the viscous heating to the thermal conductivity of the fluid and is observed only when the channel width is sufficiently large, where the macroscopic transport becomes significant. Thus, the solidification is thought to be due to the macroscopic heat and mass transport induced by the *viscous heating* generated in the fluid regime by the high-speed shear flow. Unexpectedly, shear bands also appear near wall when the solidification occurs.

1. Introduction

Lubrication is a fundamental and classical problem in the fluid dynamics. However, the behaviors become very complicated in complex mechanical systems and extreme environments, where the complex rheology of structured fluid, the chemical reaction at complex interface, and the phase transition in high-speed lubrication are mainly concerned.[1, 2, 3] Computer simulations for those complicated lubrication systems are still challenging and important both from academic and practical points of views.

The computational fluid dynamics (CFD) approaches based on the macroscopic models are useful unless the constitutive relations between macroscopic quantities are prescribed. However, the constitutive relations are unknown for the complex fluids in general.

The molecular-dynamics (MD) simulation is useful even for the complex fluids which cannot be addressed by the CFD approach because any complicated phenomena are autonomously reproduced once their molecular models are appropriately specified in principle.[4, 5, 6]

In this paper, we carry out the large-scale MD simulation of the high-speed lubrication of a Lennard-Jones fluid, which may involve the phase transition. Apart from the conventional MD approaches utilizing the thermostat algorithms to artificially control the fluid temperature, our MD simulation concerns the *viscous heating* generated by the high-speed shear flows as they are naturally observed in real lubrication systems.

The MD simulations for the thermal lubrication in nano channel have been studied by various researchers.[7, 8, 9, 10] However, in this study, we consider the channels which widen much larger than the molecular scale, where the macroscopic transport becomes important. Thus, our study is concerned with the thermal lubrication which is practically important in micro-mechanical systems.

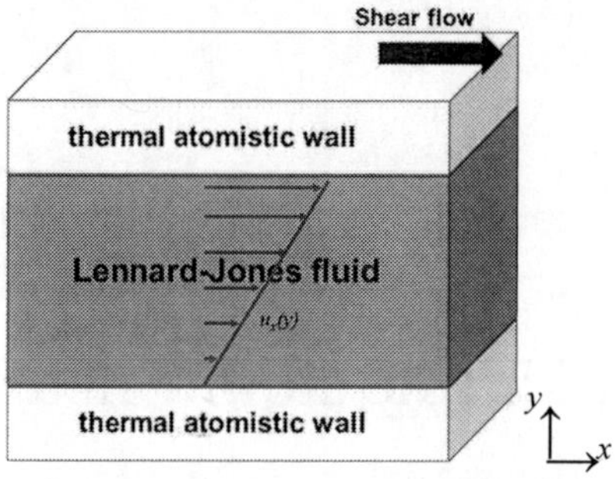

Figure 1. The schematic of the problem. The LJ fluid is sandwiched by atomistic walls kept at a constant temperature T^w. The boundary-driven shear flows are considered in this paper. The x-axis is set in the flow direction and the y-axis is set in the direction perpendicular to the parallel walls. In the x-direction, the periodic boundary condition is considered. The linear flow velocity profile for the iso-thermal fluid is also drawn in the figure.

Incidentally, the drawback of MD simulation comparing to the CFD approach is the huge computational cost. To overcome this difficulty, the development of the multiscale method is currently very active research field.[11, 12, 13, 14, 15] We have also recently developed a multiscale method called Synchronized Molecular-Dynamics (SMD) method and applied it to the thermal lubrication of polymeric fluid with a uniform density, where the mass transports are neglected.[16, 17, 18, 19, 20, 21, 22] Although the multiscale approaches are powerful and promising for complex fluids, in order to further develop the multiscale methods, we need more first-principle results obtained by the full MD simulation. Thus, in this paper, we concern the full MD simulation which can reproduce the full macroscopic transport, i.e., the mass, momentum, and heat transfers, autonomously.

2. Problem and model

We consider a Lennard-Jones (LJ) fluid between parallel plates, which is shown in Fig. 1. The fluid domain is ranged over $y \in (0, H)$ and the wall domains are ranged over $y \in [-W, 0]$ (the lower wall) and $y \in [H, H+W]$ (the upper wall). Both of the fluid and walls are composed of LJ particles which interact each other via the LJ potential

$$U(r) = \begin{cases} 4\varepsilon\left[\left(\frac{\sigma}{r}\right)^{12} - \left(\frac{\sigma}{r}\right)^{6}\right], & (0 < r < r_c), \\ 0, & (r_c \leq r). \end{cases} \tag{1}$$

Here, r_c is the cut-off parameter and ε and σ are the units of energy and length of the LJ particles, respectively. The wall and fluid particles are considered to be the same in the size and mass.

The wall particles are also connected to the face-centered cubic (FCC) lattice structure $\{\boldsymbol{r}_i^w\}$ by a spring potential, and the temperature of wall particles is kept at a constant T^w by the Langevin thermostat algorithm.

Thus, the dynamics of LJ particles are described as

$$m\ddot{\boldsymbol{r}}_i(t) = -\sum_j \frac{\partial U(|\boldsymbol{r}_{ij}|)}{\partial \boldsymbol{r}_{ij}}, \tag{2}$$

for the fluid particles (i.e., $r_{y_i} \in (0, H)$), and

$$m\ddot{\boldsymbol{r}}_i(t) = -\sum_j \frac{\partial U(|\boldsymbol{r}_{ij}|)}{\partial \boldsymbol{r}_{ij}} - k_s(\boldsymbol{r}_i - \boldsymbol{r}_i^w) - \gamma\dot{\boldsymbol{r}}_i + \boldsymbol{R}(t), \tag{3}$$

Boundary-driven		
Case	H	V^w
A	56	3.0
B	168	1.0
C	252	3.0
D	756	1.0
E	67	2.5
F	84	2.0
G	302	2.0
H	378	2.0
I	112	1.5
J	168	3.0

Table 1. Parameter values used in the boundary-driven shear flows. In this proceedings, only the results of Case C are mainly presented.

for the wall particles (i.e., $r_{y_i} \in [-W, 0] \cup [H, H+W]$), where $R_\alpha(t)$ $(\alpha = x, y, z)$ is the white Gaussian noise, which satisfies,

$$< R_\alpha(t) R_\beta(t-s) >= 2mk_B T^w \gamma \delta_{\alpha\beta} \delta(s). \tag{4}$$

Here, $\boldsymbol{r}_i$ represents the position of the ith particle, $\boldsymbol{r}_{ij}$ is defined as $\boldsymbol{r}_{ij} = \boldsymbol{r}_i - \boldsymbol{r}_j$, m is the mass of LJ particle, k_s is the spring constant, γ is the damping coefficient, and k_B is the Boltzmann constant. Note that the summation $\sum_j$ counts for both of fluid- and wall-particles.

The cut-off length $r_c = 2.8$, spring constant $k_s = 10$ and damping coefficient $\gamma = 0.1$ are fixed. Hereafter, we measure the quantities by the units of mass m, energy ε, length σ, and time $\tau = \sqrt{m\sigma^2/\varepsilon}$.

We remark that the temperature of the fluid is not artificially controlled by any thermostat algorithms but varies spatially and temporally due to the macroscopic transport of mass, momentum, and heat between the parallel plates walls with a constant temperature T^w. We also note that the temperature of the moving wall (the upper wall) is controlled only by the y and z components of thermal velocity while the x velocity is given by a constant velocity $v_x = V^w$.

The LJ fluid is initially in a uniform liquid state whose density $\rho_0 = 0.844$ and temperature $T_0 = 0.722$ are fixed near the triple point of the LJ potential. This initial state of LJ fluid is produced by a long-time quiescent MD simulation of the system shown in Fig. 1.

The width of channel H, which is shown in Table 1, is much larger than the molecular size. Thus, the inhomogeneous distributions of macroscopic quantities, e.g., density, velocity, and temperature, may be created due to the macroscopic transport of mass, momentum, and heat between the parallel plates.

3. Results

The lengths of simulation box in x- and z-directions, $L_x = L_z = 16.8$ are fixed while the width of channel H varies as shown in Table 1. The thickness of slab wall $W \simeq 4.1$ is fixed. Thus, for example, the number of particles is about 6.2×10^4 for Case C. The MD simulations are performed by using the LAMMPS software package [23, 24], in which Eq.(2) is time-integrated by the velocity-verlet method with the time-step size $\Delta t = 0.005$.

The channel width H is uniformly divided into 20 bins and, in each bin, the local macroscopic quantities, i.e., density ρ, flow velocity u_x, temperature T, and stress $p_{\alpha\beta}$ are calculated in the stationary state after a long time τ_0 has passed (i.e., $\tau_0 = 1.53 \times 10^8 \Delta t$ is taken for Case D in

Table 1 and $\tau_0 = 6 \times 10^7$ for the other cases). The local macroscopic quantities are also time-averaged in each 100,000 time-steps interval, in which instantaneous macroscopic quantities are sampled at every 10 time steps (i.e., 10,000 samples are averaged for each local macroscopic quantity). The standard deviations of the instantaneous local macroscopic quantities are at most 0.0044 for the density, 0.023 for the velocity, 0.018 for the temperature, and 0.15 for the normal stress. See also Fig. 2.

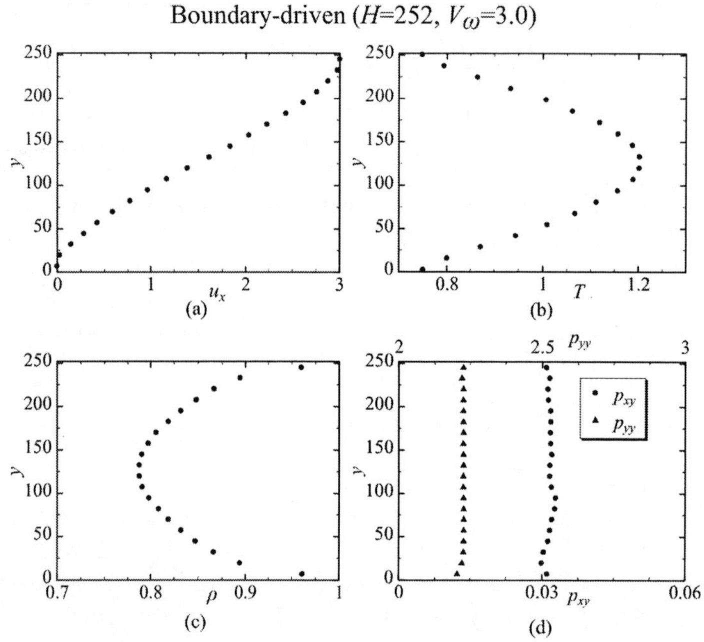

Figure 2. The spatial distributions of velocity (a), temperature (b), density (c), and shear and normal stresses (d) for Case C in Table 1. The standard deviations of the instantaneous local macroscopic quantities are at most 0.023 for u_x, 0.018 for T, 0.0044 for ρ, and 0.15 for p_{yy}.

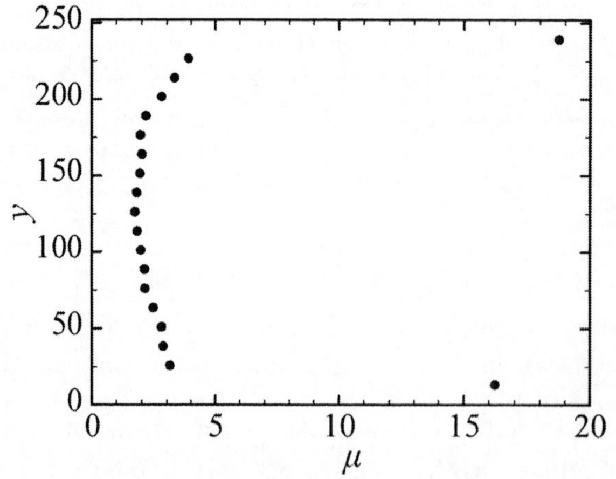

Figure 3. The spatial distribution of local viscosity μ, which is calculated as the ratio of local shear stress p_{xy} to the local share rate $\dot{\gamma}$ (=$\partial u_x/\partial y$), i.e., $\mu = p_{xy}/\dot{\gamma}$, for Case C in Table 1.

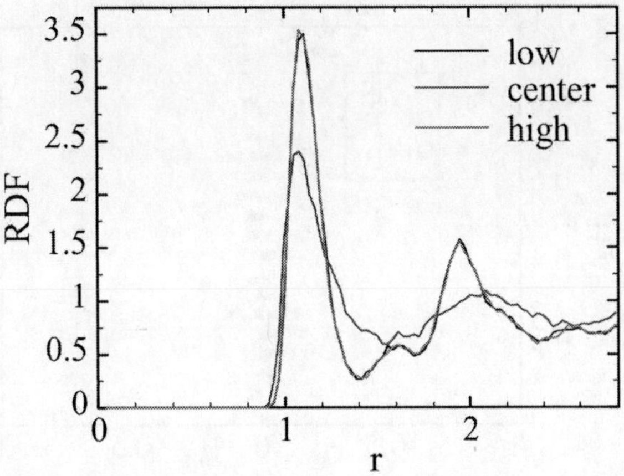

Figure 4. The radial distribution functions (RDF) in the upper, middle, and lower regions at time $t = \tau_0$. The small second peaks of RDF in the upper and bottom regions correspond to the characteristic of the RDF in FCC lattice structure. The local density in the upper and lower regions (solid phase) is $\rho = 0.96$ while that in the middle region is $\rho = 0.79$. See also Fig. 2(c) for the local density profile.

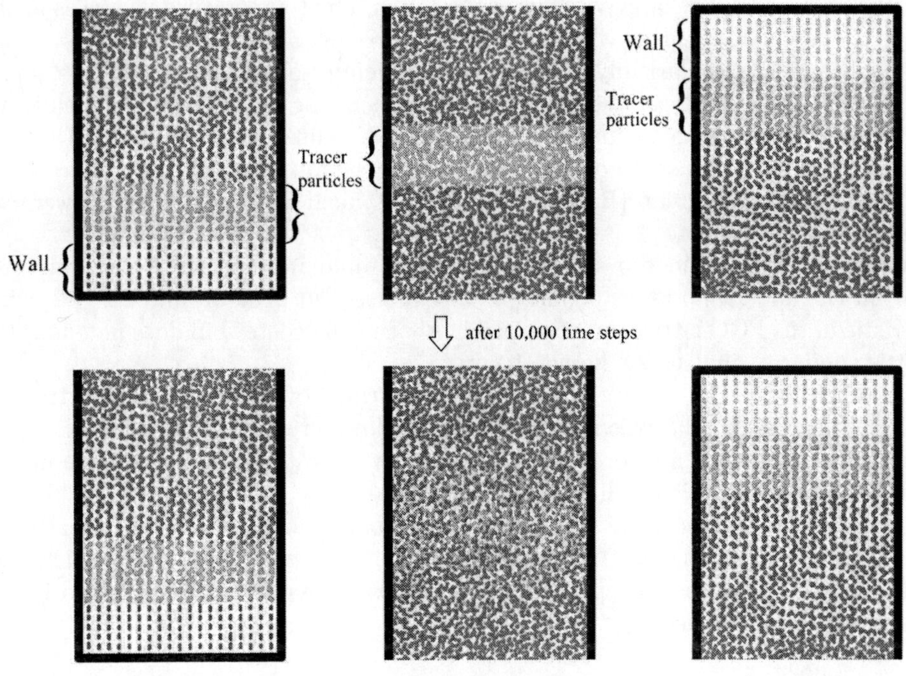

Figure 5. The snapshots of molecules contained in the lower (a), middle (b), and upper (b) regions at two different time steps. The movements of colored molecules in upper figures after 10,000 time steps are shown in the lower figures. The shear banding layer near walls broadens out about 15σ's thickness in Figs. (a) and (c).

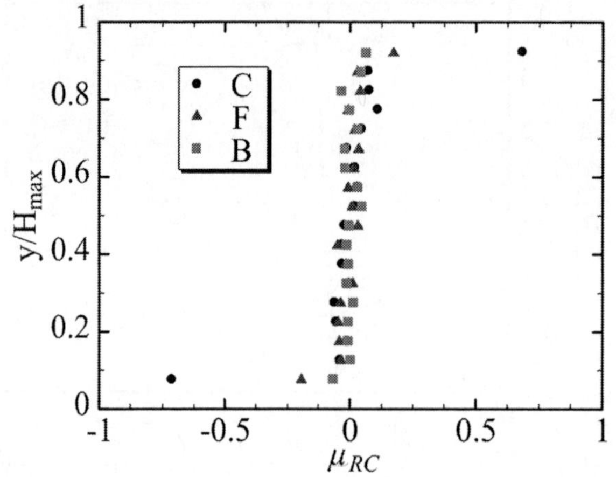

Figure 6. The spatial distributions of the rate of change of local viscosity defined by Eq. 5 for case C, F, and B in Table 1.

Figures 2 and 3 show the spatial distributions of local macroscopic quantities (i.e., velocity, temperature, density, and stress) and local viscosity for Case C in Table 1, respectively.

In Fig. 2, it is seen that both of the normal and shear stresses, p_{yy} and p_{xy} are uniformly distributed between the upper and bottom walls. This confirms that the local stresses are balanced so that the flow velocity is in the stationary state. On the other hand, the other macroscopic quantities vary largely between the walls. The temperature increases in the middle region due to the viscous heating while the temperatures near walls are kept at constant $T \simeq T_w$. Thus, the local density remarkably increases in the vicinity of walls.

Interestingly, the velocity gradient becomes much smaller in the vicinity of walls, where the local viscosities are remarkably increased. See Fig. 3. Thus, the thickening of local viscosity occurs in the vicinity of walls. In order to investigate the local material state, we also calculate the radial distribution function (RDF) of LJ particles included in the upper, lower, and center bins in Fig. 4.

Surprisingly, we can find the solidification of LJ fluid in the vicinity of walls in Fig. 4. The RDF in the upper and lower regions exhibit the second peak around $r \simeq 1.7$, which is the characteristic of the FCC lattice structure, while that in the center bin does not show the second peak but is similar to that of the liquid structure.

The solidification in the vicinity of wall is also confirmed by the movements of tracer particles in Fig. 5. Remarkably, the tracer particles in the lower region do not diffuse in the lateral direction (y direction) but are confined in the vicinity of wall while those in the middle region diffuse in the lateral direction as liquid molecules.

Furthermore, unexpectedly, we can observe shear bands in the vicinities of walls. The shear banding layers broaden out about 15σ's thickness over the walls, where the particles are closely packed because they are strongly pushed toward the wall by the dilation of fluid in the middle region due to the viscous heating.

4. Discussion

In this section, we discuss the non-dimensional parameter which is relevant to the solidification mechanism. We classify the state of LJ fluid near wall by using the rate of change of local viscosity defined by

$$\mu_{RC} = \frac{\partial \log \mu}{\partial y}. \tag{5}$$

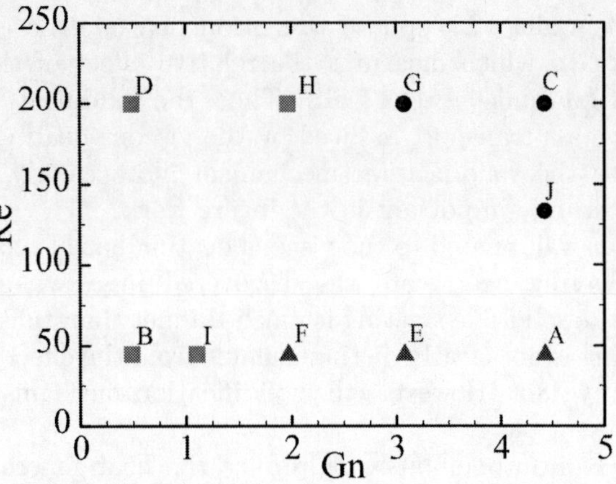

Figure 7. The diagram of the local state near wall in terms of two non-dimensional parameters defined by Eqs. (6) and (7). The results for the parameter sets in Table 1 are shown. The symbols ○, △, and □ represent the solid, semi-solid, and liquid state, respectively.

Figure 6 shows the distributions of μ_{RC} for case C, F, and B in Table 1. The rate of change of viscosity is quite small, $|\mu_{RC}| \ll 1$ in the region except for the vicinity of walls. Remarkably, it exhibits jumps only in the vicinity of walls for Case C, where the solidification occurs.

We qualitatively distinguish the states of LJ fluid near wall for all cases shown in Table 1 by using the maximum values of $|\mu_{RC}|$; say, the solid state for $|\mu_{RC}| \geq 0.4$, semi-solid state for $0.1 < |\mu_{RC}| < 0.4$, and liquid state for $|\mu_{RC}| \leq 0.1$ in this study.

Figure 7 is the diagram of state near wall in terms of two non-dimensional parameters defined as,

$$Re = \frac{H}{\mu_0/\rho_0 V_M} \frac{\rho_0 V_M H}{\mu_0} \propto V_M H, \tag{6}$$

$$Gn = \frac{\mu_0 V_M^2/H^2}{\lambda_0 \Delta T/H^2} = \frac{\mu_0 V_M^2}{\lambda_0 \Delta T} \propto V_M^2. \tag{7}$$

Here V_M is the maximum flow velocity, ΔT is the characteristic temperature difference, and μ_0 and λ_0 are the viscosity and thermal conductivity of LJ fluid in the initial state, which is calculated as $\mu_0 = 3.2$ and $\lambda_0 = 6.5$, respectively, by the equilibrium MD simulation. We also remark that the choice of ΔT is arbitrary and $\Delta T = 1$ is fixed in this paper.

The first non-dimensional parameter Re, the Reynolds number measures the relative width of channel H to viscous length $\mu_0/(\rho_0 V_M)$ and the second non-dimensional parameter Gn measures the relative effect of viscous heating to thermal conductivity.

It is seen that Gn is more relevant to characterize the transition from liquid to solid states than Re. However, the solid states do not appear at small Reynolds numbers even if Gn is large. The solidification only occurs when the channel width is sufficiently large. This fact may indicate that the large system size is required for the LJ fluid to be segregated in two different phases.

5. Concluding remarks and perspective

We discovered a novel solidification and shear banding phenomenon near wall occurring in the high-speed shear flow between the parallel plates by the molecular-dynamics simulation. The solidification and shear banding near wall is only observed when the channel width is sufficiently large, where the highly-inhomogeneous distributions of macroscopic quantities are created between the walls. The phase transition near wall is characterized by the non-dimensional parameter Gn, which measures the relative effect of the viscous heating to the thermal conductivity of LJ fluid between walls. Thus, the solidification near wall is thought to be due to the heat and mass transport induced by the viscous heating in the fluid regime.

In order to elucidate the solidification mechanism more clearly, the effect of the lattice structure of wall atoms will be important in the future work.

The solidification near wall related to the viscous heating has also been discussed in Ref. [25], where the solid-like layering near wall was discovered in a two-dimensional nanochannel. Actually, in the paper, the solid-like layering is much thinner than that in Fig. 5 and the effect of the macroscopic transport is not clearly distinguishable from the effect of the wall atoms because the channel width is only 48σ. However, the solidification mechanism seems to be similar to that observed in the present paper.

The MD simulation can autonomously reproduce the heat-generating shear flows coupling with phase transition as they may occur in real high-speed lubrication systems. Importantly, this study demonstrates that the application of MD simulation to fluid dynamics problem can unveil a hidden physical phenomenon, which cannot be ever addressed by any conventional CFD approaches, even in the fundamental flow problems.

Acknowledgements
This study was financially supported by JSPS KAKENHI Grant Number 16K17554 and 17H01083.

References

[1] R. B. Bird, R. C. Armstrong, and O. Hassager, *Dynamics of polymeric liquids* Vol. 1 (John Wiley and Sons, New York, 1987).
[2] Y. Hori, *Hydrodynamics lubrication* (Springer-Verlag, Tokyo, 2006).
[3] C. J. Pipe, T. S. Majmudar, and G. H. McKinley, "High shear rate viscometry", Rheol Acta **47**, 621 (2008).
[4] M. P. Allen and D. J. Tildesley, *Computer Simulation of Liquids*, (Oxford University Press, New York, 1989).
[5] D. J. Evans and G. Morris, *Statistical mechanics of nonequilibrium liquids*, (Cambridge university press, New York, 2008).
[6] G. E. Karniadakis, A. Beskok, and N. Aluru, *Microflows and nanoflows: fundamentals and simulation* (Springer, New York, 2005).
[7] R. Khare, J. de Pablo, and A. Yethiraj, "Molecular simulation and continuum mechanics study of simple fluids in non-isothermal planar couette flows", J. Chem. Phys. **107**, 2589 (1997).
[8] J. S. Hansen and J. T. Ottesen, "Molecular dynamics simulations of oscillatory flows in microfluidic channels", Microfluid Nanofluid **2**, 301 (2006).
[9] B. H. Kim, A. Beskok, and T. Cagin, "Viscous heating in nanoscale shear driven liquid flows", Microfluid Nanofluid **9**, 31 (2010).
[10] J. Ghorbanian and A. Beskok, "Temperature profiles and heat fluxes observed in molecular dynamics simulations of force-driven liquid flows", Phys. Chem. Chem. Phys. **19**, 10317 (2017).
[11] W. E and B. Engquist, "The heterogeneous multi-scale methods", Commun Math. Sci. **1**, 87 (2003).
[12] I. G. Kevrekidis, C. W. Gear, J. M. Hyman, P. G. Kevrekidis, O. Runborg, and C. Theodoropoulos, "Equation-free, coarse-grained multiscale computation: enabling microscopic simulations to perform system-level analysis", Comm. Math. Sci. **1**, 715 (2003).
[13] W. Ren and W. E, "Heterogeneous multiscale method for the modeling of complex fluids and micro-fluidics", J. Comput. Phys. **204**, 1 (2005).
[14] I. G. Kevrekidis and G. Samaey, "Equation-free multiscale computation: algorithms and applications", Annu. Rev. Phys. Chem. **60**, 321 (2009).
[15] M. K. Borg, D. A. Lockerby, and J. M. Reese, "Multiscale method for micro/nano flows of high aspect ration", J. Comput. Phys. **233**, 400 (2013).
[16] S. Yasuda and R. Yamamoto, "A model for hybrid simulation of molecular dynamics and computational fluid dynamics", Phys. Fluids **20**, 113101 (2008).
[17] S. Yasuda and R. Yamamoto, "Rheological properties of polymer melt between rapidly oscillating plates: an application of multiscale modeling", Europhys. Lett. **86**, 18002 (2009).

[18] S. Yasuda and R. Yamamoto, "Multiscale modeling and simulation for polymer melt flows between parallel plates", Phys. Rev. E **81**, 036308 (2010).
[19] S. Yasuda and R. Yamamoto, "Dynamic rheology of a supercooled polymer melt in nonuniform oscillating flows between rapidly oscillating plates", Phys. Rev. E **84**, 031501 (2011).
[20] T. Murashima, S. Yasuda, T. Taniguchi, and R. Yamamoto, "Multiscale modeling for polymeric flow: particle-fluid bridging scale methods", J. Phys. Soc. Jpn. **82**, 012001 (2013).
[21] S. Yasuda and R. Yamamoto, "Synchronized molecular-dynamics simulation via macroscopic heat and momentum transfer", Phys. Rev. X **4**, 041011 (2014).
[22] S. Yasuda and R. Yamamoto, "Synchronized molecular-dynamics simulation for the thermal lubrication of a polymeric liquid between parallel plates", Comput. Fluids **124**, 185 (2016).
[23] S. Plimpton, "Fast parallel algorithms for short-range molecular dynamics", J. Comput. Phys., **117**, 1–19 (1995).
[24] http://lammps.sandia.gov
[25] M. Priya and Y. Rabin, "Molecular dynamics simulations of shear-induced thermophoresis and non-Newtonian flow in compressible fluids", arXiv: 1703.1046.

Groundstates of liquid crystals with colloids: a project for undergraduate students.

Joan Adler[1], **Priel Aharonian**[1], **Oshri Halimi**[1],

[1] Physics Department, Technion, Haifa, Israel, 32000

E-mail: phr76ja@technion.ac.il

Abstract. Although simulated annealing has become a useful tool for optimization of many systems, its initial raison d'etre of achieving the groundstate structure for a spin or atomic/molecular condensed system remains important. Such modelling, whether using analog models such as glass beads or by invoking simple computer models can be suited to undergraduate projects. In this paper we discuss the application of simulated annealing to find the groundstate of a system of liquid crystals (LC) with suspended colloids. These systems are expected to have interesting conductive behaviour, relevant to applications for television and computer screens. In our first stage, a pure LC system was simulated in python and vizualized by undergraduates and presented on an educational website. In the next stage colloid(s) were added, and the original code modified accordingly. Interesting effects such as ordering around the colloid have been seen and will be described. In the final stage and in order to study larger samples, the code was rewritten in C++ and several algorithmic modifications were made. Speed up factors between 100 and more than 1000 were obtained, and fascinating closed cells surrounding the colloids were observed.

1. Introduction

This report is part of the Technion Computational Physics Group's presentation of educational websites with downloadable code on many diverse topics, with special emphasis on Condensed Matter/Statistical Physics topics and model visualization. Earlier presentations [1, 2, 3] cover a wide range of topics. Each project includes physical background, downloadable code and instructions for implementation, and is designed to introduce a topic at a level suitable for other senior undergraduates.

2. Simulated annealing for groundstate search

How do crystals reach their groundstates in nature? The groundstate is the state with the lowest energy, there could be several equivalent ones, or a situation of almost minimum energy which encompasses several domains. Let us consider the example of diamonds, which are naturally formed from molten carbon in high temperature and pressure conditions in volcanoes as they slowly cool or anneal. Rapid cooling or quenching would lead to many small crystal domains. An animated demonstration simulated annealing computer code, using a simple two-dimensional Lennard-Jones (LJ) interaction, between two atoms i and j, a distance r_{ij} apart, where σ is the separation of the particles when the potential is zero and ϵ is the depth of the potential at its minimum,

$$V_{LJ} = 4\epsilon\left[\left(\frac{\sigma}{r_{ij}}\right)^{12} - \left(\frac{\sigma}{r_{ij}}\right)^{6}\right] \quad (1)$$

was presented in [4] to illustrate this. The calculation used simple Metropolis annealing, with a slow lowering of the temperature until an ordered state was found. The code in [4] has an interactive command-line interface which enables dynamic change to the temperature or lowering rate, as well as a visualization window to view the current state. This approach enables the student to self-learn about the optimal cooling rate. A variation for the same model using a master-slave genetic algorithm was given in [5]. For these calculations "reflecting" boundary conditions were used and these will be discussed further below. It should be noted that we are not discussing any behaviour at higher temperatures or phase transitions, (there could be one as the temperature is lowered and the system solidifies) just the groundstate formation and structure. Another model of this type that demonstates the application of a simulated annealing algorithm is the telescope phasing problem, [6, 7]. These latter models illustrate the importance of scheduling the decrease of "temperature" (noise) appropriately, but we will not discuss this further here.

Another way to educate students about annealing and groundstate search is analog demonstrations [8, 9]. These can be used to demonstrate the annealing process, a full discussion is given in [8] and examples for defects and grain boundaries in [9]. This approach has proved successful for extracurricular work with highschool and young undergraduate students and also as preliminary step for beginning graduate students. It requires a collection of beads or glass balls (such as those used for chemical filters) and transparent boxes. In earlier days one had to use a photocopier to take the images, today a smartphone can be used and so for the present project larger boxes (from empty CD disks Fig. 1) in Figs. 1 were photographed "as is" (Fig. 2), equivalent to a simulated quench then shook to imitate movement as kinetic energy, a simulated anneal and photographed again (Fig. 3). Insight towards a good choice of scheduling can be obtained from the analog systems as well as from discussion of annealing in nature.

The cases, computer and analog, with spherical particles were easy because with spherical particles the domains dissolve with relatively little effort, however once the particles become elongated groundstate search becomes far more difficult. We can take the intuition from the spherical analog systems to the computer to begin our simulations, but we may need additional stages.

Figure 1. The box.

Figure 2. Disorder.

Figure 3. Groundstate.

3. Experimental motivation

The present series of studies began when the Technion RBNI Nanotechnology Institute had a visit by Professor Shunsuke Kobayashi from Japan, who spoke about how colloids could change the properties of liquid crystals. In discussions following his talk, we decided to begin projects containing educational material with downloadable code to encorage modelling these systems.

4. Visualization and potentials for liquid crystals with or without colloids

Liquid crystal groundstates have long range order in at least one direction, and the molecular objects usually are non-spherical molecules. The Technion Computational Physics/SimPhoNy vizualization software, AViz, [10, 11, 12] had a LC option, (proposed by Mike Allen, quite a while ago) that had been used for quadrupole simulations of elongated molecules but never for liquid crystals. (A quadrupole Pa_3 ordered state simulated by O. Cohen [13], shown in Fig. 4.) The molecules are shown as cylinders. Within AViz there is the possibility to use color to reinforce the perception of spin direction, and it works equally well on cylinders. As opposed to atoms, that only have a type and location, spins have type, location and a direction with x, y and z components, and can be drawn as pins or cones. The LC option has a location and direction and can have additional properties such as angular deviation from some selected direction, or movement of center from initial location. They can be drawn as lines or cylinders of varying lengths and thicknesses. At this point in time, AViz enables only one type of object (atom/spin/liquid crystal) so the introduction of colloids required a workaround. The colloids are indicated by stars of cylinders joined at the center, which has the added visual advantage over a larger blob of being able to view order thru their branches. However, it is important to note that this is merely a visual trick - in the simulations the particles have the correct spherical or spheriodal shape. A cartoon graph of energy versus configuration space for a simple two dimensional LC system surrounding a colloid is shown in Fig. 5.

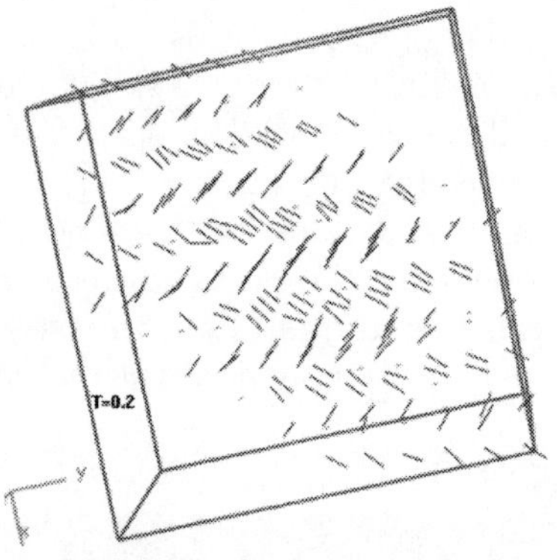

Figure 4. Quadrupolar order with AViz [13].

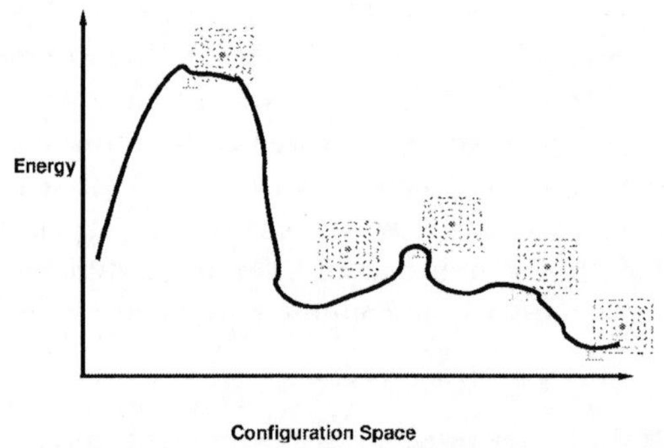

Figure 5. Cartoon of energy vs. configuration space.

A LC potential [14] that has the same general functional form as the LJ was developed by Gay and Berne [GB], for this anisotropic case. Each of parameters σ, r_{ij} and ϵ are generalized since there are now orientational degrees of freedom as well as the translational ones and shape anisotropy. There is a wonderful website and educational project about generalised LJ/GB potentials developed by the Concord Consortium [15], which discusses potentials beyond the usual LJ/GB as well as mixed systems. This site is molecular dynamics oriented to study dynamic behaviour rather than groundstate search that is our main concern, but the potentials are, of course, identical. We refer the reader to their excellent exposition for detailed information. In brief, a unit vector, $\hat{u}_i$, a length l_i, breadth d_i and an electric dipole moment p_i are added for each molecule, as well as χ and α which are functions of the length/breadth ratios. The parameters are now functions of these, with details in [15, 20, 21]. In addition there are parameters that can be chosen to include the possible dependence of the well depth and end to end versus side by side values.

There are many excellent research articles concerning liquid crystal simulation and experiment with and without colloids. We refer the interested reader to [16, 17, 18, 19] where more sophisticated methods and models are explored.

5. Groundstates of liquid crystals

The project to present educational sites for liquid crystal groundstate search was begun by Hila Glanz and Izrael Nadir with a python code for the GB potential (including generalization for other potentials) and an interesting webinterface for presenting results as shown in Figs 6 and 7. We used the "color to indicate direction of the molecule feature" of AViz graphics. Each of the screenshots of the interface includes pre-calculated results for the energy and a visualization of the state for selected cases. This project was described in [20]. The present code is also presented on a website [21], where details of the similar and distinct aspects of the two projects are described in depth. Both codes use the same, rather basic simple Metropolis annealing selected for the Lennard Jones potential of [4], but do not have the interactive option since they require much longer run times.

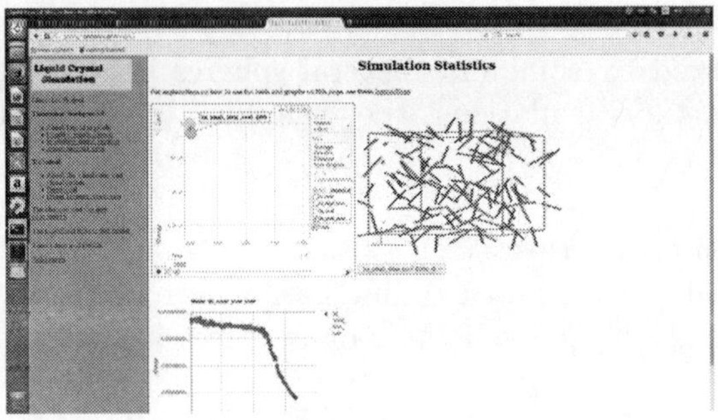

Figure 6. Screenshot of the interface at high temperature, with AViz [20].

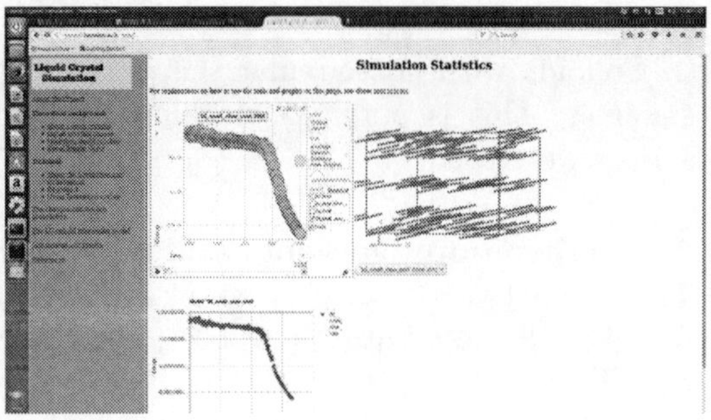

Figure 7. Screenshot of the interface at low temperature after annealing.

6. Groundstates of liquid crystals with colloids

Pure LCs are relatively easy to anneal into groundstates. The addition of colloids complicates this greatly. Some liquid crystals are themselves colloids, but we study the case of small liquid crystal molecules with a few large colloids. Once colloids are introduced, the groundstate stuctures must satisfy minimal energy of the LCs as well as of their order around the colloids. The first decision is whether one requires the LCs to orient parallel or perpendicular to the colloids. For the 2d simulations parallel was selected, with an intuitive explanation that when you push 1 rod-like object towards a big sphere-like object the rod-like object will try to be parallel in order not to be squashed, all the other rod-like objects are more affected by the rod-like object near them than the sphere-like object which is much further away. The rod-like objects had a rejection between themselves and so tried to fill the space until they reached very close to the colloid. Therefore there was also a rejection between rods and the colloid. The start and end configurations for two dimensional systems with a central colloid are shown in Fig. 8.

In [20] simple periodic boundary conditions were used, since the system was homogeneous. The addition of colloids breaks this symmetry and thus boundary conditions must be re-evaluated. The novel "reflecting" conditions of [4], are not practical or needed here. They were introduced because a simple Lennard-Jones system has a very high vapour pressure and the default situation with a sample is that the atoms fly away. (This is not the case with low vapour pressure metals [25].) However totally free boundaries can cause problems also with liquid crystals and so periodic boundary conditions were selected. These obviously mean we have a "lattice" of structures and voids, but the structures are at a substantial distance from each other. Despite the differences in models and boundary conditions from [4] the same simple Metropolis annealing is used.

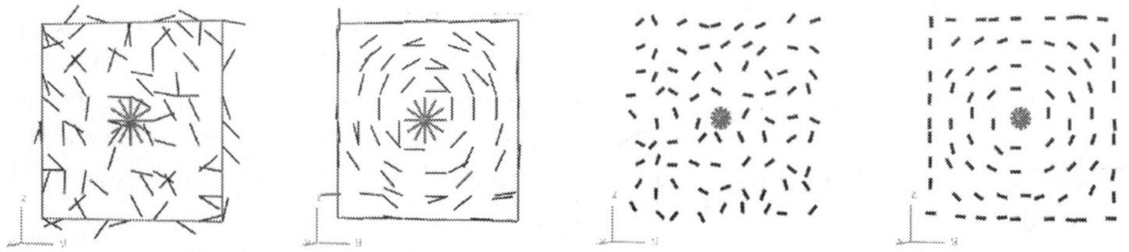

Figure 8. Screenshot of the interface at low temperature after annealing; rightmost images with different choices of periodic boundary conditions.

Colloids with non-circular shapes were visualized by combining several spheres in rows or squares. This is purely a visualization trick, in the calculations, the true ovoid or circular shapes were used.

7. Further improvements and three dimensional cases

The interactive interface shown in Figs.6 and 7 and Python based coding in [20] have been very useful for demonstration purposes, but three dimensions and colloids took an excessive time to equilibrate.

Thus, in order to proceed further a new C++ code was prepared, validated and then run, with some examples given in Fig. 9. This code is many times more efficient and we are continuing its exploration. Details of the speedup are in Table 1, where the times are averages over all temperatures. Note that in addition to the language change there are also some algorithmic changes and bug fixes in the new code. In the Python code the potential of the whole system was calculated from scratch at each Monte Carlo step! Even if we wanted to calculate the potential of a certain radius, to find this radius we would have had to run through all the molecules to check if they were in the radius. The C++ implementation has a lookup table that saves all the potentials thus we need to calculate only the potentials needed for this change. In addition we added a discretization of the 3d space so we could know in O(1) (on average) the neighbor molecules. The time for one iteration (moving one molecule) is linear in the number of molecules; or to be more accurate O(1) is replaced by O(m), where "m" is the number of grid cells that potential covers until it decays. Since "m" depends only on the potential parameters we can keep "m" fixed and extend the system to contain more and more LC molecules, keeping the iteration time unchanged.

The final structures around the colloids have similarities in 2d and 3d, with circular/spherical symmetries given enough development time. However, in our 3d example we selected an attraction between the colloid and LC and expected and found a "porcupine" structure. The periodic potential and periodic boundary conditions of the LC were already considered in the earlier implementation. In the latest one a periodic potential for the colloids in adjacent systems was added. As expected we found similar results because the system is large relative to the colloid

Table 1. Comparison of times for C++ and Python codes.

Model	Steps	Python time	C++ time
5×5 molecules	100,000	21.25 mins	9.5 secs
$5 \times 5 \times 5$ molecules	200,000	3.5 hours	52 secs
$11 \times 11 \times 11$ molecules	500,000	6 days	8.33 mins

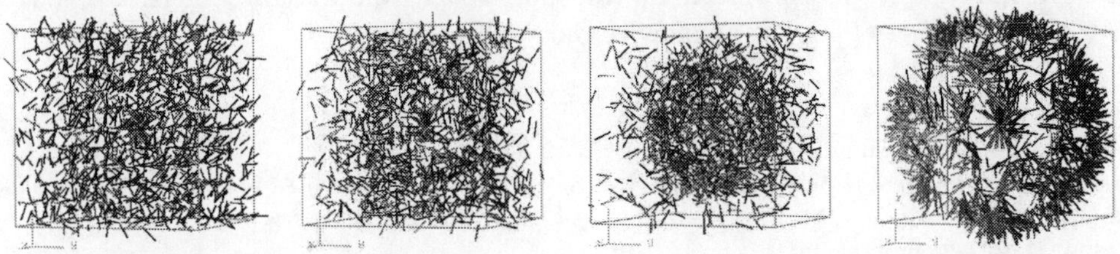

Figure 9. 3d sample images (l to r) A few steps after initial random state, about 25 percent done, 35 percent, final configuration. The color of the liquid crystal molecules is selected according to their angle to guide the eye.

and the potential of adjacent colloids decays substantially when we look inside our system. The final code can be found on Github [23], and some comments on [21]. We note that for all cases a simple slow temperature reduction has been selected.

One important issue at the level beyond educational codes with simulated annealing when the algorithm is molecular dynamics is the selection of ensemble such that there are minimal constraints on the structural order. This has been studied intensively by Keiko Aoki and we refer the interested reader to her series of papers [22].

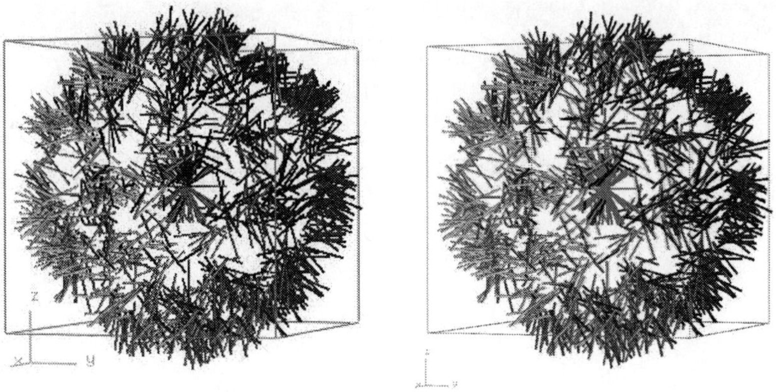

Figure 10. An enlargement of the final state shown in Fig. 9 and an analglyphic stereo version [24]. An animation can be found at: http://phycomp.technion.ac.il/~phr7ja/newmovie.gif.

8. Conclusions

This project began as a modest addition to the Technion Computational Physics groups series of educational projects on the simulation of liquid crystals, but was extended beyond the usual scope in order to create large three dimensional models. The structure of the final(?) states in the three dimensional colloid samples as shown in Fig. 10 appears to be very interesting and will be explored further in the future.

Acknowledgments

Support from SimPhoNy, an FP7 program of the European Commission (FP7-NMP-2013-SMALL-7) with grant number 604005 is acknowledged with thanks.

References

[1] Mazvovsky D, Halioua G and Adler J *Physics Procedia* 2012 **34** 1-5
[2] Adler J, Artzi Y, ben Bashat L, Izraeli T Y, Kreif M, Lavi I, Leibenzon A, Levi A, Schlesinger I, Toledano E, Peretz U, Weisler Y and Alon Yagil 2014 *Journal of Physics: Conference Series* **510** 012003
[3] Adler J, Nissim G, Kiswani A 2017 *IOP Conf. Series: Journal of Physics: Conf. Series* **905** 012017
[4] Silverman A and Adler J 1992 *Computers in Physics* **6** 277
[5] Adler J and A. Berengoltz A 2000 in *Recent Developments in Computer Simulation Studies in Condensed Matter Physics* **XIII**, Springer, edited by Landau D, Lewis SP and Schuttler B 71
[6] Ribak E, Adler J and Lipson SG, 1990 J. Phys. A **23** L809-14
[7] Paykin I, Yacobi L, Adler J and Ribak E, 2015 *Phys. Rev. E* **91** 023302
[8] Adler J, 1999 *Am. J. Physics* **67** 1145
[9] https://bananushka.github.io/aviz/index.html
[10] http://phony1.technion.ac.il/~aviz
[11] Adler J 2003 *Computers in Science and Engineering* **5** 61
[12] https://github.com/simphony/AViz
[13] Adler J and Cohen O 2010 *Physics Procedia,* **6** 2
[14] Gay JG and Berne BJ 1981 *J. Chem. Phys.* **74** 3316
[15] http://mw.concord.org/modeler/articles/beyondlj.pdf
[16] Forman CJ, Fejer SN, Chakrabarti D, Barker PD, and Wales DJ 2013 *J Phys Chem B* **117** 7918
[17] Milller MA, Shepherd JJ and Wales DJ 2008 *Mol Phys* **106** 1655
[18] Bakker HE, Dussi S, Droste BL, Besseling TH, Kennedy CL, WiegantEI, Bing Liu, Imhof A, Dijkstraa M and van Blaaderen A 2016 *Soft Matter* **12** 9238
[19] Tkalec U, Ravnik M, Copar S, Zumer S and Musevic I 2011 *Science* **333** 62
[20] Adler J, Glanz H and Izrael N 2014 *Physics Procedia* **57c** 2
[21] http://phony1.technion.ac.il/~oshrih (with account cpstudent and password cppw16)
[22] Aoki KM and Ohnishi S 2015 *Molecular Crystals and Liquid Crystals* **612:1** 64
[23] https://github.com/alexoshri/LC_periodic
[24] Peled D, Silverman A and Adler J 2013 *IOP Conference Series* **454** 012076
[25] Adham Hashibon, Ph. D. thesis 2002

IOP Conf. Series: Journal of Physics: Conf. Series **1136** (2018) 012029 doi:10.1088/1742-6596/1136/1/012029

Concept formulation and university teaching methodology for dynamic chaos

A K Imanbayeva[1], R N Syzdykova[2] and A A Temirbayev[1]

[1] IETP Al Farabi Kazakh National University, 71 al-Farabi Avenue, Almaty 050040, Kazakhstan

[2] Almaty University of Power Engineering and Telecommunications, 126 Baytursynov Street, Almaty 050013, Kazakhstan

akmaral@physics.kz

Abstract. Chaos is a typical property of dynamical systems that is widespread in all areas of modern physics. We can say that it is a very rare phenomenon when systems demonstrate only regular behavior. Researches on chaos have highlighted the study of a whole series of issues that have general physical and general scientific significance in a new way and provided new impulses. These changes have caused a number of methodological issues that need to be addressed in the process of development of methods of teaching modern concepts and phenomena. First of all, it is important to carefully select the material and adapt it in order to achieve a combination of scientific character and accessibility for students. This paper presents methodological techniques of teaching dynamic chaos on the basis of the authors' experience. This technique enhances formulation of modern ideas on physical picture of the world, skills of working with nonlinear models of physics, and develop practical skills of working with computer models.

1. Introduction

One of the most important discoveries in modern science is a dynamic chaos. Dynamic or deterministic chaos is a non-periodic oscillation in non-linear deterministic systems and shows high sensitivity to the initial conditions [1-2]. From the very beginning of the formation of dynamic chaos as a scientific direction, a great interest was shown in exploring this phenomenon in a wide range of applied problems from physics, astronomy, radio electronics, telecommunications, biology, in particular, neurodynamics, to financial analysis. The reasons for the continued interest lie both in studying the fundamental properties of chaos, and in finding ways of application of this phenomenon. Research on chaos supplemented the existing picture of the world in terms of natural science. Previously it was believed that world structure in general terms can be explained on the basis of a small number of fundamental principles. All the rest was perceived as the getting consequences. Studies of chaos have shown that obtaining the consequences can lead to profound conceptual changes in science [2-4]. The work of Mushkin [5] provides a review of how a new language and system for describing dynamic chaos has developed, with emphasis on attributes that have a physical and general scientific significance. It is also important to note that chaos studies have interdisciplinary character.

IOP Conf. Series: Journal of Physics: Conf. Series **1136** (2018) 012029 doi:10.1088/1742-6596/1136/1/012029

This present article is motivated by the need to understand the student's conceptualization of dynamic chaos and highlight the challenges to the formulation of teaching chaos at undergraduate level. The aim of the paper is to provide justifications for the chosen methodology in the authors' experience.

2. Concept of dynamic chaos

Dynamic chaos is one of the basic concepts of nonlinear physics [6-7]. The Nobel Prize winner V.L. Ginzburg paid special attention to non-equilibrium processes, chaos, attractors and underlined their importance and fundamental character in science, noting that the study of dynamic chaos can be called nonlinear physics and is undoubtedly the current trend of physics development [1].

In the literature there are several different definitions of the concept of chaos. We proceed from the following definition that chaos is defined as some random process that is observed in dynamic systems that are not affected by noise or any random forces. We note the main properties of the chaotic system [1]:

– unpredictability (exponential instability),
– indecomposability (transitivity),
– element of irregularity (cycle density).

The theory of chaos is usually considered as part of the theory of dynamical systems.

A set of dynamic variables is specified for a dynamic system, that describes the state of the system during the time period before and after using the initial state, as well as subject to some rule which is set by evolution operator:

$$\frac{dx_1}{dt} = F_1(x_1, \dots, x_N, \mu),$$

$$\frac{dx_2}{dt} = F_2(x_1, \dots, x_N, \mu),$$

$$\frac{dx_N}{dt} = F_N(x_1, \dots, x_N, \mu),$$

here the values $x_1, x_2, \dots, x_N$ are system variables, μ is vector of control parameters, $F_1, F_2, \dots, F_N$ are some functions. If we consider the values $x_1, x_2, \dots, x_N$ as coordinates of the point x in N-dimensional space, then we get a visual geometric representation of the state of a system in the form of this point (The state of a system is defined by specifying values of a set of measurable properties sufficient to determine all other properties. The state of a system is defined by specifying values of a set of measurable properties sufficient to determine all other properties. For fluid systems, typical properties are pressure, volume and temperature. More complex systems may require the specification of more unusual properties. As an example, the state of an electric battery requires the specification of the amount of electric charge it contains.). This point is called the image or phase point, the values $x_1, x_2, \dots, x_N$ are called phase coordinates of the point or phase variables of the system, and the state space is called system phase space. The motion of the phase point along a certain line, which is called the phase trajectory responds to the changed state of the system in time. $F_1, F_2, \dots, F_N$ determine the speed of motion of image point in the N-dimensional phase space.

Dynamic systems are classified as conservative and dissipative. A system will be conservative if its energy is constant. The dissipative system is displayed by dissipation of energy under the influence of external conditions. The main property of dissipative systems is the contraction of the phase volume. This means that within the time period, starting from the dynamic equations, the original volume decreases. It is expressed in the form of the inequality $\mathrm{Div}(v) < 0$.

An important difference between dissipative system and conservative dynamical systems is that in the latter systems only that attracting sets occur. As $t \to \infty$, all phase trajectories will converge to some subset of zero volume, which is called the attractor of dynamical system. Scientists are studying a significant number of different attractors, but for an initial view of this phenomenon, there are three options:

IOP Conf. Series: Journal of Physics: Conf. Series **1136** (2018) 012029 doi:10.1088/1742-6596/1136/1/012029

1. "Point attractor". An example of such a system is a swinging pendulum, which, with time, stops the friction force at one point. The system is "attracted" to the initial point of equilibrium.

2. "Limiting cycle". Suppose there is no friction. Then the pendulum will always fluctuate and represent a regular periodic system. Figure 1 shows this type of attractor. Motion on limit cycle reflects a complex process of energy changes in time that takes place in the system. If the external disturbance shifts the trajectory on the phase plane inside the limit cycle, then the introduced energy will exceed the dissipated one on average. Here the average value of divergence will be positive. Outside the limit cycle, the divergence is negative, which leads to the striving of phase trajectories to the limiting cycle from the outside.

3. "Strange attractor". If we randomly change the energy communicated to the pendulum through equal time intervals, the resultant motion will be different and non-periodic. However, it is limited by the maximum amplitude of the pendulum and by the laws of physics (gravitational force, etc.). The result of such a movement will be a chaotic, or strange attractor. It is usually assumed that a dynamical system possesses a strange attractor if in its phase space there is a limit set consisting of chaotic trajectories. Strange attractors are complex fractal sets attracting to themselves all trajectories from some adjacent area [8].

Here we briefly consider an example of a strange attractor of a system with chaotic dynamics on the Chua scheme (Fig.2). It is necessary to have at least three variables corresponding to the reactive elements of the circuit for describing the chaotic dynamics. In this case, these elements are two capacitors and one inductor.

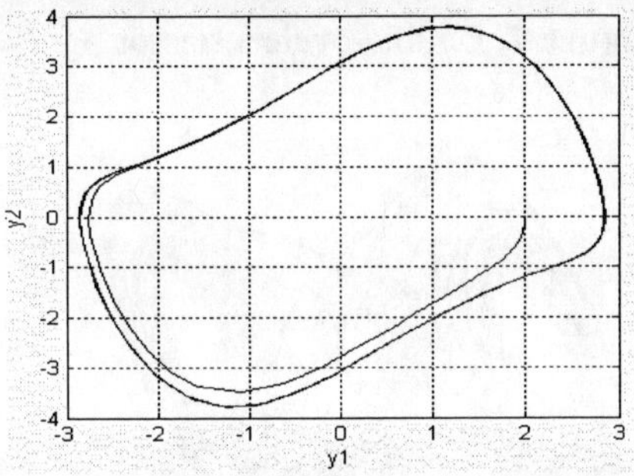

Figure 1. The limiting cycle arising in classical non-linear Van der Pol oscillator

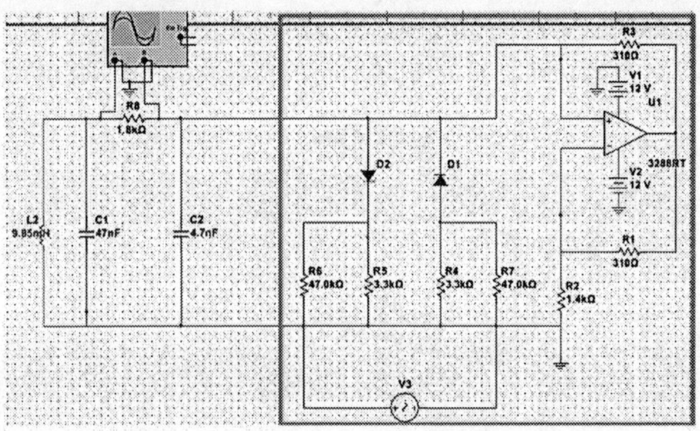

Figure 2. Strange attractor of chaotic system

The active element of the Chua scheme is highlighted with a rectangle in Fig. 2. The mathematical model of the circuit is described by the following system of equations:

$$\frac{dx}{dt} = \alpha\big(y - x - f(x)\big),$$
$$\frac{dy}{dt} = -y + x + z,$$
$$\frac{dz}{dt} = -\beta y.$$

$$f(x) = \begin{cases} bx + a - b, & x \geq 1. \\ ax, & |x| \leq 1. \\ bx - a + b, & x \leq -1. \end{cases}$$

Here we introduce the following notation for dimensionless variables and parameters $x = V_{C1}/E$, $y = V_{C2}/E$, $z = I_L R/E$, $\alpha = C2/C1$, $\beta = R^2 C2/L$, $f(x) = V_{C1}/E + RG(V_{C1})/E$, where V_{C1} and V_{C2} are the voltages on the capacitors $C1$ and $C2$, I_L is the current flowing through the coil L. We can obtain types of attractors by changing the parameter a in the range from $a = -0.1$ to $a = -1.11$ and for fixed values of the remaining parameters $b = 0.714$, $\alpha = 15.6$, $\beta = 28$.

An attractor of the "limit cycle" type is obtained for $a = -0.1$ (Fig.3). The period-doubling bifurcation is observed with a further decrease of the parameter up to $a = -1.045$. (Fig.4). The oscillations in the system become chaotic when the value of the parameter a reaches $a = -1.079$ and a random attractor of the Rössler type appears, this result is shown in the Figure 5. Finally, the merging of two spiral attractors into a single attractor of the "double curl" type occurs as the parameter decreases to $a = -1.11$ (Fig.6).

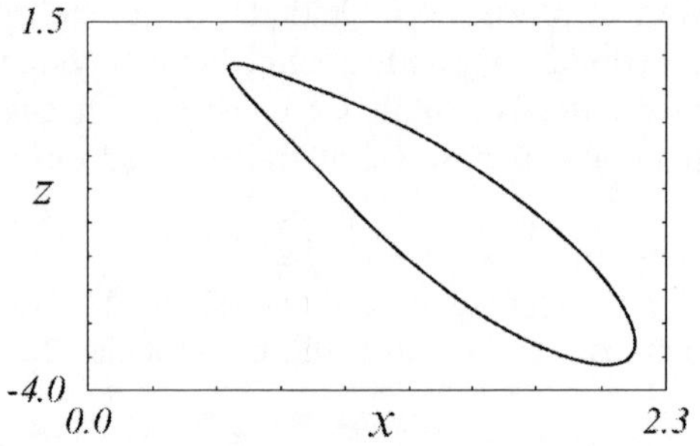

Figure 3. The attractor of the "limit cycle" type

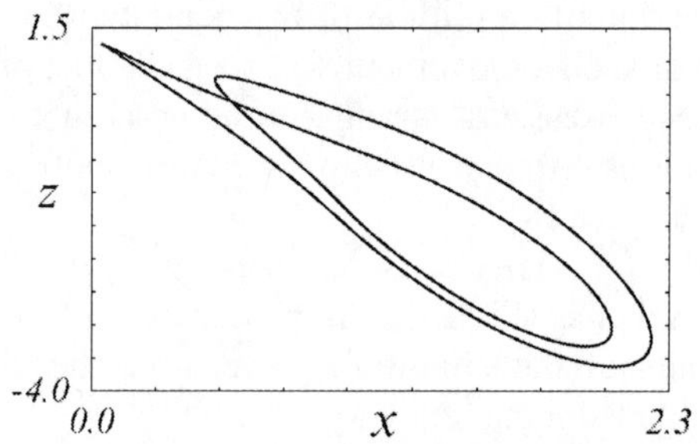

Figure 4. Double-cycle attractor

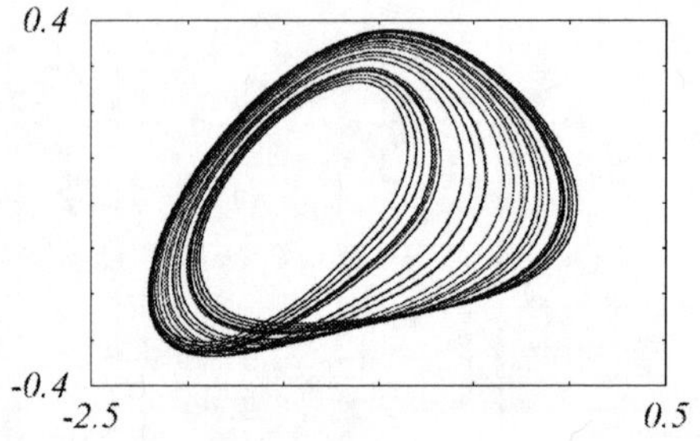

Figure 5. Chaotic attractor of the Rössler type

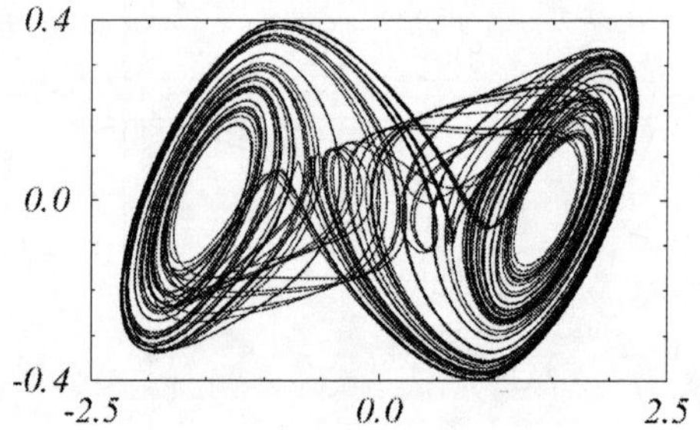

Figure 6. Chaotic attractor of the "double curl" type

3. Methodology

The phenomenon of dynamic chaos can be studied in four stages. Below are the stages of studying the phenomenon within a 15-week course consisting of 3 credits including lectures and laboratory works. This discipline is intended for bachelor students of the 3-d year of study, specialty 5B071900 "Radio engineering, electronics and telecommunications" at Al-Farabi Kazakh National University [9].

Figure 7 shows the work curriculum with the names of the topics of the lectures in the form of a flowchart. Each stage of lecture classes is accompanied by a laboratory computer practice [10]. Students conduct research within the framework of the mentioned topic and present the results. During the computer experiment the methods and models of dynamic programming, data visualization method, and analysis are used.

We also offer students the following topics for independent work. All themes of independent work are related to our scientific works: synchronization, fractal antennas, stochastic resonance effect in nonlinear dynamical systems [11-13].

All laboratory work is carried out in a computer lab. The MatLab system is used for modeling.

We consider one of these laboratory works as an example, and describe the results that students themselves should receive.

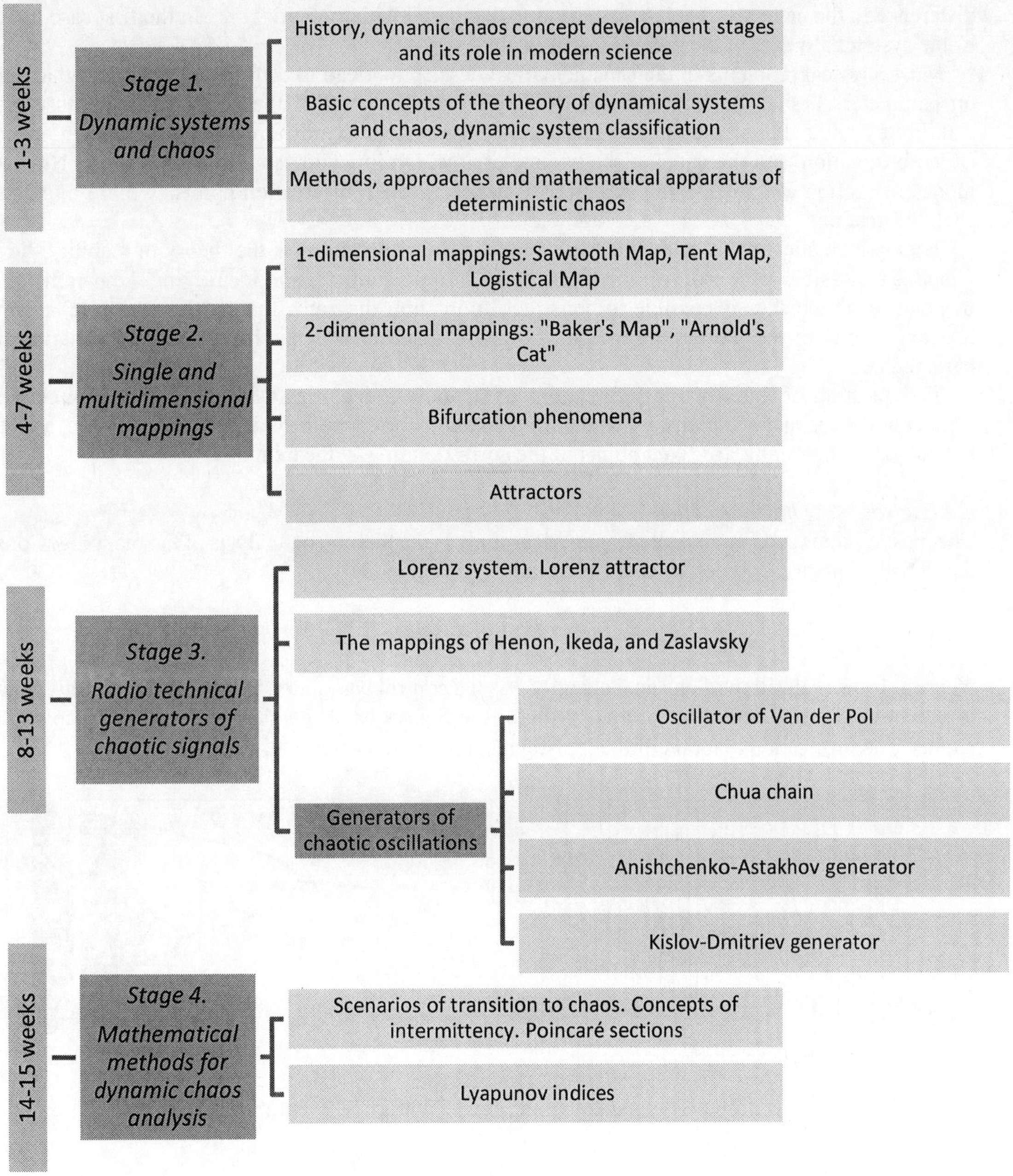

Figure 7. Block diagram of the distribution of topics for a 15-week course

4. Practical work

4.1 Aim

The study of pre-chaotic or post-chaotic changes in the dynamic system in variation of its parameters by means of bifurcation diagrams.

IOP Conf. Series: Journal of Physics: Conf. Series **1136** (2018) 012029 doi:10.1088/1742-6596/1136/1/012029

4.2 Theory

The reaction of a dynamical system to a small perturbation is determined by its state, and in some cases the perturbing factors affect the mode of system functioning insignificantly, in others they lead to a sharp difference in the nature of the disturbed motion compared with the initial one. In the first case, the state of the system is stable, in the second case it is not.

Most physical problems in their mathematical description lead to differential equations that depend on parameters beside the initial boundary conditions. Changing of the parameter can cause loss of stability by one mode of motion and transition of the system to a different state. This phenomenon is called bifurcation, and the value of the parameter at which it occurs is a bifurcation point. New stable modes of motion can arise in the system as it passes through bifurcation points. We are interested in such bifurcations.

The mathematical basis of the elementary theory of bifurcations is the theory of stability, through which it is possible to breakdown the phase space into trajectories, and identifying the parameter sets. In practice, this makes it possible to construct bifurcation diagrams that explain the mechanisms of rearrangements of the motion regimes in the phase space of a dynamical system in variation of its parameters.

The question of stability and bifurcations of periodic trajectories can be considered directly with respect to differential equations, when the limit solution corresponds to a particular solution, as well as by analyzing the stability of fixed points of the corresponding Poincaré map.

4.3 Logistic mapping bifurcation

The most obvious and convenient for numerical study is the method of analysis of the mapping. Consider the logistic equation that describes the population growth model:

$$x_{n+1} = r x_n (1 - x_n), n = 1,2, \dots N, \tag{1}$$

where x_n is the realization of physical quantity, r is the control parameter. The realization of this equation is shown in Figure 8 (the order of numbering of the figures has been changed in accordance with this paper). The bifurcation diagram of the logistic mapping is shown in Fig. 9.

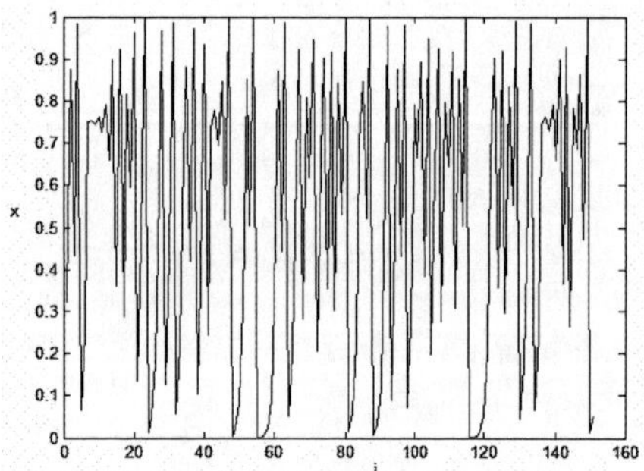

Figure 8. Realization of logistic equation

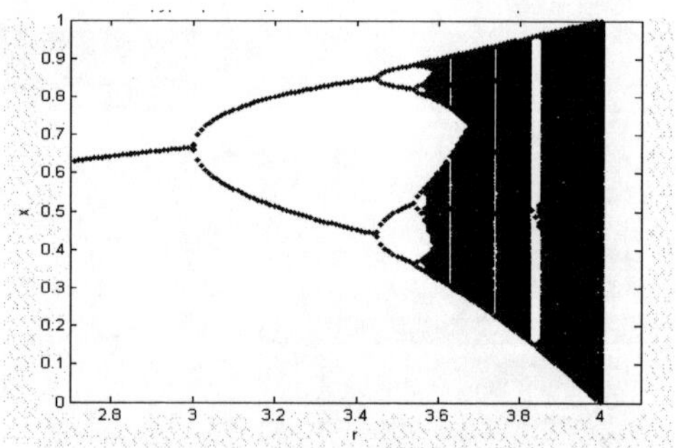

Figure 9. Bifurcation diagram of Logistic map

Bifurcation in system with inertia

The emergence of chaos results from the consecutive increase in the period of oscillations according to the law of natural row.

The system with inertia is described by the following system of equations [14]:

$$\begin{aligned} \dot{X} &= Y + (m_1 + m_2)X - XZ, \quad X \leq q, \\ \dot{X} &= Y - m_2 X - qZ, \quad X > q, \end{aligned} \tag{2}$$

$$\dot{Y} = -X,$$
$$\dot{Z} = -gZ + gF(X)X^2,$$

where $F(X)$ is the Heaviside function; m_1, m_2, q, g are the parameters of excitation, dissipation, limitation and inertia, respectively. This system differs from the Anishchenko-Astakhov's generator by its dynamic characteristics of a nonlinear amplifier, which has a linear section at $X \leq q$ and a saturation section at $X > q$. The fourth equation describes an inertial one half-wave inertial transducer. Thus, the system dynamics is determined by two mechanisms for limiting oscillations. The first mechanism is inertial-free and is related to the nonlinear character of amplifying element. The second - inertial, which is caused by the influence of voltage from the output of the inertial transformation on the steepness of the nonlinear amplifier.

In Figure 10, the fragment of bifurcation diagram of changes of maximal values $[X]$ of the oscillatory process $X(t)$ as the parameter g changes.

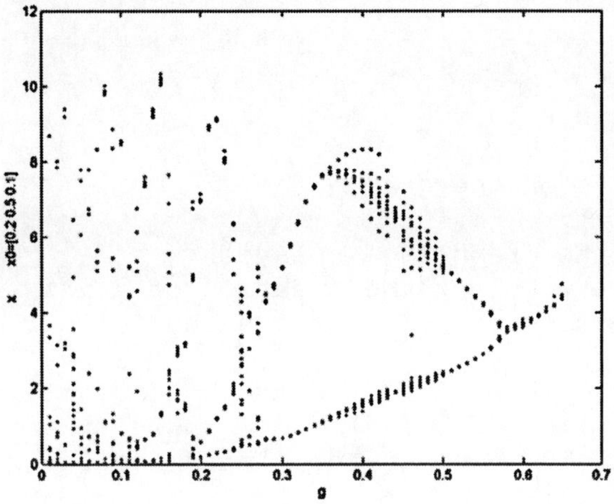

Figure 10. The bifurcation diagram of system (2)

4.4 Assignments

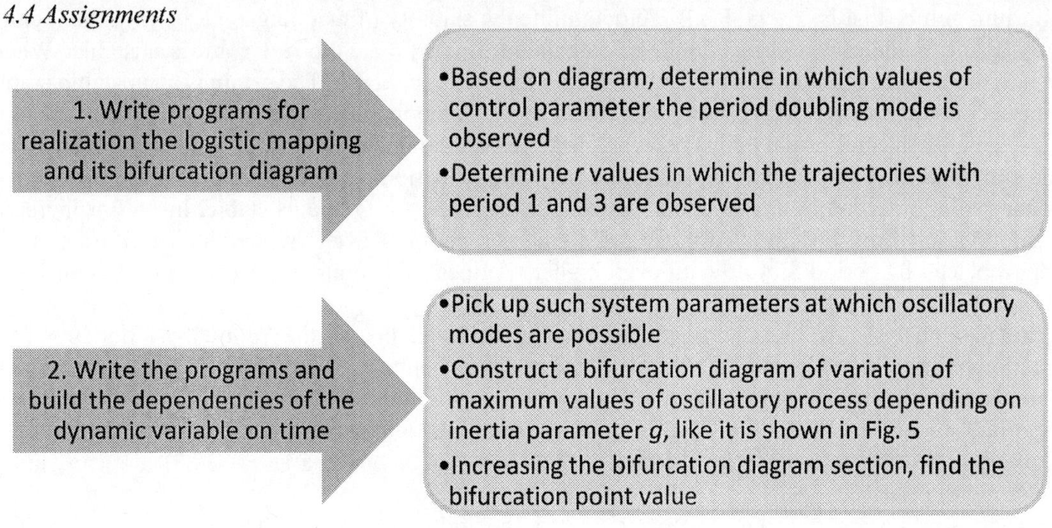

4.5 Analysis of the results of laboratory work

Here we briefly give the main results that students should obtain in the course of performing this work.

IOP Conf. Series: Journal of Physics: Conf. Series **1136** (2018) 012029 doi:10.1088/1742-6596/1136/1/012029

Below is a program in MatLab, the result of which is Figure 9.

```
% argument:
% rb - initial value of r (r is control parameter)
% rh - step change of r
% N - number of steps
% default values
rb=0.01;
rh=0.01;
N=500;
rk=rh*(N-1)+ rb; % calculate the final value of r
r=[rb:rh:rk]; % set r
x(1:N)=0.1; % initial value of x
% we skip 1000 steps for the mapping to reach the attractor
for i=1:1000
    x=(1-x).*x.*r;
end
%- now we draw points
for i=1:1000
% draws points on the graph
plot(r,x,'k.');
    % until the graph is displayed
    hold on % we block the creation of a new window
    x=(1-x).*x.*r; % we calculate the following points
end
% we finally draw points (the graph is displayed in the window)
plot(r,x,'k.');
xlim([2.7 4.1]); % we set the limits for the x axis
xlabel('r');
ylabel('x');
```

Analytical analysis of equation (1) leads to the following conclusion. When $r > 1$, there are two equilibrium points (that is, $x = rx(1–x)$). To determine the stability of mapping $x_n+1 = f(x_n)$, the incline value $|f'(x)|$ at the quiescent point should be calculated. If $|f'(x)| > 1$, the rest point is unstable. When $1 < r < 3$, the logistic equation has two quiescent points: $x = 0$, $(r – 1) / r$. The origin is an unstable point, and the second quiescent point is stable. However, in $r = 3$, the incline at $x = (r – 1)/r$ exceeds the unit ($f' = 2– r$) and both equilibrium points become unstable. In r parameter values, contained between 3 and 4, this simple difference equation describes the set of multi-periodic and chaotic motions. In $r = 3$, the stationary solution becomes unstable, but bi-cycle or bivariate orbit remains stable. In further increase in r, the bivariate orbit becomes unstable and a emerges period 4 cycle, which due to bifurcation, is rapidly replaced by period 8 cycle with even higher r values. This doubling process period continues until r reaches the value $r = 3.56994$…

Graphical analysis of bifurcation diagram leads the students to the following. Students find bifurcation points of doubling period. It is observed at values below $r = 3.57$. Section where each branch of a tree splits into two. Starting with $r < 3$, they can see a trajectory with a period 1. To see longer trajectories, students mark the first 30-50 iterations with dots, and subsequent iterations with another symbol. Chaotic trajectories can be detected at $3.57 < r < 4.0$. In the area $r = 3.83$, one can find a trajectory with period 3.

In analyzing system (2), the students come to the following results. The development of the oscillatory process occurs at the parameters $m_1 = 1.6$, $m_2 = 0.1$, $q = 0.05$. A small change in the values $[X]$ on bifurcation diagram corresponds to the regular system dynamics, the random scatter of points –

to chaotic dynamics. In the interval $g \in [0.57; 0.65]$, the dynamics of the system is characterized by a stable limit cycle. The interval $g \in [0.56; 0.57]$ corresponds to chaotic oscillations. This can be seen in Figure 11.

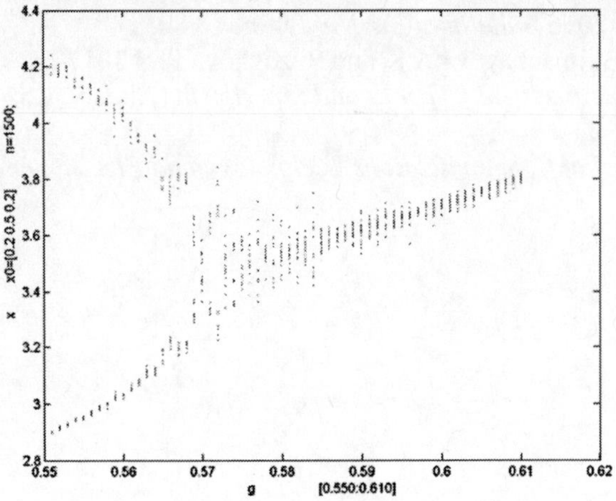

Figure 11. Increased section of bifurcation diagram in Figure 5

The point $g = 0.56$ is a bifurcation point. In each bifurcation, the distances between critical values of the changing parameter are reduced, zones of chaotic oscillations are denser.

5. Conclusion

Modernized teaching using computers and other innovations requires careful formulation and implementation of various methods. In this regard, it is necessary to pay attention to the developing resources of the content of physical education, in particular, the content of individual concepts of modern science. Also it is necessary to note the effectiveness of the use of significant scientific and theoretical material, which form the basis for the formation of professional competence.

Every year lecture material and computer lab works are continuously improved and updated. In our opinion, the proposed methodology makes it possible to achieve a better formation of students' knowledge of dynamic chaos, formation of the required skills in scientific knowledge of modern physics.

Acknowledgements
This work has been supported by the Ministry of Education and Science of Kazakhstan under Grant number 3837/GF4.

References

[1] Loskutov A 2010 *Phys. Usp.* **53** 1257
[2] Boeing G 2016 *Systems* **4** 37
[3] Motter A E and Campbell D K 2013 *Physics Today* **66** 27
[4] Doebeli M and Ispolatov I 2014 *Evolution* **68** 1365
[5] Mukhin P P 2010 *History of Science and Engineering* **4** 46
[6] Smirnov O 2010 *Nonlinear physics* (Moscow: Sputnik+) p 289
[7] Zhanabaev Z Zh 1997 *Lectures on nonlinear physics. Tutorial* (Almaty: Kazakh University Press) p 71 (in Russian)
[8] Kuznetsov S P 2006 *Dynamical Chaos* (Moscow: Fizmatlit) p 295 (in Russian)
[9] Imanbayeva A K, Zhanabaev Z Zh and Saymbetov A K 2016 *Proc. 9th Int. Conf. of Education, Research and Innovation* (Spain: IATED) p 3449

IOP Conf. Series: Journal of Physics: Conf. Series **1136** (2018) 012029 doi:10.1088/1742-6596/1136/1/012029

[10] Zhanabayev Z Zh, Imanbayeva A K and Almasbekov N Ye 2005 *Computer modeling in radiophysics and electronics. Tutorial* (Almaty: Kazakh University Press) p 144 (in Russian)
[11] Temirbayev A A, Zhanabaev Z Zh, Nalibayev Y, Naurzbayeva A Z and Imanbayeva A K 2014 *Communications in computer and information science* **438** 30
[12] Temirbayev A A, Imanbayeva A K, Karibayev B A, Namazbayev T A, Kapurnova S T and Tleubayeva I S 2016 *Bulletin of KazNu. Physics* **3** 80
[13] Tokmyrzaeva Zh B, Imanbayeva A K and Syzdykova R N 2017 *Proc. 10th Int. Conf. Chaos and structures in nonlinear systems. Theory and experiment* (Almaty: Kazakh University Press) p 328 (in Russian)
[14] Savel'ev S V 2013 *J of Communications Technology and Electronics* **58** 3 221